AF824055

Marker-Assisted Breeding in Legumes for Drought Tolerance

Muhammad Azhar Nadeem
Faheem Shehzad Baloch
Filippos Bantis • Sajid Fiaz
Muhammad Aasim
Editors

Marker-Assisted Breeding in Legumes for Drought Tolerance

Springer

Editors
Muhammad Azhar Nadeem
Department of Biotechnology
Faculty of Science
Mersin University
Mersin, Türkiye

Faheem Shehzad Baloch
Department of Biotechnology
Faculty of Science
Mersin University
Mersin, Türkiye

Filippos Bantis
Department of Agriculture
University of Western Macedonia
Florina, Greece

Sajid Fiaz
Institute of Molecular Biology
and Biotechnology
The University of Lahore
Lahore, Pakistan

Muhammad Aasim
Faculty of Agricultural Sciences and
Technology
Sivas University of Science and Technology
Sivas, Sivas, Türkiye

ISBN 978-981-96-4111-6 ISBN 978-981-96-4112-3 (eBook)
https://doi.org/10.1007/978-981-96-4112-3

© The Editor(s) (if applicable) and The Author(s), under exclusive license to Springer Nature Singapore Pte Ltd. 2025, Corrected Publication 2026

This work is subject to copyright. All rights are solely and exclusively licensed by the Publisher, whether the whole or part of the material is concerned, specifically the rights of translation, reprinting, reuse of illustrations, recitation, broadcasting, reproduction on microfilms or in any other physical way, and transmission or information storage and retrieval, electronic adaptation, computer software, or by similar or dissimilar methodology now known or hereafter developed.
The use of general descriptive names, registered names, trademarks, service marks, etc. in this publication does not imply, even in the absence of a specific statement, that such names are exempt from the relevant protective laws and regulations and therefore free for general use.
The publisher, the authors and the editors are safe to assume that the advice and information in this book are believed to be true and accurate at the date of publication. Neither the publisher nor the authors or the editors give a warranty, expressed or implied, with respect to the material contained herein or for any errors or omissions that may have been made. The publisher remains neutral with regard to jurisdictional claims in published maps and institutional affiliations.

This Springer imprint is published by the registered company Springer Nature Singapore Pte Ltd.
The registered company address is: 152 Beach Road, #21-01/04 Gateway East, Singapore 189721, Singapore

If disposing of this product, please recycle the paper.

Preface

Legumes are essential to world agriculture, providing substantial benefits to food security, nutrition, and sustainable farming practices. Their ability to fix atmospheric nitrogen reduces the reliance on synthetic fertilizers, thereby supporting soil health and ecosystem balance. Legume production encounters significant hurdles from environmental conditions, particularly drought. The increasing frequency and intensity of drought events threaten yields, especially in regions heavily dependent on legumes for sustenance and economic stability.

This book entitled "Marker-Assisted Breeding in Legumes for Drought Tolerance," edited by Muhammad Azhar Nadeen, Faheem Shehzad Baloch, Filippos Bantis, Sajid Fiaz, and Muhammad Aasim, is aimed to address the above-discussed challenges by presenting a comprehensive resource for leveraging molecular tools to improve drought tolerance in legumes. Marker-assisted breeding (MAB) and conventional and advanced genomic techniques have revolutionized crop enhancement initiatives. Marker-assisted breeding expedites the breeding process by facilitating accurate identification and selection of preferred attributes, enabling the development of resilient cultivars adapted to a changing climate.

This book targeted a diverse reader, comprising academics, plant breeders, graduate students, and policymakers. Researchers will discover the most recent insights into genomics-based breeding strategies, while plant breeders will gain from practical case studies the use of marker-assisted selection in diverse legume crops. Graduate students will acquire a robust foundation in molecular breeding techniques, while policymakers will comprehend the wider implications of implementing these strategies for achieving food and nutritional security.

The book is structured into thematic sections to ensure a logical progression from fundamental concepts to sophisticated applications. The initial chapters provide the foundation by examining the effects of abiotic conditions on legume cultivation, particularly emphasizing drought. They also examine regional challenges, particularly in the Mediterranean region and Europe, where legumes hold both cultural and economic importance. The following chapters explore traditional breeding methods and their historical impact on drought tolerance, transitioning to the examination of contemporary genomic tools. Tools like genome-wide association studies (GWAS) offer a framework for comprehending the genetic foundations of drought tolerance and for pinpointing markers associated with key characteristics.

The main body of the book comprises species-specific case studies that emphasize the implementation of marker-assisted breeding in prominent legumes, such as common bean, pea, faba bean, lentil, soybean, and mung bean. These chapters include comprehensive descriptions of breeding projects, encompassing marker identification to field assessment of drought-resistant cultivars. Subsequent chapters discuss the cultivation of underutilized crops, including lablab and neglected legumes, highlighting their capacity to improve resilience and sustainability in arid settings. The concluding chapter outlines prospects, emphasizing the necessity of incorporating genomes, phenomics, and bioinformatics into breeding operations. It advocates for worldwide cooperation and investment in research to optimize the advantages of marker-assisted breeding for global food security. The last section highlights prospects, emphasizing the necessity of incorporating genomes, phenomics, and bioinformatics into breeding operations. It advocates for worldwide cooperation and investment in research to optimize the advantages of marker-assisted breeding for global food security.

This book would not have been possible without the invaluable contributions of leading experts in the fields of plant genetics and breeding. Each chapter reflects the author's extensive experience and dedication to advancing agricultural science. We express our profound appreciation to the contributors for imparting their experience and to the reviewers for their insightful criticism, which improved the quality of this work. *Marker-Assisted Breeding in Legumes for Drought Tolerance* goes beyond being a mere technical manual; it serves as a strategic framework for addressing one of the most critical challenges in agriculture. This work aims to equip the scientific community with cutting-edge knowledge and tools, fostering innovation and advancing sustainable solutions for legume cultivation under drought stress.

I am assured that this book will serve as an indispensable resource for individuals committed to enhancing the resilience and productivity of legume crops, thus significantly contributing to the future of sustainable agriculture.

Mersin, Türkiye — Muhammad Azhar Nadeem
Mersin, Türkiye — Faheem Shehzad Baloch
Florina, Greece — Filippos Bantis
Lahore, Pakistan — Sajid Fiaz
Sivas, Sivas, Türkiye — Muhammad Aasim

Contents

1 **Abiotic Stress and Legume Production** 1
Uttarayan Dasgupta, Somsole Bharath, and Suman Dutta

2 **Drought Stress and Sustainable Legume Production** 23
Abdel Rahman M. Tawaha, Hiba Alatrash, Yaman Jabbour, Abdel Razzaq Al-Tawaha, Ali M. Qaisi, Rania Jammal, Arun Karnwal, Mohamad Shatnawi, Palani Saranraj, and Jana Rammal

3 **Legume Production and Problems in the Mediterranean Region** 41
Taufiq Nawaz, Shah Fahad, Nitish Joshi, Tanzeel Ur Rahman, Muhammad Nasir Rasheed Khan, Shah Hassan, and Shah Saud

4 **Drought Stress and Production of Legumes in Europe** 77
Aleksandra Ilić, Ana Uhlarik, and Zlatica Mamlić

5 **Legume Breeding for Drought Tolerance Experience from the Past Using Classical Breeding Approaches** 107
Shumaila Ijaz, Javed Iqbal, Banzeer Ahsan Abbasi, Zakir Ullah, Akhtar Munir, Tabassum Yaseen, Muhammad Majeed, Noshin Ilyas, Amin Fathi, Sobia Kanwal, and Tariq Mahmood

6 **Genome-Wide Association Studies (GWAS) in Pulses: A Genomic Perspective to Drought** 133
Hafiz Muhammad Ahmad, Muhammad Farhan Khan Pasha, Shazia Arif, Saira Ishaq, Syed Hamza Mahfooz, Sumaira Hafeez, and Sadaf Zahra

7 **Marker-Assisted Breeding in Legumes for Drought Tolerance** 151
Aiman Hina, Ahtsham Ul Hassan, Muhammad Tajammal Khan, Adeela Haroon, Asim Abbasi, Awais Hanif, Hayssam M. Ali, and Nader R. Abdelsalam

8 **Marker-Assisted Breeding in Common Bean for Drought Tolerance** 179
Xubo Ke, Jia Yao, Xinyue Gu, Zhihan Jiang, and Pei Xu

9 Marker-Assisted Breeding of Pea for Drought Tolerance 195
Md. Mobinur Rahman, Jaber Bin Azim, Md. Yousuf Ali, Touhidul Islam, Md. Parvez Kabir, and Akbar Hossain

10 Marker-Assisted Breeding in Faba Bean for Drought Tolerance 213
Zakir Ullah, Javed Iqbal, Banzeer Ahsan Abbasi, Shumaila Ijaz, Akhtar Munir, Tabassum Yaseen, Muhammad Majeed, Noshin Ilyas, Amin Fathi, Sobia Kanwal, and Tariq Mahmood

11 Marker-Assisted Breeding in Lentil for Drought Tolerance 269
Afifa Younas, Sajid Fiaz, Nadia Riaz, Madiha Rashid, Zahra Noreen, Aasma Tufail, Maha Noor, Sundas Fatima, and Hurmat Mehdi Mirrani

12 Marker-Assisted Breeding of Soybean for Drought Tolerance 301
Jaber Bin Azim, Md. Mobinur Rahman, Md. Yousuf Ali, Touhidul Islam, Md. Parvez Kabir, Shahin Imran, and Akbar Hossain

13 Marker-Assisted Breeding in Mung Bean (*Vigna radiata* L.) for Drought Stress 323
Meliha F. Sarıkaya, Muhammed Tatar, Amjad Ali, Muhammad T. Altaf, Parnaz Mortazavi, Sarmad A. Qureshi, Mehmet Bedir, and Kağan Kökten

14 Lablab (*Lablab Purpureus* L. Sweet): Current Developments in Breeding and Prospects for Drought Tolerance, Improved Yield, and Sustainable Production in Sub-Saharan Africa (SSA) 345
Julius S. Missanga and Pavithravani B. Venkataramana

15 Breeding Advancements in Neglected Legume Species for Drought Tolerance 373
Suman Dutta, Debashis Paul, Sampa Saha, and Sandeep Bagri

16 Future Perspectives in Legume Breeding for Sustainable Agriculture 399
Muhammad Tanveer Altaf, Amjad Ali, Waqas Liaqat, Parnaz Mortazavi, Sarmad Ali Qureshi, Muhammed Tatar, Amna Jamil, Muhammad Sheraz, Muhammad Aneeq Ur Rahman, Sabeen Rehman Soomro, Salma Naimatullha Soomro, Muhammad Azhar Nadeem, and Faheem Shehzad Baloch

Correction to: Future Perspectives in Legume Breeding for Sustainable Agriculture C1
Muhammad Tanveer Altaf, Amjad Ali, Waqas Liaqat, Parnaz Mortazavi, Sarmad Ali Qureshi, Muhammed Tatar, Amna Jamil, Muhammad Sheraz, Muhammad Aneeq Ur Rahman, Sabeen Rehman Soomro, Salma Naimatullha Soomro, Muhammad Azhar Nadeem, and Faheem Shehzad Baloch

Editors and Contributors

About the Editors

Muhammad Azhar Nadeem is working as Associate Professor of plant breeding and genetics at Mersin University, Mersin, Turkiye. He is currently engaged in research activities involving genetic diversity assessment and genome-wide association studies (GWAS) to identify genomic regions and validate them for marker-assisted breeding. Dr. Nadeem is focusing on performing GWAS in legumes, especially in common bean, lentil, and other legumes to identify genomic regions associated with agronomic, mineral elements, cooking, and quality traits. He has more than 75 SCI/SCIE publications in world-renowned and prestigious journals and also published more than 10 book chapters on important aspects of molecular genetics about plant species. He serves as an editorial board member of several international journals. Dr. Nadeem is an active member of various COST actions as well. Dr. Nadeem served as editor and co-editor of four other books. He has a good number of citations (over 3000) as per Google Scholar with an h-Index of 25.

Faheem Shehzad Baloch is working as Professor of plant genetics and breeding at Mersin University, Turkey. He has more than 11 years of teaching and research experience in plant phenomics, genetics and genomics, and biotechnology. He led and participated in many projects funded by national and international organizations. He supervised and co-supervised MSC, PhD, and postdoc candidates from various countries. He is serving as a Management Committee and active member of various COST actions and played important roles in different work packages of various COST actions. He has more than 120 SCI publications in Web of Science (WOS) and edited 9 books and written over 19 book chapters on important aspects of molecular genetics in relation to plant species. He has an extensive array of citations with over 4300 times as per Google Scholar with an h-Index of 35. He serves as an editorial board member of several impacted journals.

Filippos Bantis is an Assistant Professor at the Vegetable and Greenhouse Crops, School of Agriculture, University of Western Macedonia, Florina, 53100, Greece. Dr. Bantis has undertaken as Principal Investigator 2 research projects funded by individual companies and has also participated or is participating in 12 European and national research projects as well as in 4 joint action projects (COST Actions). He is involved in a patent entitled "Method for the production of improved grafted watermelon seedlings using artificial lighting." He has 33 publications in peer-reviewed international journals and there are 547 citations in his published work. He has also 19 presentations at international and 11 presentations at national conferences, is a member of the editorial board of two international journals, and is a guest editor for 2 special issues in an international journal.

Sajid Fiaz is an Assistant Professor working at the Institute of Molecular Biology and Biotechnology, The University of Lahore, Pakistan. Dr. Fiaz is also serving as adjunct scholar at Henan Institute of Science and Technology, China. Dr. Fiaz's research interests include marker-assisted selection for agronomic traits, mutation breeding for biotic and abiotic stress resistance, Quantitative trait loci (QTL) mapping and genome editing for yield and quality traits in cereals. Dr. Fiaz is actively supervising undergraduate and postgraduate students. He has published more than 162 research, 10 review articles, and 12 book chapters with various prestigious international scientific journals. He is working as academic editor with several high-notch scientific magazines.

Muhammad Aasim is a Professor at the Department of Plant Protection, Sivas University of Science and Technology, Sivas, Turkey since 2020. He completed his graduation and masters from Pakistan with honors and received gold medal from Bahauddin Zakariya University (BZU), Multan (graduation), and silver medal from University of Agriculture Faisalabad (UAF) (Master). He worked in multinational pesticide companies from 1998 to 2003 in Pakistan. He started his PhD in 2004 and was awarded PhD in 2010 from Ankara University with honor. After completing PhD, he worked as Assistant Professor from 2011 to 2013 and Associate Professor from 2013 to 2015 at the Department of Biology, Karamanoğlu Mehmetbey University, Karaman, Türkiye. He also served at the Department of Biotechnology, Necmettin Erbakan University, Konya, Türkiye, as Associate Professor from 2016 to 2020. He is currently engaged in research activities involving plant biotechnology, plant tissue culture, genetic transformation, phytoremediation, stress physiology, plant-based biopesticides, medicinal and aquatic plants, and nanoparticles. He is also working as an active consultant for plant tissue culture labs. He has published 95 SCIE articles, and cumulatively more than 100 international research articles. He is co-editor of 1 book and wrote more than 30 international book chapters since 2018. Currently, he is working on application of artificial intelligence (AI) in plant biotechnology and plant sciences and published research articles in world-renowned journals in 2022 and 2023. Currently, his h-index is 24 with more than 1690 citations on Google Scholar.

Contributors

Asim Abbasi Department of Entomology, University of Agriculture, Faisalabad, Pakistan

Banzeer Ahsan Abbasi Department of Botany, Rawalpindi Women University, Rawalpindi, Pakistan

Nader R. Abdelsalam Agricultural Botany Department, Faculty of Agriculture (Saba Basha), Alexandria University, Alexandria, Egypt

Hafiz Muhammad Ahmad Department of Bioinformatics and Biotechnology, Government College University, Faisalabad, Pakistan

Hiba Alatrash Department of field Crops, Faculty Agricultural Engineering, University of Aleppo, Aleppo, Syria

Amjad Ali Department of Plant Protection, Faculty of Agricultural Sciences and Technologies, Sivas University of Science and Technology, Sivas, Turkey

Hayssam M. Ali Department of Botany and Microbiology, College of Science, King Saud University, Riyadh, Saudi Arabia

Md. Yousuf Ali Division of Maize Breeding, Bangladesh Wheat and Maize Research Institute, Dinajpur, Bangladesh

Yousuf Ali Division of Maize Breeding, Bangladesh Wheat and Maize Research Institute, Dinajpur, Bangladesh

Muhammad T. Altaf Department of Field Crops, Faculty of Agriculture, Recep Tayyip Erdoğan University, Rize/Pazar, Turkey

Muhammad Tanveer Altaf Department of Field Crops, Faculty of Agriculture, Recep Tayyip Erdoğan University, Rize/Pazar, Turkey

Abdel Razzaq Al-Tawaha National Agricultural Research Center, Baqa, Jordan

Shazia Arif Department of Plant Breeding and Molecular Genetics, University of Poonch, Rawalkot, Pakistan

Jaber Bin Azim Division of Maize Breeding, Bangladesh Wheat and Maize Research Institute, Dinajpur, Bangladesh

Sandeep Bagri ICAR-Central Institute for Cotton Research, Regional Station, Sirsa, Haryana, India

Faheem Shehzad Baloch Department of Biotechnology, Faculty of Science, Mersin University, Mersin, Turkey

Department of Plant Resources and Environment, Jeju National University, Jeju, South Korea

Mehmet Bedir Department of Plant Production and Technologies, Faculty of Agricultural Sciences and Technologies, Sivas University of Science and Technology, Sivas, Turkey

Somsole Bharath Division of Genetics, ICAR- Indian Agricultural Research Institute, New Delhi, India

Uttarayan Dasgupta Division of Genetics, ICAR- Indian Agricultural Research Institute, New Delhi, India

Suman Dutta School of Agriculture and Rural Development, Ramakrishna Mission Vivekananda Educational and Research Institute, Howrah, West Bengal, India

Shah Fahad Department of Biology/Microbiology, South Dakota State University, Brookings, SD, USA

Amin Fathi Department of Agronomy, Ayatollah Amoli Branch, Islamic Azad University, Amol, Iran

Sundas Fatima Department of Botany, Division of Science and Technology, University of Education, Lahore, Pakistan

Sajid Fiaz Department of Plant Breeding and Genetics, The University of Haripur, Haripur, Pakistan

School of Life Sciences, Henan Institute of Science and Technology, Xinxiang, Henan, China

Institute of Molecular Biology and Biotechnology, The University of Lahore, Lahore, Pakistan

Xinyue Gu Key Laboratory of Specialty Agri-product Quality and Hazard Controlling Technology, College of Life Sciences, China Jiliang University, Hangzhou, People's Republic of China

Sumaira Hafeez Department of Plant Breeding and Molecular Genetics, University of Poonch, Rawalkot, Pakistan

Awais Hanif Department of Plant Breeding and Genetics, University of Agriculture, Faisalabad, Pakistan

Adeela Haroon Department of Botany, The Women University Multan, Multan, Pakistan

Ahtsham Ul Hassan Centre of Agricultural Biochemistry and Biotechnology (CABB), University of Agriculture, Faisalabad, Pakistan

Shah Hassan Department of Agricultural Extension Education & Communication, The University of Agriculture, Peshawar, Khyber Pakhtunkhwa, Pakistan

Aiman Hina Soybean Research Institute, Ministry of Agriculture (MOA) Key Laboratory of Biology and Genetic Improvement of Soybean (General), MOA National Centre for Soybean Improvement, State Key Laboratory for Crop Genetics and Germplasm Enhancement, Nanjing Agricultural University, Nanjing, China

Akbar Hossain Division of Soil Science, Bangladesh Wheat and Maize Research Institute, Dinajpur, Bangladesh

Shumaila Ijaz School of Biomedical Engineering, Shenzhen University Medical School, Shenzhen University, Shenzhen, China

Aleksandra Ilić Institute of Field and Vegetable Crops, National Institute of the Republic of Serbia, Novi Sad, Serbia

Noshin Ilyas Department of Botany, PMAS Arid Agriculture University, Rawalpindi, Pakistan

Shahin Imran Department of Agronomy, Khulna Agricultural University, Khulna, Bangladesh

Javed Iqbal Department of Botany, Bacha Khan University, Charsadda, Khyber Pakhtunkhwa, Pakistan

Saira Ishaq Department of Food Science and Technology, University of Poonch, Rawalkot, Pakistan

Touhidul Islam Division of On-Farm Research, Bangladesh Wheat and Maize Research Institute, Dinajpur, Bangladesh

Yaman Jabbour Department of field Crops, Faculty Agricultural Engineering, University of Aleppo, Aleppo, Syria

Amna Jamil Department of Horticultural Sciences, The Islamia University of Bahawalpur, Bahawalpur, Pakistan

Rania Jammal Department of Clinical Pharmacy, Jordan University of Science and Technology, Irbid, Jordan

Zhihan Jiang Key Laboratory of Specialty Agri-product Quality and Hazard Controlling Technology, College of Life Sciences, China Jiliang University, Hangzhou, People's Republic of China

Nitish Joshi Department of Agronomy, Horticulture & Plant Science, South Dakota State University, Brookings, SD, USA

Md. Parvez Kabir Division of Soil Science, Bangladesh Wheat and Maize Research Institute, Dinajpur, Bangladesh

Sobia Kanwal Department of Biology and Environmental Sciences, Allama Iqbal Open University, Islamabad, Pakistan

Arun Karnwal Department of Microbiology, School of Bioengineering and BioSciences, Lovely Professional University, Phagwara, India

Department of Microbiology, Graphic Era (Deemed to be University), Dehradun, Uttarakhand, India

Xubo Ke Key Laboratory of Specialty Agri-product Quality and Hazard Controlling Technology, College of Life Sciences, China Jiliang University, Hangzhou, People's Republic of China

Muhammad Nasir Rasheed Khan Department of Agronomy, Abdul Wali Khan University Mardan, Khyber Pakhtunkhwa, Pakistan

Muhammad Tajammal Khan Division of Science and Technology, Department of Botany, University of Education, Lahore, Pakistan

Kağan Kökten Department of Plant Production and Technologies, Faculty of Agricultural Sciences and Technologies, Sivas University of Science and Technology, Sivas, Turkey

Waqas Liaqat Department of Field Crops, Faculty of Agriculture, Institute of Natural and Applied Sciences, Çukurova University, Adana, Turkey

Syed Hamza Mahfooz Barani Agriculture Research Institute, Chakwal, Pakistan

Tariq Mahmood Department of Plant Sciences, Faculty of Biological Sciences, Quaid-i-Azam University, Islamabad, Pakistan

Muhammad Majeed Department of Botany, University of Gujrat, Punjab, Pakistan

Zlatica Mamlić Institute of Field and Vegetable Crops, National Institute of the Republic of Serbia, Novi Sad, Serbia

Hurmat Mehdi Mirrani Department of Botany, Division of Science and Technology, University of Education, Lahore, Pakistan

Julius S. Missanga Department of Biology, University of Dodoma (UDOM), Dodoma, Tanzania

Parnaz Mortazavi Department of Plant Production and Technologies, Faculty of Agricultural Sciences and Technologies, Sivas University of Science and Technology, Sivas, Turkey

Akhtar Munir Department of Chemistry, Quaid-I-Azam University, Islamabad, Pakistan

Department of Chemistry, University of Sialkot, Sialkot, Pakistan

Muhammad Azhar Nadeem Department of Biotechnology, Faculty of Science, Mersin University, Mersin, Türkiye

Taufiq Nawaz Department of Biology/Microbiology, South Dakota State University, Brookings, SD, USA

Maha Noor Department of Botany, Division of Science and Technology, University of Education, Lahore, Pakistan

Zahra Noreen Department of Botany, Division of Science and Technology, University of Education, Lahore, Pakistan

Muhammad Farhan Khan Pasha Pakistan Oilseed Department, Ministry of National Food Security and Research, Islamabad, Pakistan

Debashis Paul ICAR-Central Institute for Cotton Research, Regional Station, Sirsa, Haryana, India

Ali M. Qaisi Department of Pharmaceutical Sciences, School of Pharmacy, University of Jordan, Amman, Jordan

Sarmad A. Qureshi Department of Plant Production and Technologies, Faculty of Agricultural Sciences and Technologies, Sivas University of Science and Technology, Sivas, Turkey

Md. Mobinur Rahman Division of Maize Breeding, Bangladesh Wheat and Maize Research Institute, Dinajpur, Bangladesh

Muhammad Aneeq Ur Rahman Biotechnology Research Institute, Graduate School of Chinese Academy of Agricultural Sciences, Beijing, China

Tanzeel Ur Rahman Department of Hotel Management and Tourism, University of Swabi, Khyber Pakhtunkhwa, Pakistan

Jana Rammal Lebanese University, Faculty of Agronomy and Veterinary, Beirut, Lebanon

Madiha Rashid Department of Botany, Division of Science and Technology, University of Education, Lahore, Pakistan

Department of Botany, Lahore College for Women University, Lahore, Pakistan

Nadia Riaz Department of Botany, Lahore College for Women University, Lahore, Pakistan

Sampa Saha ICAR-National Bureau of Plant Genetic Resources, New Delhi, India

Palani Saranraj Department of Microbiology, Sacred Heart College (Autonomous), Tirupattur, Tamil Nadu, India

Meliha F. Sarıkaya Department of Plant Production and Technologies, Faculty of Agricultural Sciences and Technologies, Sivas University of Science and Technology, Sivas, Turkey

Shah Saud College of Life Science, Linyi University, Linyi, Shandong, People's Republic of China

Mohamad Shatnawi Department of Biotechnology, Faculty of Agricultural Technology, Al-Balqa Applied University, Amman, Jordan

Muhammad Sheraz Fodder and Forrage Program, Crop Sciences Institute, NARC, Islamabad, Pakistan

Sabeen Rehman Soomro Department of Plant Production and Technologies, Faculty of Agricultural Sciences and Technologies, Sivas University of Science and Technology, Sivas, Turkey

Salma Naimatullha Soomro Department of Plant Production and Technologies, Faculty of Agricultural Sciences and Technologies, Sivas University of Science and Technology, Sivas, Turkey

Muhammed Tatar Department of Plant Protection, Faculty of Agricultural Sciences and Technologies, Sivas University of Science and Technology, Sivas, Turkey

Abdel Rahman M. Tawaha Department of Biological Sciences, Al Hussein bin Talal University, Maan, Jordan

Aasma Tufail Department of Botany, Division of Science and Technology, University of Education, Lahore, Pakistan

Ana Uhlarik Institute of Field and Vegetable Crops, National Institute of the Republic of Serbia, Novi Sad, Serbia

Zakir Ullah Department of Plant Sciences, Faculty of Biological Sciences, Quaid-i-Azam University, Islamabad, Pakistan

Pavithravani B. Venkataramana Department of Sustainable Agriculture, Nelson Mandela African Institution of Science and Technology (NM-AIST), Arusha, Tanzania

Pei Xu Key Laboratory of Specialty Agri-product Quality and Hazard Controlling Technology, College of Life Sciences, China Jiliang University, Hangzhou, People's Republic of China

Jia Yao Key Laboratory of Specialty Agri-product Quality and Hazard Controlling Technology, College of Life Sciences, China Jiliang University, Hangzhou, People's Republic of China

Tabassum Yaseen Department of Botany, Bacha Khan University, Charsadda, Khyber Pakhtunkhwa, Pakistan

Afifa Younas Department of Botany, Division of Science and Technology, University of Education, Lahore, Pakistan

Department of Botany, Lahore College for Women University, Lahore, Pakistan

Sadaf Zahra Department of Bioinformatics and Biotechnology, Government College University, Faisalabad, Pakistan

Abiotic Stress and Legume Production 1

Uttarayan Dasgupta, Somsole Bharath, and Suman Dutta

Abstract

Because they fix nitrogen, improve soil fertility, and provide vital nutrients, legumes are critical to world agriculture. However, environmental stressors, especially abiotic ones like drought, salt, high temperatures, and nutritional deficits, frequently reduce their output. This chapter explains stress response mechanisms and mitigation techniques while thoroughly examining the impact of abiotic stress on the production of legumes. It is essential to comprehend how abiotic stress affects legumes. Reduced yield and quality are the result of essential processes like photosynthesis and nitrogen absorption being disrupted by drought stress. Growth is disrupted by salinity stress because it modifies metabolic pathways and water balance. Hot or cold, extreme temperatures hinder development and reproduction, which affects seed set and yield. Deficits in some nutrients, particularly potassium, phosphate, and nitrogen, further limit productivity by interfering with vital physiological and biochemical functions. Intricate coping mechanisms have evolved by legumes to withstand abiotic stress. Key adaptive responses include osmotic adjustment, antioxidant defense mechanisms, hormone modulation, and gene expression regulation. These systems control hormonal signaling, scavenge reactive oxygen species, preserve cellular homeostasis, and alter the expression of genes that respond to stress. There are several ways to mitigate abiotic stress in the development of legumes. Conventional and molecular methods are used in stress tolerance breeding to create resilient cultivars. Important strategies include genetic engineering, biological methods that use beneficial microbes, and efficient crop

U. Dasgupta · S. Bharath
Division of Genetics, ICAR- Indian Agricultural Research Institute, New Delhi, India

S. Dutta (✉)
School of Agriculture and Rural Development, Ramakrishna Mission Vivekananda Educational and Research Institute, Howrah, West Bengal, India

© The Author(s), under exclusive license to Springer Nature Singapore Pte Ltd. 2025
M. A. Nadeem et al. (eds.), *Marker-Assisted Breeding in Legumes for Drought Tolerance*, https://doi.org/10.1007/978-981-96-4112-3_1

management techniques. Other crucial tactics include adopting precision agricultural technology, controlling soil health, increasing water usage efficiency, and putting climate-smart agriculture policies into effect. Moreover, capacity-building programs and post-harvest management techniques enable farmers to successfully respond to stress-related difficulties. In conclusion, addressing abiotic stress in legume production is vital for ensuring sustainable agriculture and food security. Through interdisciplinary research and collaborative efforts, resilient legume varieties and sustainable practices can be developed to mitigate the adverse effects of abiotic stress, ensuring productivity and resilience in the face of changing environmental conditions.

Keywords

Legumes · Abiotic stress · Mitigation strategies · Environmental challenges · Crop resilience · Stress response mechanisms

1.1 Introduction

Legumes belong to the Leguminosae/Fabaceae family, which includes a diverse array of species such as beans, peas, lentils, soybeans, and peanuts. These plants are renowned for their nutrient-rich seeds or pods, which play a crucial role in human and animal nutrition, as well as in sustainable agricultural practices. Legumes provide an excellent source of essential amino acids and are a significant source of protein (Messina 1999). They have a high dietary fiber content, which aids in proper digestion and blood sugar regulation (Mudryj et al. 2014). They are also a good source of many vitamins and minerals, including potassium, zinc, iron, and folate (Messina 1999; Ali et al. 2022). In many parts of the world, especially in poor nations, legumes are a staple diet because they offer an inexpensive and wholesome source of protein, promoting food security and thwarting malnutrition (Foyer et al. 2016; Akibode and Maredia 2011). Through a symbiotic interaction with rhizobia bacteria in their root nodules, legumes may fix atmospheric nitrogen, negating the need for synthetic nitrogen fertilizers (Peoples et al. 2009). This capacity to fix nitrogen contributes to increased soil fertility and the adoption of sustainable farming methods (Unkovich et al. 2008). Additionally, they can be added to crop rotation systems to improve soil health and disrupt pest and disease cycles (Unkovich et al. 2008; Abbas et al. 2023). In addition to their nutritional benefits, legumes have a lower carbon footprint compared to animal proteins (Nadeem et al. 2021). They can be cultivated as cash crops for many farmers in developing countries (Daryanto et al. 2015) and are often used as animal feed and green manure, making them indispensable to human civilization. Worldwide, legumes are primarily grown as pulses or oilseeds, and in some cases, as green manure crops. In 2022, global pulse production reached 120 million tons, including beans, chickpeas, cowpeas, Bambara beans, broad and horse beans (dry), lentils, lupins, dry peas, pigeon peas, string beans, and other pulses (FAOSTAT 2024). Additionally, significant quantities of other legumes, such as soybeans and groundnuts, were produced, totaling 348.856 and 54.238 million tons, respectively (FAOSTAT 2024).

Unfavorable environmental circumstances or other elements that have a detrimental effect on a plant's growth, development, and production are referred to as abiotic stress. Extreme temperatures both high and low drought, floods, salinity, heavy metals, and radiation are some of these stressors. Unfavorable circumstances have the potential to impede a number of physiological processes in plants, including water absorption, photosynthesis, and nutrient uptake. This can lead to a decrease in crop yields and an impact on food security. In light of climate change and other environmental issues, improving agricultural productivity and guaranteeing a sustainable food supply depend heavily on comprehending and reducing the impacts of abiotic stress. These stresses, which might include a range of environmental elements like temperature, water availability, soil conditions, light intensity, and pollution, are considered nonliving or nonbiological. There has been a recent increase in the occurrence of abiotic stress extremities owing to climate change (Onyekachi et al. 2019). Crop yields have been significantly impacted by climate change in recent years, with yield declines observed in many places as a result of numerous climate-related issues. According to a study by Battisti et al. (2018), soybean yields in Brazil decreased by 9.1% on average between 1980 and 2015 as a result of climate change. The scientists ascribed the decrease in yield to increased temperatures, altered precipitation patterns, and an increase in the frequency of extreme weather occurrences. The aim of this chapter is to explore the importance of legumes in nutrition, sustainable agriculture, and environmental health. It highlights the nutrient-rich seeds and pods of various legume species, emphasizing their role in providing essential amino acids, proteins, dietary fiber, vitamins, and minerals. The chapter also discusses the socioeconomic benefits of legumes, particularly in developing countries, and their contribution to food security. Additionally, it examines the ability of legumes to fix atmospheric nitrogen, enhancing soil fertility and promoting sustainable farming. The chapter also addresses the impact of abiotic stress on legume production, especially in the context of climate change, and its implications for agricultural productivity and food security.

1.1.1 Effect of Abiotic Stress on Legume Production

The development of legumes is severely hampered by abiotic challenges, which include high temperatures, drought, salt, floods, and heavy metal pollution. These conditions interfere with vital physiological functions including photosynthesis, nutrient uptake, and water absorption (Altaf et al. 2023). These adverse circumstances may cause legume crops to develop more slowly, produce fewer beans, and have lower-quality beans. Climate change is making these pressures more frequent; therefore it is critical to understand how they affect legumes to create resilient agriculture techniques. To maintain legume output and guarantee food security, countermeasures including cultivating stress-tolerant cultivars and enhancing soil management are crucial. Details of different legumes with their most important abiotic stresses are given in Table 1.1.

Table 1.1 List of major cultivated legumes and their most important abiotic stresses

S. no.	Legume crop	*Scientific name*	Affected abiotic stresses and tolerance	References
1.	French bean	*Phaseolus vulgaris*	Drought, Al toxicity susceptible	Souter et al. (2017) and Thung and Rao (1999)
2.	Chickpea	*Cicer arietinum* L.	Drought, terminal heat stress, low temperature, water-logging, salinity susceptible	Jha et al. (2014) and Arriagada et al. (2022)
3.	Lentil	*Lens culinaris*	Heat, cold, drought, salinity, flooding & submergence susceptible	Singh et al. (2022a) and Aski et al. (2023)
4.	Soybean	*Glycine max* (L.) Merr.	Drought, heat, cold, salinity, flood, acidity, mineral toxicity susceptible	Ratnaparkhe et al. (2022)
5.	Mungbean/ greengram	*Vigna radiate* L.	Drought, heat, salinity, cadmium toxicity susceptible	Pathak et al. (2023), Chaudhary et al. (2022) and Manu et al. (2022)
6.	Blackgram	*Vigna mungo*	Drought, heat, salinity, cold, flooding and submergence susceptible	Manu et al. (2022)
7.	Lathyrus	*Lathyrus sativus*	Flood, drought, low temperature, salinity tolerant	Patto et al. (2006) and Gonçalves et al. (2022)
8.	Fababean	*Vicia faba*	Cold, heat, drought, acidity, salinity susceptible	Maalouf et al. (2022)
9.	Indian bean/ hyacinth bean	*Lablab purpureus*	Heat, drought, salinity susceptible	Rai et al. (2018)
10.	Cowpea	*Vigna unguiculata* (L.) Walp.	Heat, cold, drought, salinity, Al toxicity susceptible	Veeranagappa et al. (2022) and Sahay et al. (2022)
11.	Tepary bean	*Phaseolus acutifolius* A.Gray	Cold, drought tolerant	Souter et al. (2017) and Buitrago-Bitar et al. (2021)
12.	Moth bean	*Vigna acontifolia* (Jacq.) Maréchal	Drought, heat, salinity tolerant	Gayacharan et al. (2023) and Tiwari et al. (2018)
13.	Rice bean	*Vigna umbellata* (Thunb.) Ohwi et Ohashi	Drought, heavy metal toxicity susceptible	Atta et al. (2021, 2022)
14.	Horsegram	*Macrotyloma uniflorum*	Drought tolerant	Jha et al. (2022)
15.	Pea	*Pisum sativum*	Heat, cold, drought, salinity, waterlogging susceptible	Gondalia et al. (2022)

(continued)

Table 1.1 (continued)

S. no.	Legume crop	*Scientific name*	Affected abiotic stresses and tolerance	References
16.	Pigeon pea	*Cajanus cajan*	Terminal drought, waterlogging, salinity, Al toxicity susceptible	Nandini et al. (2022)
17.	Groundnut	*Arachis hypogaea* L.	Drought, heat, salinity, cold, heavy metal toxicity susceptible	Aravind et al. (2022) and Kadirimangalam et al. (2022)

1.1.2 Drought Stress

A state of water deficit in plants caused by insufficient soil water supply caused by either low rainfall, high rates of evapotranspiration, or a combination of the two is known as drought stress. It happens when the amount of water evaporating or transpiring is greater than the amount of water absorbed by the roots of the plant (Farooq et al. 2009). Legumes cultivated by mankind give varying responses to water/drought stress, and some are tolerant up to a level, but some are extremely susceptible, but in all cases, there is a significant reduction in their final yield. Reduced assimilate translocation and carbon fixation, restricted flowering time and its impact on reproductive organs, pollen grain sterility, fewer pods, poorer grain set, and decreased sink activity are all associated with this (Nadeem et al. 2019a, b). Legume growth and development is impacted by drought in a number of ways, including germination, the development of shoots and roots, photosynthesis, and the reproductive stage. Drought is becoming one of the most unpredictable and uncontrollable factors restricting crop productivity and negatively affecting legume crops as a result of global climate change. Grain legumes are particularly susceptible to drought as they are mostly cultivated in rain-fed areas without sufficient infrastructure for irrigation (Khatun et al. 2021). The effects of drought on grain legumes are summarized in Table 1.2.

1.1.3 Salinity Stress

Salinity affects capacity of grain legumes for biological nitrogen fixation, germination and growth, and reproductive stage, all of which have disastrous effects on plants. Salinity can be a major obstacle as early as the germination period, preventing the embryo from absorbing water. It can also prevent the growth of the embryo axis by preventing starch from being mobilized (Farooq et al. 2015; Farooq et al. 2017a, b). Legume plants subjected to salinity stress may experience stunted development, reduced plant height, and decreased biomass accumulation (Essa 2002). It can also result in leaf chlorosis, necrosis, and early senescence, which lowers the ability of plants to photosynthesize (Essa 2002; Sehrawat et al. 2013). The nodulation process and the symbiotic interaction between legumes and rhizobia bacteria can both be adversely affected by salinity stress (Elsheikh and Wood 1995; Zahran

Table 1.2 Effect of drought stress on grain legumes

Effects	Target areas in plants	References
Morphological effects	Reduction in plant growth, leaf area, pod number due to impaired cell division, and cell expansion and differentiation	Taiz et al. (2015) and Khatun et al. (2021)
Physiological effects	Increase in leaf temperature, decrease in chlorophyll content, lower photosynthetic rate, reduction in transpiration, decrease in leaf-water potential	Chowdhury et al. (2017), Anjum et al. (2011), Fathi and Tari (2016), Makbul et al. (2011) and Khatun et al. (2021)
Morpho-physiological effects	Reduced germination, reduction in flowering and grain-filling period, reduced pod production, and high pod abortion. Reduction in nutrient levels in grains and deterioration of quality	Shrestha et al. (2006), Ghanbari et al. (2013), Khatun et al. (2021) and Farooq et al. (2017a, 2017b)
Physio-biochemical effects	ROS buildup and oxidative stress	Farooq et al. (2009) and Khatun et al. (2021
Molecular effects	Differential expression of certain genes like LEA proteins and lipid transfer proteins	Gao et al. (2008)

1999). Excessive levels of salt can also prevent root nodules from forming and functioning, which lowers nitrogen fixation and impacts plant growth and output (Elsheikh and Wood 1995; Sehrawat et al. 2013). High salt levels can lead to fewer pods, smaller seeds, and reduced seed weight, affecting the overall yield and quality of the crop. It can cause critical nutrients to be poorly absorbed and translocated, which can result in toxicities or nutrient deficits in legumes (Munns and Tester 2008; Shabala and Cuin 2008). In addition to inducing oxidative stress, high salt levels can potentially harm cells and interfere with metabolic activities by producing more reactive oxygen species (ROS) (Hasanuzzaman et al. 2013; Sehrawat et al. 2013).

1.1.4 Heat Stress

When plants are subjected to temperatures over their ideal range for an extended length of time, it can cause a variety of physiological and metabolic disruptions that can have a detrimental effect on growth, development, and productivity. Elevated temperatures have the potential to impede bean seed germination and seedling establishment (Kumar et al. 2011). Heat stress can harm embryos and decrease mobilization of seed reserves, which can have an impact on germination. It can result in reduced biomass accumulation, decreased leaf area, and stunted development (Awasthi et al. 2014). Elevated temperatures may result in leaf burning, chlorosis, and early senescence, thereby diminishing the plant's ability to photosynthesize (Kumar et al. 2011). Reduced pod and seed set and poor flower development, pollination, and fertilization in legumes can result from heat stress (Awasthi et al. 2014). Increased temperatures in the reproductive stage can lead to ovule abortion, pod dropping, and pollen sterility, all of which can diminish yield (Kumar et al. 2011). It can also cause reactive oxygen species (ROS) to be produced, which can damage cell membranes and cause oxidative stress (Awasthi et al. 2014). Elevated temperatures have the potential to compromise membrane stability, impacting the movement of water, nutrients, and other vital components (Kumar et al. 2011). Reduced nodulation and nitrogen fixation are two detrimental effects of heat stress on the symbiotic connection between legumes and nitrogen-fixing rhizobia bacteria (Awasthi et al. 2014). Rhizobia's ability to survive and function can be hampered by high temperatures, which reduces the amount of nitrogen available to the plant (Kumar et al. 2011).

1.1.5 Cold Stress

When a specific plant species or cultivar experiences a temperature decrease below its minimal threshold, it is known as cold stress. This can cause a variety of physiological, biochemical, and molecular changes that can hinder the plant's ability to grow, develop, and produce. In legumes, low temperatures can impede or postpone seed germination and seedling emergence. In addition to reducing the mobilization of seed reserves and harming the embryo, cold stress can also have an adverse effect on early seedling growth and germination. Legume plants subjected to cold stress may experience slowed development, decreased leaf area, and decreased biomass accumulation. It can reduce a plant's ability to photosynthesize by causing leaf chlorosis, necrosis, and early senescence. In legumes, cold stress can have a detrimental effect on flower development, pollination, and fertilization, which can diminish pod and seed set. Low temperatures during the reproductive stage can lead to pod dropping, ovule abortion, and pollen sterility, all of which can reduce production. Reactive oxygen species (ROS) can be produced by cold stress, which can cause oxidative stress and membrane damage in legumes. Chilling stress can interfere with the stability of membranes, which can impact the movement of nutrients, water, and other vital components.

1.1.6 Flooding and Submergence Stress

Flooding stress is the state that results from a plant's roots being partly or fully submerged in water for an extended length of time, which causes hypoxia or an oxygen shortage in the root zone. This may happen as a result of heavy rains, inadequate soil drainage, or other factors that produce soggy conditions. Plant growth and development can be impeded by flooding and submersion, which reduces biomass accumulation and production. This is mostly because of the restricted oxygen supply in the root zone, which has an impact on a number of physiological functions (Linkemer et al. 1998; Bharadwaj et al. 2023). Waterlogging can impair the uptake of essential nutrients, such as nitrogen, phosphorus, and potassium, from the soil due to restricted root growth and function (Palta et al. 2010). Extended periods can also result in early leaf senescence and chlorosis (yellowing) of the leaves, which lowers the plant's ability to photosynthesize (Bharadwaj et al. 2023). It can result in hypoxia, or an oxygen shortage, in the root zone, which can harm roots and impede their ability to grow and produce biomass (Bacanamwo and Purcell 1999). Legumes' nodulation process and nitrogen fixation, which are essential to their growth and productivity, can be negatively impacted by flooding and submersion (Bacanamwo and Purcell 1999). Circumstances that are wet can provide ideal conditions for soil-borne pathogens, making legumes more vulnerable to illness (Bharadwaj et al. 2023). The degree and length of the flooding or submergence stress, the stage at which the legume crop is growing, and the particular species or cultivar can all affect these consequences.

1.1.7 Mineral Toxicity Stress

Exposure of plants to elevated concentrations of specific minerals can lead to mineral toxicity stress, which can have detrimental effects on their growth, development, and production. Heavy metals like cadmium (Cd) and lead (Pb), as well as common minerals like manganese (Mn) and aluminum (Al), can all accumulate to hazardous amounts in soil. This stress is more common in acidic soils because the minerals become more soluble there and are therefore more accessible to plants (Brunetti et al. 2011; Clemens 2006). One of the main obstacles in acidic soils ($pH < 5.5$) is aluminum toxicity. Plants that are stunted and have poor nutrient and water uptake are the result of Al^{3+} ions in the soil solution inhibiting root growth and function. Al poisoning mostly damages the apex of the root, preventing cell division and elongation (Kochian et al. 2015). As a result, the root surface area is decreased, which restricts the plant's capacity to search the soil for water and nutrients. Plants may be poisoned by too much Mn, especially in soggy or acidic soils. Chlorosis, or the yellowing of leaves, and necrotic patches, particularly on older leaves, are signs of magnesium poisoning. The absorption and use of other vital nutrients, such as iron (Fe), magnesium (Mg), and calcium (Ca), are hampered by high Mn levels (Millaleo et al. 2010). Reactive oxygen species (ROS) are produced when heavy metals like cadmium and lead are present, and this can lead to oxidative damage to cellular constituents such as proteins, lipids, and nucleic acids. Chromosome abnormalities, mutations, and DNA damage can all be brought on by heavy metals, which may have long-term consequences for the growth and development of plants (Gallego et al. 2012; Hossain et al. 2012; Singh et al. 2022a, b).

1.2 Mechanisms of Abiotic Stress Response in Legumes

Legumes exhibit remarkable adaptability to harsh environmental conditions through a multitude of responses. These reactions involve complex interactions among genes, fostering communication and synergy across diverse pathways. At various levels ranging from the entirety of the plant to intricate cellular and molecular mechanisms, these adaptations manifest through physiological and biochemical adjustments. Each response is finely tuned to the specific stressors encountered, ensuring survival and growth. This intricate web of adaptation showcases the sophistication of plant biology and highlights the dynamic nature of plant-environment interactions, ultimately contributing to the resilience and success of plant life in diverse ecological settings. A few of the mechanisms are described below.

1.2.1 Osmotic Adjustment

Accumulating osmoprotectants, including proline, is a typical response in numerous plants including legumes under diverse stresses, such as water scarcity (Verbruggen and Hermans 2008; Yoshiba et al. 1997). Plants utilize δ(1)-pyrroline-5-carboxylate (P5C) and δ (1)-pyrroline-5-carboxylate synthetase (P5CS) and δ (1)-pyrroline-5-carboxylate reductase (P5CR) to generate proline from L-glutamic acid (L-Glu). Similar to many other metabolites, proline can be converted back to L-Glu via P5C, where two enzymes are involved: P5C dehydrogenase (EC 1.5.1.12) and proline dehydrogenase (oxidase) (ProDH; EC 1.5.99.8) (Kim and Nam 2013). Enhancing proline accumulation has been employed to improve drought resistance in plants, such as legumes. Soybean plants engineered to overexpress a cDNA for P5CR show increased proline accumulation and serve as an excellent model for studying proline's role in enhancing drought resistance (De Ronde et al. 2004). Additionally, altering proline levels impacts the concentrations of various other amino acids during stress, indicating that their metabolic pathways are coordinately regulated (Simon-Sarkadi et al. 2005). Adjusting trehalose metabolism in *Medicago truncatula* appears to be a promising method to boost resistance to water shortage. This method helps plants recover from extreme water scarcity and efficiently utilize short rain periods during prolonged droughts. Ion balance is crucial for adapting to salt stress in tolerant plants. In *Medicago truncatula*, the TN1.11 genotype, which is salt-tolerant, shows greater root growth under saline conditions and accumulates more Na^+ and Cl^- ions in shoots than roots. In comparison, the Jemalong A17 genotype, which is moderately salt-sensitive, exhibits different ion distribution. This suggests a connection between ion transport and growth (Zahaf et al. 2012). Linked to oxidative stress, the buildup of osmoprotectants such as amino acids and sugars in the cytoplasm of pea plants contributes to maintaining high turgor pressure and stabilizing proteins and membranes, thus mitigating stress-induced damage. A metabolomic study of drought-stressed pea plants revealed notable elevations in proline, valine, myo-inositol, GABA, threonine, homoserine, and trigonelline levels (Sánchez et al. 1998).

1.2.2 Antioxidant Defense System

Under stress, plants generate increased levels of reactive oxygen species (ROS), reactive nitrogen species (RNS), and reactive sulfur species (RSS) (Matamoros and

Becana 2021). These must be regulated by antioxidant enzymes and metabolites to prevent toxicity and facilitate their role in signaling. Hydrogen peroxide (H_2O_2), superoxide radical (O^{2-}), nitric oxide (NO), and hydrogen sulfide (H_2S) are crucial for plant acclimation to abiotic stress (Choudhury et al. 2017). The signaling capacity and bioactivity of ROS, RSS, and RNS largely arise from their ability to induce post-translational alterations of protein amino acid side chains. These post-translational alterations can alter protein function, leading to changes in activity or even protein degradation (Friso and van Wijk 2015). ROS are produced in plants under various environmental stresses due to photorespiration and an imbalance between carbon dioxide and electrons from light reactions. This imbalance occurs during drought, cold, heat, high light intensity, UV radiation, or salinity stress (Shao et al. 2007). In leaves, ROS are primarily generated in the apoplast by NADPH oxidases (also called respiratory burst oxidase homologues, RBOHs) and by other oxidases and peroxidases. Additionally, ROS are produced in the chloroplasts, mitochondria, and peroxisomes (Choudhury et al. 2017). In plant cells, the system that scavenges reactive oxygen intermediates enzymatically includes superoxide dismutase (SOD), catalase (CAT), ascorbate peroxidase (APX), monodehydroascorbate reductase (MDAR), dehydroascorbate reductase (DHAR), glutathione peroxidase (GPX), and glutathione reductase (GR), all part of the water-water cycle (Asada 2000). Rasool et al. (2013) observed that chickpea genotypes SKUA-06 and SKUA-07 show salinity tolerance due to effective enzymatic antioxidants. SOD, CAT, APX, and GR help manage reactive oxygen species (ROS), maintaining redox balance and cellular integrity. In cells and organelles, stress-induced redox changes typically cause an oxidative shift. However, how these redox state variations lead to specific responses is still unclear. Factors like the concentration, exposure time, subcellular location, and types of ROS, RNS, and RSS are important. Loss of homeostasis of these reactive molecules in each cell compartment may modify susceptible amino acids, primarily sulfur-containing ones like cysteine and methionine, but also tyrosine, lysine, and arginine. These modifications can change enzyme and transcription factor activities, affecting metabolic processes, signaling pathways, and gene expression (Matamoros and Becana 2021).

1.2.3 Hormonal Regulation

Plants can detect environmental changes by creating nonheritable modifications, which show the plant's morphological and physiological flexibility, or by creating heritable changes (Llanes et al. 2015). These reactions are controlled by phytohormones, which are compounds from plant biosynthetic pathways. Phytohormones can act at their production site or, after being transported, in other parts of the plant. These regulators are crucial for adapting to abiotic stresses by controlling physiological responses (Javid et al. 2011). Abscisic acid (ABA), a key phytohormone, plays a central role in managing these responses. The quick buildup of ABA during stress and its regulation of various adaptive responses make it known as "the stress hormone," aiding plant survival under adverse conditions (Davies and Zhang 1991).

Early research showed that ABA functions as a long-distance signal for water stress, indicating soil drying (Hartung 1999). In legumes, ABA is quickly transported from roots to shoots, where it builds up in the apoplast of leaf guard cells in response to drought. In lupins subjected to water deficit, ABA production increases, which correlates with reduced stomatal conductance (Jia et al. 1996). Numerous studies highlight the importance of this phytohormone in legumes and indicate that understanding ABA responses can enhance crop productivity. The role of ABA is typically linked to inhibition processes (Peng et al. 2006; Finkelstein and Rock 2002). However, recent findings indicate it may also promote growth in developing tissues. Drought stress in soybean was shown to raise ABA levels in flowers and pods, correlating with fewer pods setting (Westgate et al. 1996). ABA buildup prompts H_2O_2 production, which signals stress tolerance. Research shows ABA levels rise before H_2O_2 appears, with both playing roles in closing stomata and inhibiting their opening (Lü et al. 2013).

1.2.4 Gene Expression Regulation

Identifying specific genes linked to plant resilience against environmental challenges is pivotal in comprehending their responses and enhancing crop durability. Recent strides in genomics provide an array of tools, such as sequencing entire genomes and transcriptomes, as well as utilizing gene expression microarrays and RNA-seq technology. These methodologies allow for the identification and isolation of genes responsive to environmental stresses across various legume crops. Understanding the function of these genes elucidates the intricate molecular mechanisms underlying stress tolerance. Ultimately, this knowledge aids in the development of legume varieties better equipped to thrive under unfavorable environmental conditions, thereby improving productivity and resilience (Kudapa et al. 2013). Researchers examined the global effects of salt stress and recovery in *Medicago truncatula* at the transcriptional level. This analysis identified 328 genes associated with root growth recovery from salt stress in the salt-sensitive R108 genotype. One key gene identified was MtZPT2-1, a TFIIIA-like transcription factor involved in adjusting root adaptation to salt stress (Merchan et al. 2007). A comparative analysis of salt stress regulation in the root apexes and whole roots of the Jemalong A17 genotype was done using two methods: quantitative real-time RT-PCR for transcription factors in whole roots and Mt16K+ microarray for root apexes. The study identified 824 differentially expressed genes under early salt stress, including 84 transcription factors. Notably, transcription factors showed different regulation in root apexes compared to whole roots (Gruber et al. 2009). In soybean roots, a digital gene expression profile experiment revealed a greater ratio of upregulated to downregulated genes in the salt-tolerant *G. soja* compared to the salt-sensitive *G. max* genotype. This study identified 490 salt-responsive genes, including many transcription factors, involved in various steps of the salt regulatory pathway (Ali et al. 2012). Recent advancements in proteomics and metabolomics have provided an integrated perspective on salt tolerance mechanisms in legumes. These studies have enhanced our understanding of how plants adapt to salt stress. Increasing evidence indicates that transcription factors play a crucial role in regulating plant responses to salt stress (Song et al. 2012). The freezing tolerance response in *M. truncatula* at the molecular level involves the expression of the CBF/DREB1 gene family. This family includes transcriptional activators that bind to the C-repeat/dehydration-responsive element found in many cold-responsive gene promoters. Four CBF genes, MtCBF1-4, have been identified in *M. truncatula*. MtCBF2 and MtCBF3 are quickly induced by low temperatures, and a transient increase in MtCBF1 transcripts has also been observed under these conditions (Pennycooke et al. 2008; Zhang et al. 2010; Li et al. 2011).

1.3 Breeding Strategies for Abiotic Stress Tolerance in Legumes

The development of cultivars that can tolerate a range of environmental stresses, including drought, salt, flooding, severe temperatures, and biotic challenges, is the goal of breeding for stress tolerance in legumes. In this chapter we have mentioned some of the breeding strategies used to improve stress tolerance in legume cultivars.

1.3.1 Conventional Breeding

Phenotypic Selection: Parameters including yield, biomass, and stress-related morphological or physiological features, parent lines, and progenies are chosen based on how well they function under stress (Muñoz-Perea et al. 2006).

Hybridization and Pedigree Breeding: Creating stress-tolerant lines by crossing parents who can withstand stress and moving generations forward through pedigree selection (Gaur et al. 2012).

1.3.2 Marker-Assisted Selection

Quantitative Trait Loci (QTL) Mapping: Utilizing molecular markers to identify genomic areas (QTLs) linked to stress tolerance features, then utilizing these QTLs for multiple exposure syndrome (MAS) (Varshney et al. 2014).

Genome-Wide Association Studies (GWAS): These studies identified important alleles for MAS by linking molecular markers with stress tolerance features in various germplasm collections (Kale et al. 2015).

1.3.3 Genomic Selection

Whole-Genome Prediction Models: Jarquín et al. (2014) describe the use of high-density molecular markers and phenotypic data to create prediction models that aid in the selection of superior genotypes based on their genomic estimated breeding values (GEBVs).

1.3.4 Transgenic Approaches

Overexpression or Silencing of Stress-Responsive Genes: To increase stress tolerance, add or modify the expression of genes implicated in stress tolerance pathways (Bhatnagar-Mathur et al. 2007; Dita et al. 2006).

Pyramiding Several Stress Tolerance Genes: This genetic engineering technique combines several stress tolerance genes into a single genotype (Bhatnagar-Mathur et al. 2007).

1.3.5 Integrated Approaches

Integrating Molecular and Conventional Breeding: Using phenotypic selection along with molecular markers, QTLs, and GS to hasten the creation of stress-tolerant cultivars (Varshney et al. 2014).

Taking Advantage of Natural Genetic Resources: Investigating and applying genetic diversity in landraces and wild relatives of legumes for features related to stress tolerance (Berger et al. 2012; Mallikarjuna et al. 2011).

These methods of breeding have been effectively used to create cultivars that are resistant to stress in a variety of legume crops, including peanuts, chickpeas, soybeans, common beans, and lentils. The intricacy of the stress tolerance features, marker data, and genetic resource availability are major factors that influence the choice of breeding technique.

1.4 Challenges and Future Prospects

There are several important obstacles in the way of breeding legumes to withstand abiotic stress. Stress tolerance characteristics are complicated; they entail complex physiological and biochemical pathways and are frequently regulated by several genes (Araujo et al. 2015). Phenotypic selection and hybridization processes become labor-intensive and time-consuming as a result of complex inheritance of the traits. Furthermore, because of the diversity of the environment, it might be challenging to recognize and precisely measure stress-related features in the field. Because stress tolerance is polygenic, marker-assisted selection and genomic selection need large amounts of molecular data and sophisticated statistical models, which can be resource-intensive and may not always yield accurate findings (Varshney et al. 2014). Furthermore, the broad use of transgenic techniques is limited by regulatory obstacles and public acceptability difficulties. Genetic resources, such as varied germplasm collections, are essential yet sometimes scarce, particularly for legume landraces and their wild cousins. This restricts the genetic variation that breeders may use to create cultivars that can withstand stress. Ultimately, there is still more work to be done in terms of transferring laboratory results into real-world applications, which calls for interdisciplinary cooperation and large financial commitments.

Despite these obstacles, the development of stress-tolerant legume cultivars appears to have a bright future because of advances in breeding methods and a broader comprehension of plant genetics. The accurate identification of stress tolerance alleles made possible by molecular approaches such as GWAS and QTL mapping can greatly improve conventional breeding methods (Veeranagappa et al. 2022). A significant tool for anticipating and choosing superior genotypes is provided by the application of high-density molecular markers and sophisticated prediction models in genomic selection, which may hasten the breeding process. With the development of CRISPR-Cas9 and other gene editing technologies that allow for precise and targeted alterations to increase stress tolerance, transgenic techniques remain promising (Dutta et al. 2024). Furthermore, novel stress tolerance features can be introduced into cultivated variety by using the inherent genetic diversity seen in wild relatives and landraces. Using phenotypic, molecular, and genomic data in integrated breeding methods should result in more resilient stress-tolerant cultivars (Maity et al. 2023). By offering real-time data and maximizing resource utilization, the use of digital tools and precision agricultural technology may help these breeding efforts even more. These cutting-edge breeding techniques will be crucial to maintaining the resilience and sustainability of legume production as climate change continues to have an influence on agriculture.

1.5 Conclusion

Globally, abiotic stressors such as salinity, drought, flooding, and extremely high temperatures provide serious obstacles to the development of legumes. These stressors have the potential to negatively affect several facets of legume development, productivity, and growth, which could result in significant yield losses and lowered crop quality. Developing practical methods to lessen the negative consequences of abiotic stresses requires an understanding of the physiological, biochemical, and molecular mechanisms underpinning plant responses to these challenges. One promising strategy to solve the problem of abiotic stress is to breed for stress tolerance. Conventional breeding methods have made it possible to identify and use important genetic resources carrying stress-tolerant features, thanks to the combination of contemporary molecular tools and biotechnological procedures. By precisely locating and adding favorable alleles or genes linked to stress tolerance, marker-assisted selection, genomic selection, and transgenic techniques have sped up the production of stress-tolerant legume cultivars. However, a multidisciplinary approach is required due to the intricacy of abiotic stress tolerance features and the intricate interconnections among various stress causes. Complementing traditional breeding methods with cutting-edge genomics, transcriptomics, proteomics, and metabolomics approaches can offer a thorough grasp of the genetic and molecular processes behind stress tolerance. Furthermore, investigating how symbiotic relationships with advantageous microorganisms like rhizobia and plant growth-promoting bacteria can improve stress tolerance is promising for the development of sustainable legumes. The development of legume cultivars that are climate-resilient and capable of withstanding the difficulties presented by abiotic stresses necessitates ongoing research endeavors, worldwide collaborations, and the utilization of state-of-the-art technology. Reaching this objective will help ensure food security and agricultural sustainability as well as improve smallholder farmers' livelihoods and encourage the preservation of priceless genetic resources.

References

Abbas A, Ali A, Hussain A, Ali A, Alrefaei AF, Naqvi SAH et al (2023) Assessment of genetic variability and evolutionary relationships of *Rhizoctonia solani* inherent in legume crops. Plan Theory 12(13):2515

Akibode S, Maredia M (2011) Global and regional trends in production, trade and consumption of food legume crops. Department of Agriculture, Food, and Resource Economics, Michigan State University

Ali Z, Zhang DY, Xu ZL, Xu L, Yi JX, He XL, Huang YH, Liu XQ, Khan AA, Trethowan RM, Ma HX (2012) Uncovering the salt response of soybean by unraveling its wild and cultivated functional genomes using tag sequencing. PLoS One 7(11):e48819

Ali A, Altaf MT, Nadeem MA, Karaköy T, Shah AN, Azeem H, Baloch FS, Baran N, Hussain T, Duangpan S, Aasim M (2022) Recent advancement in OMICS approaches to enhance abiotic stress tolerance in legumes. Front Plant Sci 13:952759

Altaf MT, Liaqat W, Iqbal J, Baig MMA, Ali A, Nadeem MA, Baloch FS (2023) Exploring Omics approaches to enhance stress tolerance in soybean for sustainable bioenergy production. In: Biotechnology and Omics approaches for bioenergy crops. Singapore, Singapore, pp 123–154

Anjum SA, Xie X, Wang L, Saleem MF, Man C, Lei W (2011) Morphological, physiological and biochemical responses of plants to drought stress. Afr J Agric Res 6(9):2026–2032

Araujo SS, Beebe S, Crespi M, Delbreil B, González EM, Gruber V, Lejeune-Henaut I, Link W, Monteros MJ, Prats E, Rao I (2015) Abiotic stress responses in legumes: strategies used to cope with environmental challenges. Crit Rev Plant Sci 34(1–3):237–280

Aravind B, Nayak SN, Choudhary RS, Gandhadmath SS, Prasad PVV, Pandey MK et al (2022) Integration of genomics approaches in abiotic stress tolerance in groundnut (Arachis hypogaea L.): an overview. In: Genomic designing for abiotic stress resistant oilseed crops, pp 149–197

Arriagada O, Cacciuttolo F, Cabeza RA, Carrasco B, Schwember AR (2022) A comprehensive review on chickpea (Cicer arietinum L.) breeding for abiotic stress tolerance and climate change resilience. Int J Mol Sci 23(12):6794

Asada K (2000) The water-water cycle as alternative photon and electron sinks. Philos Trans R Soc Lond Ser B Biol Sci 355(1402):1419–1431

Aski MS, Dikshit HK, Mishra GP, Yadav PS, Iquebal MA, Sarika et al (2023) Current and future strategies in breeding lentil for abiotic stresses. In: Legumes: physiology and molecular biology of abiotic stress tolerance. Singapore, Singapore, pp 235–257

Atta K, Pal AK, Jana K (2021) Effects of salinity, drought and heavy metal stress during seed germination stage in ricebean [Vigna umbellata (Thunb.) Ohwi and Ohashi]. Plant Physiol Rep 26(1):109–115

Atta K, Adhikary S, Mondal S, Mukherjee S, Pal A, Mondal S et al (2022) A review on stress physiology and breeding potential of an underutilized, multipurpose legume: Rice bean (Vigna umbellata). In: Developing climate resilient grain and forage legumes, pp 235–253

Awasthi R, Kaushal N, Vadez V, Turner NC, Berger J, Siddique KHM, Nayyar H (2014) Individual and combined effects of transient drought and heat stress on carbon assimilation and seed filling in chickpea. Funct Plant Biol 41(11):1148–1167. https://doi.org/10.1071/FP13340

Bacanamwo M, Purcell LC (1999) Soybean root morphological and anatomical traits associated with acclimation to flooding. Crop Sci 39(1):143–149

Battisti R, Sentelhas PC, Parker PS, Nendel C, Gil MDS, Farias JR Basso CJ (2018) Assessment of crop-management strategies to improve soybean resilience to climate change in Southern Brazil. Crop and Pasture Sci 69(2), pp. 154–162.

Berger JD, Milroy SP, Turner NC, Siddique KH, Imtiaz M, Malhotra R (2012) Chickpea evolution has selected for contrasting phenological mechanisms among different habitats. Euphytica 185(1):1–15

Bhatnagar-Mathur P, Devi MJ, Reddy DS, Lavanya M, Vadez V, Serraj R et al (2007) Stress-inducible expression of at DREB1A in transgenic peanut (Arachis hypogaea L.) increases transpiration efficiency under water-limiting conditions. Plant Cell Rep 33(12):2139–2152

Bharadwaj B, Mishegyan A, Nagalingam S, Guenther A, Joshee N, Sherman SH, Basu C (2023) Physiological and genetic responses of lentil (Lens culinaris) under flood stress. Plant Stress, 7, p.100130.

Brunetti G et al (2011) Amendments for the remediation of soils contaminated by heavy metals. J Soils Sediments 11(4):634–643

Buitrago-Bitar MA, Cortés AJ, López-Hernández F, Londoño-Caicedo JM, Muñoz-Florez JE, Muñoz LC, Blair MW (2021) Allelic diversity at abiotic stress responsive genes in relationship to ecological drought indices for cultivated tepary bean, Phaseolus acutifolius A. Gray, and its wild relatives. Genes 12(4):556

Chaudhary S, Priya M, Jha UC, Pratap A, HanumanthaRao B, Singh I et al (2022) Approaches toward developing heat and drought tolerance in mungbean. In: Developing climate resilient grain and forage legumes, pp 205–234

Choudhury FK, Rivero RM, Blumwald E, Mittler R (2017) Reactive oxygen species, abiotic stress and stress combination. Plant J Cell Mol Biol 90(5):856–867

Chowdhury JA, Karim MA, Khaliq QA, Ahmed AU, Mondol ATMAI (2017) Effect of drought stress on water relation traits of four soybean genotypes

Clemens S (2006) Toxic metal accumulation, responses to exposure and mechanisms of tolerance in plants. Biochimie 88(11):1707–1719

Daryanto S, Wang L, Jacinthe PA (2015) Global synthesis of drought effects on food legume production. PLoS One 10(6):e0127401. https://doi.org/10.1371/journal.pone.0127401

Davies WJ, Zhang J (1991) Root signals and the regulation of growth and development of plants in drying soil. Annu Rev Plant Biol 42(1):55–76

De Ronde JA, Cress WA, Krüger GH, Strasser RJ, Van Staden J (2004) Photosynthetic response of transgenic soybean plants, containing an Arabidopsis P5CR gene, during heat and drought stress. J Plant Physiol 161(11):1211–1224

Dita MA, Rispail N, Prats E, Rubiales D, Singh KB (2006) Biotechnology approaches to overcome biotic and abiotic stress constraints in legumes. Euphytica 147(1):1–24

Dutta S, Paul D, Das A, Kumar S, Roy A, Barua A, Chinnusamy V (2024) Directed mutagenesis in plants through genome editing using guide RNA library. Nucleus:1–13

Elsheikh EAE, Wood M (1995) Nodulation and N2 fixation by soybean inoculated with salt-tolerant rhizobia or salt-sensitive bradyrhizobia. Soil Biol Biochem 27(4–5):657–661. https://doi.org/10.1016/0038-0717(95)98644-9

Essa TA (2002) Effect of salinity stress on growth and nutrient composition of three soybean (Glycine max L. Merrill) cultivars. J Agron Crop Sci 188(2):86–93. https://doi.org/10.1046/j.1439-037X.2002.00537.x

FAO (2024). https://www.fao.org/faostat/en/#data/QCL. Accessed on 07.06.2024

Farooq M, Wahid A, Kobayashi N, Fujita D, Basra SMA (2009) Plant drought stress: effects, mechanisms and management. Agron Sustain Dev 29(1):185–212. https://doi.org/10.1051/agro:2008021

Farooq M, Hussain M, Wakeel A, Siddique KHM (2015) Salt stress in maize effects resistance mechanisms and management: a review. Agron Sustain Dev 35:461e481. https://doi.org/10.1007/s13593-015-0287-0

Farooq M, Gogoi N, Barthakur S, Baroowa B, Bharadwaj N, Alghamdi SS, Siddique KH (2017a) Drought stress in grain legumes during reproduction and grain filling. J Agron Crop Sci 203(2):81–102

Farooq M, Gogoi N, Hussain M, Barthakur S, Paul S, Bharadwaj N et al (2017b) Effects, tolerance mechanisms and management of salt stress in grain legumes. Plant Physiol Biochem 118:199–217

Fathi A, Tari DB (2016) Effect of drought stress and its mechanism in plants. Int J Life Sci 10(1):1–6

Finkelstein RR, Rock CD (2002) Abscisic acid biosynthesis and response. In: The Arabidopsis book, vol 1. American Society of Plant Biologists

Foyer CH, Lam HM, Nguyen HT, Siddique KH, Varshney RK, Colmer TD et al (2016) Neglecting legumes has compromised human health and sustainable food production. Nat Plants 2(8):1–10. https://doi.org/10.1038/nplants.2016.112

Friso G, van Wijk KJ (2015) Posttranslational protein modifications in plant metabolism. Plant Physiol 169(3):1469–1487

Gallego SM, Pena LB, Barcia RA, Azpilicueta CE, Iannone MF, Rosales EP et al (2012) Unravelling cadmium toxicity and tolerance in plants: insight into regulatory mechanisms. Environ Exp Bot 83:33–46

Gao WR, Wang XS, Liu QY, Peng H, Chen C, Li JG et al (2008) Comparative analysis of ESTs in response to drought stress in chickpea (C. arietinum L.). Biochem Biophys Res Commun 376(3):578–583

Gaur PM, Jukanti AK, Varshney RK (2012) Impact of genomic technologies on chickpea breeding strategies. Agronomy 2(3):199–221

Gayacharan, Chandora R, Rana JC (2023) Moth bean (Vigna aconitifolia): a minor legume with major potential to address global agricultural challenges. Front Plant Sci 14:1179547

Ghanbari AA, Shakiba MR, Toorchi M, Choukan R (2013) Nitrogen changes in the leaves and accumulation of some minerals in the seeds of red, white and chitti beans ('Phaseolus vulgaris') under water deficit conditions. Aust J Crop Sci 7(5):706–712

Gonçalves L, Rubiales D, Bronze MR, Vaz Patto MC (2022) Grass pea (Lathyrus sativus L.) – a sustainable and resilient answer to climate challenges. Agronomy 12(6):1324

Gondalia N, Vashi R, Barot V, Sharma F, Anishkumar PK, Chatterjee M et al (2022) Genomic designing for abiotic stress tolerance in pea (Pisum sativum L.). In: Genomic designing for abiotic stress resistant pulse crops. Springer, Cham, pp 45–113

Gruber V, Blanchet S, Diet A, Zahaf O, Boualem A, Kakar K, Alunni B, Udvardi M, Frugier F, Crespi M (2009) Identification of transcription factors involved in root apex responses to salt stress in Medicago truncatula. Mol Genet Genomics: MGG 281(1):55–66

Hartung W (1999) Abscisic acid-a hormonal long distance stress signal in plants under drought and salt stress. Handbook of crop stress

Hasanuzzaman M et al (2013) Physiological, biochemical, and molecular mechanisms of heat stress tolerance in plants. Int J Mol Sci 14(5):9643–9684. https://doi.org/10.3390/ijms14059643

Hossain MA, Piyatida P, da Silva JAT, Fujita M (2012) Molecular mechanism of heavy metal toxicity and tolerance in plants: central role of glutathione in detoxification of reactive oxygen species and methylglyoxal and in heavy metal chelation. J Bot 2012:1–37

Jarquín D, Crossa J, Lacaze X, Du Chene P, Lorgeou J, Piraux J et al (2014) A reaction norm model for genomic selection using high-dimensional genomic and environmental data. Theor Appl Genet 127(3):595–607

Javid MG, Sorooshzadeh A, Moradi F, Modarres Sanavy SAM, Allahdadi I (2011) The role of phytohormones in alleviating salt stress in crop plants. Aust J Crop Sci 5(6):726–734

Jha UC, Chaturvedi SK, Bohra A, Basu PS, Khan MS, Barh D (2014) Abiotic stresses, constraints and improvement strategies in chickpea. Plant Breed 133(2):163–178

Jha UC, Nayyar H, Parida SK, Siddique KH (2022) Horse gram, an underutilized climate-resilient legume: breeding and genomic approach for improving future genetic gain. In: Developing climate resilient grain and forage legumes. Singapore, Singapore, pp 167–178

Jia W, Zhang J, Zhang DP (1996) Metabolism of xylem-delivered ABA in relation to ABA flux and concentration in leaves of maize and Commelina communis. J Exp Bot 47(8):1085–1091

Kadirimangalam SR, Bagudam R, Mathew A, Deshmukh D, Pasupuleti J (2022) Breeding groundnut cultivars for resilience to climate change effects. In: Developing climate resilient grain and forage legumes, pp 141–165

Kale SM, Jaganathan D, Ruperao P, Chen C, Punna R, Kudapa H et al (2015) Prioritization of candidate genes in "QTL-hotspot" region for drought tolerance in chickpea (Cicer arietinum L.). Sci Rep 5(1):1–16

Khatun M, Sarkar S, Era FM, Islam AM, Anwar MP, Fahad S et al (2021) Drought stress in grain legumes: effects, tolerance mechanisms and management. Agronomy 11(12):2374

Kim GB, Nam YW (2013) A novel Δ(1)-pyrroline-5-carboxylate synthetase gene of Medicago truncatula plays a predominant role in stress-induced proline accumulation during symbiotic nitrogen fixation. J Plant Physiol 170(3):291–302

Kochian LV, Piñeros MA, Liu J, Magalhaes JV (2015) Plant adaptation to acid soils: the molecular basis for crop aluminum resistance. Annu Rev Plant Biol 66:571–598

Kudapa H, Ramalingam A, Nayakoti S, Chen X, Zhuang WJ, Liang X, Kahl G, Edwards D, Varshney RK (2013) Functional genomics to study stress responses in crop legumes: progress and prospects. Funct Plant Biol: FPB 40(12):1221–1233

Kumar S, Kaur R, Kaur N, Bhandhari K, Kaushal N, Gupta K, Bains TS Nayyar H (2011) Heat-stress induced inhibition in growth and chlorosis in mungbean (Phaseolus aureus Roxb.) is partly mitigated by ascorbic acid application and is related to reduction in oxidative stress. Acta Physiologiae Plantarum 33, pp. 2091–2101.

Li D, Zhang Y, Hu X, Shen X, Ma L, Su Z, Wang T, Dong J (2011) Transcriptional profiling of Medicago truncatula under salt stress identified a novel CBF transcription factor MtCBF4 that plays an important role in abiotic stress responses. BMC Plant Biol 11:109

Linkemer G, Board JE, Musgrave ME (1998) Waterlogging effects on growth and yield components in late-planted soybean. Crop Sci 38(6):1576–1584

Llanes A, Devinar G, Luna V (2015) Role of abscisic acid in legumes under abiotic stress. In: Legumes under environmental stress: yield, improvement and adaptations, pp 145–160

Lü D, Wang W, Miao C (2013) ATHK1 acts downstream of hydrogen peroxide to mediate ABA signaling through regulation of calcium channel activity in Arabidopsis guard cells. Chin Sci Bull 58:336–343

Maalouf F, Khater LA, Babiker Z, Mohamed A (2022) Genetic and genomic research for abiotic stresses in Faba bean. In: Genomic designing for abiotic stress resistant pulse crops. Springer, Cham, pp 249–270

Maity A, Paul D, Lamichaney A, Sarkar A, Babbar N, Mandal N, Dutta S, Maity PP, Chakrabarty Chakrabarty SK (2023) Climate change impacts on seed production and quality: current knowledge, implications, and mitigation strategies. Seed Sci Technol 51(1):7–38

Makbul S, Güler NS, Durmuş N, Güven S (2011) Changes in anatomical and physiological parameters of soybean under drought stress. Turk J Bot 35(4):369–377

Mallikarjuna N, Pande S, Jadhav DR, Sastri DC, Rao JN (2011) Introgression of disease resistance genes from Arachis kempff-mercadoi into cultivated groundnut. Plant Breed 130(2):242–245

Manu B, Biradar R, Sabale PR, Kumar K, Aski MS, Mohite N et al (2022) Genomic designing for abiotic stress tolerance in mungbean and urdbean. In: Genomic designing for abiotic stress resistant pulse crops. Springer, Cham, pp 271–343

Matamoros MA, Becana M (2021) Molecular responses of legumes to abiotic stress: post-translational modifications of proteins and redox signaling. J Exp Bot 72(16):5876–5892

Merchan F, de Lorenzo L, Rizzo SG, Niebel A, Manyani H, Frugier F, Sousa C, Crespi M (2007) Identification of regulatory pathways involved in the reacquisition of root growth after salt stress in Medicago truncatula. Plant J Cell Mol Biol 51(1):1–17

Messina V (1999) Legumes and soybeans: overview of their nutritional profiles and health effects. Am J Clin Nutr 70(3):439s–450s. https://doi.org/10.1093/ajcn/70.3.439s

Millaleo R, Reyes-Díaz M, Ivanov AG, Mora ML, Alberdi M (2010) Manganese as essential and toxic element for plants: transport, accumulation and resistance mechanisms. J Soil Sci Plant Nutr 10(4):470–481

Mudryj AN, Yu N, Aukema HM (2014) Nutritional and health benefits of pulses. Appl Physiol Nutr Metab 39(11):1197–1204. https://doi.org/10.1139/apnm-2013-0557

Munns R, Tester M (2008) Mechanisms of salinity tolerance. Annu Rev Plant Biol 59:651–681. https://doi.org/10.1146/annurev.arplant.59.032607.092911

Muñoz-Perea CG, Terán H, Allen RG, Wright JL, Westermann DT, Singh SP (2006) Selection for drought resistance in dry bean landraces and cultivars. Crop Sci 46(5):2111–2120

Nadeem M, Li J, Yahya M, Sher A, Ma C, Wang X, Qiu L (2019a) Research progress and perspective on drought stress in legumes: a review. Int J Mol Sci 20(10):2541

Nadeem M, Li J, Yahya M, Wang M, Ali A, Cheng A et al (2019b) Grain legumes and fear of salt stress: focus on mechanisms and management strategies. Int J Mol Sci 20(4):799

Nadeem MA, Habyarimana E, Karaköy T, Baloch FS (2021) Genetic dissection of days to flowering via genome-wide association studies in Turkish common bean germplasm. Physiol Mol Biol Plants 27(7):1609–1622

Nandini B, Venkatesh, Reddy UG, Mallikarjuna BP, Manu B, Vaijayanthi PV et al (2022) Genomic design for abiotic stress resistance in pigeonpea. In: Genomic designing for abiotic stress resistant pulse crops. Springer, Cham, pp 169–248

Onyekachi OG, Boniface OO, Gemlack NF, Nicholas N (2019) The effect of climate change on abiotic plant stress: a review. Abiotic Biot Stress Plants 17:71–83

Palta JA, Ganjeali A, Turner NC, Siddique KHM (2010) Effects of transient subsurface waterlogging on root growth, plant water relations, and yields of chickpea. Plant Soil 333(1–2):13–27

Pathak R, Panchariya P, Choudhary M, Solanki K, Rani R, Kakani RK, Kalia RK (2023) Morphophysiological and molecular diversity in mung bean (Vigna radiata L.). In: Legumes: physiology and molecular biology of abiotic stress tolerance, pp 115–147

Patto MV, Skiba B, Pang ECK, Ochatt SJ, Lambein F, Rubiales D (2006) Lathyrus improvement for resistance against biotic and abiotic stresses: from classical breeding to marker assisted selection. Euphytica 147:133–147

Peng YB, Zou C, Wang DH, Gong HQ, Xu ZH, Bai SN (2006) Preferential localization of abscisic acid in primordial and nursing cells of reproductive organs of Arabidopsis and cucumber. New Phytol 170(3):459–466

Pennycooke JC, Cheng H, Stockinger EJ (2008) Comparative genomic sequence and expression analyses of Medicago truncatula and alfalfa subspecies falcata COLD-ACCLIMATION-SPECIFIC genes. Plant Physiol 146(3):1242–1254

Peoples MB, Brockwell J, Herridge DF, Rochester IJ, Alves BJR, Urquiaga S et al (2009) The contributions of nitrogen-fixing crop legumes to the productivity of agricultural systems. Symbiosis 48(1–3):1–17. https://doi.org/10.1007/BF03179980

Rai KK, Rai N, Rai SP (2018) Recent advancement in modern genomic tools for adaptation of Lablab purpureus L to biotic and abiotic stresses: present mechanisms and future adaptations. Acta Physiol Plant 40(9):164

Rasool S, Ahmad A, Siddiqi TO, Ahmad P (2013) Changes in growth, lipid peroxidation and some key antioxidant enzymes in chickpea genotypes under salt stress. Acta Physiol Plant 35:1039–1050

Ratnaparkhe MB, Satpute GK, Kumawat G, Chandra S, Kamble VG, Kavishwar R et al (2022) Genomic designing for abiotic stress tolerant soybean. In: Genomic designing for abiotic stress resistant oilseed crops. Springer, Cham, pp 1–73

Sahay G, Mehta BK, Dikshit N, Shashikumara P, Bhargavi HA, Priyadarshini P (2022) Developing climate-resilient cowpea (Vigna unguiculata [L.] Walp.) through genomics-assisted breeding approaches. In: Developing climate resilient grain and forage legumes. Springer, Singapore, pp 77–103

Sánchez FJ, Manzanares M, de Andres EF, Tenorio JL, Ayerbe L (1998) Turgor maintenance, osmotic adjustment and soluble sugar and proline accumulation in 49 pea cultivars in response to water stress. Field Crop Res 59(3):225–235

Sehrawat N et al (2013) Effect of salinity stress on mungbean [Vigna radiata (L.) Wilczek] during consecutive summer and spring seasons. J Agric Sci Belgrade 58(1):23–32. https://doi.org/10.2298/JAS1301023S

Shabala S, Cuin TA (2008) Potassium transport and plant salt tolerance. Physiol Plant 133(4):651–669. https://doi.org/10.1111/j.1399-3054.2007.01008.x

Shao HB, Chu LY, Lu ZH, Kang CM (2007) Primary antioxidant free radical scavenging and redox signaling pathways in higher plant cells. Int J Biol Sci 4(1):8–14

Shrestha R, Turner NC, Siddique KHM, Turner DW, Speijers J (2006) A water deficit during pod development in lentils reduces flower and pod numbers but not seed size. Aust J Agric Res 57(4):427–438

Simon-Sarkadi L, Kocsy G, Várhegyi A, Galiba G, de Ronde JA (2005) Genetic manipulation of proline accumulation influences the concentrations of other amino acids in soybean subjected to simultaneous drought and heat stress. J Agric Food Chem 53(19):7512–7517

Singh A, Dikshit HK, Mishra GP, Aski M, Kumar S, Sarker A (2022a) Breeding for abiotic stress tolerance in lentil in genomic era. In: Genomic designing for abiotic stress resistant pulse crops. Springer, Cham, pp 145–167

Singh B, Padhy AK, Ambreen H, Yadav M, Bhardwaj S, Singh G et al (2022b) Understanding abiotic stress responses in lentil under changing climate regimes. In: Developing climate resilient grain and forage legumes. Singapore, Singapore, pp 179–204

Song Y, Ji D, Li S, Wang P, Li Q, Xiang F (2012) The dynamic changes of DNA methylation and histone modifications of salt responsive transcription factor genes in soybean. PLoS One 7(7):e41274

Souter JR, Gurusamy V, Porch TG, Bett KE (2017) Successful introgression of abiotic stress tolerance from wild tepary bean to common bean. Crop Sci 57(3):1160–1171

Taiz L, Zeiger E, Møller IM, Murphy A (2015) Plant physiology and development

Thung M, Rao IM (1999) Integrated management of abiotic stresses. In: Common bean improvement in the twenty-first century, pp 331–370

Tiwari B, Kalim S, Tyagi N, Kumari R, Bangar P, Barman P et al (2018) Identification of genes associated with stress tolerance in moth bean [Vigna aconitifolia (Jacq.) Marechal], a stress hardy crop. Physiol Mol Biol Plants 24:551–561

Unkovich M, Herridge D, Peoples M, Cadisch G, Boddey B, Giller K et al (2008) Measuring plant-associated nitrogen fixation in agricultural systems. ACIAR Monogr 136:258

Varshney RK, Thudi M, Nayak SN, Gaur PM, Kashiwagi J, Krishnamurthy L et al (2014) Genetic dissection of drought tolerance in chickpea (Cicer arietinum L.). Theor Appl Genet 127(2):445–462

Veeranagappa P, Manu B, Prasad G, Blair MW, Hickok D, Naveena NL et al (2022) Advanced breeding strategies for abiotic stress tolerance in cowpea. In: Genomic designing for abiotic stress resistant pulse crops, pp 115–144

Verbruggen N, Hermans C (2008) Proline accumulation in plants: a review. Amino Acids 35(4):753–759

Westgate ME, Passioura JB, Munns R (1996) Water status and ABA content of floral organs in drought-stressed wheat. Funct Plant Biol 23(6):763–772

Yoshiba Y, Kiyosue T, Nakashima K, Yamaguchi-Shinozaki K, Shinozaki K (1997) Regulation of levels of proline as an osmolyte in plants under water stress. Plant Cell Physiol 38(10):1095–1102

Zahaf O, Blanchet S, de Zélicourt A, Alunni B, Plet J, Laffont C, de Lorenzo L, Imbeaud S, Ichanté JL, Diet A, Badri M, Zabalza A, González EM, Delacroix H, Gruber V, Frugier F, Crespi M (2012) Comparative transcriptomic analysis of salt adaptation in roots of contrasting Medicago truncatula genotypes. Mol Plant 5(5):1068–1081

Zahran HH (1999) Rhizobium-legume symbiosis and nitrogen fixation under severe conditions and in an arid climate. Microbiol Mol Biol Rev 63(4):968–989. https://doi.org/10.1128/MMBR.63.4.968-989.1999

Zhang G, Chen M, Chen X, Xu Z, Li L, Guo J, Ma Y (2010) Isolation and characterization of a novel EAR-motif-containing gene GmERF4 from soybean (Glycine max L.). Mol Biol Rep 37(2):809–818

Drought Stress and Sustainable Legume Production

2

Abdel Rahman M. Tawaha, Hiba Alatrash, Yaman Jabbour, Abdel Razzaq Al-Tawaha, Ali M. Qaisi, Rania Jammal, Arun Karnwal, Mohamad Shatnawi, Palani Saranraj, and Jana Rammal

Abstract

Drought stress is a considerable obstacle to sustainable legume production, as legumes exhibit heightened sensitivity to water deficiency, which can profoundly affect their growth, yield, and quality. Drought stress diminishes soil moisture,

A. R. M. Tawaha (✉)
Department of Biological Sciences, Al Hussein bin Talal University, Maan, Jordan

H. Alatrash · Y. Jabbour
Department of field Crops, Faculty Agricultural Engineering, University of Aleppo, Aleppo, Syria

A. R. Al-Tawaha
National Agricultural Research Center, Baqa, Jordan

A. M. Qaisi
Department of Pharmaceutical Sciences, School of Pharmacy, University of Jordan, Amman, Jordan

R. Jammal
Department of Clinical Pharmacy, Jordan University of Science and Technology, Irbid, Jordan

A. Karnwal
Department of Microbiology, School of Bioengineering and BioSciences, Lovely Professional University, Phagwara, India

Department of Microbiology, Graphic Era (Deemed to be University), Dehradun, Uttarakhand, India

M. Shatnawi
Department of Biotechnology, Faculty of Agricultural Technology, Al-Balqa Applied University, Amman, Jordan

P. Saranraj
Department of Microbiology, Sacred Heart College (Autonomous), Tirupattur, Tamil Nadu, India

J. Rammal
Lebanese University, Faculty of Agronomy and Veterinary, Beirut, Lebanon

© The Author(s), under exclusive license to Springer Nature Singapore Pte Ltd. 2025
M. A. Nadeem et al. (eds.), *Marker-Assisted Breeding in Legumes for Drought Tolerance*, https://doi.org/10.1007/978-981-96-4112-3_2

constraining water availability for essential plant processes, hence impacting nutrient absorption, photosynthesis, and nitrogen fixation—a crucial activity in legumes owing to their symbiotic association with nitrogen-fixing bacteria. The process of nitrogen fixation, which is an essential function of legumes, is inhibited when there is a drought. In order for the nitrogen-fixing bacteria known as Rhizobium to successfully form nodules on the roots of legumes, they require an adequate amount of moisture. Nitrogen fixation is reduced, which results in nutrient deficiencies. These deficiencies diminish the plant's ability to produce healthy seeds, which in turn reduces the yield. When it comes to the health of the soil, drought causes disruptions in the microbial activity and the availability of nutrients, which in turn makes the plant's stress worse and has an effect on the soil's fertility over the long term. When faced with these obstacles, sustainable legume production in drought conditions requires integrated approaches. These techniques include breeding for drought-resistant cultivars, effective water management practices such as drip irrigation, and soil conservation methods to conserve moisture. It is possible for these measures to help offset the negative impacts of drought on legumes, so assuring more resilient output in environments with limited water resources.

Keywords

Legume · Nitrogen fixation · Water stress · Plant growth

2.1 Introduction

The agricultural sector is the foundation of a nation's expanding economy because of the significant role it plays in satisfying the demands for food on a worldwide scale. On the other hand, the industry is confronted with a multitude of obstacles, such as the escalation of global temperatures, the reduction of arable land, the excessive application of fertilizers and pesticides, the assaults of plant infections, and other environmental stresses. According to Baweja et al. (2020), the entirety of these elements works to impede agricultural output and sustainability. Climate change is increasing the incidence of abiotic stresses such as drought, heavy rainfall, and extreme temperatures, which adversely affect crop growth and important traits, ultimately reducing yields (Srivastava and Kumar 2020). Drought stress, in particular, poses a significant threat to agriculture. Plants encounter several environmental challenges, with drought stress becoming progressively more critical. Meteorological models predict an expansion of drought-prone regions, exacerbating the effects on agriculture. Legumes, essential sources of fiber, oils, and protein, significantly contribute to the world economy and are integral to sustainable agriculture due to their ability to fix atmospheric nitrogen through symbiosis with nitrogen-fixing bacteria called rhizobia (Ali et al. 2022). This nitrogen intake enhances the biosphere, promoting soil fertility and diminishing the need on synthetic fertilizers. Comprehending the molecular, metabolic, physiological, and agronomic reactions of legumes to drought is crucial for formulating creative techniques to enhance their drought resilience and guarantee agricultural sustainability.

Agronomic crops like chickpea (*Cicer arietinum*), lentil (*Lens culinaris*), faba bean (*Vicia faba*), cowpea (*Vigna unguiculata*), soybean (Glycine max), lupin (*Genistoid clade*), and common bean (*Phaseolus vulgaris*), are all part of the Fabaceae (Leguminosae) family (Nadeem et al. 2018). This family, with over 800 genera and 20,000 species, is the third-largest family of flowering plants and ranks as the most crucial crop family after cereals (Lewis et al. 2005; Gepts et al. 2005; Bitocchi et al. 2017). Due to the intensity and length of the stress, drought-induced crop yield losses frequently transcend those caused by other causes (Farooq et al. 2009; Hasegawa et al. 2013). Water shortage is a major environmental limitation that leads to a reduction in plant production. The impacts of drought stress on plant growth, phenology, water and nutrient dynamics, photosynthesis, assimilate distribution, and respiration are significant, and it is essential to understand these effects in order to design strategies for adaptation (Barnabás et al. 2008; Jaleel et al. 2009). According to Osakabe et al. (2014), drought resistance mechanisms act on morphological, physiological, and molecular levels. These mechanisms also include a variety of management strategies that aim to mitigate the negative consequences of drought. Plants exhibit a wide range of physiological and biochemical responses at the cellular and whole-plant levels in order to endure drought stress (Chaves et al. 2003; Flexas and Medrano 2002). These responses include a reduction in leaf area, stem development, and root expansion, as well as a disruption of water relations and a decrease in water-use efficiency. Stomatal closure, membrane degradation, and poor enzyme activity, particularly those linked to CO_2 fixation and ATP generation, are the primary factors that primarily contribute to a reduction in the amount of CO_2 that is assimilated by the leaf (Lawlor and Cornic 2002).

According to Apel and Hirt (2004), an increase in the flow of metabolites through photorespiration magnifies the effects of oxidative stress by producing reactive oxygen species (ROS), which have the potential to cause damage to macromolecules and to inhibit growth. Plants have a number of different mechanisms that allow them to tolerate drought. These mechanisms include minimizing water loss by increasing their diffusive resistance, maximizing water intake by having deep and extensive root systems, and reducing transpirational loss by having smaller and thicker leaves (Serraj et al. 2004). The control of nutrients is also very important. Potassium helps with osmotic adjustment, and silicon strengthens the root endodermis, which improves the water balance in the cell (Ashraf et al. 2011). According to Sakamoto and Murata (2002), low-molecular-weight osmolytes such as glycinebetaine, proline, and organic acids provide significant contributions to the maintenance of cellular functioning in the presence of drought. According to Gill and Tuteja (2010), growth regulators such salicylic acid, auxins, gibberellins, cytokinins, and abscisic acid assist modify plant responses to drought. On the other hand, polyamines, citrulline, and certain enzymes function as antioxidants to alleviate the damage caused by reactive oxygen species (ROS). At the molecular level, a number of drought-responsive genes and transcription factors have been identified as contributors to drought tolerance (Shinozaki and Yamaguchi-Shinozaki 2007). These genes and transcription factors include dehydration-responsive element-binding genes, aquaporins, late embryogenesis abundant proteins, and dehydrins. Mass screening, breeding, marker-assisted selection, the exogenous application of hormones and osmoprotectants, and genetic engineering are some of the strategies that can be utilized to improve drought resistance (Bartels and Sunkar 2005). Legumes are especially valuable for several reasons:

1. *Food-System Level*: They are essential for human and animal nutrition, particularly in developing countries, as they are rich sources of protein (17–40%), vitamins (e.g., folic acid and vitamin B), and minerals (e.g., calcium, zinc, iron) (Tharanathan and Mahadevamma 2003; Zhu et al. 2005).
2. *Production-System Level*: Legumes improve soil properties and fertility by fixing atmospheric nitrogen, increasing soil organic carbon, humus content, and phosphorus availability (Jensen et al. 2012). They also help mitigate climate change by reducing greenhouse gas emissions such as nitrous oxide (N_2O) and carbon dioxide (CO_2) (Lemke et al. 2007; Fabio et al. 2017).
3. *Cropping System Level*: Legumes play a role in controlling weeds, pests, and diseases (Robson et al. 2002; Seymour et al. 2012). Like other crops, legumes face multiple abiotic stresses, with drought being one of the most limiting factors for productivity (Micheletto et al. 2007; Farooq et al. 2016). Since legumes are primarily grown in rainfed areas, many global climate models predict an increase in drought frequency and intensity, underscoring the risk of water scarcity (Mittal et al. 2014).

2.2 Drought Stress

Climate change intensifies the negative impacts of drought stress on agriculture. Drought stress is a major cause of crop production losses and a serious threat to global food security, as it hinders plant growth and development, limiting yield and productivity (Chaves and Oliveira 2004). Drought stress in agricultural fields arises from multiple factors, including (1) insufficient water supply from rainfall or irrigation, which reduces soil moisture; (2) extreme temperatures; (3) water loss from plants through transpiration; (4) dry winds; (5) high light intensity; and (6) elevated soil salinity, indicated by high electrical conductivity (Chaves and Oliveira 2004; Trujillo et al. 2008; Dai 2012). Drought has four categories provided below.

2.2.1 Meteorological Drought

This type of drought occurs due to factors like dry winds and rising temperatures and is typically characterized by a significant decrease in atmospheric humidity or precipitation. Meteorological drought begins and ends quickly, responding directly to changes in weather conditions. According to Wilhite and Glantz (1985), meteorological drought occurs when there is a considerable and persistent shortage in precipitation relative to the average in a given region. This results in decreased atmospheric moisture and frequently leads to a variety of environmental stresses because of the lower moisture content in the atmosphere. According to Dai (2012), it is typically the initial stage in the development of drought, and it frequently leads to successive droughts in the agricultural and hydrological sectors as a result of the decrease in water levels in soil, rivers, and reservoirs. Indicators such as the Standardized Precipitation Index (SPI), which evaluates the deviation from the average rainfall, and the Palmer Drought Severity Index (PDSI), which takes into consideration both precipitation and temperature, are utilized in the process of evaluating meteorological drought (McKee et al. 1993; Palmer 1965).

The frequency, intensity, and length of meteorological droughts are impacted by climate events such as the El Niño–Southern Oscillation (ENSO), which has the potential to disrupt regular patterns of precipitation and result in severe droughts in a number of different locations. According to the Intergovernmental Panel on Climate Change (IPCC), climate change is anticipated to make meteorological droughts more frequent and severe in certain regions. This is because altering weather patterns would result in less predictable rainfall and longer dry periods. According to Allen et al. (2010), the increased frequency of droughts places additional pressure on ecosystems and water resources, which in turn has an impact on agriculture, forestry, and on the availability of drinking water. Meteorological drought frequently results in what is known as a "flash drought," which is characterized by quick depletion of soil moisture brought on by high temperatures, dry winds, and bright sunshine (Otkin et al. 2016). This phenomenon causes severe stress to agricultural production. The affects on soil, water reservoirs, and vegetation may endure longer, which may have repercussions for crop yields and water availability (Wilhite et al. 2007). Although the meteorological drought may end as soon as regular rainfall patterns return, the implications on these areas may last for a longer period of time. According to Van Loon (2015), it is essential to monitor meteorological drought through the use of dependable climatic indices and early warning systems in order to reduce the socioeconomic costs of drought and to assist vulnerable regions in adapting to the increase in the risk of drought.

2.2.2 Agricultural Drought

This type of drought involves a decrease in soil moisture due to reduced atmospheric precipitation. It negatively impacts plant growth, productivity, and animal health and often occurs as a consequence of meteorological drought.

2.2.3 Hydrological Drought

A hydrological drought is defined as an extended period of time when surface or groundwater levels fall below normal, impacting the availability of water for agricultural, industrial, and household uses. It occurs due to the lack of the ground or surface water. Many activates increase hydrological drought such as, human activities, farming, and energy. Hydrological drought, in contrast to meteorological drought, is caused by chronic water system imbalances brought about by things like inadequate precipitation, over-extraction of groundwater, and decreased river flow (Wilhite and Glantz 1985). How long and how well the affected region's water bodies and aquifers hold water determine how severe a hydrological drought would be (Van Loon 2015). The depletion of water storage areas like rivers, reservoirs, and aquifers is a sign of hydrological drought, which typically follows meteorological dryness by some time. Soil moisture levels fall first, streamflow and lake levels fall second, while groundwater depletion occurs last during dry periods (Tallaksen and van Lanen 2004). This development demonstrates how ecosystems and water resources can be affected by hydrological drought for a long time after precipitation levels have normalized (Van Loon and Laaha 2015). Hydrological drought can be worsened by human activities like damming, irrigation, and changes in land use, which disrupt the patterns of natural water flow and put additional stress on aquatic ecosystems (Vörösmarty et al. 2000). Reduced water quality owing to stagnation and higher expenses for water delivery and irrigation are only two of the many diverse impacts of hydrological drought. Because water shortage impacts crop irrigation, which in turn reduces yields and increases food insecurity, agricultural systems are especially at risk (Wilhite et al. 2007). Sustainable water management strategies that consider both natural and human-made elements are necessary in light of the environmental and economic impacts of hydrological drought (Vörösmarty et al. 2000).

2.2.4 Socioeconomic Drought

This type of drought arises when water demand exceeds supply. Contributing factors include population growth and excessive demand for certain goods, which place additional strain on limited water resources. It significantly impacts the phenological, physiological, and morphological traits of plants (Farooq et al. 2009; Bhargava and Sawant 2013) and can also reduce both the quantity and quality of yields and yield components (Zlatev and Lidon 2012; Nezhadahmadi et al. 2013).

2.3 Effect of Drought Stress on Legumes

Drought is a significant abiotic stress that reduces legume productivity and acts as a barrier to crop production. Since legumes typically rely on rainfall, they are especially vulnerable to drought stress, particularly during the grain-filling and reproductive growth stages (Savita and Saurabh 2020). Turk et al. (2004) conducted controlled experiments to assess germination rates, seedling growth, and early-stage development of three lentil cultivars under different levels of moisture stress. The results showed that moisture stress substantially reduced both germination rates and seedling growth in all cultivars. However, the extent of this impact varied, indicating differences in drought tolerance among the cultivars (Turk et al. 2004). Tawaha and Turk et al. (2004) investigate optimal seeding management practices for field pea (*Pisum sativum*) in semi-arid Mediterranean environments, where water scarcity is a frequent challenge. The study examines the effects of different seeding rates and planting dates on field pea growth, yield, and water-use efficiency. Through field experiments, Tawaha and Turk et al. (2004) found that both seeding rate and planting time significantly influenced field pea productivity. Higher seeding rates improved yield by increasing plant density, though excessive rates could lead to competition for limited resources (Turk et al. 2004). Early planting generally resulted in better yield outcomes, as it allowed the crop to take advantage of residual soil moisture from winter rains before the onset of summer drought (Turk et al. 2004). Tawaha et al. (2005) examine how chickpea (*Cicer arietinum*) adapts to various cultural practices in a Mediterranean-type environment, where crop productivity is challenged by seasonal rainfall and frequent drought. The study evaluates the effects of different sowing dates, seeding rates, and planting depths on chickpea growth, yield, and adaptability. Results indicate that planting time significantly affects chickpea performance, with early sowing leading to higher yields due to increased soil moisture from winter rains (Tawaha et al. 2005). The study also identifies optimal seeding rates that ensure sufficient plant density while minimizing resource competition (Tawaha et al. 2005). Additionally, planting depth plays an important role, as deeper sowing enhances access to residual soil moisture but must be carefully managed to prevent reduced seedling emergence (Tawaha et al. 2005).

2.3.1 Physiological Traits

Yield losses in crops due to drought often result from changes in various morphological, physiological, and biochemical processes, particularly a reduction in photosynthetic rates. Drought may reduce yields by limiting photosynthesis or leaf expansion rates (Fang et al. 2010) and by lowering stomatal resistance and transpiration rates (Llorens et al. 2003; Sardans et al. 2008), or through a combination of these factors. These impacts directly diminish photosynthetic efficiency and, consequently, plant productivity. The process of growth covers a variety of processes, including cell division, cell expansion, and differentiation. Each of these processes involves intricate genetic, physiological, ecological, and morphological occurrences, as well as their interactions. At the end of the day, the quality and quantity of plant development are ultimately determined by these processes, which are controlled by restricted water availability. Because it is intimately linked to turgor pressure, cell growth is very vulnerable to drought. A decrease in turgor pressure can considerably affect cellular development. Cell growth is highly sensitive to moisture deprivation. Due to the fact that drought stress affects both root and shoot growth, it ultimately results in a reduction in the total growth and development of plants. On the other hand, grain legumes that are drought-resistant tend to produce root systems that are deeper than their counterparts that are more vulnerable to drought. In a field trial, for instance, a drought-resistant soybean landrace displayed a greater number of fibrous roots and explored a bigger soil volume. This was an example of a drought-resistant soybean. As the severity of drought stress increases, the cell wall becomes more stiff and less flexible. This results in a decrease in cell volume and pressure, which has a negative impact on the plant's ability to grow and develop. The size of the leaves and the number of leaves are two key elements that are influenced by this. The mesophyll cells in the leaves become dehydrated as a result of drought, and when there is a substantial lack of water, the roots may constrict, which results in changes in the amount of substances that are deposited in the leaves. Within the first stages, water stress limits cell proliferation, which ultimately leads to a reduction in leaf development. When it comes to several species of grain legumes, the development of leaf area is more susceptible to the effects of a lack of soil moisture than leaf abscission. As a result of increased synthesis and heightened sensitivity to stress hormones, plants tend to shed their leaves when they are under water stress. Furthermore, there is a high association between drought-induced reductions in leaf area development and the amount of leaf senescence. This correlation is a result of several factors. Both cowpea and pigeon pea experience senescence and abscission of older basal leaves when they are subjected to dryness throughout the flowering and pod-filling stages of their development. This rise in leaf senescence is made worse by a reduction in nitrogen uptake, which in turn leads to a loss in chlorophyll synthesis and a reduction in radiation efficiency. It would appear that the combined impacts of drought and monocarpy contribute to comparable patterns of acropetal leaf senescence in cowpea, which in turn accelerates the process of senescence when drought circumstances are present.

2.3.2 Nitrogen Fixation

Legume crops have the unique ability to fix atmospheric nitrogen through symbiosis with rhizobium bacteria, a process known as symbiotic nitrogen fixation (SNF). This process can fulfill the nitrogen needs of the plant, potentially eliminating the requirement for nitrogen fertilizers, making it essential for sustainable agriculture (Worrall and Roughley 1976). However, drought stress significantly impacts nitrogen fixation in legumes (Serraje 2003). Several hypotheses explain the inhibition of nitrogen fixation under drought conditions. Reduced oxygen permeability is thought to restrict nodule activity (Minchin 1997; Denison 1998), while decreased carbon flux to nodules also contributes to the decline in nitrogen fixation (Arrese-Igor et al. 1999). During drought, nodule sucrose synthase activity declines sharply, limiting the carbon influx required for bacteroid respiration (Gonza'lez et al. 1995; Gordon et al. 1997). Gonza'lez et al. (1995) identified sucrose synthase as the first enzyme affected by drought in soybean nodules, while Gordon et al. (1997) found a positive correlation between sucrose synthase activity and nitrogen fixation in drought-stressed soybean nodules. A nitrogen feedback mechanism may also regulate nitrogen fixation under drought condition. Drought stress reduces nitrogenase activity, an indicator of nitrogen fixation efficiency, making it a useful metric for assessing drought effects on legumes (Streeter 2003). A study on common bean (*Phaseolus vulgaris* L.) found that drought stress decreased nodule fresh weight, nitrogenase activity, ammonia production, and the activity of ammonia-assimilating enzymes. Drought also caused structural changes in nodules, altered intercellular glycoprotein content, and damaged bacteroid membranes. Drought stress impacts symbiotic nitrogen fixation by affecting root and nodule formation and function (Zahran 1999). For example, drought reduces root length and dry biomass accumulation in soybean (Thu et al. 2014). Drought-induced changes in root architecture include modifications in root depth, branching density, root angle, and increased root-to-shoot biomass allocation (Franco et al. 2011; Fenta et al. 2014). Studies have reported that soybean nodule numbers decrease under drought (Fernandez-Luquen et al. 2008; Márquez-García et al. 2015). In faba bean, drought reduced the number of infection threads, external nodule diameter, and lenticel presence in soybean nodules (Davis and Imsande 1988).

2.3.3 Yield and Yield Components

The impact of drought on crop yield is complex, affecting processes such as fertilization, gametogenesis, embryogenesis, and grain formation (Farooq et al. 2014). The flowering and reproductive stages are particularly susceptible to drought stress, as it hinders pollen tube growth and pollen grain germination (Samarah et al. 2009; Gusmao et al. 2012; Phillips et al. 2018). Drought also reduces the number of flowers, pods, and seeds, leading to an overall decline in yield (Pushpavalli et al. 2015). For instance, drought stress has been shown to decrease the number of pods per plant (Demirta et al. 2010; Li et al. 2013; Mak et al. 2014). Kokubun et al. (2001) reported a 92.7% reduction in pod number when drought occurred at the start of pod development and an 81.6% reduction during pod elongation. Additionally, drought stress reduces seed numbers, with the greatest declines occurring during flowering (Demirta et al. 2010; He et al. 2016; Maleki et al. 2013). In many seed crops, the most drought-sensitive stages are stem elongation and flowering (Kakaei et al. 2010; Zebarjadi et al. 2012). In chickpea, drought stress accounts for 40–45% of global yield losses (Ahmed et al. 2005), with some estimates placing these losses at around 33% (Kashiwagi et al. 2015). Drought reduces chickpea grain yield by 12%, aboveground biomass by 13%, plant height by 17%, and 100-seed weight (Muruiki et al. 2018; Arif et al. 2021). Chickpea yield declines significantly, with sensitivity varying based on growth stage. Ali et al. (1998) found that exposure to drought from flowering to maturity substantially reduced seed yield per plant. For soybean, drought stress decreases yield and yield components, with effects depending on growth stage, duration, and intensity of drought. The pod set and seed-filling stages are especially vulnerable, and drought at these stages shortens the seed-filling period and reduces grain weight (Liu et al. 2005). Kobraei et al. (2011) reported that drought stress reduced soybean yield by 43–44%. Lentil is also highly sensitive to drought, which can reduce yield by 6–54% (Fouad et al. 2011). In a study on the impact of drought on lentil traits, Rahimi et al. (2016) found that drought reduced all traits except leaf width, with the greatest decreases in pod weight (54%), seed yield (45.3%), and pod number per plant (42.2%). Hosseini et al. (2011) reported a 52% decrease in lentil seed yield under drought conditions during flowering. In grass pea, drought stress reduces seed yield by 19–39%, pod number per plant by 39.96%, seeds per pod by 20.51%, thousand-seed weight by 3.1%, biological yield by 25.73%, pod length by 10.53%, and branch number per plant by 10.27% (Boukecha et al. 2017). Pea yield also declines under drought, particularly during the anthesis stage (Mafakheri et al. 2010). In mungbean, drought during pod development reduces the number of seeds per pod and pod number, thus decreasing yield (Mirzaei et al. 2014). For canola (*Brassica napus*), drought after flowering reduces pod number per plant by shortening the flowering period, reproductive growth duration, and flower fertility (Nasri et al. 2008).

2.4 Drought Tolerance Mechanisms

Various factors influence a crop's ability to withstand drought stress, with some factors pertaining directly to the drought itself—such as its severity and duration—while others relate to the crop, including species, age, and developmental stage (Madhava et al. 2006). Crops have naturally evolved resistance mechanisms to mitigate the negative effects of drought stress, employing strategies, such as escaping, avoiding, and tolerating drought; however, the effectiveness of these mechanisms varies among species (Ashraf et al. 2009; Madhava et al. 2006; Salehi-lisar et al. 2012). Generally, plants with a perennial growth habit and deep root systems exhibit greater drought tolerance compared to annuals with shallow root systems (Farooq et al. 2009).

2.4.1 Escaping

This strategy refers to the ability of plants to complete their life cycles before water deficiency occurs and affects their performance (Siddique et al. 1993). This mechanism relies on a successful alignment between phenological development and the periods of soil moisture availability, particularly in environments where the growing season is shorter and terminal drought stress is prevalent (Farooq et al. 2014). Early flowering and maturity are critical traits for legumes to effectively escape drought stress (Thomson et al. 1997). Certain legume species, such as cowpea and common bean, can alleviate the effects of temporary drought by producing new organs during the recovery period after stress (Calvache et al. 1997).

2.4.2 Avoiding

This mechanism primarily relies on morphological adaptations and physiological processes that help maintain high water potential in plants by preventing tissue dehydration. This is achieved by reducing transpiration through various strategies, including decreasing leaf area, increasing leaf thickness, reducing the number of stomata, and developing glaucous surfaces on the aerial parts of the plant. Additionally, maintaining effective water uptake from the soil is facilitated by an extensive and robust root system (Serraj et al. 2003; Kashiwag et al. 2006; Sicher et al. 2012; Goufo et al. 2017). The root system is a crucial trait, as a deep and thick root structure enables plants to access water from deeper and more distant soil sources (Akhtar and Nazir 2013; Franco 2011). Generally, plants employing the avoidance mechanism tend to be smaller in size, as the energy expended in developing these traits can lead to lower overall crop production (Khan et al. 2011; Salehi-lisar et al. 2012).

2.4.3 Tolerating

Plants can endure drought stress by making osmotic adjustments at the cellular level, enhancing cell wall structure, and expanding their root systems (Khan et al. 2015; Blum 2017).

2.5 Drought Management

There is increasing international apprehension regarding the inadequacy of existing drought-management strategies, which primarily rely on crisis intervention. These reactive strategies concentrate on mitigating the effects of drought rather than tackling the underlying vulnerabilities that intensify these effects (Wilhite et al. 2014). By emphasizing immediate relief initiatives, such programs frequently overlook essential preventive actions, hence failing to foster resilience against forthcoming drought occurrences (Sivakumar et al. 2010). This reactive approach allocates resources to mitigate immediate effects, such as agricultural losses and water scarcity, without establishing policies that could diminish long-term vulnerability (Pulwarty and Sivakumar 2014). Conversely, proactive drought management—focused on preparedness, mitigation, and adaptation—provides a more sustainable approach by detecting and mitigating risks prior to the onset of droughts (UNCCD 2019). Transitioning from crisis-oriented solutions to a proactive, risk-based strategy is crucial for enhancing resilience in at-risk areas and reducing drought-related impacts (Wilhite et al. 2014; Vörösmarty et al. 2000). In addition to natural resistance, drought stress can be reduced by drought management strategies such as agronomic strategies, breeding approach.

2.5.1 Agronomic Strategy

Various strategies can enhance plant tolerance to drought stress, including changes in planting timing, plant geometry, nutrient management, zero tillage, mulching, and deep ploughing during the rainy season (Siddique et al. 2012; Jha et al. 2015;

Johnsona et al. 2018). The primary goal of these management practices is to improve the efficiency with which crops utilize available water through transpiration while reducing water loss from evaporation and weed competition. Additionally, the application of gypsum can enhance soil structure, thereby promoting water infiltration and root growth (Hamza and Anderson 2003).

2.5.2 Breeding Approaches

Breeding techniques such as classical breeding, genetic engineering, genome editing, and molecular breeding play a crucial role in enhancing crop varieties, not only for higher yields but also for tolerance to biotic (pests and pathogens) and abiotic stresses like drought and temperature extremes (Kumar et al. 2020; Munawar et al. 2020; Baloch et al. 2024). Numerous breeding approaches have been employed to improve drought tolerance in legumes. For instance, Hamwieh et al. (2013) identified 12 quantitative trait loci (QTLs) related to seedling drought tolerance in chickpea, including NCPGR-50, TR-50, SCEA19, TAA-58, H6C-07, H5E-02, H5G-01, H1B-04, TA-113, and H1F-21. Additionally, Li et al. (2017) reported that GmFDL19 plays a significant role in enhancing drought tolerance in soybean. Iuchi et al. (2002) found that VuNCED1 is involved in the synthesis of abscisic acid (ABA) in cowpea during drought conditions. Furthermore, Cai et al. (2015) successfully employed CRISPR/Cas9-mediated genetic engineering to study the efficiency of sgRNAs in soybean hairy roots.

2.6 Conclusions

Legumes are sensitive to water deficiency, which can impact their growth, yield, and quality, making drought stress a major impediment to sustainable legume production. Due to their symbiotic relationship with nitrogen-fixing bacteria, legumes need water for nutrient absorption, photosynthesis, and nitrogen fixation, which is reduced by drought stress. Drought hinders legume nitrogen fixing. Rhizobium, a nitrogen-fixing bacteria, needs moisture to form nodules on legume roots. Reduced nitrogen fixation causes nutrient deficits. Due to these shortcomings, the plant produces fewer healthy seeds, reducing output. Drought disrupts microbial activity and nutrient availability, which worsens plant stress and lowers soil fertility over time. Sustainable legume cultivation under drought requires coordinated techniques. These include drought-resistant cultivar breeding, drip irrigation, and soil conservation to conserve moisture. These strategies can help legumes withstand drought and produce more in water-scarce areas.

Contribution Statement A.M. Tawaha and H. Alatrasha conceptualized the chapter structure, drafted the main content, and conducted the primary literature review. Y. Jabbour, A. Al-Tawaha, and A.M. Qaisi reviewed the manuscript and provided critical feedback to ensure clarity and coherence throughout the chapter. R. Jammal,

A. Karnwal, M. Shatnawi, P. Saranraj, and J. Rammal supervised the overall development of the chapter, offered expert insights, and finalized the draft for submission. All authors reviewed and approved the final version of the chapter for publication.

References

Ahmed FP, Gaur PM, Croser J (2005) Chickpea (Cicer arietinum). In: Singh R, Jauhar P (eds) Genetic resources, chromosome engineering, and crop improvement—grains legumes. CRC Press, Boca Raton, pp 185–214

Akhtar I, Nazir N (2013) Effect of water logging and drought stress in plants. Int J Water Res Environ Sci 2:34–40

Allen MS, Voelker Linton JA (2010) Rates of particle size reduction and passage are faster for legume compared with cool-season grass, resulting in lower rumen fill and less effective fiber. J Anim Sci 88(2):711–720.

Ali Y, Haq MA, Ahmad N, Alam SS (1998) Differences in drought tolerance in ten chickpea genotypes. I- some studies on yield and yield components. Pakistan J Biol Sci 1(3):199–201

Ali A, Altaf MT, Nadeem MA, Karaköy T, Shah AN, Azeem H, Baloch FS, Baran N, Hussain T, Duangpan S, Aasim M (2022) Recent advancement in OMICS approaches to enhance abiotic stress tolerance in legumes. Front Plant Sci 13:952759

Apel K, Hirt H (2004) Reactive oxygen species: metabolism, oxidative stress, and signal transduction. Annu Rev Plant Biol 55:373–399

Arif A, Parveen N, Waheed MQ, Atif RM, Waqar I, Shah TM (2021) A comparative study for assessing the drought-tolerance of chickpea under varying natural growth environments. Front Plant Sci 11:607869

Arrese-Igor C, Gonza'lez EM, Gordon AJ, Minchin FR, Ga'lvez L, Royuela M, Cabrerizo PM, Aparicio-Tejo PM (1999) Sucrose synthase and nodule nitrogen fixation under drought and other environmental stresses. Symbiosis 27:189–212

Ashraf M, Ozturk M, Athar HR (2009) Salinity and water stress, improving crop efficiency. Springer, Dordrecht

Ashraf M, Akram NA, Al-Qurainy F, Foolad MR (2011) Drought tolerance: roles of organic osmolytes, growth regulators, and mineral nutrients. Environ Exp Bot 68(3):301–312

Baloch FS, Ali A, Tajibayev D, Nadeem MA, Ölmez F, Aktaş H et al (2024) Stripe rust resistance gene Yr15 in Turkish and Kazakhstan wheat germplasms and the potential of Turkish wild emmer for stripe rust breeding. Genet Resour Crop Evol 71(6):2699–2719

Barnabás B, Jäger K, Fehér A (2008) The effect of drought and heat stress on reproductive processes in cereals. Plant Cell Environ 31(1):11–38

Bartels D, Sunkar R (2005) Drought and salt tolerance in plants. Crit Rev Plant Sci 24(1):23–58

Baweja P, Kumar S, Kumar G (2020) Fertilizers and pesticides: their impact on soil health and environment. Soil Health:265–285

Bhargava S, Sawant K (2013) Drought stress adaptation: metabolic adjustment and regulation of gene expression. Plant Breed 132:21–32

Bitocchi E, Rau D, Bellucci E, Rodriguez M, Murgia ML, Gioia T et al (2017) Beans (Phaseolus ssp.) as a model for understanding crop evolution. Front Plant Sci 8:722

Blum A (2017) Osmotic adjustment is a prime drought stress adaptive engine in support of plant production: osmotic adjustment and plant production. Plant Cell Environ 40:4–10. https://doi.org/10.1111/pce.12800

Boukecha S, et al (2017). Drought tolerance in some populations of grass pea (Lathyrus sativus L.). Legume Research 40(3):346–352.

Cai Y, Chen L, Liu X, Sun S, Wu C, Jiang B, Han T, Hou W (2015) CRISPR/Cas9-mediated genome editing in soybean hairy roots. PLoS One 10:e0136064
Calvache M, Reichardt K, Bacchi OOS, Dourado-Neto D (1997) Deficit irrigation at different growth stages of the common bean (Phaseolus vulgaris L., cv. Imbabello). Sci Agric 54:1–16
Chaves MM, Oliveira MM (2004) Mechanisms underlying plant resilience to water deficits: prospects for water saving agriculture. J Exp Bot 55:2365–2384
Chaves MM, Maroco JP, Pereira JS (2003) Understanding plant responses to drought—from genes to the whole plant. Funct Plant Biol 30(3):239–264
Dai A (2012) Drought under global warming: a review. Wiley Interdiscip Rev Clim Change 2:45–65
Davis LC, Imsande J (1988) Direct test for altered gas exchange rates in water-stressed soybean nodules. Ann Bot 61:169–177
Demirta C, Yazgan S, Candogan BN, Sincik M (2010) Quality and yield response of soybean (Glycine max L. Merrill) to drought stress in sub–humid environment. Afr J Biotechnol 9:6873–6881
Denison RF (1998) Decreased oxygen permeability: a universal stress response in legume root nodules. Bot Acta 111:191–192
Fabio S, Albino M, Angelica G, Michele P (2017) Multiple benefits of legumes for agriculture sustainability: an overview. Chem Biol Technol Agric 4:2
Fang X, Turner NC, Yan G, Li F, Siddique KHM (2010) Flower numbers, pod production, pollen viability, and pistil function are reduced and flower and pod abortion increased in chickpea (Cicer arietinum L.) under terminal drought. J Exp Bot 61:335–345
Farooq M, Wahid A, Kobayashi N, Fujita D, Basra SMA (2009) Plant drought stress: effects, mechanisms and management. Agron Sustain Dev 29:185–212
Farooq M, Hussain M, Siddique KH (2014) Drought stress in wheat during flowering and grain-filling periods. Crit Rev Plant Sci 33:331–349
Farooq M, Gogoi N, Barthakur S, Baroowa B, Bharadwaj N, Alghamdi SS (2016) Drought stress in grain legumes during reproduction and grain filling. J Agron Crop Sci 203:81–102
Fenta BA, Beebe SE, Kunert KJ, Burridge JD, Barlow KM, Lynch PJ et al (2014) Field phenotyping of soybean roots for drought stress tolerance. Agronomy 4:418–435
Fernandez-Luquen F, Dendooven L, Munive A, Corlay-Chee L, Serrano- Covarrubias LM, Espinosa-Victoria D (2008) Micro-morphologyof common bean (Phaseolusvulgaris L.) nodules undergoing senescence. Acta Physiol Plant 30:545–552
Flexas J, Medrano H (2002) Drought-inhibition of photosynthesis in C3 plants: stomatal and non-stomatal limitations revisited. Ann Bot 89(2):183–189
Fouad M, Imtiaz M, Kumar S, Malhotra R (2011) Breeding food legumes for enhanced drought and heat tolerance to cope with climate change. In: Solh M, Saxena MC (eds) Food security and climate change in dry areas. Proceedings of International Conference, Syria, ICARDA, pp 244–254
Franco JA (2011) Root development under drought stress. Technol Knowl Transf e-Bull 2:1–3
Franco JA, Bañón S, Vicente MJ, MirallesJ M-SJJ (2011) Root development inhorticultural plants grown under abiotic stress conditions-a review. J Hortic Sci Biotechnol 86:543–556
Gepts P, Beavis WD, Brummer EC, Shoemaker RC, Stalker HT, Weeden NF, Young ND (2005) Legumes as a model plant family. Genomics for food and feed report of the cross-legume advances through genomics conference. Plant Physiol 137:1228–1235
Gill SS, Tuteja N (2010) Reactive oxygen species and antioxidant machinery in abiotic stress tolerance in crop plants. Plant Physiol Biochem 48(12):909–930
Gonza'lez EM, Gordon AJ, James CL, Arrese-Igor C (1995) The role of sucrose synthase in the response of soybean nodules to drought. J Exp Bot 46:1515–1523
Gordon AJ, Minchin FR, Skøt L, James CL (1997) Stress-induced declines in soybean N2 fixation are related to nodule sucrose synthase activity. Plant Physiol 114:937–946
Goufo P, Moutinho-Pereira JM, Jorge TF, Correia CM, Oliveira MR, Rosa EAS et al (2017) Cowpea (Vigna unguiculata L.Walp.) metabolomics: osmoprotection as a physiological strategy for drought stress resistance and improved yield. Front Plant Sci 8:586. https://doi.org/10.3389/fpls.2017.00586

Gusmao M, Siddique KHM, Flower K, Nesbitt H, Veneklaas EJ (2012) Water deficit during the reproductive period of grass pea (Lathyrus sativus L.) reduced grain yield but maintained seed size. J Agron Crop Sci 198:430–441

Hamwieh A, Imtiaz M, Maize I, Malhotra RS (2013) Multi-environment QTL analyses for drought-related traits in a recombinant inbred population of chickpea (Cicer arietinum L.). Theor Appl Genet 126:1025–1038

Hamza MA, Anderson W (2003) Response of soil properties and grain yield to deep ripping and gypsum application in a compacted loamy sand soil contrasted with a sandy clay loam in Western Australia. Aust J Agric Res 54:273–282

Hasegawa PM, Bressan RA, Zhu JK, Bohnert HJ (2013) Plant cellular and molecular responses to high salinity. Annu Rev Plant Physiol Plant Mol Biol 51(1):463–499

He J, Du Y, Wang T, Turner NC, Yang R, Jin Y, Xi Y, Zhang C, Cui T, Fang X et al (2016) Conserved water use improves the yield performance of soybean. Agric Water Manag 179:236

Hosseini FS, Nezami A, Parsa M, Hajmohammadnia K (2011) Effects of supplementary irrigation on yield and yield components of lentil (Lens culinaris Medik.) cultivars in Mashhad climate. J Water Soil 25(3):625–633

Iuchi S, Kobayashi M, Yamaguchi-Shinozaki K, Shinozaki KA (2002) Stress-inducible gene for9-cis-epoxycarotenoid dioxygenase involved in abscisic acid biosynthesis under water stress in drought-tolerant cowpea. Plant Physiol 123:553–562

Jaleel CA, Manivannan P, Wahid A, Farooq M, Somasundaram R, Panneerselvam R (2009) Drought stress in plants: a review on morphological characteristics and pigments composition. Int J Agric Biol 11(1):100–105

Jensen ES, Peoples MB, Boddey RM, Gresshoff PM, Hauggaard-Nielsen H, Alves BJ, Morrison MJ (2012) Legumes for mitigation of climate change and the provision of feedstock for biofuels and biorefineries. A review. Agron Sustain Dev 32:329–364

Jha S, Sehga VK, Rao YVS (2015) Effect of sowing direction and crop geometry on water use e_ciency and productivity of Indian mustard (Brassica juncea L.) in semi arid region of India. J Oilseed Brassica 6:257–264

Johnsona YK, Fredrick OA, Josiah MK, Sijali IV (2018) Effects of tillage practices on water use efficiency and yield of different effects of tillage practices on water use efficiency and yield of different drought tolerant common bean varieties in Machakos county, Eastern Kenya. Am Sci Res J Eng Technol Sci 40:217–234

Kakaei M, Kahrizi D, Ebadi AG (2010) Study of drought response extremes wheat varieties via seed storage constitutive proteins. Am J Sci Res 12:32–35

Kashiwag J, Krishnamurthy L, Crouch JH, Serraj R (2006) Variability of root characteristics and their contribution to seed yield in chickpea (Cicer arietinum L.) under terminal drought stress. Field Crop Res 95:171–181

Kashiwagi J, Krishnamurthy L, Purushothaman R, Upadhyaya HD, Gaur PM, Gowda CLL, Ito O, Varshney RK (2015) Scope for improvement of yield under drought through the root traits in chickpea (Cicer arietinum L.). Field Crop Res 170:47–54

Khan MA, Iqbal M, Jameel M, Nazeer W, Shakir S, Aslam MT, Iqbal B (2011) Potentials of molecular based breeding to enhance drought tolerance in wheat (Triticum aestivum L.). Afr J Biotechnol 10:11340–11344

Khan MS, Kanwal B, Nazir S (2015) Metabolic engineering of the chloroplast genome reveals that the yeast ArDH gene confers enhanced tolerance to salinity and drought in plants. Front Plant Sci 6:725. https://doi.org/10.3389/fpls.2015.00725

Kobraei S, Etminan A, Mohammadi R, Kobraee S (2011) Effects of drought stress on yield and yield components of soybean. Ann Biol Res 2(5):504–509

Kokubun M, Shimada S, Takahashi M (2001) Flower abortion caused by preanthesis water deficit is not attributed to impairment of pollen in soybean. Crop Sci 41:1517–1521

Kumar S, Gupta P, Choukri H, Siddique KHM (2020) Efficient breeding of pulse crops. In: Gosal SS, Wani SH (eds) Accelerated plant breeding, vol 3. Springer Nature, Switzerland

Lawlor DW, Cornic G (2002) Photosynthetic carbon assimilation and associated metabolism in relation to water deficits in higher plants. Plant Cell Environ 25(2):275–294

Lemke RL, Zhong Z, Campbell CA, Zentner RP (2007) Can pulse crops play a role in mitigating greenhouse gases from North American agriculture? Agron J 99:1719–1725

Lewis G, Schrire B, Mackinder B, Lock M (2005) Legumes of the world. Royal Botanic Gardens, Kew

Li D, Liu H, Qiao Y, Wang Y, Cai Z, Dong B, Shi C, Liu Y, Li X, Liu M (2013) Effects of elevated CO2 on the growth, seed yield, and water use e_ciency of soybean (Glycine max. L.) under drought stress. Agric Water Manag 129:105–112

Li Y, Chen Q, Nan H, Li X, Lu S, Zhao X, Liu B, Gu C, Kong F, Cao D (2017) Overexpression of GmFDL19 enhances tolerance to drought and salt stresses in soybean. PLoS One 12:e0179554

Liu X, Jin J, Herbert SJ, Zhang Q, Wang G (2005) Yield components, dry matter, LAI and LAD of soybeans in Northeast China. Field Crop Res 93:85–93

Llorens et al (2003) Erica multiflora, and Globularia alypum, to experimentally drier and warner conditions. Physiol Plant 119(2):231–243

Madhava RKV, Raghavendra AS, Janardhan Reddy K (eds) (2006) Physiology and molecular biology of stress tolerance in plants. Springer, Dordrecht

Mafakheri A, Siosemardeh B, Bahramnejad A, Struik PC, Sohrabi E (2010) Effect of drought stress on yield, proline and chlorophyll contents in three chickpea cultivars. Aust J Crop Sci 4(8):580–585

Mak M, Babla M, Xu S, Carrigan AO, Liu X, Gong Y, Holford P, Chen Z (2014) Leaf mesophyll K+, H+ and Ca2+ fluxes are involved in drought-induced decrease in photosynthesis and stomatal closure in soybean. Environ Exp Bot 98:1–12

Maleki A, Naderi A, Naseri R, Fathi A, Bahamin S, Maleki R (2013) Physiological performance of soybean cultivars under drought stress. Bull Environ Pharmacol Life Sci 2:38–44

Márquez-García B, Shaw D, Cooper JW, Karpinska B, Quain MD, MakgopaE M et al (2015) Redox markers for drought-induced nodule senescence, a process occurring after drought-induced senescence of the lowest leaves in soybean (Glycinemax Merr.). Ann Bot 116:497–510

McKee TB, Doesken NJ, Kleist J (1993) The relationship of drought frequency and duration to time scales. In: Proceedings of the 8th conference on applied climatology, vol 17, pp 179–183

Micheletto S, Rodriguez-uribe L, Hernandez R, Richins RD, Curry J, Connell MAO (2007) Comparative transcript profiling in roots of Phaseolus acutifolius and P. vulgaris under water deficit stress. Plant Sci 173:510–520

Minchin FR (1997) Regulation of oxygen diffusion in legume nodules. Soil Biol Biochem 29:881–888

Mirzaei A, Naseri R, Vafa P, Moradi M (2014) Effects of drought stress on qualitative and quantitative traits of mungbean. Int J Agric Biol Eng 8:148–152

Mittal N, Mishra A, Singh R, Kumar P (2014) Assessing future changes in seasonal climatic extremes in the Ganges river basin using an ensemble of regional climate models. Clim Chang 123:273–286

Munawar S, Ul Qamar MT, Mustafa G, Khan MS, Joyia FA (2020) Role of biotechnology in climate resilient agriculture. In: Fahad S et al (eds) Environment, climate, plant and vegetation growth. Springer Nature Switzerland

Muruiki JM, et al (2018) Drought stress in chickpea: Physiological, breeding, and omics insights. In: Varshney RK, Thudi M (Eds.), Chickpea Breeding and Management (pp. 201–218). Springer.

Nadeem MA, Habyarimana E, Çiftçi V, Nawaz MA, Karaköy T, Comertpay G et al (2018) Characterization of genetic diversity in Turkish common bean gene pool using phenotypic and whole-genome DArTseq-generated silicoDArT marker information. PLoS One 13(10):e0205363

Nasri M, Zahedi H, Tohidi MHR, Ghooshchi F, Paknejad F (2008) Investigation of water stress on macroelements in canola genotypes leaf (Brassica napus). Am J Agric Biol Sci 3(4):669–672

Nezhadahmadi A, Hossain Prodhan Z, Faruq G (2013) Drought tolerance in wheat. Sci World J 2013:1–12

Osakabe Y, Osakabe K, Shinozaki K, Tran L-SP (2014) Response of plants to water stress. Front Plant Sci 5:86

Otkin JA, Anderson MC, Hain CR, Mladenova IE (2016) Examining rapid onset drought development using the thermal infrared–based evaporative stress index. J Hydrometeorol 17(2):675–689
Palmer WC (1965) Meteorological drought. U.S. Department of Commerce Weather Bureau Research Paper No. 45
Phillips BB, Shaw RF, Holland MJ, Fry EL, Bardgett RD, Bullock JM, Osborne JL (2018) Drought reduces floral resources for pollinators. Glob Chang Biol 24:3226–3235
Pulwarty RS, Sivakumar MVK (2014) Information systems in a changing climate: early warnings and drought risk management. Weather Clim Extrem 3:14–21
Pushpavalli R, Zaman-allah M, Turner NC, Baddam R, Rao MV, Vadez V (2015) Higher flower and seed number leads to higher yield under water stress conditions imposed during reproduction in chickpea. Funct Plant Biol 42:162–174
Rahimi MH, Houshm S, Khodambashi M, Shiran B, Mohammady S (2016) Effect of drought stress on agro-morphological traits of lentil (lens Culinaris Medik.) recombinant inbred lines. Bangladesh J Agril Res 41(2):207–219
Robson MC, Fowler SM, Lampkin NH, Leifert C, Leitch M, Robinson D et al (2002) The agronomic and economic potential of breakcrops for ley/arable rotations in temperate organic agriculture. Adv Agron 77:369–427
Sakamoto A, Murata N (2002) The role of glycine betaine in the protection of plants from stress: clues from transgenic plants. Plant Cell Environ 25(2):163–171
Salehi-lisar SY, Motafakkerazad R, Hossain MM, Rahman IMM (2012) In: Rahman IMM (ed) Water stress in plants: causes, effects and responses, water stress. InTech
Samarah NH, Haddad N, Alqudah AM (2009) Yield potential evaluation in chickpea genotypes under late terminal drought in relation to the length of reproductive stage. Ital J Agron 3:111–117
Sardans et al (2008) Drought and warming-induced changes in P and K concentration and accumulation in plant biomass and soil in a Mediterranean shrubland. Plant Soil 306:261–271
Savita AT, Saurabh KS (2020) Drought stress tolerance in legume crops. In: Hasanuzzaman M (ed) Agronomic crops. Springer Nature Singapore Pte Ltd, pp 149–155
Serraj R, Bidinger FR, Chauhan YS, Seetharama N, Nigam SN, Saxena NP (2003) Management of drought in ICRISAT cereal and legume mandate crops. In: Water productivity in agriculture: limits and opportunities for improvement. CAB International, Wallingford, pp 127–144
Serraj R, Sinclair TR, Purcell LC (2004) Symbiotic N_2 fixation response to drought. J Exp Bot 55(394):425–434
Serraje R (2003) Effects of drought stress on legume symbiotic nitrogen fixation: physiological mechanisms. Indian J Exp Biol:1136–1141
Seymour M, Kirkegaard JA, Peoples MB, White PF, French RJ (2012) Break-crop benefits to wheat in Western Australia-insights from over three decades of research. Crop Pasture Sci 63:1–16
Shinozaki K, Yamaguchi-Shinozaki K (2007) Gene networks involved in drought stress response and tolerance. J Exp Bot 58(2):221–227
Sicher RC, Timlin D, Bailey B (2012) Responses of growth and primary metabolism of water-stressed barley roots to rehydration. J Plant Physiol 169:686–695. https://doi.org/10.1016/j.jplph.2012.01.002
Siddique KHM, Walton GH, Seymour M (1993) A comparison of seed yields of winter grain legumes in Western Australia. Aust J Exp Agric 33:915–922
Siddique KHM, Johansen C, Alghamdi SS (2012) Innovations in agronomy for food legumes. A review. Agron Sustain Dev 32:45–64
Sivakumar MVK et al (2010) Managing weather and climate risks in agriculture. Springer Science & Business Media
Srivastava S, Kumar P (2020) Abiotic stress responses and tolerance mechanisms for sustaining crop productivity in sugarcane. In: Hasanuzzaman M (ed) Agronomic crops, vol 29. Springer Nature Singapore Pte Ltd
Streeter JG (2003) Effects of drought on nitrogen fixation in soybean root nodules. Plant Cell Environ 26:1199–1204

Tallaksen LM, van Lanen HAJ (2004) Hydrological drought: processes and estimation methods for streamflow and groundwater. Elsevier
Tawaha AM, Turk MA (2004) Field pea seeding management for semi-arid Mediterranean conditions. J Agron Crop Sci 190(2):86–92
Tawaha AR, Turk MA, Lee KD (2005) Adaptation of chickpea to cultural practices in a Mediterranean type environment. Res J Agric Bio Sci 1:152–157
Tharanathan RN, Mahadevamma S (2003) Grain legumes—a boon to human nutrition. Trends Food Sci Tech 14:507–518
Thomson BD, Siddique KHM, Barr MD, Wilson JM (1997) Grain legume species in low rainfall Mediterranean type environments. I. Phenology and seed yield. Field Crop Res 54:173–187
Thu NB, Nguyen QT, Hoang XL, Thao NP, Tran LS (2014) Evaluation of drought tolerance of the Vietnamese soybean cultivars provides potential resources for soybean production and genetic engineering. BioMed Res Int 9:1
Trujillo LE, Sotolongo M, Menendez C, Ochogavia ME, Coll Y, Hernandez I, Borras-Hidalgo O, Thomma BPHJ, Vera P, Hernandez L (2008) SodERF3, a novel sugarcane ethylene responsive factor (ERF), enhances salt and drought tolerance when overexpressed in tobacco plants. Plant Cell Physiol 49:512–525
Turk MA, Taw ARM, Lee KD (2004) Seed germination and seedling growth of three lentil cultivars under moisture stress. Asian J Plant Sci 3(3):394–397
UNCCD (2019) Drought resilience, adaptation and management policy framework. United Nations Convention to Combat Desertification
Van Loon AF (2015) Hydrological drought explained. Wiley Interdiscip Rev Water 2(4):359–392
Van Loon AF, Laaha G (2015) Hydrological drought severity explained by climate and catchment characteristics. J Hydrol 526:3–14
Vörösmarty CJ, et al (2000) Global water resources: Vulnerability from climate change and population growth. Science 289(5477):284–288. https://doi.org/10.1126/science.289.5477.284.
Wilhite DA, Glantz MH (1985) Understanding: the drought phenomenon: the role of definitions. Water Int 10(3):111–120
Wilhite DA, Sivakumar MV, Pulwarty R (2007) Managing drought risk in a changing climate: the role of national drought policy. Weather Clim Extrem 3:4–13
Wilhite DA, Sivakumar MVK, Pulwarty RS (2014) Managing drought risk in a changing climate: the role of national drought policy. Weather Clim Extrem 3:4–13
Worrall VS, Roughley RJ (1976) Effect of moisture stress on infection of Trifolium Subterraneum L. by RhizobiumTrifolii Dang. J Exp Bot 27:1233–1241
Zahran HH (1999) Rhizobium-legume symbiosis and nitrogen fixation under severe conditions and in an arid climate. Microbiol Mol Biol Rev 63:968–989
Zebarjadi A, Mirany T, Kahrizi D, Ghobadi M, Nikoseresht R (2012) Assessment of drought tolerance in some bread wheat genotypes using drought resistance indices. Biharean Biol 6:94–98
Zhu H, Choi HK, Cook DR, Shoemaker RC (2005) Bridging model and crop legumes through comparative genomics. Plant Physiol 137:1189–1196
Zlatev Z, Lidon FC (2012) An overview on drought induced changes in plant growth, water relations and photosynthesis. Emir J Food Agric 24:57–72

3 Legume Production and Problems in the Mediterranean Region

Taufiq Nawaz, Shah Fahad, Nitish Joshi,
Tanzeel Ur Rahman, Muhammad Nasir Rasheed Khan,
Shah Hassan, and Shah Saud

Abstract

Legume production in the Mediterranean region faces unique challenges and opportunities influenced by its diverse climatic and soil conditions. This chapter examines the cultivation practices, yield statistics, economic significance, and prevalent issues associated with major legumes including chickpeas (*Cicer arietinum*), lentils (*Lens culinaris*), faba beans (*Vicia faba*), and peas (*Pisum sativum*). Factors such as optimal sowing times, irrigation techniques, soil preparation, and fertilization methods are critical in enhancing productivity and mitigating environmental impacts. Weed, pest, and disease control strategies are imperative for sustaining crop health and maximizing yields. Furthermore, the

T. Nawaz · S. Fahad
Department of Biology/Microbiology, South Dakota State University, Brookings, SD, USA
e-mail: taufiq.nawaz@jacks.sdstate.edu

N. Joshi
Department of Agronomy, Horticulture & Plant Science, South Dakota State University, Brookings, SD, USA

T. U. Rahman
Department of Hotel Management and Tourism, University of Swabi, Khyber Pakhtunkhwa, Pakistan
e-mail: tanzeel.htm13@gmail.com

M. N. R. Khan
Department of Agronomy, Abdul Wali Khan University Mardan, Khyber Pakhtunkhwa, Pakistan

S. Hassan
Department of Agricultural Extension Education & Communication, The University of Agriculture, Peshawar, Khyber Pakhtunkhwa, Pakistan

S. Saud (✉)
College of Life Science, Linyi University, Linyi, Shandong, People's Republic of China

© The Author(s), under exclusive license to Springer Nature Singapore Pte Ltd. 2025

M. A. Nadeem et al. (eds.), *Marker-Assisted Breeding in Legumes for Drought Tolerance*, https://doi.org/10.1007/978-981-96-4112-3_3

potential for expanding legume cultivation into marginal lands and leveraging export opportunities is discussed, highlighting the need for sustainable practices and innovation in agricultural systems. Collaboration and knowledge sharing among Mediterranean countries are identified as key drivers for advancing legume production in the region. This chapter underscores the interdisciplinary approach required to address the complex interplay of agronomic, economic, and environmental factors in Mediterranean legume agriculture.

Keywords

Legume production · Mediterranean region · Economic significance · Soil preparation · Fertilization · Sustainability · Climate resilience · Export opportunities

3.1 Introduction

Legumes, such as beans, lentils, chickpeas, and peas, are crucial to the agricultural systems and diets of the Mediterranean region due to their multifaceted significance. They contribute to food security and nutrition, as well as to soil fertility through biological nitrogen fixation (Maxted and Bennett 2001). The Mediterranean climate, with its mild, wet winters and hot, dry summers, provides a unique environment for legume cultivation, presenting both advantages and challenges. Legumes also play a key role in sustainable agriculture, as they release fewer greenhouse gases, sequester carbon in soils, and reduce the need for fossil energy inputs (Stagnari et al. 2017). In the dryland farming systems of the Mediterranean, forage legumes are particularly important for animal production, with vetches, chicklings, and annual medics being the most significant species (Ates et al. 2014). The conservation and use of legume species is crucial for food security and the preservation of genetic diversity in the Mediterranean region (Vishnyakova et al. 2019).

Legumes, such as chickpeas and lentils, are crucial for sustainable agriculture in the Mediterranean due to their ability to fix nitrogen and improve soil health (Okumu et al. 2023). They also play a key role in diversifying and intensifying agriculture, particularly in the face of climate change (Vasconcelos et al. 2020). However, their potential is underutilized in Europe, where they occupy less than 4% of arable lands. To address this, there is a need for research on the constraints of current legume systems, new uses for legumes, and the promotion of agroecology (Voisin et al. 2014). In the Mediterranean, climate change is a major challenge, affecting rain events, floods, and droughts, and necessitating mitigation strategies for sustainable crop production (Dos Santos et al. 2022; Dawar et al. 2023).

Legume production in the Mediterranean faces a range of challenges, including biotic and abiotic stresses, and socio-economic barriers (Toker and Mutlu 2011). These challenges are exacerbated by climate change, necessitating adaptive strategies (Araújo et al. 2015). Recent research has focused on developing stress-tolerant legume varieties through biotechnological tools (Chandrashekharaiah et al. 2021). These efforts are crucial for ensuring the resilience and sustainability of legume farming in the region. Recent research has highlighted the significant impact of climate change on legume production in the Mediterranean region, necessitating the development of effective management strategies and technological innovations (Hossain 2020). These strategies include irrigation, changes in sowing date, and cultivar choice (Marteau-Bazouni et al. 2024), as well as the integration of agronomic and biotechnological approaches, and advanced genome editing tools to develop drought-tolerant legume cultivars (Nadeem et al. 2019). Additionally, the management of Mediterranean soil resources under global change, including the implementation of changes in agricultural practices, soil water management, and vegetal material, is crucial for the resilience and sustainability of legume farming in the region (Lagacherie et al. 2018; Nawaz et al. 2024b, c, d, e).

3.2 Major Legumes Crops in the Mediterranean Region

3.2.1 Chickpeas (*Cicer arietinum*)

Chickpea cultivation in the Mediterranean region is influenced by a range of traditional and modern practices. Chickpea yields in the Mediterranean region vary significantly due to differences in climatic conditions, soil types, and farming practices. On average, chickpea yields range from 0.8 to 1.2 tonnes per hectare, although improved management and modern cultivars can achieve yields up to 2.5 tonnes per hectare (Antichi et al. 2023). Nunes et al. (2023) emphasize the importance of careful irrigation, particularly during flowering and pod formation, to avoid fungal issues. Malhotra et al. (1997) and Siddique et al. (2012) both highlight the potential benefits of irrigation, with Malhotra's study in Syria showing a significant increase in seed yield with irrigation, and Basha's review recommending two irrigations at specific growth stages. However, Pacucci et al. (2006) found that supplementary irrigation only increased the yield of certain chickpea genotypes in Southern Italy, suggesting that the effectiveness of irrigation may vary by region and genotype.

Chickpeas are a significant crop in the Mediterranean region, contributing to both rural livelihoods and national economies (Merga and Haji 2019). They are a crucial source of income for smallholder farmers, providing financial stability and employment opportunities in rural areas. The crop's high protein content and nutritional value make it a vital component of local diets, supporting food security and nutritional health (Fikre et al. 2020). In the global market, Mediterranean countries play a pivotal role in chickpea trade, exporting significant quantities to meet the growing demand in regions such as South Asia, the Middle East, and Europe (Korbu et al. 2016). The economic value of chickpeas is further enhanced by their use in various processed foods, including hummus, falafel, and flour, which cater to diverse culinary traditions and preferences (Ali et al. 2022b). Additionally, chickpea farming contributes to sustainable agriculture through nitrogen fixation, improving soil health and reducing the need for chemical fertilizers, thus promoting environmental sustainability (Muehlbauer and Sarker 2017).

3.2.2 Lentils (*Lens culinaris*)

Lentils (*Lens culinaris*) are well-suited to the Mediterranean region, particularly in low to medium rainfall areas, where early sowing can lead to higher yields (Siddique et al. 1998). However, they face challenges in diverse organic Mediterranean environments, with genotype selection being crucial for balancing yield and stability (Baxevanos et al. 2024). Effective weed management, particularly removing weeds at 30 and 90 days after seeding, is essential for optimal growth and yield (Turk and Tawaha 2002). Rainfall and temperature significantly impact lentil seed yield, with late vegetative growth being particularly sensitive to March rainfall (Erskine and El Ashkar 1993).

Lentil yields in the Mediterranean region exhibit variability influenced by climatic conditions, soil quality, and agricultural practices. On average, lentil yields range from 0.6 to 1.0 tonnes per hectare, with potential yields reaching up to 1.5 tonnes per hectare under optimal conditions. Leading producers in the Mediterranean include Turkey, Syria, and Spain, which collectively contribute a significant portion of the region's lentil output (Antichi et al. 2023). The timing of lentil harvest is crucial, as it affects both the physical properties of the seeds and the quality of the produce. Gürsoy (2020) found that delayed harvest time led to decreased moisture content, seed dimensions, and bulk density, while increasing sphericity, true density, porosity, and hardness of lentil seeds. However, Erskine et al. (1991) and Bansal et al. (1994) both noted that mechanical harvesting methods, while efficient, can result in significant harvest losses compared to hand harvesting. Khatun et al. (2009) further emphasized the importance of the harvesting stage, with seeds collected at the stage when most of the pods were light brown showing the highest germination percentage and seedling vigor.

3.2.3 Faba Beans (*Vicia faba*)

Faba beans, a traditional Mediterranean crop, are best planted in late autumn or early winter to take advantage of the wet season (Tedone et al. 2022). No-tillage systems have been found to enhance nitrogen fixation and improve soil quality, reducing the need for synthetic fertilizers. Intra-row spacing of 10–12 cm is optimal for seed yield in Mediterranean conditions (Yucel 2013). Faba beans rely more on nitrogen fixation than other legumes, leading to higher nitrogen benefits for subsequent crops (Ruisi et al. 2017). The choice of tillage system and residual nitrogen levels can influence faba bean yield, with conventional tillage being more productive in rainy years (López-Bellido et al. 2008). Faba bean production in the Mediterranean region is significant, with key producers including Egypt, Morocco, and Italy. The region's production figures are influenced by climatic variability, soil fertility, and agricultural practices. On average, faba bean yields range from 1.5 to 3.0 tonnes per hectare, with potential yields reaching up to 4.0 tonnes per hectare under optimal conditions. Egypt is one of the largest producers and consumers of faba beans, contributing substantially to the region's total output (Antichi et al.

2023). Faba beans, also known as broad beans, have a rich history and cultural significance in the Mediterranean region and the Middle East (Pasqualone et al. 2020). They are a vital source of protein and other nutrients, playing a crucial role in food security and nutrition in these areas (Askar 1986; Alam et al. 2022). The beans are consumed in various forms, including soups, falafel, and ful medames, and are a staple in the diet of countries like Egypt, Morocco, and Italy (Pasqualone et al. 2020). Faba beans are also a significant export commodity, with potential for further expansion in food and nutritional security (Dhull et al. 2022).

3.2.4 Peas (*Pisum sativum*)

Peas (*Pisum sativum*) are a valuable crop in the Mediterranean region due to their adaptability to the local climate and soil conditions (Karkanis et al. 2016). They are typically planted in early spring or late winter, with a sowing depth of 2.5 to 5 cm and row spacing of 20–30 cm. The crop's yield and quality can be influenced by factors such as seeding rate, date, and depth, as well as phosphorus fertilization (Tawaha and Turk 2004). The development of high-yielding pea varieties in the region is also important, with a focus on early maturation and long preflowering periods (Al-Karaki 1999). Pea productivity in the Mediterranean region varies widely, influenced by factors such as soil quality, climate, and farming practices. Average yields range from 1.0 to 2.5 tonnes per hectare, with potential yields reaching up to 3.5 tonnes per hectare under optimal conditions. Leading producers in the region include France, Spain, and Italy, each contributing significantly to the total output (Antichi et al. 2023). Optimal germination of peas occurs at soil temperatures between 10 °C and 20 °C. Crop rotation, particularly following cereals or grasses, is beneficial for peas, reducing disease pressure and improving soil fertility. Minimal tillage practices are encouraged to preserve soil moisture and structure (Lenssen et al. 2018). Peas are known for their nitrogen-fixing capabilities, enhancing soil fertility for subsequent crops. Effective weed management is crucial, often achieved through mechanical cultivation and selective herbicides (Saeed et al. 2024). Pest and disease management focuses on controlling aphids, powdery mildew, and root rot through integrated pest management strategies (Grünwald et al. 2004). Ascochyta blight, a common disease in peas, can be managed through a combination of strategies, including time of sowing and the use of wheat stubble from the preceding crop (Bennett et al. 2019). Pea production in Europe has declined but is being revived due to its agronomic advantages, including support for sustainable cropping systems and reduced nitrogen fertilizer use (Karkanis et al. 2016). The Mediterranean region is a key player in the global pulse market, with peas being a significant part of this sector (Joshi and Rao 2016). The region's production and use of field peas, both for human consumption and animal feed, have been discussed in the context of European cropping systems (Karkanis et al. 2016). The adaptability of peas in this region, along with their potential to support sustainable cropping systems and reduce nitrogen fertilizer use, has been highlighted. The rise in the export of processed pea products, such as pea protein and pea-based snacks, aligns with global trends towards convenience foods and plant-based protein sources (Sozer et al. 2017). This trend is further supported by the increasing integration of pulse proteins, including those from peas, into the plant protein ingredient market (Singhal et al. 2016).

3.3 Types of Soil in the Mediterranean Regions

The Mediterranean region is characterized by diverse soil types, which significantly influence agricultural practices and crop productivity (Fig. 3.1). The predominant soil types include:

3.3.1 Calcareous Soils

The presence of calcium carbonate in Mediterranean soils, derived from limestone or chalk, can lead to alkalinity and affect nutrient availability (Carreira et al. 1996). This can result in low organic matter and available nitrogen, as well as reduced availability of phosphate, iron, and zinc (Wahba et al. 2019). The high pH level in these soils can also lead to micronutrient deficiencies, particularly zinc and iron, which can hamper crop production (Rashid and Ryan 2004). Therefore, the management of these soils, including efficient water use and fertilization, is crucial for sustainable agriculture in Mediterranean areas (Sgouras et al. 2007).

3.3.2 Terra Rossa

Red, clayey soils rich in iron oxides are a common feature of Mediterranean regions, with specific properties and management needs. These soils, classified as

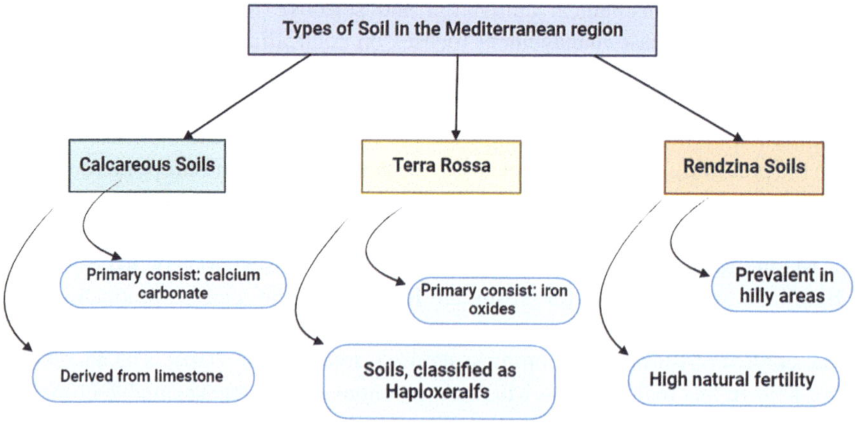

Fig. 3.1 Characteristics and types of soil in the Mediterranean region

Haploxeralfs and Rhodoxeralfs, have a pH range of 5.2 to 7.69 and a texture ranging from clay to sandy clay loam (Sgouras et al. 2007). They are typically found on well-drained sites and are affected by water deficit during the summer period (Noulas et al. 2009). The predominant iron oxide responsible for their red coloration is hematite (Aydinalp and Cresser 2008). To maintain their fertility, regular fertilization and the use of nonacidifying fertilizers are recommended (Aydinalp and Cresser 2008). These soils are also affected by erosion and may require practices such as rotation schemes, minimum tillage, liming, and proper irrigation systems to sustain crop production and mitigate erosion risk (Noulas et al. 2009).

3.3.3 Rendzina Soils

The formation and properties of shallow, rocky soils on limestone and marl are influenced by their parent material, with specific characteristics such as high natural fertility and limited root penetration (Mitkova and Markoski 2019; Hewitt et al. 2021). These soils are prevalent in hilly and mountainous areas, where they are subjected to strong hillslope processes and may accumulate organic carbon (Egli and Poulenard 2016). The specialized root system of shallow-soil endemics, which invest more in deep roots, may be an adaptation to these challenging soil conditions (Poot and Lambers 2008).

3.3.4 Mediterranean Brown Earths

The development and fertility of woodland and shrubland soils are influenced by factors such as root density, organic matter incorporation, and mineralogy (Grieve 1980; Tripathi et al. 2006). These soils, while suitable for a wide range of crops with appropriate management, are often fragile and prone to erosion (Binkley and Fisher 2019). The impact of different land uses on the physical, chemical, and biological properties of these soils varies, with arable cropping and pine plantations leading to a decline in soil structure and biological properties (Sparling et al. 2000; Irfan et al. 2021).

3.3.5 Fertilization Methods Used in the Mediterranean Region

Effective fertilization is crucial to optimizing crop yields in the Mediterranean region, where soil fertility can vary widely. Key fertilization methods (Fig. 3.2) include:

I. *Chemical Fertilizers:* Chemical fertilizers, commonly used in conventional farming, are applied based on soil tests to meet crop nutritional needs (Gerenfes 2021). However, their use can lead to environmental and health concerns, such as nutrient leaching, soil salinity, and water eutrophication. Sustainable agricul-

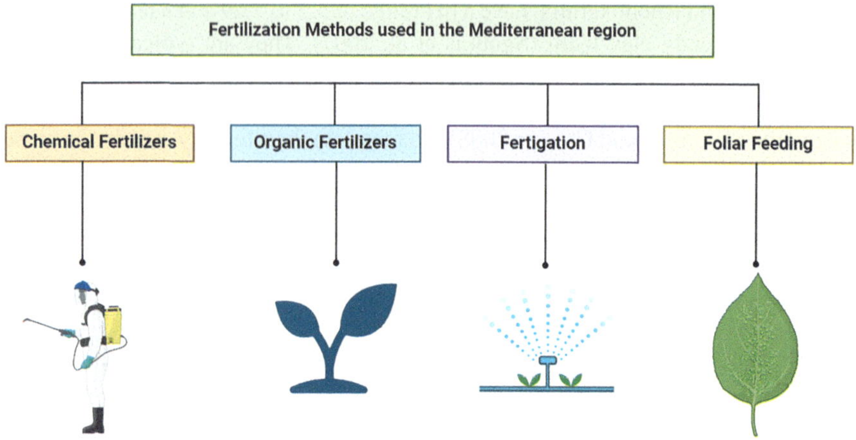

Fig. 3.2 Various fertilization methods used in the Mediterranean region

ture, including organic farming, emphasizes the use of organic materials as soil amendments and sources of plant nutrients (Hue and Silva 2000). Nanofertilizers, a potential substitute for chemical fertilizers, are engineered materials that can increase nutrient uptake efficiency and reduce environmental impact (Nongbet et al. 2022).

II. *Organic Fertilizers:* Organic fertilizers, such as compost, manure, and green manure, play a crucial role in improving soil structure and fertility by adding organic matter (Assefa and Tadesse 2019). These fertilizers release nutrients slowly and enhance microbial activity, promoting long-term soil health (Kumar and Pawar 2018). However, the impact on soil properties depends on the quantity and quality of the materials applied, with long-term applications of bulky organic materials being particularly effective in increasing soil organic carbon and improving soil quality (Bhogal et al. 2018).

III. *Integrated Nutrient Management (INM):* The concept of Integrated Nutrient Management (INM) has been widely recognized as a sustainable approach to soil fertility and crop productivity (Vanlauwe and Giller 2006). By combining organic and inorganic fertilizers, INM optimizes nutrient availability and maintains soil health, addressing both immediate crop needs and long-term sustainability (Wu and Ma 2015). This approach has been particularly effective in improving soil properties, enhancing crop yields, and reducing environmental pollution (Kumar et al. 2024). Furthermore, INM has been shown to increase the activity of soil microorganisms and improve soil physical, chemical, and biological properties, making it a key practice for sustainable agriculture (Panta and Parajulee 2021).

IV. *Fertigation:* Fertigation, the application of water-soluble fertilizers through irrigation systems, has been shown to significantly increase crop productivity and nutrient uptake (Shedeed et al. 2009). This method is particularly effective in areas with limited rainfall or erratic precipitation, as it allows for precise nutrient delivery and reduces nutrient losses (Kafkafi and Tarchitzky 2011). In Israel, where 80% of the irrigated land uses fertigation, this technique has been crucial in the production of high-quality crops with minimal pollution (Imas 1999). Furthermore, fertigation can increase water and nutrient use efficiency, making it a valuable tool for sustainable agriculture (Yu et al. 2003).

V. *Foliar Feeding:* Foliar application of micronutrients, such as iron, zinc, boron, copper, manganese, chlorine, molybdenum, and nickel, can effectively address specific deficiencies and improve crop health (Patel et al. 2022). This method supplements soil fertilization and provides a quick nutrient boost during critical growth stages (Fageria et al. 2009). It is particularly beneficial in vegetable production, where it can enhance growth, yield, and quality attributes (Krishnasree et al. 2021). The application of both macro- and micronutrients is crucial for the proper growth and development of crops, with foliar application being particularly effective for micronutrients (Kashyap et al. 2022).

3.3.6 Organic Versus Conventional Farming Practices

Organic farming emphasizes sustainable agricultural practices, avoiding synthetic chemicals and focusing on natural inputs (Fig. 3.3). Key features include:

I. *Soil Health:* Organic farming practices, such as composting, manure application, crop rotations, and cover crops, play a crucial role in enhancing soil fertility (Nair and Delate 2016). These practices improve soil structure, increase organic matter content, and promote beneficial soil organisms (Jamar et al. 2006). Crop rotation, in particular, is highlighted as a central tool in maintaining and developing soil fertility in organic systems (Watson et al. 2002). The use of legume-based multi-annual rotations and on-farm manures is emphasized, as they replenish nutrient elements and increase biological activity (Watson 2008). Composting, crop rotation, and cover crops are also identified as key practices in organic vegetable production, with the potential to improve soil quality, break pest and disease cycles, and increase yields (Nair and Delate 2016).

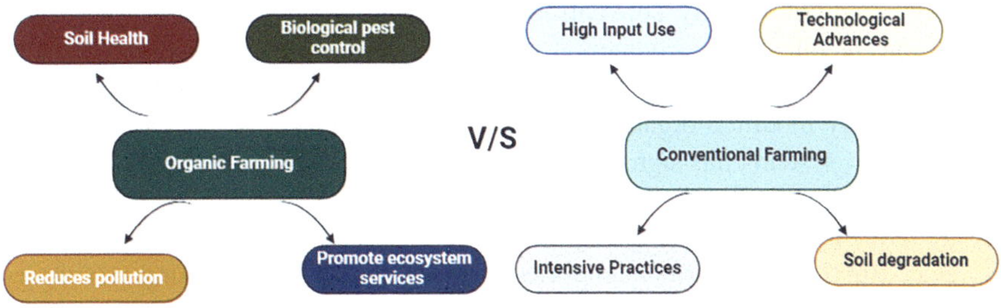

Fig. 3.3 Characteristics of organic and conventional farming

II. *Pest and Disease Management:* Organic farmers employ a range of methods to manage pests and diseases, including biological control, crop rotations, and resistant varieties (Saxena 2021). These methods are used to prevent and control pests and diseases without the use of synthetic pesticides, which are prohibited in organic farming (Irfan et al. 2022). The emphasis on variety and diversity in organic farming is crucial for effective pest and disease management (Stoleru and Sellitto 2016). These practices not only contribute to sustainable agriculture but also promote natural predator populations and biodiversity (Wyss et al. 2005).

III. *Environmental Impact:* Organic farming has been shown to have a positive impact on the environment, particularly in reducing pollution and conserving water resources (Sivaranjani and Rakshit 2019). It achieves this by minimizing the use of synthetic chemicals, which can leach into water sources and cause pollution. Additionally, organic farming practices enhance biodiversity and promote ecosystem services (Kühling and Trautz 2013). This is achieved through the use of biological methods for pest and disease control, which minimizes negative impacts on natural ecosystems. The positive effects of organic farming on biodiversity and ecosystem services are further supported by Winqvist et al. (2012) and Underwood et al. (2011), who both found that organic farming enhances species richness and abundance and supports sustainable food production.

3.3.7 Conventional Farming

Conventional farming relies on synthetic inputs to maximize crop yields and efficiency (Fig. 3.3). Key features include:

I. *High Input Use:* Conventional farming's heavy reliance on synthetic fertilizers, pesticides, and herbicides has led to significant ecological and health concerns (Degeronimo 2015). While these practices have increased agricultural productivity, they have also resulted in soil and water contamination, loss of biodiversity, and pest resistance. The need for agricultural training and capacity building, as well as the promotion of commercial organic farming, has been highlighted as potential solutions (Kharel and Marahatta 2020). Despite their role in increasing productivity, conventional agrochemicals have been associated with various health and environmental risks (Jastrzębska et al. 2022).

II. *Technological Advances:* The integration of technological innovations such as genetically modified organisms (GMOs), precision agriculture, and advanced irrigation systems has significantly improved farming systems. Precision agriculture, in particular, has been shown to enhance sustainable crop production by reducing input waste and optimizing crop health and stress status (Naresh et al. 2024). This is further supported by the use of soil-less farming techniques like hydroponics, aquaponics, and aeroponics, which have been found to be more efficient than traditional soil-based farming (Kumar et al. 2020). The adoption of precision farming tools, including those for small-scale farming systems in developing countries, can lead to significant environmental benefits (Finger et al. 2019).

III. *Intensive Practices:* Conventional farming practices, while capable of higher yields and more efficient land use, can also contribute to soil degradation, water pollution, and loss of biodiversity if not managed sustainably (Tal 2018). Sustainable management is crucial to mitigate these negative impacts, with a focus on reducing the use of harmful chemicals and integrating crop production with soil protection and landscape ecological structures (Lindang et al. 2021). Organic agriculture, with its emphasis on soil fertility, biodiversity preservation, and reduced environmental impact, presents a potential alternative to conventional farming. However, the implications of reduced productivity and the need for further research and investment in organic farming should be considered (Gomiero et al. 2011).

IV. *Yield and Efficiency*: A comparison of organic and conventional farming practices reveals that while conventional farming initially achieves higher yields, organic farming focuses on long-term soil health and sustainability (Yadav et al. 2024). This is supported by a quantitative review which found that organic farming had lower yields but also lower intensity of soil use, resulting in a productivity gap of 29% to 44% (Alvarez 2022). Furthermore, organic farming has been found to have a lower environmental impact, with improved soil characteristics, higher water holding capacity, and greater biodiversity (Gomiero et al. 2011).

V. *Environmental Impact:* Organic farming is generally associated with a lower environmental footprint, promoting soil health, water conservation, and biodiversity (Raj et al. 2024). It has been found to have positive impacts on the environment per unit of area, with higher soil organic matter content and lower nutrient losses (Tuomisto et al. 2012). Organic agriculture also enhances soil fertility, reduces soil loss, and increases water holding capacity, leading to larger yields in water-scarce conditions (Gomiero et al. 2011). However, the environmental performance of organic farming can vary depending on the assessment method and the specific environmental indicator being considered (Schader et al. 2012).

VI. *Market Demand:* Consumer preferences for organic products are influenced by a variety of factors, including price sensitivity, attitudes toward food, and the perceived health and environmental benefits of organic foods (Stolz 2011). These preferences have contributed to the growth of the organic market and the maintenance of premium prices for organic products (Islam 2014). However, the potential for excess supplies and changes in consumer attitudes may lead to increased competition and pressure on prices, narrowing the price gap between organic and conventional products (Hallam 2003). Despite this, the presence of cosmetic defects and the influence of store choice and household characteristics can still impact the choice of organic produce (Thompson 1998; Shah et al. 2023).

3.3.8 Sowing and Irrigation

I. *Chickpeas* (*Cicer arietinum*)

Research consistently shows that winter sowing of chickpeas in the Mediterranean region, particularly in late autumn or early winter, is beneficial for the crop's growth and yield. Singh et al. (1997) found that winter-sown chickpeas had a 70% higher seed yield compared to spring-sown crops, with a longer growing period contributing to this increase. Yau (2005) further supported this, recommending sowing in December or January for optimal yield. López-Bellido et al. (2008) and Özdemir and Karadavut (2003) both found that winter sowing, particularly in late autumn, led to significantly higher yields compared to traditional late winter or spring sowing. These findings highlight the importance of sowing chickpeas in the cooler months to avoid heat stress and maximize the use of residual soil moisture (Singhal et al. 2023).

II. *Lentils* (*Lens culinaris*)

The timing of lentil sowing is crucial for optimal growth and yield. Early sowing, particularly in low to medium rainfall regions, allows for longer growth periods, greater dry matter production, and higher seed yields (Siddique et al. 1998). However, the specific timing varies by location, with the end of April to mid-May identified as the optimum sowing time in New South Wales (Maphosa et al. 2023). In a rainfed Mediterranean environment, early winter sowing has been found to significantly increase seed and straw yields (Silim et al. 1991). In the UK, lentil has been shown to have potential as a grain legume crop, with stable and high yields predicted for both spring and autumn sowings (Andrews et al. 2001; Shahab et al. 2023).

III. *Faba Beans* (*Vicia faba*)

Faba beans are best sown in late autumn or early winter to take advantage of winter rains and avoid high temperatures and drought conditions. This timing allows for establishment and cold acclimation, optimizing hardiness and yield potential (Landry et al. 2016). However, the success of this approach can vary depending on location, with mild winter temperatures potentially affecting hardening against late frosts (Annicchiarico and Iannucci 2007). The CROPGRO model has been adapted to simulate faba bean growth and yield, predicting high seed yields with early winter sowing in some regions (Boote et al. 2002). In New Zealand, faba beans sown from early March to late April have been shown to be suitable for forage, with a sowing-to-harvest window allowing for late spring feed (Martini et al. 2012).

IV. *Peas* (*Pisum sativum*)

Research has shown that the timing of sowing significantly impacts the growth and yield of peas. Silim et al. (1985) found that autumn-sown peas can produce similar seed yields to spring sowings when winter survival is adequate, with mid-November and early March being the optimum sowing dates. Similarly, Sümer (2023) reported that early sowing resulted in higher dry matter content and growing degree day (GDD) values, while late sowing led to a shorter period from pod set to maturation. However, Castillo et al. (1994) noted that early sowing can lead to lower seed vigor, as indicated by higher conductivity and hollow heart incidence. Knott and Belcher (1998) further emphasized the importance of sowing date, recommending mid-November as the optimum time for sowing winter peas to achieve reliable yields.

3.4 Irrigation Techniques and Water Management

Effective irrigation and water management are crucial for optimizing crop yields and ensuring sustainable water use in the Mediterranean region (Fig. 3.4), characterized by its hot, dry summers and irregular rainfall.

I. *Drip Irrigation:* Drip irrigation is a highly efficient method for delivering water to the root zone of plants, particularly beneficial for chickpeas, lentils, faba beans, and peas (Ali et al. 2000). It reduces water loss due to evaporation and runoff, making it ideal for water-scarce regions (Yang et al. 2023). However, it is important to manage water stress during flowering and pod development, as these crops are intolerant of excessive soil water (Farah et al. 1988). Drip irrigation can significantly increase crop yields and water use efficiency while reducing fertilizer leaching and soil salinity (Yang et al. 2023). Further research is needed to understand the specific factors that affect water use efficiency in these crops (Cooper et al. 1988).

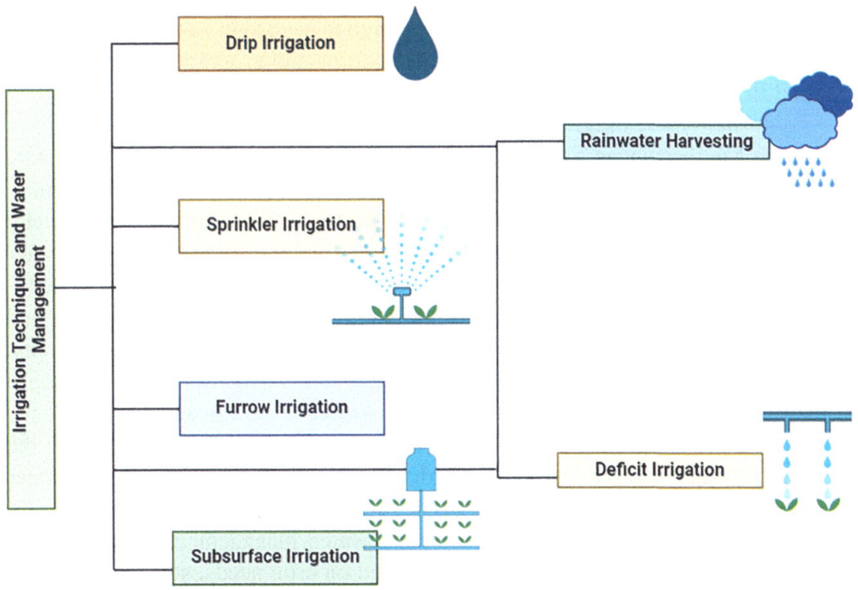

Fig. 3.4 Major irrigation techniques and water management techniques in Mediterranean region

II. *Sprinkler Irrigation:* Sprinkler irrigation, a method suitable for legumes, mimics natural rainfall and allows for even water distribution (Kalpana and Selvi 2008). However, it may result in higher evaporation losses compared to drip irrigation (Trout and Kincaid 2007). Proper design and management of sprinkler systems can improve moisture distribution and reduce wind drift and evaporation losses (Arunadevi et al. 2021). The contribution of droplet evaporation to total evaporation losses during sprinkler irrigation is still a controversial issue (Uddin et al. 2010).
III. *Furrow Irrigation:* Furrow irrigation, while less efficient than pressurized systems, can be effective with proper management. Factors such as furrow length, shape, and soil type can significantly impact its performance (Horst et al. 2007; Karim and Karim 2020). In particular, furrow length reduction, change in shape, and cut-off irrigation can improve water distribution and soil productivity (Khalifa and Ghannam 2021). These findings underscore the importance of considering these factors in the design and management of furrow irrigation systems for faba beans and peas.
IV. *Subsurface Irrigation:* Subsurface irrigation, particularly through methods like subsurface drip irrigation (SDI) and porous clay pipe irrigation, has been shown to significantly improve water use efficiency and crop yield in various studies (Camp 1998). These methods reduce evaporation and surface runoff, delivering water directly to the root zone and minimizing water quality hazards (Lamm 2002). They are particularly beneficial for legumes, especially in sandy or well-drained soils (Adnan, Khan et al. 2023). However, the high cost and complexity of installation and maintenance may limit their widespread adoption (Siyal et al. 2011). Further research is needed to explore the feasibility of subsurface irrigation in different regions and cropping systems (Douh and Boujelben 2011).
V. *Rainwater Harvesting:* Rainwater harvesting has been a longstanding practice in the Mediterranean region, with ancient civilizations using cisterns to store rainwater for sustainability (Mays 2014). Modern studies have further demonstrated the effectiveness of rainwater harvesting systems in addressing water scarcity, with a focus on domestic non-potable use and community water demand (Kakoulas et al. 2022). The potential for rainwater harvesting in buildings with green roofs has also been explored, with a focus on water quality and runoff management (Monteiro et al. 2016). In a case study at Ege University, the potential for rainwater harvesting to supplement irrigation needs was highlighted, with the harvested rainwater providing a significant portion of the required water (Ullah et al. 2022). These studies collectively underscore the value of rainwater harvesting in the Mediterranean region for enhancing water sustainability and reducing dependence on groundwater.
VI. *Scheduling and Management:* Effective water management in agriculture requires precise irrigation scheduling based on crop needs, soil moisture levels, and weather conditions (Aguilar et al. 2015; Ullah et al. 2023a). This can be achieved through the use of soil moisture sensors and evapotranspiration (ET) models, which help determine the optimal timing and amount of irriga-

tion (Surve and Patel 2024). Soil moisture sensors, in particular, are crucial for providing real-time data on soil moisture levels, enabling growers to avoid over- or under-watering (Maughan et al. 2015). However, the uncertainty associated with both ET estimates and soil moisture measurements underscores the need for careful management, especially in deficit irrigation strategies (Sayde et al. 2008).

VII. *Deficit Irrigation:* Deficit irrigation, a water-saving technique, can be effectively applied to legumes like chickpeas and lentils, which are relatively drought-tolerant (Laita et al. 2024). This method can optimize water use efficiency and maintain acceptable yield levels, making it particularly useful in water-scarce regions (Amiri et al. 2021). For example, in a study on lentils, a 20% reduction in water use did not significantly impact fruit yield or quality, and even increased water productivity by 33%. Similarly, in another study, a 10% reduction in grain yield was observed with a 20% reduction in water use, but this was accompanied by a 27% decrease in biological yield, leading to a 20% increase in harvest index (Amiri et al. 2023). These findings suggest that deficit irrigation can be a valuable tool for managing water resources and enhancing crop production in arid areas (Javed et al. 2024).

In conclusion, choosing the right irrigation schedules and sowing dates is essential for successful legume farming in the Mediterranean region. Enhancing overall agricultural resilience, conserving water resources, and ensuring sustainable crop output are all achieved by combining efficient irrigation techniques with smart water management strategies.

3.5 Weed Control

Effective weed management is essential for optimizing legume production in the Mediterranean region, as weeds compete with crops for nutrients, water, and light.

3.5.1 Cultural Practices

I. *Crop Rotation*: Crop rotations that include legumes, such as clover, lucerne, and vetch-oats, can effectively disrupt the life cycle of weeds, reducing their prevalence in subsequent cereal crops (Böhm et al. 2020; Maikštėnienė et al. 2009). This is particularly important in the context of intensive farming, where long-term crop rotations have been shown to increase crop productivity and reduce weed prevalence (Butkevičienė et al. 2021; Ullah et al. 2024). However, the use of legumes in rotation regimes may have short-term effects on cereal yield (Palmer 1998).

II. *Intercropping*: Research has consistently shown that growing legumes alongside other crops can suppress weed growth through various mechanisms. Bilalis et al. (2010) found that maize-legume intercropping increased soil canopy cover, reducing available light for weeds and leading to a decrease in weed density and dry matter. Ekeleme et al. (2003) identified specific legume cover crops, such as lablab and velvetbean, that effectively suppressed weeds by providing ground cover. Fisk et al. (2001) demonstrated that annual legume cover crops, including burr medic and barrel medic, reduced both winter and summer annual weed populations. Elsalahy et al. (2019) further supported these findings, showing that only-legume cover crop mixtures, particularly those with alsike clover and black medic, can suppress weeds, with the effectiveness being influenced by seed density and environmental conditions (Ahmad et al. 2022). These studies collectively highlight the potential of legumes to suppress weed growth when grown alongside other crops (Fahad et al. 2023).

3.5.2 Mechanical Methods

I. *Tillage*: Tillage is a common practice for weed control, but it can also have negative effects on soil structure and moisture conservation (Jones et al. 1990). Conservation tillage, which aims to balance short-term crop profitability with long-term soil health, is a potential solution (Onasanya et al. 2021). However, the impact of different tillage systems on weed ecology, herbicide activity, and persistence needs further investigation (Chauhan et al. 2006). Studies have shown that minimum tillage systems can increase organic matter content and water-stable aggregate content, leading to better soil structure and moisture conservation (Rusu et al. 2013).

II. *Hand Weeding*: Mechanical weeding, particularly with the use of manually drawn rotor weeders, has been found to be more cost-effective and efficient than manual weeding in small fields (Rahaman et al. 2022). This is further supported by the development of a novel manual mechanical weeder that improves weeding efficiency in row crops (Sundaram et al. 2021; Khan et al. 2023a). However, the use of herbicides is less effective than hand-weeding, which is still widely practiced due to its cost efficiency and minimal soil disturbance (Abebe 2024). Therefore, while mechanical weeding may be more practical in some cases, the choice of weeding method should be based on a careful consideration of factors such as cost, efficiency, and environmental impact.

3.5.3 Chemical Methods

I. *Use of Herbicides*: A range of studies have demonstrated the effectiveness of selective herbicides in controlling weeds without harming legume crops. Kousta et al. (2024) found that pre-emergence herbicides like flumioxazin, benfluralin, and terbuthylazine plus pendimethalin were particularly effective in controlling weeds in grain legumes, with minimal phytotoxicity. Similarly, Cox and Harrington (2005) identified several pre-emergence and post-emergence herbicides that were safe for use in desmanthus seed crops, providing effective weed control (Ali et al. 2022b). However, Kanatas and Gazoulis (2022) noted that chickpea had lower tolerance for pre-emergence herbicides like pendimethalin and s-metolachlor, suggesting a need for caution in their use (Bawazeer et al. 2021). Lastly, Schreiber (1960) highlighted the potential of pre-emergence herbicides in aiding the establishment of forage legumes, further underscoring the importance of selective herbicides in weed management.

3.5.4 Pest Control

Pest management is crucial to prevent yield losses and ensure healthy crop development. Common pests affecting legumes in the Mediterranean include aphids, weevils, and caterpillars.

3.5.4.1 Cultural Practices

I. *Resistant Varieties*: Research has shown that planting pest-resistant legume varieties can significantly reduce the impact of pest infestations. Rubiales et al. (2015) and Fattah et al. (2023) both emphasize the importance of developing resistant legume varieties to combat a range of pests and diseases. Sharma et al. (2017) further support this, highlighting the need to utilize wild relatives of crops and genetic engineering to enhance resistance to insect pests (Nawaz et al. 2025). These studies collectively underscore the potential of pest-resistant legume varieties in mitigating the impact of pest infestations (Saljoqi et al. 2022).
II. *Sanitation Method*: Research consistently shows that maintaining field hygiene and removing crop residues can significantly reduce pest habitats and sources of infestation (Pearsons and Tooker 2017; Cloyd and Herrick 2022). These practices can promote soil community diversity, facilitate the activity of arthropod predators, and enhance natural pest control, ultimately leading to lower pest pressure and crop injury (Bianchi et al. 2006). Therefore, implementing these strategies as part of an integrated pest management program is crucial for sustainable pest regulation in agricultural landscapes.

3.5.4.2 Biological and Chemical Control

1. *Natural Enemies*: Encouraging beneficial insects, such as ladybugs and parasitic wasps, can effectively control pest populations naturally (Gredler 2001). These insects can be attracted to crops by planting flowering plants and perennial grasses around farms. The use of parasitoid wasps, in particular, has been shown to be effective in targeting the life cycle stages of pests, leading to sustainable pest control (Faithpraise et al. 2013; Khan et al. 2022c). However, the feasibility of managing insect-pest problems through the release of parasites in host ecosystems needs further examination.
2. *Biopesticides*: *Bacillus thuringiensis* (Bt) is a naturally occurring bacterium that produces insecticidal proteins, making it a safe and effective pest management option (Kumar et al. 2008). Bt-based products have been widely used in agriculture and forestry due to their specificity and environmental friendliness (Naz

et al. 2023). However, these products have limitations, such as a narrow activity spectrum and instability in certain conditions. To address these limitations, new strains with improved insecticidal activity and methods to increase toxin persistence have been developed (Sanchis 2011). Bt has also been used to create transgenic plants with pest resistance (Nawaz et al. 2023a). The use of Bt and other insect-pathogenic bacteria in integrated pest management is a growing area of research and development (Ruiu 2015).

3. *Chemical Insecticides*: The use of selective insecticides in integrated pest management (IPM) is crucial for sustainable agriculture, as they minimize harm to beneficial insects and the environment (Hardy 2012). However, the impact of these insecticides on natural enemies must be carefully considered, as they play a key role in biological control of pests (Fadlelmawla et al. 2021). The selectivity of a pesticide, which refers to its harmlessness for beneficial organisms, is influenced by its physiological selectivity and application procedures (Blumel et al. 1999). To address the negative effects of traditional insecticides, the development of greener alternatives, such as pyrethroids, neonicotinoids, and biological insecticides, is crucial (López et al. 2005; Ullah et al. 2023b).

3.5.5 Disease Control

Disease management is vital to prevent significant yield losses and maintain crop health. Common diseases affecting legumes in the Mediterranean include ascochyta blight, Fusarium wilt, and powdery mildew.

3.5.5.1 Cultural Practices

I. *Disease-Free Seeds*: Certified disease-free seeds play a crucial role in minimizing the risk of introducing pathogens into the field (Shepard and Claflin 1975). This is particularly important in low-income countries, where seed degeneration is a significant issue (Thomas-Sharma et al. 2017). The effectiveness of seed potato certification in controlling potato diseases has been demonstrated over a century-long program in Wisconsin. However, the sustainability of this program may be challenged by soilborne diseases (Frost et al. 2013).

II. *Proper Seed Spacing*: Research consistently shows that adequate plant spacing can significantly impact disease development by improving air circulation and reducing humidity levels. Copes and Scherm (2005) found that increasing plant spacing led to higher evaporation and reduced the number of hours within the temperature range favorable for disease development. Similarly, Sconyers et al. (2005) observed a decrease in stem rot severity and incidence as plant spacing increased. Both studies highlight the role of plant spacing in creating a microclimate less conducive to disease. Gnanesh et al. (2023) further support these findings, demonstrating that balanced fertilizer application and wider plant spacing can reduce disease severity. These studies collectively underscore the importance of plant spacing in disease management (Khan et al. 2023b).

3.5.5.2 Chemical and Biological Control

I. *Fungicides*: Selective fungicides are effective in controlling fungal diseases, but their overuse can lead to the development of resistant pathogen strains (Gossen et al. 2014). To prevent this, it is crucial to rotate fungicides with different modes of action (Damicone and Smith 2009). This approach, along with other resistance management strategies such as using higher or lower dose rates, mixtures or alternations, and molecular diagnostics, can help maximize the effective life of these plant protection products (Corkley et al. 2022; Khan et al. 2023c). Additionally, the use of PCR-based molecular techniques can aid in the early detection of resistant strains, contributing to integrated disease control and reduced pesticide inputs in the agricultural environment (Ishii 2006).

II. *Integrated Disease Management (IDM)*: The integration of cultural, biological, and chemical methods in disease management is a key strategy for sustainable and effective control (Musheer et al. 2023). This approach is particularly important in developing countries, where the implementation of integrated disease management (IDM) requires a supportive policy environment. In nursery settings, the use of biological control, cultural practices, and chemical control methods has been found to be effective in managing pests and diseases (Thapa and Juyal 2022). However, the adoption of these practices in India faces challenges, including the need for more effective biocontrol methods and the development of new technologies (Kumar 2014). The successful development and testing of an IDM package for multiple tomato diseases further underscores the importance of this holistic approach (Panday 2005).

To summarize, the successful management of weeds, pests, and diseases in the production of legumes necessitates the integration of cultural, mechanical, biological, and chemical techniques (Nawaz et al. 2024b, c, d, e). The Mediterranean region's agricultural resilience is improved and sustainable crop production is guaranteed by the application of integrated management solutions that are customized to address particular difficulties (Nawaz et al. 2023a).

3.6 Future Prospects and Recommendations

3.6.1 Untapped Areas for Legume Cultivation

The Mediterranean region presents significant opportunities for expanding legume cultivation into untapped areas. Several strategies can facilitate this expansion:

3.6.2 Utilization of Marginal Lands

I. *Drought-Resistant Varieties:* Research and development efforts have made significant progress in the creation of drought-tolerant legume varieties, which can thrive in arid and semi-arid regions (Coba de la Pena and Pueyo 2012). These varieties are crucial for utilizing underused lands in these regions (Nawaz et al.

2024b, c, d, e). The use of biotechnological approaches, such as transgenic technology, has been particularly effective in enhancing the tolerance of legumes to environmental stresses, including drought (Serraj et al. 2003). The International Crops Research Institute for the Semi-Arid Tropics (ICRISAT) has played a key role in this research, identifying genetic sources of drought tolerance and developing screening methods for specific traits (Bhatnagar-Mathur et al. 2009). The integration of various approaches, including agronomic and biotechnological strategies, is essential for the development of drought-tolerant legume cultivars (Nadeem et al. 2019).

II. *Soil Health Improvement:* Legumes, with their nitrogen-fixing ability, play a crucial role in improving soil fertility in degraded or marginal lands, making them more suitable for subsequent agricultural use (Yuvaraj et al. 2020). This is achieved through the symbiotic relationship between legumes and nitrogen-fixing bacteria, which enriches the soil with essential nutrients and reduces the need for synthetic inputs (Meena et al. 2018; Nawaz et al. 2024f). The use of legumes in reclamation of marginal soils is particularly promising, with research focusing on the selection of tolerant rhizobial strains and legume varieties. However, while legumes can significantly increase soil fertility and forage productivity, they may have lower seed production compared to other crops (Nohong et al. 2024; Khan et al. 2022b).

3.6.3 Adaptation to Climate Change

I. *Climate-Resilient Cultivation Practices*: The cultivation of legumes faces numerous challenges, including genetic, socio-economic, institutional, and environmental factors (Abobatta et al. 2021). However, implementing climate-smart agricultural practices, such as water-saving irrigation techniques and soil moisture conservation methods, can enhance the resilience of legume cultivation to climate variability (Fahad et al. 2024). These practices, when combined with improved seeds and inorganic fertilizers, can significantly increase yields and reduce the probability of low yields, even under critical weather stress (Arslan et al. 2015; Nawaz et al. 2023b). Furthermore, the adoption of conservation agriculture, stress-adapted legume varieties, and drought-tolerant maize can further enhance the resilience of smallholder farmers to climate change, leading to increased productivity and income (Makate et al. 2019). However, the adoption of these innovations is hindered by constraints such as technology transfer and resource limitations (Siddique et al. 2012). Therefore, there is a need for targeted interventions and policy efforts to promote the adoption of these climate-smart agricultural practices, particularly in resource-poor farming situations (Khan et al. 2022a; Nawaz et al 2025b).

II. *New Growing Zones*: The potential for legume cultivation is shifting due to climate change, with some regions becoming more suitable for these crops (Nawaz et al. 2024b, c, d, e) However, there are significant challenges to global legume cultivation, including genetic, socio-economic, and institutional constraints. To address these challenges, there is a need for the development of legume varieties adapted to changing climatic conditions (Abobatta et al. 2021). Geovisualization applications can be used to map the fundamental climate niche of legumes, aiding in the identification of new potential cultivation areas (Peter et al. 2020). It is also crucial to breed new varieties with improved resistance to abiotic and biotic stresses, taking into account the predicted impacts on productivity and the number of people who depend on the crop (Lane and Jarvis 2007; Nawaz et al. 2024a).

3.6.4 Land Reclamation and Agroforestry

I. *Rehabilitation of Degraded Lands*: Legumes, with their ability to fix nitrogen and improve soil structure, are crucial in land reclamation projects (Abiala et al. 2018). They have been shown to be effective in restoring degraded and salt-affected soils, making them suitable for reclamation of marginal lands (Layek et al. 2018). The symbiotic association of shrubby legumes, in particular, has been found to enhance soil fertility and plant performance in degraded soils (Pérez-Fernández et al. 2016). Therefore, incorporating legumes into land reclamation projects can indeed restore soil health and increase arable land availability.
II. *Integration into Agroforestry*: Legume cultivation, when combined with tree planting, can have a range of positive impacts. Legumes, with their nitrogen-fixing abilities, can enhance soil quality and reduce the need for chemical fertilizers (Everwand et al. 2017). They also have the potential to enhance carbon sequestration, thereby contributing to biodiversity and reduced greenhouse gas emissions (Abberton 2010). Tree legumes, in particular, offer a variety of benefits, including soil erosion control, habitat provision, and biological nitrogen fixation (Dubeux et al. 2017). Furthermore, legumes can play a crucial role in sustainable agriculture, providing high-quality food and feed, reducing greenhouse gas emissions, and saving fossil energy inputs (Stagnari et al. 2017). Therefore, the combination of legume cultivation and tree planting can lead to enhanced biodiversity, improved soil quality, and additional economic benefits (Fahad et al. 2024).

3.6.5 Export Opportunities

Expanding legume cultivation in the Mediterranean region can lead to significant export opportunities, driven by global market trends and increasing demand for legumes.

I. *Health and Nutrition Trends*: The increasing awareness of the health benefits of legumes, such as their high protein and fiber content, is driving demand in markets like Europe, North America, and Asia (Schmidt and Oliveira 2023). Mediterranean legumes, in particular, are well-positioned to cater to these health-conscious consumers due to their nutritional value and potential for use in a variety of food products (Velásquez-Barreto et al. 2020). However, there are still challenges to be addressed, such as the need for convenient preparation methods and the development of novel products to meet consumer preferences (Affrifah et al. 2023).

II. *Plant-Based Diets*: The rising popularity of plant-based diets has created a demand for legumes as a key protein source, particularly in the Mediterranean region (Hernández-López et al. 2022). Legumes are not only a sustainable protein source but also offer numerous health benefits, including a lower risk of cardiovascular diseases and type 2 diabetes (Yanni et al. 2023). The Mediterranean diet, which is known for its sustainability, can be further enhanced by the production and consumption of legumes (Carlini et al. 2024). The use of legumes in high-quality food production can also contribute to environmental sustainability and human health (Alandia et al. 2020). Therefore, Mediterranean producers have a significant opportunity to tap into the growing demand for legumes by supplying high-quality legumes and legume-based products (Ahmad et al. 2024).

3.6.6 Strategic Export Markets

I. *European Union:* The EU is a key market for Mediterranean legumes due to its proximity and strong trade relations. However, the impact of trade liberalization on these products is limited, as Mediterranean countries already have preferential access and face significant non-tariff trade barriers. The Mediterranean basin is a diverse region, with varying levels of tariff sensitivity and non-tariff trade resistance. Italy plays a crucial role in the trade of fruits and vegetables from the Mediterranean, with special relationships with Turkey and Egypt (Mulazzani and Malorgio 2009).

II. *Middle East and North Africa (MENA)*: The MENA region's traditional consumption of legumes presents a significant market opportunity, particularly for the Mediterranean region, which can leverage its logistical advantages to efficiently supply this market (Hernández-López et al. 2022). However, there are challenges to be addressed, including the need for efficient supply chains (Moslemi et al. 2016) and trade logistics (Devlin and Yee 2005). The region also needs to implement measures to benefit from trade-enhancing initiatives (Alonso-Gamo et al. 1997). Despite these challenges, the potential for the Mediterranean region to capitalize on the demand for legumes in the MENA market is evident.

3.7 Conclusion

In summary, the future of legume production in the Mediterranean hinges on several key points. Expanding cultivation into untapped areas and capitalizing on export opportunities, driven by global demand and health trends, are crucial. Sustainability efforts, including the use of drought-resistant varieties and organic farming practices, will be essential for adapting to climate change and preserving soil health. Innovation in cultivation techniques and value-added products can further enhance market competitiveness. Moving forward, fostering collaboration among Mediterranean countries and sharing knowledge on best practices in pest management, disease control, and sustainable farming will be pivotal. Emphasizing these strategies will not only bolster the region's agricultural resilience but also position Mediterranean legume producers as leaders in the global market.

References

Abberton M (2010) Enhancing the role of legumes: potential and obstacles. Integr Crop Manag 11:177–187

Abebe D (2024) Effect of hand frequency weeding on crop growth parameters and yield. J Food Nutrit 3(1):2836–2276

Abiala MA, Abdelrahman M, Burritt DJ, Tran LSP (2018) Salt stress tolerance mechanisms and potential applications of legumes for sustainable reclamation of salt-degraded soils. Land Degrad Dev 29(10):3812–3822

Abobatta WF, El-Hashash EF, Hegab RH (2021) Challenges and opportunities for the global cultivation and adaption of legumes. J Appl Biotechnol Bioeng 8(5):160–172

Adnan M, Khan MA, Basir A, Fahad S, Nasar J, Imran, Alharbi S, Ghoneim AM, Yu G-H, Saleem MH (2023) Biochar as soil amendment for mitigating nutrients stress in crops. In: Sustainable agriculture reviews 61: Biochar to improve crop production and decrease plant stress under a changing climate. Springer, pp 123–140

Affrifah NS, Uebersax MA, Amin S (2023) Nutritional significance, value-added applications, and consumer perceptions of food legumes: a review. Legume Sci 5(4):e192

Aguilar J, Rogers D, Kisekka I (2015) Irrigation scheduling based on soil moisture sensors and evapotranspiration. Kansas Agri Exp Stat Res Rep 1(5):20

Ahmad K, Din Z u, Ullah H, Ouyang Q, Rani S, Jan I, Alam M, Rahman Z, Kamal T, Ali S (2022) Preparation and characterization of bio-based nanocomposites packaging films reinforced with cellulose nanofibers from unripe banana peels. Starch-Stärke 74(5–6):2100283

Ahmad R, Liaquat M, Sammi S, Al-Hawadi JS, Jahangir M, Mumtaz A, Khan I, Okla MK, Alaraidh IA, AbdElgawad H (2024) Physicochemical and nutritional profiles of wild adlay (Coix lacryma-jobi Linn) accessions by GC, FTIR, and spectrophotometer. Food Chem X:101418

Alam H, Zamin M, Adnan M, Ahmad N, Nawaz T, Saud S, Basir A, Liu K, Harrison MT, Hassan S (2022) Evaluating the resistance mechanism of Atriplex leucoclada (Orache) to salt and water stress; a potential crop for biosaline agriculture. Front Plant Sci 13:948736

Alandia G, Pulvento C, Sellami MH, Hoidal N, Anemone T, Nigussie E et al (2020) Grain legumes may enhance high-quality food production in Europe. In: Emerging research in alternative crops. Springer, Cham, pp 25–53

Ali M, Dahan R, Mishra JP, Saxena NP (2000) Towards the more efficient use of water and nutrients in food legume cropping. In: Linking research and marketing opportunities for pulses in the 21st century: proceedings of the third international food legumes research conference. Springer Netherlands, pp 355–368

Ali A, Altaf MT, Nadeem MA, Karaköy T, Shah AN, Azeem H, Baloch FS, Baran N, Hussain T, Duangpan S, Aasim M (2022a) Recent advancement in OMICS approaches to enhance abiotic stress tolerance in legumes. Front Plant Sci 13:952759

Ali Y, Nawaz T, Ahmed N, Junaid M, Kanwal M, Hameed F, Ahmed S, Ullah R, Shahab M, Subhan F (2022b) Maize (Zea mays) response to abiotic Stress. In: Maize genetic resources-breeding strategies and recent advances, IntechOpen

Al-Karaki GN (1999) Phenological development–yield relationships in dry pea in semiarid Mediterranean conditions. J Agron Crop Sci 182(2):73–78
Alonso-Gamo MP, Fennell MS, Sakr MK (1997) Adjusting to new realities: MENA, the Uruguay round, and the EU-Mediterranean initiative. International Monetary Fund
Alvarez R (2022) Comparing productivity of organic and conventional farming systems: a quantitative review. Arch Agron Soil Sci 68(14):1947–1958
Amiri SR, Salimi K, Ziaei SM (2021) The effect of deficit irrigation on yield and water use efficiency of lentil (Lens culinaris Medik.). Env Stresses Crop Sci 14(1):75–83
Amiri S, Ziaei SM, Zakeri N (2023) Managing drought risk on lentil using deficit irrigation in arid zones. Romanian Agri Res 40:483
Andrews M, McKenzie BA, Joyce A, Andrews ME (2001) The potential of lentil (Lens culinaris) as a grain legume crop in the UK: an assessment based on a crop growth model. Ann Appl Biol 139(3):293–300
Annicchiarico P, Iannucci A (2007) Winter survival of pea, faba bean and white lupin cultivars in contrasting Italian locations and sowing times, and implications for selection. J Agric Sci 145(6):611–622
Antichi D, Pampana S, Tramacere LG, Biarnes V, Stute I, Kadžiulienė Ž et al (2023) An experimental dataset on yields of pulses across Europe. Sci Data 10(1):708
Arslan H, Ekin Z, Hatipoglu H (2018) Performances of sesame genotypes (Sesamum indicum L.) with different seed shell colors in semi-arid climate conditions. Fresenius Environ Bull 27
Araújo SS, Beebe S, Crespi M, Delbreil B, González EM, Gruber V et al (2015) Abiotic stress responses in legumes: strategies used to cope with environmental challenges. Crit Rev Plant Sci 34(1–3):237–280
Arunadevi K, Nongkynrih I, Ramachandran J (2021) Wind drift evaporation loss and soil moisture distribution under sprinkler irrigated Blackgram (Vigna mungo L.). Int J Bio-resource Stress Manag 12(Dec 6):731–736
Askar A (1986) Food science: Faba beans (Vicia faba L.) and their role in the human diet. Food Nutr Bull 8(3):1–11
Assefa S, Tadesse S (2019) The principal role of organic fertilizer on soil properties and agricultural productivity-a review. Agri Res and Tech Open Access J 22(2):556192
Ates S, Feindel D, El Moneim A, Ryan J (2014) Annual forage legumes in dryland agricultural systems of the West Asia and North Africa Regions: research achievements and future perspective. Grass Forage Sci 69(1):17–31
Aydinalp C, Cresser MS (2008) Red soils under Mediterranean type of climate: their properties use and productivity. Bulg J Agri Sci 14:576–582
Bansal RK, Monroe GE, Dahan R, El Gharras O, Bahri A (1994) Mechanization of lentil harvesting in Morocco. Appl Eng Agric 10(5):641–646
Bawazeer S, Rauf A, Nawaz T, Khalil AA, Javed MS, Muhammad N, Shah MA (2021) Punica granatum peel extracts mediated the green synthesis of gold nanoparticles and their detailed in vivo biological activities. Green Process Synth 10(1):882–892
Baxevanos D, Kargiotidou A, Noulas C, Kouderi AM, Aggelakoudi M, Petsoulas C et al (2024) Lentil cultivar evaluation in diverse organic Mediterranean environments. Agronomy 14(4):790
Bennett SJ, Lee RC, Sharma A (2019) Management of field peas to reduce Ascochyta blight and maximise yield
Bhatnagar-Mathur P, Rao JS, Vadez V, Sharma KK (2009) Transgenic strategies for improved drought tolerance in legumes of semi-arid tropics. J Crop Improv 24(1):92–111
Bhogal A, Nicholson FA, Rollett A, Taylor M, Litterick A, Whittingham MJ, Williams JR (2018) Improvements in the quality of agricultural soils following organic material additions depend on both the quantity and quality of the materials applied. Front Sustain Food Syst 2:9. https://doi.org/10.3389/fsufs.2018.00009
Bianchi FJ, Booij CJH, Tscharntke T (2006) Sustainable pest regulation in agricultural landscapes: a review on landscape composition, biodiversity and natural pest control. Proc R Soc B Biol Sci 273(1595):1715–1727

Bilalis D, Papastylianou P, Konstantas A, Patsiali S, Karkanis A, Efthimiadou A (2010) Weed-suppressive effects of maize–legume intercropping in organic farming. Int J Pest Manag 56(2):173–181

Binkley D, Fisher RF (2019) Ecology and management of forest soils. Wiley

Blumel S, Matthews GA, Grinstein A, Elad Y (1999) Pesticides in IPM: selectivity, side-effects, application and resistance problems. In: Developments in plant pathology. Springer, Dordrecht, pp 150–167

Böhm H, Dauber J, Dehler M, Gallardo DAA, de Witte T, Fuss R et al (2020) Crop rotations with and without legumes: a review

Boote KJ, Mínguez MI, Sau F (2002) Adapting the CROPGRO legume model to simulate growth of faba bean. Agron J 94(4):743–756

Butkevičienė LM, Skinulienė L, Auželienė I, Bogužas V, Pupalienė R, Steponavičienė V (2021) The influence of long-term different crop rotations and monoculture on weed prevalence and weed seed content in the soil. Agronomy 11(7):1367

Camp CR (1998) Subsurface drip irrigation: a review. Trans ASAE 41(5):1353–1367

Carlini B, Lucini C, Velázquez J (2024) The role of legumes in the sustainable Mediterranean diet: analysis of the consumption of legumes in the Mediterranean population over the last ten years a PRISMA statement methodology. Sustain For 16(7):3081

Carreira JA, Arevalo JR, Niell FX (1996) Soil degradation and nutrient availability in fire-prone Mediterranean shrublands of southeastern Spain. Arid Land Res Manag 10(1):53–64

Castillo AG, Hampton JG, Coolbear P (1994) Effect of sowing date and harvest timing on seed vigour in garden pea (Pisum sativum L.). N Z J Crop Hortic Sci 22(1):91–95

Chandrashekharaiah PS, Paul V, Kushwaha S, Sanyal D, Dasgupta S (2021) Biotechnological approaches for enhancing stress tolerance in legumes. In: Sustainable agriculture reviews 51: legume agriculture and biotechnology, vol 2. Springer, Cham, pp 247–293

Chauhan BS, Gill GS, Preston C (2006) Tillage system effects on weed ecology, herbicide activity and persistence: a review. Aust J Exp Agric 46(12):1557–1570

Cloyd RA, Herrick NJ (2022) The case for sanitation as an insect pest management strategy in greenhouse production systems. J Entomol Sci 57(3):315–322

Coba de la Pena T, Pueyo JJ (2012) Legumes in the reclamation of marginal soils, from cultivar and inoculant selection to transgenic approaches. Agron Sustain Dev 32:65–91

Cooper PJM, Campbell GS, Heath MC, Hebblethwaite PD (1988) Factors which affect water use efficiency in rainfed production of food legumes, and their measurement. In: World crops: cool season food legumes: A global perspective of the problems and prospects for crop improvement in pea, lentil, faba bean and chickpea. Springer, Dordrecht, pp 813–829

Copes WE, Scherm HARALD (2005) Plant spacing effects on microclimate and Rhizoctonia web blight development in container-grown Azalea. HortScience 40(5):1408–1412

Corkley I, Fraaije B, Hawkins N (2022) Fungicide resistance management: maximizing the effective life of plant protection products. Plant Pathol 71(1):150–169

Cox KG, Harrington KC (2005) Selective herbicide strategies for use in Australian desmanthus seed crops. Trop Grassl 39(3):171–181

Damicone J, Smith D (2009) Fungicide resistance management. Oklahoma Cooperative Extension Service

Dawar K, Mian IA, Khan S, Zaman A, Danish S, Liu K, Harrison MT, Saud S, Hassan S, Nawaz T (2023) Alleviation of cadmium toxicity and fortification of zinc in wheat cultivars cultivated in Cd contaminated soil. S Afr J Bot 162:611–621

Devlin J, Yee P (2005) Trade logistics in developing countries: the case of the Middle East and North Africa. World Econ 28(3):435–456

Dhull SB, Kidwai MK, Siddiq M, Sidhu JS (2022) Faba (broad) bean production, processing, and nutritional profile. In: Dry beans and pulses: production, processing, and nutrition. Wiley, pp 359–381

Dos Santos MM, Da Silva RS, Soares MA, Lopes ÉMG, Shabani F (2022) Global risks of Bedellia somnulentella (Lepidoptera: Bedelliidae) invasion: a modeling exercise using a mechanistic model. CLIMEX Theor Appl Climatol 149(1):401–411

Douh B, Boujelben A (2011) Improving water use efficiency for a sustainable productivity of agricultural systems with using subsurface drip irrigation for maize (Zea mays L.). J Agric Sci Technol B 1:881–888
Dubeux JCB, Muir JP, Apolinário VXDO, Nair PR, Lira MDA, Sollenberger LE (2017) Tree legumes: an underexploited resource in warm-climate silvopastures. Rev Bras Zootec 46(8):689–703
Egli M, Poulenard J (2016) Soils of mountainous landscapes. In: International encyclopedia of geography: people, the earth, environment and technology: people, the earth, environment and technology. Wiley, pp 1–10
Ekeleme F, Akobundu IO, Fadayomi RO, Chikoye D, Abayomi YA (2003) Characterization of legume cover crops for weed suppression in the moist savanna of Nigeria. Weed Technol 17(1):1–13
Elsalahy H, Döring T, Bellingrath-Kimura S, Arends D (2019) Weed suppression in only-legume cover crop mixtures. Agronomy 9(10):648
Erskine W, El Ashkar F (1993) Rainfall and temperature effects on lentil (Lens culinaris) seed yield in Mediterranean environments. J Agric Sci 121(3):347–354
Erskine W, Diekmann J, Jegatheeswaran P, Salkini A, Saxena MC, Ghanaim A, El Ashkar F (1991) Evaluation of lentil harvest systems for different sowing methods and cultivars in Syria. J Agric Sci 117(3):333–338
Everwand G, Cass S, Dauber J, Williams M, Stout J (2017) Legume crops and biodiversity. In: Legumes in cropping systems. CABI, Wallingford, pp 55–69
Fadlelmawla M, Abdelbagi A, Hammad A, Hur JH (2021) Effects of nitrogenous fertilization and some insecticides on the natural enemies of transgenic Bt cotton pests. EntomoBrasilis 14:5
Fageria NK, Filho MB, Moreira A, Guimarães CM (2009) Foliar fertilization of crop plants. J Plant Nutr 32(6):1044–1064
Fahad S, Ali I, Hussain I, Ahmad D, Saud S, Dawar K, Danish S, Datta R, Hassan S, Liu H (2023) Modulation of maize growth, yield and soil enzymes activities by introducing wheat straw mulching and tillage practices. Plant Soil 1–21:699–719
Fahad S, Adnan M, Zhou R, Nawaz T, Saud S (2024) Biochar-assisted remediation of contaminated soils under changing climate. Elsevier, Amsterdam
Faithpraise F, Chatwin C, Obu J, Young R, Birch P (2013) Targeting the life cycle stages of the Diamond Back Moth (Plutella Xylostella) with three different parasitoid wasps. World Acad Sci Eng Technol Int J Biol Veterinar Agri Food Eng 8(5):516–524
Farah SM, Arar A, Miller DE (1988) Water requirements and the irrigation management of pea, lentil, faba bean and chickpea crops. In: World crops: Cool season food legumes: a global perspective of the problems and prospects for crop improvement in pea, lentil, faba bean and chickpea. Springer Netherlands, Dordrecht, pp 271–278
Fattah A, Arrahman A, Wahditiya AA, Yasin M, Widiarta IN, Nugraha Y (2023) Effect of the number of rows and cultivars of soybeans on damage intensity of pest and predator populations in corn-Soybean intercropping, South Sulawesi Indonesia. Legum Res 46(8):1087–1091
Fikre A, Desmae H, Ahmed S (2020) Tapping the economic potential of chickpea in sub-Saharan Africa. Agronomy 10(11):1707
Finger R, Swinton SM, El Benni N, Walter A (2019) Precision farming at the nexus of agricultural production and the environment. Ann Rev Resour Econ 11:313–335
Fisk JW, Hesterman OB, Shrestha A, Kells JJ, Harwood RR, Squire JM, Sheaffer CC (2001) Weed suppression by annual legume cover crops in no-tillage corn. Agron J 93(2):319–325
Frost KE, Groves RL, Charkowski AO (2013) Integrated control of potato pathogens through seed potato certification and provision of clean seed potatoes. Plant Dis 97(10):1268–1280
Gerenfes D (2021) Review on laboratory mineral soil analysis and soil mineral management in organic farming. Bioorg Chem 6(1):1–6
Gnanesh BN, Arunakumar GS, Tejaswi A, Supriya M, Pappachan A, Harshitha MM (2023) Molecular diagnostics of soil-borne and foliar diseases of mulberry: present trends and future perspective. In: The mulberry genome. Springer International Publishing, Cham, pp 215–241

Gomiero T, Pimentel D, Paoletti MG (2011) Environmental impact of different agricultural management practices: conventional vs. organic agriculture. Crit Rev Plant Sci 30(1–2):95–124
Gossen BD, Carisse O, Kawchuk LM, Van Der Heyden H, McDonald MR (2014) Recent changes in fungicide use and the fungicide insensitivity of plant pathogens in Canada. Can J Plant Pathol 36(3):327–340
Gredler G (2001) Encouraging beneficial insects in your garden
Grieve IC (1980) Some contrasts in soil development between grassland and deciduous woodland sites. J Soil Sci 31(1):137–145
Grünwald NJ, Chen W, Larsen RC (2004) Pea diseases and their management. In: Diseases of fruits and vegetables: volume II: diagnosis and management. Springer Netherlands, Dordrecht, pp 301–331
Gu L, Nawaz T, Qiu Y, Wu Y, Zhou R (2023) Photosynthetic conversion of CO2 and H2O to long-chain terpene alcohol by genetically engineered N2-fixing cyanobacteria. Elsevier, Photosynthesis, pp 451–461
Gürsoy S (2020) Some physical properties of lentil seeds affected by harvest time. Agri Sci Technol (1313–8820) 12(3):264–271
Hallam D (2003) The organic market in OECD countries: past growth, current status and future potential. Organic agriculture: sustainability, markets and policies. 179–186
Hardy MC (2012) Using selective insecticides in sustainable IPM. CABI Rev 2011:1–7
Hernández-López I, Ortiz-Solà J, Alamprese C, Barros L, Shelef O, Basheer L et al (2022) Valorization of local legumes and nuts as key components of the Mediterranean diet. Food Secur 11(23):3858
Hewitt AE, Balks MR, Lowe DJ, Hewitt AE, Balks MR, Lowe DJ (2021) Melanic soils. In: The soils of Aotearoa New Zealand. Springer, Cham, pp 101–111
Horst MG, Shamutalov SS, Goncalves JM, Pereira LS (2007) Assessing impacts of surge-flow irrigation on water saving and productivity of cotton. Agric Water Manag 87(2):115–127
Hossain A, Sab AE, Barutcular C, Bhatt R, Cig F, Seydosoglu S, ... Saneoka H. (2020). Sustainable crop production to ensuring food security under climate change: a Mediterranean perspective. Aust J Crop Sci 14(3):439–446
Hue NV, Silva JA (2000) Organic soil amendments for sustainable agriculture: organic sources of nitrogen, phosphorus, and potassium. In: Plant nutrient management in Hawaii's soils, approaches for tropical and subtropical agriculture. College of Tropical Agriculture and Human Resources, University of Hawaii, Manoa, pp 133–144
Imas P (1999) Recent techniques in fertigation of horticultural crops in Israel. In: Recent trends in nutrition management in horticultural crops. Dapoli, Maharashtra
Irfan M, Mudassir M, Khan MJ, Dawar KM, Muhammad D, Mian IA, Ali W, Fahad S, Saud S, Hayat Z (2021) Heavy metals immobilization and improvement in maize (Zea mays L.) growth amended with biochar and compost. Sci Rep 11(1):18416
Irfan M, Dawar K, Fahad S, Mehmood I, Alamri S, Siddiqui MH, Saud S, Khattak JZK, Ali S, Hassan S (2022) Exploring the potential effect of Achnatherum splendens L.–derived biochar treated with phosphoric acid on bioavailability of cadmium and wheat growth in contaminated soil. Environ Sci Pollut Res 29(25):37676–37684
Ishii H (2006) Impact of fungicide resistance in plant pathogens on crop disease control and agricultural environment. Jpn Agri Res Quart 40(3):205–211
Islam S (2013) Marketing organic foods through conventional retail outlets
Jamar L, Leclercq V, Hardy B, Vanwindekens F, Stilmant D, Huyghebaert B (2006) Crop yields, nutrient flows and soil status in four organic vegetable cropping systems with high fertility-building crop diversity–a long-term experiment in Belgium
Jastrzębska M, Kostrzewska M, Saeid A (2022) Conventional agrochemicals: pros and cons. In: Smart agrochemicals for sustainable agriculture. Academic Press, pp 1–28
Javed U, Kumar P, Hussain S, Nawaz T, Fahad S, Ashraf S, Ali K (2024) Geospatial analysis of soil resistivity and hydro-parameters for groundwater assessment. Discover Geoscience 2(1):3

Jones OR, Allen RR, Unger PW (1990) Tillage systems and equipment for dryland farming. In: Advances in soil science: dryland agriculture: strategies for sustainability, vol 13. Springer, New York, pp 89–130

Joshi PK, Rao PP (2016) Global and regional pulse economies: current trends and outlook, vol 144. The International Food Policy Research Institute

Kafkafi U, Tarchitzky J (2011) Fertigation: a tool for efficient fertilizer and water management. International Fertilizer Industry Association, Paris, pp 141–141

Kakoulas DA, Golfinopoulos SK, Koumparou D, Alexakis DE (2022) The effectiveness of rainwater harvesting infrastructure in a Mediterranean Island. Water 14(5):716

Kalpana R, Selvi RV (2008) Effect of irrigation layouts on growth and yield of legumes–a review. Agric Rev 29(2):139–144

Kanatas PJ, Gazoulis I (2022) The integration of increased seeding rates, mechanical weed control and herbicide application for weed management in chickpea (Cicer arietinum L.). Phytoparasitica 50(1):255–267

Karim SM, Karim TH (2020) Enhancing furrow irrigation performance and water productivity through better design and water management in a cracked soil. Iraqi J Agric Sci 51(5):1276

Karkanis A, Ntatsi G, Kontopoulou CK, Pristeri A, Bilalis D, Savvas D (2016) Field pea in European cropping systems: adaptability, biological nitrogen fixation and cultivation practices. Notulae Botanicae Horti Agrobotanici Cluj-Napoca 44(2):325–336

Kashyap C, Bainade SP, Kumar V, Singh A (2022) Foliar application of macro and micronutrient in field crops and their effect on growth, yield, quality and economics. Pharma Innov J 11(5):970–976

Khalifa RM, Ghannam M (2021) Characterizing water distribution of furrow irrigation to improve soil productivity in North Nile delta of Egypt. Menoufia J Soil Sci 6(3):87–110

Khan FU, Khan AA, Li K, Xu X, Adnan M, Fahad S, Ahmad R, Khan MA, Nawaz T, Zaman F (2022a) Influences of long-term crop cultivation and fertilizer management on soil aggregates stability and fertility in the Loess Plateau, Northern China. J Soil Sci Plant Nutr 22(2):1446–1457

Khan MA, Basir A, Fahad S, Adnan M, Saleem MH, Iqbal A, Amanullah, Al-Huqail AA, Alosaimi AA, Saud S (2022b) Biochar optimizes wheat quality, yield, and nitrogen acquisition in low fertile calcareous soil treated with organic and mineral nitrogen fertilizers. Front Plant Sci 13:879788

Khan MR, Huang C, Ullah R, Ullah H, Qazi IM, Nawaz T, Adnan M, Khan A, Su H, Ren L (2022c) Effects of various polymeric films on the pericarp microstructure and storability of longan (cv. Shixia) fruit treated with propyl disulfide essential oil from the neem (Azadirachta indica) plant. Polymers 14(3):536

Khan MA, Basir A, Adnan M, Fahad S, Ali J, Mussart M, Mian IA, Ahmad M, Saleem MH, Naseem W (2023a) Biochar to improve crops yield and quality under a changing climate. In: Sustainable agriculture reviews 61: Biochar to improve crop production and decrease plant stress under a changing climate. Springer, pp 57–73

Khan MS, Ali S, Ali N, Khan SA, Akbar R, Ansari MJ, Fahad S, Datta R, Danish S, Hassan S (2023b) Response of okra (Abelmoschus esculentus L.) F5: 6 population of for earliness and yield traits. Pak J Bot 55(2):689–695

Khan SA, Sohni S, Akhtar K, Bakhsh EM, Nawaz T, Khan SB (2023c) Lignocellulose biomatrix zero-valent cobalt nanoparticles: a dip-catalyst for organic pollutants degradation. Ind Crop Prod 198:116694

Kharel S, Marahatta D (2020) Conventional farming practices around Kathmandu valley: an appraisal. Nepalese J Develop Rural Stud 17:90–103

Khatun A, Kabir G, Bhuiyan MAH (2009) Effect of harvesting stages on the seed quality of lentil (Lens culinaris L.) during storage. Bangladesh J Agric Res 34(4):565–576

Knott CM, Belcher SJ (1998) Optimum sowing dates and plant populations for winter peas (Pisum sativum). J Agric Sci 131(4):449–454

Korbu L, Damte T, Fikre A (2016) Harnessing chickpea value chain for nutrition security and commercialization of smallholder Agriculture in Africa

Kousta A, Katsis C, Tsekoura A, Chachalis D (2024) Effectiveness and selectivity of pre-and post-emergence herbicides for weed control in grain legumes. Plan Theory 13(2):211
Krishnasree RK, Raj SK, Chacko SR (2021) Foliar nutrition in vegetables: a review. J Pharmacogn Phytochem 10(1):2393–2398
Kühling INSA, Trautz DIETER (2013) The role of organic farming in providing ecosystem services. Int J Environ Rural Develop 4(1):175–178
Kumar S (2014) Plant disease management in India: Advances and challenges. Afr J Agric Res 9(15):1207–1217
Kumar V, Pawar K (2018) A review on soil health and fertility management in organic agriculture through green manuring. J Pharmacogn Phytochem 7(1S):3213–3217
Kumar S, Chandra A, Pandey KC (2008) Bacillus thuringiensis (Bt) transgenic crop: an environment friendly insect-pest management strategy. J Environ Biol 29(5):641–653
Kumar AR, Divya TM, Jayasudha BSK, Sudha PN (2020) Precision agriculture: a review on its techniques and technologies. Int Res J Mod Eng Technol Sci 2(09):1326–1332
Kumar A, Chandel N, Barkha B (2024) Organic farming vs. integrated nutrient management: a comparative review of agricultural productivity and sustainability. Int J Plant Soil Sci 36(6):460–473
Lagacherie P, Alvaro-Fuentes J, Annabi M, Bernoux M, Bouarfa S, Douaoui A et al (2018) Managing Mediterranean soil resources under global change: expected trends and mitigation strategies. Reg Environ Chang 18:663–675
Laita M, Sabbahi R, Azzaoui K, Nasri H, Messaoudi Z, Benkirane R, Aithaddou H (2024) Optimizing water use and crop yield with deficit irrigation techniques: a comprehensive overview and case study from Morocco. Multidiscip Rev 7(4):2024074–2024074
Lamm FR (2002) Advantages and disadvantages of subsurface drip irrigation. In International meeting on advances in drip/micro irrigation, Puerto de La Cruz, Tenerife, Canary Islands, pp 1–13
Landry EJ, Coyne CJ, McGee RJ, Hu J (2016) Adaptation of autumn-sown faba bean germplasm to southeastern Washington. Agron J 108(1):301–308
Lane A, Jarvis A (2007) Changes in climate will modify the geography of crop suitability: agricultural biodiversity can help with adaptation
Layek J, Das A, Mitran T, Nath C, Meena RS, Yadav GS et al (2018) Cereal+ legume intercropping: an option for improving productivity and sustaining soil health. In: Legumes for soil health and sustainable management. Springer, pp 347–386
Lenssen AW, Sainju UM, Jabro JD, Allen BL, Stevens WB (2018) Dryland pea production and water use responses to tillage, crop rotation, and weed management practice. Agron J 110(5):1843–1853
Lindang HU, Subbiah VK, Rodrigues KF, Budiman C (2021) Isolation, identification, and characterization of phosphate solubilizing bacteria, Paenibacillus sp., from the soil of Danum Valley Tropical Rainforest, Sabah, Malaysia. Biodivers J Biol Divers 22(10). https://doi.org/10.13057/biodiv/d221030
López Ó, Fernández-Bolaños JG, Gil MV (2005) New trends in pest control: the search for greener insecticides. Green Chem 7(6):431–442
López-Bellido FJ, López-Bellido RJ, Khalil SK, López-Bellido L (2008) Effect of planting date on winter kabuli chickpea growth and yield under rainfed Mediterranean conditions. Agron J 100(4):957–964
Maikštėnienė S, Arlauskienė A, Velykis A, Satkus A (2009) Enhancement of competitive ability of cereals towards weeds by means of crop rotations. Zemdirbyste-Agriculture 96:23–34
Makate C, Makate M, Mango N, Siziba S (2019) Increasing resilience of smallholder farmers to climate change through multiple adoption of proven climate-smart agriculture innovations. Lessons from Southern Africa. J Environ Manag 231:858–868
Malhotra RS, Singh KB, Saxena MC (1997) Effect of irrigation on winter-sown chickpea in a Mediterranean environment. J Agron Crop Sci 178(4):237–243
Maphosa L, Preston A, Richards MF (2023) Effect of sowing date and environment on phenology, growth and yield of lentil (Lens culinaris Medikus.) Genotypes. Plan Theory 12(3):474

Marteau-Bazouni M, Jeuffroy MH, Guilpart N (2024) Grain legume response to future climate and adaptation strategies in Europe: a review of simulation studies. Eur J Agron 153:127056
Martini MY, McKenzie BA, Moot DJ, Hill GD (2012). Dry matter accumulation of faba bean sown at different sowing dates in Canterbury
Maughan T, Allen LN, Drost D (2015) Soil moisture measurement and sensors for irrigation management
Maxted N, Bennett SJ (eds) (2001) Plant genetic resources of legumes in the Mediterranean, vol 39. Springer Science & Business Media
Mays LW (2014) Use of cisterns during antiquity in the Mediterranean region for water resources sustainability. Water Sci Technol Water Supply 14(1):38–47
Meena RS, Das A, Yadav GS, Lal R (eds) (2018) Legumes for soil health and sustainable management. Springer
Merga B, Haji J (2019) Economic importance of chickpea: production, value, and world trade. Cogent Food Agric 5(1):1615718
Mitkova T, Markoski M (2019) A review to the genesis, evolution and classification of the soils formed on limestones and dolomites in the Republic of Macedonia. Contributions Sect Nat Math Biotech Sci 40(1):63
Monteiro CM, Calheiros CS, Pimentel-Rodrigues C, Silva-Afonso A, Castro PM (2016) Contributions to the design of rainwater harvesting systems in buildings with green roofs in a Mediterranean climate. Water Sci Technol 73(8):1842–1847
Moslemi A, Hilmola OP, Vilko J (2016) Risks in emerging markets: logistics services in the Mediterranean region. Marit Bus Rev 1(3):253–272
Muehlbauer FJ, Sarker A (2017) Economic importance of chickpea: production, value, and world trade. In: The chickpea genome. Springer, Cham, pp 5–12
Mulazzani L, Malorgio G (2009) Market dynamics and commercial flows in the Mediterranean area: triangular effects among the EU, the MCPs and Italy in the fruit and vegetable sector. New Medit 8(1):37–45
Musheer N, Jamil A, Choudhary A (2023) Solo and combined applications of fungicides and bio-agents to reduce severity of Fusarium oxysporum and induce antioxidant metabolites in Ocimum tenuiflorum L. J Plant Pathol 105(1):237–251
Nadeem M, Li J, Yahya M, Sher A, Ma C, Wang X, Qiu L (2019) Research progress and perspective on drought stress in legumes: a review. Int J Mol Sci 20(10):2541
Nair A, Delate K (2016) Composting, crop rotation, and cover crop practices in organic vegetable production. In: Organic farming for sustainable agriculture. Springer, Cham, pp 231–257
Naresh R, Singh NK, Sachan P, Mohanty LK, Sahoo S, Pandey SK, Singh B (2024) Enhancing sustainable crop production through innovations in precision agriculture technologies. J Sci Res Rep 30(3):89–113
Nawaz T, Gu L, Fahad S, Saud S, Jiang Z, Hassan S, Harrison MT, Liu K, Khan MA, Liu H (2023a) A comprehensive review of the therapeutic potential of cyanobacterial marine bioactives: unveiling the hidden treasures of the sea. Food Energy Secur 12(5):e495
Nawaz T, Junaid M, Kanwal M, Ahmed S, Ahmed N, Ullah R, Adnan M, Saeed M, Romman M, Fahad S (2023b) Bio-fertilizer effects on plant-parasitic nematodes. In: Biofertilizers for sustainable soil management. CRC Press, pp 279–289
Nawaz T, Joshi D, Nelson D, Saud S, Abdelsalam NR, Abdelhamid MM, ... Fahad S (2024a) Harnessing the potential of nitrogen-fixing cyanobacteria: a rich bio-resource for sustainable soil fertility and enhanced crop productivity. Environ Technol Innov 103886
Nawaz T, Gu L, Fahad S, Saud S, Bleakley B, Zhou R (2024b) Exploring sustainable agriculture with nitrogen-fixing cyanobacteria and nanotechnology. Molecules 29(11):2534
Nawaz T, Gu L, Fahad S, Saud S, Harrison MT, Zhou R (2024c) Sustainable protein production through genetic engineering of cyanobacteria and use of atmospheric N2 gas. Food Energy Secur 13(2):e536
Nawaz T, Gu L, Fahad S, Saud S, Hassan S, Harrison MT, Liu K, Zhou R (2024d) Unveiling the antioxidant capacity of fermented foods and food microorganisms: a focus on cyanobacteria. J Umm Al-Qura Univ Appl Sci 10(1):232–243

Nawaz T, Gu L, Gibbons J, Hu Z, Zhou R (2024e) Bridging nature and engineering: protein-derived materials for bio-inspired applications. Biomimetics 9(6):373
Nawaz T, Saud S, Gu L, Khan I, Fahad S, Zhou R (2024f) Cyanobacteria: harnessing the power of microorganisms for plant growth promotion, stress alleviation, and phytoremediation in the era of sustainable agriculture. Plant Stress 11:100399
Nawaz T, Nelson D, Saleem A, Bibi M, Joshi N, Fahad S, ... Khan I (2025a) Genetic Engineering for Crop Biofortification. Crop Biofortification: Biotechnological Approaches for Achieving Nutritional Security Under Changing Climate, 263–294
Nawaz T, Nelson D, Fahad S, Saud S, Aaqil M, Adnan M, ... Raihan MJ (2025b). Impact of elevated CO2 and temperature on overall agricultural productivity. In Challenges and solutions of climate impact on agriculture (pp. 163–202). Academic Press
Naz T, Iqbal MM, Fahad S, Akhtar J, Saqib M, Alamri S, Siddiqui MH, Saud S, Khattak JZK, Ali S (2023) Bio-fortification of two wheat cultivars with iron and zinc through their soil and foliar application in salt-factored soil: growth, ionic, physiological, and biochemical modifications. J Plant Growth Regul 42(9):5727–5745
Nohong B, Islamiyati R, Semaun R (2024) Increasing the fertility and productivity of marginal land by planting leguminous plants. Indian J Agric Res 58(2):301–305
Nongbet A, Mishra AK, Mohanta YK, Mahanta S, Ray MK, Khan M et al (2022) Nanofertilizers: a smart and sustainable attribute to modern agriculture. Plan Theory 11(19):2587
Noulas C, Karyotis T, Charoulis A, Massas I (2009) Red Mediterranean soils: nature, properties, and management of Rhodoxeralfs in Northern Greece. Commun Soil Sci Plant Anal 40(1–6):633–648
Nunes BHDN, Cardeal IRP, Leite ODC, Freitas GAD, Milhomem RM, Nunes HV et al (2023) Revised approach, conditions, and cultivation practices of Cicer arietinum L. in the Cerrado. Int J Plant Soil Sci 35(23):250–256
Okumu OO, Otieno HM, Okeyo GO (2023) Production systems and contributions of grain legumes to soil health and sustainable agriculture: a review. Arch Agric Environ Sci 8(2):259–267
Onasanya OO, Hauser S, Necpalova M, Salako FK, Kreye C, Tariku M et al (2021) On-farm assessment of cassava root yield response to tillage, plant density, weed control and fertilizer application in southwestern Nigeria. Field Crop Res 262:108038
Özdemir S, Karadavut U (2003) Comparison of the performance of autumn and spring sowing of chickpeas in a temperate region. Turk J Agric For 27(6):345–352
Pacucci G, Troccoli C, Leoni B (2006) Effect of supplementary irrigation on yield of chickpea genotypes in a Mediterranean climate
Palmer C (1998) An exploration of the effects of crop rotation regime on modem weed floras. Environ Archaeol 2(1):35–48
Panday KK, Panday PK, Mishra KK (2005) Development and testing of an integrated disease management package for multiple diseases of tomato. Indian Phytopathol 58(3):294–297
Panta S, Parajulee D (2021) Integrated nutrient management (INM) in soil and sustainable agriculture. Int J Appl Sci Biotechnol 9(3):160–165
Pasqualone A, Abdallah A, Summo C (2020) Symbolic meaning and use of broad beans in traditional foods of the Mediterranean Basin and the Middle East. J Ethnic Foods 7(1):39
Patel JJ, Mangroliya RM, Patel NA (2022) Role of micronutrients foliar nutrition in vegetable production: a review. J Exp Agric Int 44(11):159–168
Pearsons KA, Tooker JF (2017) In-field habitat management to optimize pest control of novel soil communities in agroecosystems. Insects 8(3):82
Pérez-Fernández MA, Calvo-Magro E, Valentine A (2016) Benefits of the symbiotic association of shrubby legumes for the rehabilitation of degraded soils under Mediterranean climatic conditions. Land Degrad Dev 27(2):395–405
Peter BG, Messina JP, Lin Z, Snapp SS (2020) Crop climate suitability mapping on the cloud: a geovisualization application for sustainable agriculture. Sci Rep 10(1):15487
Poot P, Lambers H (2008) Shallow-soil endemics: adaptive advantages and constraints of a specialized root-system morphology. New Phytol 178(2):371–381

Rahaman S, Ramana C, Rao AS, Reddy BR (2022) Field Evaluation and Economic Analysis of Manual Drawn Rotor Weeder for Small Farms. Economic Affairs 67(4):415–421

Raj J, Jat S, Kumar M, Yadav A (2024) The role of organic farming in sustainable agriculture. Adv Res 25(3):128–136

Rashid A, Ryan J (2004) Micronutrient constraints to crop production in soils with Mediterranean-type characteristics: a review. J Plant Nutr 27(6):959–975

Rubiales D, Fondevilla S, Chen W, Gentzbittel L, Higgins TJ, Castillejo MA et al (2015) Achievements and challenges in legume breeding for pest and disease resistance. Crit Rev Plant Sci 34(1–3):195–236

Ruisi P, Amato G, Badagliacca G, Frenda AS, Giambalvo D, Di Miceli G (2017) Agro-ecological benefits of faba bean for rainfed Mediterranean cropping systems. Ital J Agron 12(3):865

Ruiu L (2015) Insect pathogenic bacteria in integrated pest management. Insects 6(2):352–367

Rusu T, Pacurar I, Dirja M, Pacurar HM, Oroian I, Cosma SA, Gheres M (2013) Effect of tillage systems on soil properties, humus and water conservation. Agric Sci 4(05):35

Saeed S, Ullah S, Amin F, Al-Hawadi JS, Okla MK, Alaraidh IA, AbdElgawad H, Liu K, Harrison MT, Saud S (2024) Salicylic acid and Tocopherol improve wheat (Triticum aestivum L.) Physio-biochemical and agronomic features grown in deep sowing stress: a way forward towards sustainable production. BMC Plant Biol 24(1):1–14

Saljoqi A-U-R, Amin S, Salim M, Nawaz T, Anjum F (2022) Pesticide residue analysis of three different pesticides used against Helicoverpa Armigera (Hubner) in tomato crop. Sarhad J Agric 38:448

Sanchis V (2011) From microbial sprays to insect-resistant transgenic plants: history of the biospesticide Bacillus thuringiensis. A review. Agron Sustain Dev 31:217–231

Saxena S (2021) A review study on pest control in organic systems. Asian J Res Bus Econ Manag 11(11):112–117

Sayde CG, Khoury L, Gitelman A, English M (2008) Optimizing estimates of soil moisture for irrigation scheduling. In: 2008 Providence, Rhode Island, June 29–July 2, 2008. American Society of Agricultural and Biological Engineers, p 1

Schader C, Stolze M, Gattinger A (2012) Environmental performance of organic farming. In: Green technologies in food production and processing. Springer, New York, pp 183–210

Schmidt HDO, Oliveira VRD (2023) Overview of the incorporation of legumes into new food options: an approach on versatility, nutritional, technological, and sensory quality. Food Secur 12(13):2586

Schreiber MM (1960) Pre-emergence herbicides on alfalfa and birdsfoot trefoil. Weeds 8(2):291–299

Sconyers LE, Brenneman TB, Stevenson KL, Mullinix BG (2005) Effects of plant spacing, inoculation date, and peanut cultivar on epidemics of peanut stem rot and tomato spotted wilt. Plant Dis 89(9):969–974

Serraj R, Bidinger FR, Chauhan YS, Seetharama N, Nigam SN, Saxena NP (2003) Management of drought in ICRISAT cereal and legume mandate crops. In: Water productivity in agriculture: Limits and opportunities for improvement. CAB Inernational, Wallingford, pp 127–144

Sgouras ID, Tsadilas CD, Barbayiannis N, Danalatos N (2007) Physicochemical and mineralogical properties of red Mediterranean soils from Greece. Commun Soil Sci Plant Anal 38(5–6):695–711

Shah S, Chen C, Sun Y, Wang D, Nawaz T, El-Kahtany K, Fahad S (2023) Mechanisms of nitric oxide involvement in plant-microbe interaction and its enhancement of stress resistance. Plant Stress 10:100191

Shahab M, Roberto SR, Adnan M, Fahad S, Koyama R, Saleem MH, Nasar J, Saud S, Hassan S, Nawaz T (2023) Phenolic compounds as a quality determinant of grapes: a critical review. J Plant Growth Regul 42(9):5325–5331

Sharma HC, Jaba J, Vashisth S (2017) Distinguishing proof and utilization of resistance of insect pests in grain legumes: progress and limitations. In: Breeding insect resistant crops for sustainable agriculture. Springer, Singapore, pp 131–170

Shedeed SI, Zaghloul SM, Yassen AA (2009) Effect of method and rate of fertilizer application under drip irrigation on yield and nutrient uptake by tomato. Ozean J Appl Sci 2(2):139–147

Shepard JF, Claflin LE (1975) Critical analyses of the principles of seed potato certification. Annu Rev Phytopathol 13(1):271–293

Siddique KHM, Loss SP, Pritchard DL, Regan KL, Tennant D, Jettner RL, Wilkinson D (1998) Adaptation of lentil (Lens culinaris Medik.) to Mediterranean-type environments: effect of time of sowing on growth, yield, and water use. Aust J Agric Res 49(4):613–626

Siddique KH, Johansen C, Turner NC, Jeuffroy MH, Hashem A, Sakar D et al (2012) Innovations in agronomy for food legumes. A review. Agron Sustain Dev 32:45–64

Silim SN, Hebblethwaite PD, Heath MC (1985) Comparison of the effects of autumn and spring sowing date on growth and yield of combining peas (Pisum sativum L.). J Agric Sci 104(1):35–46

Silim SN, Saxena MC, Erskine W (1991) Effect of sowing date on the growth and yield of lentil in a rainfed Mediterranean environment. Exp Agric 27(2):145–153

Singh KB, Malhotra RS, Saxena MC, Bejiga G (1997) Superiority of winter sowing over traditional spring sowing of chickpea in the Mediterranean region. Agron J 89(1):112–118

Singhal A, Karaca AC, Tyler R, Nickerson M (2016) Pulse proteins: from processing to structure-function relationships. Grain Legumes 55:00

Singhal RK, Fahad S, Kumar P, Choyal P, Javed T, Jinger D, Singh P, Saha D, Md P, Bose B (2023) Beneficial elements: new players in improving nutrient use efficiency and abiotic stress tolerance. Plant Growth Regul 100(2):237–265

Sivaranjani S, Rakshit A (2019) Organic farming in protecting water quality. In: Organic farming: new advances towards sustainable agricultural systems. Springer, Berlin, pp 1–9

Siyal AA, Siyal AG, Hasini MY (2011) Crop production and water use efficiency under subsurface porous clay pipe irrigation. Pak J Agric Agric Eng Vet Sci 27(1):39–50

Sozer N, Holopainen-Mantila U, Poutanen K (2017) Traditional and new food uses of pulses. Cereal Chem 94(1):66–73

Sparling GP, Shepherd TG, Schipper LA (2000) Topsoil characteristics of three contrasting New Zealand soils under four long-term land uses. N Z J Agric Res 43(4):569–583

Stagnari F, Maggio A, Galieni A, Pisante M (2017) Multiple benefits of legumes for agriculture sustainability: an overview. Chem Biol Technol Agric 4:1–13

Stoleru V, Sellitto VM (2016) Pest control in organic systems. In Integrated Pest Management (IPM): environmentally sound pest management, InTech. https://doi.org/10.5772/64457

Stolz H, Stolze M, Hamm U, Janssen M, Ruto E (2011) Consumer attitudes towards organic versus conventional food with specific quality attributes. NJAS: Wageningen J Life Sci 58(3-4):67–72

Sümer FÖ (2023) The effects of different sowing times on the phenological characteristics and seed yield of the pea. Turkish J Field Crops 28(2):301–312

Sundaram PK, Rahman A, Singh AK, Sarkar B (2021) A novel method for manual weeding in row crops. Indian J Agric Sci 91(6):946–948

Surve V, Patel HH (2024) Sensor based irrigation management in crop production: a review. Ann Res Rev Biol 39(4):1–4

Tal A (2018) Making conventional agriculture environmentally friendly: moving beyond the glorification of organic agriculture and the demonization of conventional agriculture. Sustain For 10(4):1078

Tawaha AM, Turk MA (2004) Field pea seeding management for semi-arid Mediterranean conditions. J Agron Crop Sci 190(2):86–92

Tedone L, Alhajj Ali S, De Mastro G (2022) The effect of tillage on Faba Bean (Vicia faba L.) nitrogen fixation in durum wheat (Triticum turgidum L. subsp. Durum (Desf))-based rotation under a Mediterranean climate. Agronomy 13(1):105

Thapa R, Juyal R (2022) A study on biological control, cultural practices, and chemical control methods in nursery pest and disease management. J Surv Fish Sci 8:398–402

Thomas-Sharma S, Andrade-Piedra J, Carvajal Yepes M, Hernandez Nopsa JF, Jeger MJ, Jones RAC et al (2017) A risk assessment framework for seed degeneration: Informing an integrated seed health strategy for vegetatively propagated crops. Phytopathology 107(10):1123–1135

Thompson GD, Kidwell J (1998) Explaining the choice of organic produce: cosmetic defects, prices, and consumer preferences. AJAE 80(2):277–287
Toker C, Mutlu N (2011) 16 Breeding for abiotic stresses. In: Biology and breeding of food legumes. CAB International, p 241
Tripathi D, Verma LR, Patial KS, Singh K (2006) Characteristics, classification and suitability of soils for major crops of Kiar–Nagali micro-watershed in North-West Himalayas. J Indian Soc Soil Sci 54(2):131–136
Trout TJ, Kincaid DC (2007) On-farm system design and operation and land management. Irrigat Agric Crops 30:133–179
Tuomisto HL, Hodge ID, Riordan P, Macdonald DW (2012) Does organic farming reduce environmental impacts?–A meta-analysis of European research. J Environ Manag 112:309–320
Turk MA, Tawaha AM (2002) Lentil (Lens culinaris Medik.) response to frequencies of hand weeding. Indian. J Agric Res 36(2):137–140
Uddin J, Smith R, Hancock N, Foley JP (2010, January) Droplet evaporation losses during sprinkler irrigation: an overview. In Australian Irrigation Conference and Exhibition 2010: Proceedings
Ullah I, Dawar K, Tariq M, Sharif M, Fahad S, Adnan M, Ilahi H, Nawaz T, Alam M, Ullah A (2022) Gibberellic acid and urease inhibitor optimize nitrogen uptake and yield of maize at varying nitrogen levels under changing climate. Environ Sci Pollut Res 29:1–10
Ullah R, Junaid M, Kanwal M, Adnan M, Nawaz T, Ahmed N, Subhan F, Romman M, Saud S, Ahmad S (2023a) Status of research and applications of bio-fertilizers: global scenario. In: Biofertilizers for sustainable soil management. CRC Press, pp 255–277. https://www.taylorfrancis.com/chapters/edit/10.1201/9781003286233-14/status-research-applications-bio-fertilizersrafi-ullah-muhammad-junaid-mehwish-kanwal-muhammad-adnan-taufiq-nawaz-nazeer-ahmed-fazli-subhanmuhammad-romman-shah-saud-shakeel-ahmad-anas-iqbal-fazli-wahid-muhammad-haroon-muhammad-zaminnazish-huma-khan-jamal-nasar-shah-hassan
Ullah S, Saud S, Liu K, Harrison MT, Hassan S, Nawaz T, Zeeshan M, Nasar J, Khan I, Liu H (2023b) Germination response of Oat (Avena sativa L.) to temperature and salinity using halothermal time model. Plant Stress 10:100263
Ullah I, Ullah S, Amin F, Al-Hawadi JS, Okla MK, Alaraidh IA, AbdElgawad H, Liu K, Harrison MT, Saud S, Hassan S, Nawaz T, Zhu M, Liu H, Fahad S (2024) Germination responses of *Lens culinaris* L. seeds to osmotic potentials at cardinal temperatures using hydrothermal time model. BMC Plant Biol 24(1):502
Underwood T, McCullum-Gomez C, Harmon A, Roberts S (2011) Organic agriculture supports biodiversity and sustainable food production. J Hunger Environ Nutr 6(4):398–423
Vanlauwe B, Giller KE (2006) Popular myths around soil fertility management in sub-Saharan Africa. Agric Ecosyst Environ 116(1–2):34–46
Vasconcelos MW, Grusak MA, Pinto E, Gomes A, Ferreira H, Balázs B et al (2020) The biology of legumes and their agronomic, economic, and social impact. In: The plant family fabaceae: biology and physiological responses to environmental stresses. Springer, Singapore, pp 3–25
Velásquez-Barreto FF, Ramirez-Tixe EE, Salazar-Irrazabal MD, Salazar-Silvestre E (2020) Physicochemical properties and acceptability of three formulations containing fava bean, quinoa and corn flour extrudates. Revista de Ciencias Agrícolas 37(2):40–48
Vishnyakova MA, Aleksandrova TG, Buravtseva TV, Burlyaeva MO, Egorova GP, Semenova EV et al (2019) Species diversity of the VIR collection of grain legume genetic resources and its use in domestic breeding. Proc Appl Bot Genet Breed 180(2):109–123
Voisin AS, Guéguen J, Huyghe C, Jeuffroy MH, Magrini MB, Meynard JM et al (2014) Legumes for feed, food, biomaterials and bioenergy in Europe: a review. Agron Sustain Dev 34:361–380
Wahba M, Fawkia L, Zaghloul A (2019) Management of calcareous soils in arid region. Int J Environ Pollut Environ Modell 2(5):248–258
Watson CA, Atkinson D, Gosling P, Jackson LR, Rayns FW (2002) Managing soil fertility in organic farming systems. Soil Use Manag 18:239–247
Winqvist C, Ahnström J, Bengtsson J (2012) Effects of organic farming on biodiversity and ecosystem services: taking landscape complexity into account. Ann N Y Acad Sci 1249(1):191–203

Wu W, Ma B (2015) Integrated nutrient management (INM) for sustaining crop productivity and reducing environmental impact: a review. Sci Total Environ 512:415–427

Wyss E, Luka H, Pfiffner L, Schlatter C, Uehlinger G, Daniel C (2005) Approaches to pest management in organic agriculture: a case study in European apple orchards

Yadav MK, Kaswala AR, Dubey PK (2024) An assessment of organic and conventional farming practices for yield, pest management and soil health. Asian J Soil Sci Plant Nutrit 10(2):150–156

Yang P, Wu L, Cheng M, Fan J, Li S, Wang H, Qian L (2023) Review on drip irrigation: impact on crop yield, quality, and water productivity in China. Water 15(9):1733

Yanni AE, Iakovidi S, Vasilikopoulou E, Karathanos VT (2023) Legumes: a vehicle for transition to sustainability. Nutrients 16(1):98

Yau SK (2005) Optimal sowing time and seeding rate for winter-sown, rain-fed chickpea in a cool, semi-arid Mediterranean area. Aust J Agric Res 56(11):1227–1233

Yu JD, Ni WZ, Technique YXEN (2003) For the management of fertilizers and water-fertigation. Chinese J Soil Sci 34:148–153

Yucel DO (2013) Optimal intra-row spacing for production of local faba bean (Vicia faba L. major) cultivars in the Mediterranean conditions. Pak J Bot 45(6):1933–1938

Yuvaraj M, Pandiyan M, Gayathri P (2020) Role of legumes in improving soil fertility status. In: Legume crops-prospects, production and uses. IntechOpen, pp 1–8

Drought Stress and Production of Legumes in Europe

4

Aleksandra Ilić, Ana Uhlarik, and Zlatica Mamlić

Abstract

Legumes (fam. *Fabaceae*) are the second most important family of agricultural crops worldwide. Even though legumes are used for grain, forage, silage, or green manure they are unfairly underrepresented in European cropping systems. Legumes production in Europe has been faced with a drastic decline (around 50%) in the period between 1961 and 2020. However, an increase in the legume production area since 2010 was caused by the changes in CAP, where many European Union states offered supplemental subsidies within specific programs to motivate the cultivation of legumes. Nowadays, legume production areas in the European Union occupy less than 5% of agricultural land, but it is anticipated to reach 6.3 million tons in 2030, while cultivation area is expected to grow to 2.3–2.4 million ha. Due to the functional attributes that legumes offer, including nutritional and health provision, farmland biodiversity, and improved environmental stability, it is proposed to increase legume production in Europe as an important component in agro-ecological transitions. It is foreseen that the cultivation of pulse crops will increase due to the demand for more diversified protein sources and their better availability in Europe. Various environmental pressures, such as higher temperatures together with the changed and unpredictable distribution of precipitation, interfere with legume production. Water deficit is highlighted as the most significant environmental stressor that negatively impacts agricultural production. Drought is characterized as a prolonged period of dry weather in which some areas receive less rain than usual, and it is expected to affect at least one-third of agricultural land worldwide. Severe drought events have been documented in the European continent since the beginning of the

A. Ilić (✉) · A. Uhlarik · Z. Mamlić
Institute of Field and Vegetable Crops, National Institute of the Republic of Serbia, Novi Sad, Serbia
e-mail: aleksandra.savic@ifvcns.ns.ac.rs

© The Author(s), under exclusive license to Springer Nature Singapore Pte Ltd. 2025

M. A. Nadeem et al. (eds.), *Marker-Assisted Breeding in Legumes for Drought Tolerance*, https://doi.org/10.1007/978-981-96-4112-3_4

twenty-first century. There is a consensus that the alteration of drought events affecting Europe, caused by climate change, is obvious. Legumes are mainly grown in rain-fed agricultural systems, and they are highly sensitive to drought and heat stress during both vegetative and reproductive stages. There is a need to develop various strategies in order to increase the drought-stress tolerance of legumes and enable crop productivity to meet food demands in water-scarce scenarios.

Keywords

legume production · Europe · drought stress

4.1 Introduction

The *Fabaceae* family, commonly known as legumes, ranks as the second most significant family of agricultural crops worldwide, following closely behind the *Poaceae* family, or grasses. They contribute to 27% of the global primary crop yield and fulfill 33% of the global protein needs (Khatun et al. 2021). Legumes thrive in diverse climates and soil types, offering unique functional characteristics and providing numerous advantages in farming systems. These benefits include enhanced nutrition and health, increased biodiversity on farmland, and improved environmental sustainability. Consequently, expanding the cultivation of leguminous crops is considered vital for promoting agro-ecological transitions. Additionally, legumes are valued for their diverse uses, including grain, forage, silage, and green manure (Ali et al. 2022; Nadeem et al. 2023). It is estimated that they serve as key cash crops for approximately 700 million small-scale farmers in developing nations. Nonetheless, within European agricultural systems, legumes remain underrepresented, primarily due to the predominance of cereals and oil crops (Watson et al. 2017). Legumes contribute to the sustainability of European agri-food systems by enhancing environmental performance, resource efficiency, and protein self-sufficiency (Notz et al. 2022). For instance, grain legumes such as soybeans, chickpeas, and common beans have a protein content ranging from 20% to 45%, significantly higher than the 7–17% found in cereals. Moreover, the amino acid composition of legume proteins complements that of cereal proteins. Similar to this, legume species used for feed, such as forage peas, vetches, alfalfa, and clover, are valuable sources of protein for livestock. In addition to feeding people and domestic animals, they are increasingly used as medicinal plants for the production of various pharmaceutical preparations (Bokan et al. 2016). Legumes have exceptional agrotechnical importance in all systems of agricultural production. They are very important in crop rotation because they leave the soil in a favorable biophysical state with a high content of naturally accumulated nitrogen (Mamlić et al. 2023). Humification and mineralization of legumes increase organic mineral nitrogen in the soil (Jarak et al. 2004). *Bradyrhizobium japonicum*, *Bradyrhizobium elkani*, and *Sinorhizobium fredii* bacteria form nodules in symbiosis with soybeans, and annual nitrogen fixation is over 200 kg ha-1. *Rhizobium leguminosarum* bv. *trifolia* forms nodules on the clover root and annually fixes from 45 to 670 kg N ha-1. The symbiosis of peas, vetch, and lentils with *Rhizobium leguminosarum* bv. *viciae* fixed annually around 40 to 120 kg N ha-1. A large number of legumes are good for growing in combined and cover crops, and in the absence of high-quality manure, they are sown as intercrops. Legumes as cover crops can often provide a sufficient amount of nitrogen needed for the production of the main crop. Many scientists (Giller and Cadisch 1995; Oljača and Dolijanović 2013) are of the opinion that biological nitrogen fixation using cover crops is the only alternative to mineral nitrogen fertilizers in the function of maintaining and increasing food production in the future. Nitrogen is released into the soil when the crop is incorporated into the soil or during the decomposition of crop residues (Dozet et al. 2014). Gselman and Kramberger (2008) state that the contribution of leguminous cover crops is reflected in over 100 kg ha-1 of adopted N, with beneficial effects on the next crop in the rotation (Hanly and Gregg 2004). Tonitto et al. (2006) also state that using legumes as a cover crop for the next crop provides 10–36% of the total nitrogen needs. The enhanced utilization of legumes leads to a decrease in the necessity for mineral fertilizers, thus supporting the production of food that is safe for consumption. Simultaneously, this approach facilitates essential crop rotation and enhances the efficiency of machinery and land utilization.

4.2 Legume Production in Europe

Despite their numerous beneficial properties, legume production areas in Europe remain significantly low, occupying less than 2% of agricultural land in 2022 (EUROSTAT 2022), compared to approximately 7% of total arable land worldwide covered with legumes (FAOSTAT 2024). Cernay et al. (2015) identify the high instability of legume yields and low market competitiveness as primary reasons for the limited presence of legumes in European agricultural systems. The lower yield of legumes, compared to cereals, is attributed to the higher protein and oil content in certain crops like soybeans and lupins. Additionally, legume yield fluctuations are due to poor weed competitiveness, susceptibility to lodging, and biotic and abiotic stresses (Cernay et al. 2015; Notz et al. 2022). The low market competitiveness of legumes is closely linked to earlier versions of the Common Agricultural Policy (CAP) of the European Union, which subsidized cereals but not legumes (Martin 2014; van Loon et al. 2023). This policy is believed to have caused a 50% decline in grain legume production between 1961 and 2020. However, since 2010, changes in the CAP have prompted an increase in legume production areas, with many EU states offering additional subsidies within specific programs to encourage legume cultivation through more diverse crop rotations. For example, in 1961, the common bean was primarily cultivated as a grain legume in Europe for human consumption. Nowadays, the predominant legumes are soybean (*Glycine max* (L.) Merr.), peas (*Pisum sativum* L.), and faba bean (*Vicia faba* L.), followed by chickpea (*Cicer arietinum* L.), lupins (*Lupinus* sp.), lentils (*Lens culinaris* Medik.), vetches (*Vicia* sp.), and common bean (*Phaseolus vulgaris* L.) (Notz et al. 2022; Anwar et al. 2022; FAOSTAT 2024) (Fig. 4.1).

Over the past 50 years, Europe has experienced a substantial increase in protein demand due to population growth (Uhlarik et al. 2024). Table 4.1 illustrates the change in legume cultivation over the past 20 years, according to FAOSTAT data. Soybean (*Glycine max* (L.) Merr.), although not traditionally part of European agricultural production, has earned the title of the "plant of the twentieth century." Previously, it was known and cultivated primarily in China. In the 1930s, soybean cultivation expanded significantly in the USA, which has since become a leader in global production and trade. Since the 1970s, the area dedicated to soybean cultivation has grown considerably. The significance of soybeans lies primarily in their chemical composition; the grain contains approximately 40% protein and about 20% oil, amounting to over 60% of nutrients that can be utilized for various purposes (Đukić et al. 2020). Due to the possibility of using whole grains, especially oil and protein, soybeans are widely used not only in the food industry but also in other branches of industry, and they are of great importance in international trade. Today it is cultivated in all parts of the world. The share of Europe in the total world soybean production is 1–2%, and this plant is important only for certain countries. Until the 1980s, soybeans were grown in Europe, mainly in Romania, Bulgaria, Hungary, and Yugoslavia, i.e., in the territory of today's Serbia. Significant expansion began in 1985, primarily in Italy and France. Significant fluctuations in areas, primarily in EU countries, are the result of reduced subsidies and lower profits for producers, despite record yields. Because of all this, the leading role in soybean production in Europe is taken over by Russia and Ukraine. Soybean production in Europe is thus moving to the east, to the countries of the former USSR, where it has long been cultivated in Asian parts, as well as in the areas of southern Russia and the northern Caucasus (Hrustić et al. 2008). The four largest soybean-producing countries in

Fig. 4.1 Field pea (left) and faba bean (right) fields in Serbia. (Photos: Mamlić Z.)

Table 4.1 Change in legume cultivation in Europe within the last 20 years according to FAOSTAT data

		2012	2022	
	Legume crop	Europe	Europe	Change in the production +/– (%)
Harvested area (ha)	Soybean	3,446,289	6,255,841	+45
	Faba bean (dry + green)	296,474	688,478	+56
	Field pea (dry + green)	2,136,914	2,848,447	+26
	Common bean	251,373	191,068	−24
	Vetches	275,984	105,942	−62
	Lupins	145,066	315,851	+54
	Lentils	106,135	219,725	+52
	Chickpea	50,689	351,880	+86

Europe are the Russian Federation, Ukraine, Italy, and Serbia, followed by France and Romania (FAOSTAT 2024).

As for field pea (*Pisum sativum* L.), many commercially available products currently use pea protein isolate as their primary protein source (Abbas et al. 2023). This growing demand for plant-based products is driving the expansion of pea production and processing within the EU. Efforts to enhance the availability of legume proteins focus on increasing legume cultivation at the primary production stage and advancing food processing techniques across the EU (Etienne et al. 2021). Pea forages are known for their protein content (Uhlarik et al. 2022), although they are slightly lower in protein compared to other temperate forage legumes like alfalfa and clover (Feedipedia). Additionally, peas and vetches are crucial plant species for producing forages, which can be utilized as hay, haylage, or silage (Marković et al. 2023). However, there was previously a decline in pea production due to the effects of climate change. Peas are highly sensitive to climatic conditions during their crop cycle, with yields particularly affected by drought and high temperatures during the grain formation period. To combat these challenges, a strategy was developed to produce winter varieties in regions where this is feasible. Consequently, some countries and regions have shifted to this type of production, utilizing more winter varieties to stabilize and increase yields (Bénézit et al. 2017).

Faba bean (*Vicia faba* L.) ranks as the third most cultivated legume in Europe, covering approximately 21% of the agricultural land dedicated to grain legumes in 2022. The United Kingdom leads as the primary producer, followed by France, Lithuania, Italy, and Germany (Sepngang et al. 2020). The significant increase in faba bean production in Europe, around 56%, can be attributed to the diverse properties of faba bean cultivars and the new greening requirements of the CAP policy implemented in 2015. Faba bean is well-suited to a wide range of environments, from cool climates with short growing seasons ideal for spring cultivars in Northern Europe to more temperate climates in Southern Europe, where winter varieties are predominantly grown (De Notaris et al. 2023). Additionally, the CAP policy continued to subsidize farmers for faba bean production in ecological focus areas (EFAs), which include legume cultivation zones. Germany, for instance, saw a production increase of at least 70% within four years, largely driven by CAP greening measures (Sepngang et al. 2018). Similar trends were observed in the United Kingdom, Lithuania, Latvia, Poland, and Sweden. While faba bean is primarily used for animal feed in Europe, the growing trend towards meatless diets suggests that its role in human nutrition is expected to rise.

Chickpea (*Cicer arietinum* L.) production in Europe has seen a notable increase (Živanov et al. 2021), transitioning from a marginal crop to a thriving export commodity, particularly in countries like Spain, Italy, and Portugal, where it has gained popularity among consumers for its flavor, texture, and health benefits. This rise can be attributed to several factors. Firstly, the climatic conditions in European provinces have encouraged the cultivation of crops that are resilient to drought and require minimal inputs, making chickpeas an attractive choice (Vargas-Blandino and Cárdenas-Travieso 2021). Additionally, European agriculture faces challenges such as a deficit in protein sources for livestock and an increasing risk of drought due to projected changes in agro-climatic conditions (Neugschwandtner et al. 2015). As a drought-resistant legume, chickpea production has been on the rise in response to these challenges, offering a sustainable and resilient alternative within the modern food system (Saget et al. 2020).

White, yellow, and narrow-leafed lupins (*Lupinus* sp.), native European legumes, serve as significant sources of protein alternatives. Their seeds are rich in high-quality protein, containing up to 44%, which offers potential health benefits and supports sustainable cropping systems. Lupins are cultivated across several European countries (Poland, Germany, France, etc.). Despite their highest grain yields in certain European regions, lupin cultivation areas remain modest, and yields vary considerably. However, there has been a slight increase in cultivated areas and production across Europe (Lucas et al. 2015). According to a farmer survey, conventional farmers are motivated primarily by producing domestic protein feed and agronomic benefits, such as crop diversification and enhancing rotations. Financial incentives and lupin's drought tolerance, along with soil fertility benefits, also drive cultivation (Notz and Reckling 2022). In addition to this, lupin (*Lupinus* spp.) is one of the promising species that can offer solution for soil remediation and food security in regions susceptible to climate change (Abraham et al. 2019; Losa et al. 2021).

Lentils (*Lens culinaris* Medik.) are recognized as traditional legumes used for human consumption in Spain, France, and Italy for centuries. From years 2014 to 2018, Europe represented only 3% of lentil production worldwide. However, this trend continued to increase in the following years. Even though the areas under lentils in Europe doubled by around 50% between 2012 (106,135 ha) and 2022 (219,725 ha), they are still insignificant compared to many other legumes. The consumption of lentils in the European Union increased to around 300,000 t per year, mainly in Spain, France, and Italy, followed by Germany and the United Kingdom, which could be attributed to the foreign population that settled in these parts of Europe, still used to a traditional diet based on lentils (Sepngang et al. 2020).

The bitter vetch, hairy vetch, and other vetch species are extensively grown for forage purposes, with common vetch (*Vicia sativa* L.) being a prominent choice for ruminant animal feed due to its high protein content (Kendir et al. 2009; Aasim et al. 2011; Nguyen et al. 2020). Additionally, hairy vetch is widely used as a winter cover crop and in no-till farming, where it can fix atmospheric nitrogen and provide valuable carbon and nitrogen for rotation crops like small grains. In contrast to legumes like peas or chickpeas, whose production has increased, vetch production in Europe has significantly declined primarily due to the presence of antinutritional factors (ANFs), which restrict the utilization of vetch seeds in food and feed formulations, although common vetch seeds in ruminant diets do not adversely affect their health (Ramírez-Parra and De la Rosa 2023). Drought, exacerbated by climate change, also poses a major threat to vetch production, affecting both forage and grain yields and negatively affecting production.

Common bean (*Phaseolus vulgaris* L.) has long been part of the diet and traditional life of many European nations (Savić et al. 2020) and once was the most widely produced legume in Europe. It is well adapted to diverse ecological conditions and production systems, which is confirmed by its wide distribution (Ilić et al. 2023). Common bean distinguishes from the other legumes with the existence of great variability in terms of seed color, shape, and size (Savić et al. 2021). However, despite its compositional richness and versatility, dry bean production and consumption in Europe have been faced with serious decline. This abandonment of common bean production could be attributed to unfavorable weather conditions in recent years, while many previous producers started to rely on imported beans. According to Lioi and Piergiovanni (2013), cultivation of common bean was mainly concentrated in the Mediterranean basin (Iberian Peninsula and Italy) and Balkan countries. According to FAOSTAT (2024), the main producers of common bean in Europe nowadays are Belarus, Ukraine, and Albania. Based on future projections, increased drought events and rise in global temperatures will reduce suitable areas for bean production, including Europe.

Pedo-climatic zones significantly influence the extent of legume cultivation in Europe, as shown in Table 4.2. The primary soybean producers are located in Southern and Eastern Europe, followed by countries in Western Europe. Faba bean is predominantly grown in Western Europe, specifically in France and the United Kingdom, though its cultivation spans the entire continent. According to FAOSTAT (2024), France, Ukraine, and Germany are the largest producers of dry peas. Eastern European countries lead in the production of field peas and lupins. The United Kingdom presents the largest market for fresh peas, with additional opportunities in the Netherlands, Belgium, Germany, France, and Scandinavia (Market informtion). Dry pulses, including peas, are cultivated across nearly all EU states, with Spain dedicating the largest area at 22.5% of arable land, and average seed yields ranging from 1.01 t/ha in Norway to 4.89 t/ha in Ireland (Our World in Data). Additionally, common beans, vetches, lentils, and chickpeas are mainly grown in Eastern and Southern European countries. Overall, France, Germany, and the United Kingdom specialize in the production of soybeans, peas, and faba beans due to their colder continental climates, while Mediterranean (Italy, Spain, Greece) and Balkan

Table 4.2 The extent of legume cultivation in different European regions (FAOSTAT 2024)

Harvested area (ha)	Eastern Europe	Northern Europe	Southern Europe	Western Europe
Soybean	5,233,644	2040	687,911	332,245
Faba bean (dry + green)	53,483	376,271	99,465	159,260
Field pea (dry + green)	2,007,552	317,278	206,508	317,109
Common bean	161,331	*	29,737	*
Vetches	86,185	*	19,757	*
Lupins	251,454	4420	22,000	*
Lentils	219,665	*	60	*
Chickpea	345,303	*	6578	*

countries cultivate a more diverse array of leguminous crops (Tassoni et al. 2020). Watson et al. (2017) predict that climate change will create opportunities to cultivate less commonly grown crops in specific European regions, such as lentils in southern Germany, chickpeas in Central Europe, narrow-leafed lupins in the Nordic region, and soybeans in Central Europe.

Still, Europe is not self-sufficient in grain legume production, and most of its consumption (e.g., around 70% of high-protein livestock feed) is imported from the Americas and other parts of the world (van Loon et al. 2023; De Notaris et al. 2023). European consumers of protein, as noted by Henn et al. (2022), are increasingly open to replacing a considerable amount of meat with grain legumes. This shift appears to be driven by the European Commission's proactive promotion through the Farm to Fork strategy under the European Green Deal. This strategic focus aims to elevate plant-derived proteins in human diets, thereby driving the expansion of leguminous crop production. Responding to the escalating demand for plant-based proteins, European agricultural policies, particularly the Common Agricultural Policy (CAP), have been pivotal in encouraging farmers to diversify their crop rotations, with a specific emphasis on scaling up the cultivation of forage legumes. Collaborative endeavors spanning the entire value chain, involving stakeholders such as farmers, feed mills, processors, and policymakers, have propelled Europe's progress in bolstering the competitiveness of domestically grown fodder and grain legumes, consequently opening up new market avenues and export prospects (Hamann 2020). Noteworthy investments in research and innovation have significantly bolstered the resilience and productivity of legume crops, with a pronounced emphasis on sustainable practices that foster soil health, biodiversity, and climate resilience. Furthermore, education and awareness initiatives have played a critical role in fostering the adoption of legume-based cropping systems, highlighting their economic viability, environmental advantages, and contribution to a more robust agri-food ecosystem across Europe. The continued backing of organic legume farming, combined with collaborative efforts and knowledge-sharing initiatives, is expected to accelerate the move towards a more sustainable and diverse agricultural system in Europe. This transition could be facilitated through the intensification of current production methods or through strategic area expansion, with an eye toward maintaining harmonious land use and avoiding encroachment into natural ecosystems. Recent trends show a marked expansion in pulse cultivation areas within the EU, particularly for crops such as soybean, faba bean, lupins, and lentils, highlighting the potential for greater diversity in cropping systems (Francis and Clegg 2020). As for future projections, according to Hamann (2020), the production of pulses in the European Union is anticipated to reach 6.3 million tons in 2030, while cultivation area is expected to grow to 2.3–2.4 million ha (Hamann 2020).

4.3 Drought Stress and Its Occurrence in Europe Affecting Legume Production

The production of legumes, along with other crops, is increasingly challenged by shifting climatic conditions. Long-term global climate change is anticipated to introduce various abiotic stresses that could hinder the United Nations' goal of securing adequate food supplies for a growing population by 2050 (Redden 2013). Factors such as rising temperatures and unpredictable precipitation patterns are disrupting crop yields. The frequency of extreme weather events, including droughts, has risen, potentially affecting about a third of agricultural soils and impeding plant growth and development (Ye et al. 2018; Khatun et al. 2021). Agriculture is expected to face increased occurrences of prolonged drought, severe heat, and irregular rainfall (Kellogg and Schware 2019). Additionally, shifts in regional temperature and precipitation patterns may adversely affect local crop production, impacting food availability. Priyanka et al. (2019) define drought as an extended period of unusually dry weather in which certain regions receive significantly less rainfall than typical. While drought is a frequent and recurring event in many areas, it lacks a predictable pattern. Its manifestation can differ widely across global regions (Gliessman 2000). Drought should be understood not merely as a physical occurrence but also as a natural phenomenon that affects societies by disrupting the expected rainfall patterns, which are subject to natural climate variability. Human activities often worsen the impact of drought (Omomowo and Babalola 2021). Recent years have seen drought affecting both developed and developing nations, with notable impacts on their economies and environments, heightening societal vulnerability to this natural disaster (Luan et al. 2021). Agriculture, being an "outdoor factory," is particularly susceptible to drought, with plant production (including field crops, vegetables, fruits, and grapevines), livestock farming, and fish farming all facing significant risks. Disruptions in the food supply chain due to drought can lead to both economic and social instability (Stričević et al. 2018). Drought introduces complex stress that impacts various morpho-physiological traits throughout all stages of growth, leading to substantial economic losses (Mamlić et al. 2021). Climate change is expected to exacerbate drought stress globally in the latter half of the twenty-first century, with decreased precipitation and/or increased evaporation (Dai 2012). Simulations predict that, without adaptation measures, drought-induced yield losses may rise by approximately 16% by the end of the twenty-first century (Leng and Hall 2019).

Severe drought events have been recorded across Europe since the early twenty-first century, with clear evidence that climate change is altering the patterns of drought in the region. Europe, which spans temperate and cold climate zones, experiences varying drought causes, including rainfall deficits, reduced snow availability, and high evapotranspiration rates. Markonis et al. (2021) categorized droughts in Europe based on (i) the affected region, (ii) the underlying causes, and (iii) the timing of their peak into three types: (1) summer droughts, (2) autumn droughts, and (3) winter droughts. This classification extends beyond traditional drought typologies such as meteorological, hydrological, and agricultural droughts. For instance, "Compound warm-season droughts," prevalent in Central and Western Europe, begin in spring or early summer, driven by insufficient rainfall, high temperatures, and increased evapotranspiration, often lasting around four months and associated with heat waves (AghaKouchak et al. 2014). In contrast, "Wet-to-dry season droughts" occur in the Mediterranean region due to precipitation deficits during the colder months (late winter to early spring) and typically last about seven months (Van Loon et al. 2014). "Autumn droughts" are characterized by classic precipitation deficits and are the most common in the Mediterranean area. "Rain-to-snow season droughts," or winter droughts, result from rainfall shortages in summer and autumn, extending into reduced snowfall during winter, affecting all regions. Brunner et al. (2022) found that the distribution of dominant drought types in Europe varies geographically and temporally. Warm-season droughts have become more prevalent, especially in Central Europe over the past 15 years, with similar trends observed in the Atlantic region. In contrast, snow-influenced droughts are more common at higher altitudes, while "rain-to-snow season droughts" are declining, possibly transitioning into warm-season droughts due to earlier snowmelt (Barnett et al. 2005). Soussa et al. (2011) and Philandras et al. (2011) documented an increase in "wet-to-dry season droughts" in the Mediterranean, linked to declines in winter precipitation and an increase in compound warm-season events in Central Europe. Brunner et al. (2022) identified "rain-to-snow" and "wet-to-dry season droughts" as the most severe in terms of duration and water scarcity, with agricultural drought frequently occurring during compound warm-season droughts. Van Loon et al. (2016) projected that compound warm-season droughts will likely further escalate agricultural water demand and shortages, impacting vegetation and crop productivity in Europe. However, this trend has not been observed in Northern Europe.

Legumes are primarily cultivated in rain-fed agricultural systems globally, including Europe, and are highly susceptible to drought and heat stress, which significantly limit their production (Khatun et al. 2021; Ali et al. 2022). Drought negatively impacts legumes by reducing their growth, yield, and quality. Water stress disrupts key physiological processes such as photosynthesis and nitrogen fixation, leading to stunted growth and decreased biomass. Additionally, drought conditions can hinder root development and diminish the availability of essential nutrients, further aggravating yield losses. Extended drought periods can result in reduced seed production and quality, ultimately affecting the overall productivity and economic feasibility of legume crops. The impact of drought on yield varies depending on the growth stage at which it occurs. For instance, Daryanto et al. (2015) found that drought during the flowering stage led to a 38% reduction in yield, compared to a 27% reduction when drought occurred during the pod filling and maturity stages. Their research also revealed varying yield losses among different legume species under a 60–65% water deficit: 21% in lentils and field peas, 28% in soybeans, 40% in faba beans and chickpeas, 44% in cowpeas, and 61% in common beans. Nadeem et al. (2019) further emphasized that the extent of grain yield reduction in legumes is influenced by the frequency and severity of the drought event, the crop's developmental stage, plant biomass, and genetic variability.

Soybeans are highly sensitive to water deficits, particularly during certain developmental stages (Fig. 4.2). Yield losses in soybeans can occur due to soil water shortages at any stage of growth, with the extent of reduction influenced by the timing and severity of the stress (Miladinov et al. 2020a). Opinions differ on which stage is most vulnerable to water shortages. Some studies suggest that soybeans are most affected during flowering and pod formation (Tetsuji et al. 2004), while others argue that the seed-filling phase is the most critical (Rodrigues et al. 2012). Specifically, soybeans are known to be particularly sensitive to water deficits during flowering (R1–R2), pod formation (R3–R4), and seed development (R5–R6) (Sionit et al. 1987). According to Dybing et al. (1986), although soybean plants produce many flowers, a significant proportion abort before maturing into pods due to adverse agro-ecological conditions, primarily water shortages, which greatly impact yield (Vidić et al. 2009). Similarly, field peas exposed to drought also experience reduced yields due to flower and pod abortion. Drought stress can also affect the chemical composition of peas, including sugar and amino acid content (Sorensen et al. 2003). Furthermore, water scarcity impacts nitrogen uptake in pea plants by negatively affecting nodule performance, development, and initiation, leading to reduced nitrogen accumulation and overall dry weight of the plant (Bagheri et al. 2023). The impact of water deficit on seed germination yields mixed results in scientific studies. Some researchers indicate that water scarcity during seed development adversely affects germination, while others find no such effect. Drummond et al. (1983) observed that seed germination, as measured by the standard germination test, was lower in seeds developed under drought conditions compared to those grown with adequate water supply. Similarly, Simiciklas et al. (1989) found that

Fig. 4.2 Drought stress effect on pea (left and right) and soybean (middle). (Photos: Mamlić Z.[left and right]; Đukić V.[middle])

soybean seeds exposed to water deficit during the R5 and R6 stages exhibited reduced germination rates compared to seeds from plants with optimal water conditions, although no effect was observed during the R2 and R4 stages. Dornbos et al. (1989) reported a 5% reduction in germination using the standard germination test and a 19% reduction using the cold test for seeds affected by water deficit during the R5 stage. Vieira et al. (1992) suggested that the observed decrease in germination might be due to a combination of water deficit and other factors, rather than water shortage alone. Discrepancies in findings are often attributed to variations in testing methods, drought conditions, climatic factors, and seed moisture content.

Faba bean is highly sensitive to drought stress, despite the presence of genotypic variability that allows some varieties to produce acceptable yields (Amede and Schubert 2023; Parvin et al. 2019). Although faba beans originate from semi-arid regions, they often struggle to adapt to drought conditions due to their shallow root system, which limits their ability to access deep soil moisture (Muktadir et al. 2020). Like other legumes, drought stress in faba beans leads to the production of reactive oxygen species (ROS) in plant cells. These ROS cause damage to various cellular structures, inhibit protein synthesis, and reduce enzymatic activity. As a result, drought stress reduces leaf number and size, decreases pod count, and lowers seed weight, ultimately leading to diminished seed productivity (Siddiqui et al. 2015; Deskoy et al. 2020).

Drought and heat stress have been shown to reduce chickpea yields by damaging cell membranes and impairing photosynthesis (Asati et al. 2022). In many chickpea-growing areas, heat stress is particularly problematic during the reproductive phase. Flowering, being highly sensitive to heat, can be severely impacted, resulting in reduced flowering rates or even flower abortion, which drastically reduces yield. Temperatures exceeding 35 °C can lead to yield losses of up to 39% due to the adverse effects on anthers, pollen viability, and stigma function, ultimately decreasing fertility and pod setting. In Mediterranean and semi-arid regions, chickpea plants typically face drought and high temperatures during the pod setting and seed-filling stages.

The common bean is a legume ideally suited for cultivation in regions with moderate rainfall or irrigation, but it struggles under drought and heat conditions (Fig. 4.3). It is best adapted to temperate climates. When faced with water scarcity and high temperatures (over 30 °C during the day and 20 °C at night), common bean plants often suffer from flower abortion, the formation of parthenocarpic pods, fewer seeds per pod, and reduced seed size. These issues culminate in a notable decrease in seed yield (Rainey and Griffiths 2005). Based on future projections, increased drought events and rise in global temperatures will reduce suitable areas for bean production. Most of these drought events affect common bean during flowering and grain-filling stages when reported yields could be reduced up to 99% (Manjeru et al. 2007; Ambachew et al. 2015).

In vetch species, water constraint results in reductions in growth, yield, and water use efficiency, with common vetch showing high susceptibility to drought stress. Conversely, narbon vetch displays greater resistance due to its larger leaf area and higher relative growth rate (Haffani et al. 2014). Adapting to severe drought is vital

Fig. 4.3 Effects of drought and heat stress on lentils (left) and common bean (right). (Photos: Mamlić Z.; Ilić A.)

for white lupin in Mediterranean climates, where such stress often aligns with key reproductive phases (Pecetti et al. 2023). Wild lupin species have evolved to cope with drought by developing seeds with coats that resist water penetration. This characteristic prevents a large number of seeds from germinating in a single year, ensuring the species' continuity even when seed production is compromised by drought (Walker et al. 2011).

In addition to yield, the use of high-quality seeds is a crucial factor for enhancing agricultural production across various farming systems (Sabry 2018). Annually, over two-thirds of the world's arable land is planted with seeds, which fulfill approximately 90% of global food and agro-processing needs (Mirić and Brkić 2002). The disparity between successful seed production in some regions and failures in others underscores the significant impact of environmental factors on seed development and maturation (Vujaković et al. 2011). Environmental conditions such as temperature fluctuations, water stress, excessive rainfall, nutrient deficiencies, disease outbreaks, and insect pressures all play a role in determining seed quality during development and maturation (Miladinov et al. 2020a).

4.4 Effects of Drought Stress on Legumes and Their Mechanisms of Drought Tolerance

Drought stress generally reduces osmotic pressure and disrupts water potential in plant cells, leading to oxidative stress and cellular damage. This results in reduced photosynthesis, carbon fixation, and primary metabolism (Mladenov et al. 2023; Reddy et al. 2004). Additionally, drought stress can impair membrane integrity and enzyme activity. These disruptions can inhibit plant growth, reduce fitness, lower yield quality and quantity, and decrease the overall survival potential of the crop

(Xu et al. 2015; Zhang et al. 2019). The response of plants to water shortage depends on the timing, intensity, and duration of the drought, as well as the developmental stage of the exposed crop. According to Papathanasiou et al. (2022), legumes can experience drought stress throughout the season or at specific stages of their life cycle. Farooq et al. (2017) noted that drought can impair the productivity of grain legumes at all growth stages. In most legume species, the flowering and reproductive stages are the most critical, with drought during these stages causing significant yield losses (Fang et al. 2010). Drought conditions can lead to flower abortion, smaller flower development, low nectar quality and quantity, pollen sterility, and shortened flowering and pod-filling stages (Khatun et al. 2021).

During oxidative stress caused by drought, reactive oxygen species (ROS) are produced. At low concentrations, ROS are harmless to cells (Wang et al. 2016). However, under drought conditions, mitochondria or chloroplasts generate large amounts of ROS, which can transform into highly toxic hydroxyl radicals. These radicals destroy cell membranes, proteins, and nucleic acids and damage photosynthetic pigments if not promptly excreted (Simova-Stoilova et al. 2009). Drought stress also increases membrane lipid peroxidation, with malondialdehyde (MDA), a by-product of this process, serving as a key indicator of cell membrane damage (Jiang et al. 2001). MDA content rises under high temperatures, drought, low temperatures, and salt stress (Guo et al. 2018; Rui et al. 2018). Water scarcity in legumes impairs cell division, expansion, and differentiation, which negatively impacts plant growth and development. This impairment leads to a reduction in leaf area, plant height, and overall crop growth due to decreased turgor pressure (Farooq et al. 2008). Drought conditions exacerbate leaf senescence, especially during the flowering and pod-filling stages, primarily due to chlorophyll degradation (Sarkar et al. 2021). Additionally, drought stress reduces the number of pods per plant and decreases grain weight, ultimately leading to lower grain yields. Water shortages also impede root growth and limit the uptake of less mobile nutrients (Garg 2003). The lack of water decreases nutrient availability, impairs root absorption and translocation, and disrupts nutrient metabolism (Farooq et al. 2009a, 2009b). Consequently, plants experience stunted growth due to the depletion of essential nutrients and inorganic matter.

Water-use efficiency (WUE), which measures the ratio of dry matter production to water consumption by plants, declines under conditions of water scarcity (Anderson et al. 2016). According to Awasthi et al. (2014), drought adversely impacts the plant's photosynthetic system through its various components, including the regulation of stomatal CO_2 supply, electron transport, and the carbon-reduction cycle. This results in a decreased overall concentration of chlorophyll, which reduces the plant's ability to capture light and, consequently, lowers the rate of photosynthesis and biomass accumulation (Fathi and Tari 2016). Additionally, plants respond to drought by closing their stomata, which reduces transpiration and stomatal conductance. The process of atmospheric nitrogen fixation is particularly sensitive to water deficits, even more so than transpiration and photosynthesis (Sinclair et al. 2007). During drought conditions, nitrogen fixation is hindered by a limited oxygen supply to nodule bacteria (King and Purcell 2001) and a reduced carbon supply for these bacteria (Arrese-Igor et al. 2011). Furthermore, water scarcity leads to increased accumulation of ureides in nodules (Collier and Tegeder 2012), which can slow down nitrogen transport and impact biological nitrogen fixation. However, if water shortages are brief, nodule nitrogen fixation may resume once conditions improve.

Drought also impacts the chemical composition of grains in legumes (Miladinov et al. 2019). Farooq et al. (2018) found that the protein content in the grains of food legumes often remains stable or may even increase under drought conditions. However, other studies have reported a decrease in protein content during water scarcity, with varying responses among different protein types (Khalil and Ismail 2010). For instance, in common beans, protein accumulation in seeds is diminished due to inhibited nitrogen fixation and nutrient partitioning (Singh 2007). Alghamdi (2009) noted a reduction in albumins, globulins, and prolamins, while glutelins increased in faba beans subjected to drought. Conversely, in soybeans, the protein content generally shows a negative correlation with seed oil levels (Miladinov 2020). In soybeans, water deficits lead to a significant rise in protein content, largely due to increased nitrogen accumulation in seeds and decreased carbohydrate accumulation (Ozturk and Aydin 2004). During the seed-filling period, nitrogen uptake is often insufficient due to high demands, leading to its remobilization from vegetative tissues to the seeds (Masclaux et al. 2001). Additionally, inadequate water supply not only negatively impacts biological nitrogen fixation, the primary nitrogen source for soybeans, but also reduces the uptake of nitrogen from the soil. When water is scarce, plants absorb fewer nitrates from the soil solution, mobilizing most of the nitrogen from other plant parts into seeds and young leaves. This water deficit also results in reduced oil content due to decreased photosynthetic activity and assimilation (Stone et al. 2001).

Plants have developed three primary strategies to manage drought: (i) drought avoidance, (ii) drought escape, and (iii) drought tolerance. Drought avoidance involves maintaining a high water status under limited water conditions by reducing stomatal transpiration losses and enhancing water uptake through the roots. Drought escape pertains to a plant's ability to complete its life cycle before drought conditions arise. This strategy includes mechanisms such as rapid growth, shortening the life cycle, and seasonal development before drought onset. Tekle and Alemu (2016) highlight early flowering and maturation as effective escape mechanisms, though this often results in reduced yield. Drought tolerance refers to the ability of plants to sustain metabolic functions despite low tissue water potential (Muktadir et al. 2020). This can involve adjustments in the photosynthetic machinery, such as reducing leaf area. Additionally, legumes can develop trichomes on their leaves to lower leaf temperature through increased light reflection and to provide an extra layer of resistance against water loss via transpiration (Tiwari et al. 2020). Drought-tolerant plants may also adapt by altering root size, density, length, and growth rate (Tzortzakis et al. 2020). Other drought tolerance mechanisms identified by López-Galiano et al. (2019) include osmotic adjustment, antioxidant defense mechanisms, solute accumulation, changes in metabolic and biochemical dynamics related to stomatal closure, and an increase in root-to-shoot ratio.

Under drought conditions, plants that are tolerant to water stress adjust by decreasing the number and size of their leaves to reduce water usage and minimize yield losses. Concurrently, they enhance their root system's development, including its structure, density, and size (Khatun et al. 2021; see Fig. 4.4). For example, the distribution of roots—assessed by horizontal and vertical root length density or the amount of dry matter in different soil depths (Benjamin and Nielsen 2006)—is modified in drought-tolerant soybean varieties during periods of drought stress (Tzenova et al. 2008). Observations indicate that during seasonal drought, these plants tend to have lower root density in the dry upper soil layers but increased root density in deeper soil layers where moisture is more available. Additionally, research on drought-tolerant soybean cultivars has revealed a positive relationship between drought tolerance and factors such as dry root weight relative to plant weight, total root length relative to plant weight, and root volume relative to plant weight (Liu et al. 2005).

To mitigate transpiration, legumes reduce water absorption and evaporation by closing their stomata. This closure triggers the production of abscisic acid, which extends the initial blocking effect and inhibits both leaf development and stomatal reopening. Additionally, during drought stress, maintaining gas exchange can lead to increased stomatal conductance, indicating lower stomatal sensitivity to drought. Condon et al. (2007) suggested that stomatal conductance might be a useful criterion for identifying drought-tolerant plants. For instance, in species such as white lupine, chickpeas, faba beans, field peas, and grass pea, decreased stomatal

Fig. 4.4 Un-adapted soybean cultivar compared to drought-tolerant. (Photo: Đukić V.)

conductance has been observed in the early stages of drought stress. In drought-tolerant soybean cultivars, a larger leaf area was noted compared to less tolerant varieties, which was linked to a greater reduction in stomatal conductance and a smaller decrease in photosynthesis rate in the tolerant cultivars (Stolf-Moreira et al. 2010). This indicates that drought-tolerant cultivars can effectively reduce water loss while minimizing the impact on photosynthesis.

A key trait associated with drought tolerance is relative water content (RWC), which measures the amount of water in plant tissues under normal conditions compared to hydrated conditions (Lawlor and Cornic 2002). In drought conditions, RWC is crucial for maintaining water conservation through mechanisms such as reduced stomatal opening and decreased leaf area, as well as enhancing water uptake through root plasticity. Drought-tolerant genotypes of faba beans, common beans, and chickpeas have been effectively identified based on their RWC status (Siddiqui et al. 2015; Franca et al. 2000; Talebi et al. 2013). Another important mechanism plants use to cope with drought is osmotic adjustment. This involves the accumulation of cellular metabolites, known as osmoprotectants, which help maintain turgor pressure under lower water potential conditions. While osmoprotectants primarily function in osmoregulation, they also play roles in scavenging reactive oxygen species, preserving enzymatic activity, and maintaining membrane structure (Fathi and Tari 2016). Various molecules can act as cellular osmoprotectants, including organic acids, inorganic ions, polyols, polyamines, and carbohydrates (Muktadir et al. 2020).

Under drought stress, plants make biochemical adjustments to manage the increased production of reactive oxygen species (ROS), which can damage nucleic acids, proteins, and lipids, leading to cell death. Antioxidants are crucial for defending plants against ROS damage (Rahman et al. 2021). During drought, the levels of antioxidants such as superoxide dismutase (SOD) and catalase (CAT) typically rise, helping to lower the concentration of malondialdehyde (MDA) and support plant health (Wu et al. 2020). This boost in antioxidant activity can effectively reduce damage to membranes and lipid peroxidation. Additionally, drought conditions result in increased levels of soluble sugars and proline, which contribute to compensatory growth (Imran et al. 2021). The role of antioxidants is essential in mitigating drought stress effects by detoxifying ROS and preventing cellular oxidation, thereby aiding plant growth and productivity. Moreover, plant growth regulators, known as phytohormones, are vital for helping plants adapt to environmental changes by regulating growth, development, nutrient distribution, and resource allocation (Fahad et al. 2015).

4.5 Strategies to Mitigate Drought Stress in Legumes

There is a growing interest and ongoing projects focused on developing drought-tolerant legume varieties, including pea, vetch, and chickpea, for both grain and forage production, aligning with the imperative to ensure food security and sustainable agriculture amidst ongoing climate change challenges. Legume breeders worldwide are actively addressing drought stress challenges through innovative strategies. These include incorporating genes from wild relatives to impart traits like waxy leaves or early flowering and optimizing plant water use (Bagheri et al. 2023). Furthermore, the identification of key metabolic markers responsive to drought stress is essential for selecting resilient cultivars in breeding programs (Lahuta et al. 2022). Given the limited data on specific genes influencing legume grain yield under severe conditions, there is a pressing need for diverse approaches. These include pre-breeding to introduce genetic traits, elucidating gene functions, assessing combining abilities for drought-related traits, and employing mutation breeding to domesticate wild legume species. Conventional breeding methods have proven to be reliable for improving drought tolerance in crop species and are expected to remain the primary approach for crop enhancement. It is crucial to identify essential traits that ensure yield stability and potential under drought conditions. Utilizing modern breeding techniques and biotechnology has led to the development of some drought-tolerant legumes (Khatun et al. 2021). Key root traits in grain legumes, such as length, density, depth, and fibrous root systems, are vital for avoiding drought stress and are commonly used to assess drought resistance (Ullah and Farooq 2022). Additionally, very early-maturing cultivars have been created to avoid heat and water stress during critical growth periods, such as flowering and bean filling. Breeding for early maturation is particularly effective in increasing yield stability in regions like the Mediterranean, where summer droughts are common. While early flowering can help in avoiding drought, it is less effective in mitigating drought stress once it occurs (Leonardis et al. 2012). Recent cultivars have been developed with enhanced resistance to drought and cold, enabling earlier sowing and optimizing the use of early spring soil moisture reserves (Petrychenko 2009). Faba bean, which is more sensitive to water deficit and less to water surplus compared to pea (Stoddard et al. 2006), showed a strong correlation between yield and growing-season precipitation in recent trials, highlighting the significant impact of drought on its yield (Skovbjerg et al. 2020). The narrow-leafed lupin, which is highly sensitive to soil water deficits, delays dehydration by extending its roots deeper to access untapped water sources or by optimizing the use of existing water through improved hydraulic conductivity. Other strategies include adjusting the threshold for stomatal closure or directly manipulating leaf area, considering the variability within and between species (Palta et al. 2012).

In recent years, several strategies have been developed to mitigate water-deficit stress beyond improving cultivars. Agronomic approaches, such as applying plant growth regulators and plant-growth-promoting rhizobacteria, have proven valuable in promoting crop stand establishment under various stress factors. Seed enhancement techniques like priming and coating can improve stand establishment, seedling growth, and grain yield in legumes facing drought stress (Ullah and Farooq 2022). Additionally, adjusting sowing times, modifying plant density, and implementing efficient farm management practices are effective strategies to address drought challenges. Irrigation, when feasible, can significantly increase yields. Recent experiments demonstrated that irrigation could boost soybean yield by 41% (Karges et al. 2022), and a literature review reported faba bean yield increases of

33–53%, depending on the method used to estimate water needs (Belchew et al. 2023). With future climate change, legume production could greatly benefit from irrigation, which might become essential in some parts of Europe. Nendel et al. (2023) projected that the area at risk of drought for soybeans from 2040 to 2069 will be nearly 30% larger than from 1981 to 2010. Additionally, irrigation could improve and stabilize pea yields in more continental areas of Europe. Another promising technology for mitigating drought stress in legume crops is the use of plant growth-promoting rhizobacteria (PGPRs), which help plants thrive under drought conditions by employing various mechanisms. These comprehensive strategies and technologies are vital for enhancing drought tolerance in legume crops and ensuring sustainable agricultural practices.

Germination and initial growth are crucial for determining crop composition and uniformity, which in turn affects the established yield. Research on leguminous seeds suggests that priming seeds in specific solutions or water is the most effective method for mitigating the negative effects of drought during the germination phase (Harris et al. 2001). Miladinov et al. (2020b) discovered that priming soybean seeds with potassium nitrate, ascorbic acid, and potassium chloride solutions improved seed quality regardless of the soybean variety or water potential. Additionally, Hosseinzadeh-Mahootchi and Ghassemi-Golezani (2012) found that hydro-priming chickpea seeds increased the leaf chlorophyll content index and enhanced stomatal conductance of the plants. Saglam et al. (2010) found that hydro-priming increased the seed germination rate and seedling emergence in lentil plants. Seed priming with residual silicon-glass microparticles alleviates water stress in cowpea. Priming with extracts from carrot root, garlic cloves, or ascorbic acid reduced drought stress in faba bean and decreased carotenoids, activities of peroxidase and catalase, ascorbic acid, glucose, MDA, and soluble protein contents (Kasim et al. 2017). Seed priming with PEG improved agronomic and physiological traits in mung bean, with primed seeds showing remarkable drought resistance by enhancing antioxidant enzyme activity, such as SOD and POD (Uddin et al. 2021). Zn-priming mitigates the effects of drought stress in faba bean during the early vegetative stage by increasing total biomass (99.8%), leaf surface amylase activity (85%), soluble sugar content (54.7%), total chlorophyll content (48.7%), Zn concentration in leaves (79.8%), and decreasing MDA content (42.7%) and antioxidant activities (35.2%) under water-deficient conditions (Farooq et al., 2021). Seed priming with ascorbic acid (1 mM) or PEG-6000 (7.8 mM) significantly reduced the negative effects of drought stress on seedling growth in alfalfa by lowering MDA and H_2O_2 accumulation, and increasing total soluble proteins, soluble sugars, proline, total phenolics, and H_2O_2-scavenging enzyme activities.

4.6 Conclusion

In conclusion, addressing drought stress in legume crops requires a multifaceted approach combining advanced breeding techniques, agronomic practices, and innovative technologies. The integration of these strategies holds the potential to foster the development of resilient cultivars capable of upholding agricultural productivity and ensuring food security amidst the challenges posed by climate change. The successful implementation of these methods is of utmost importance in mitigating the adverse effects of drought on legume crops and achieving sustainable agriculture.

Acknowledgments This chapter has received funding from the Ministry of Education, Science Technological Development and Innovation of the Republic of Serbia, grant number: 451-03-66/2024-03/200032.

References

Aasim M, Şahin Demirbağ N, Khawar KM, Kendir H, Özcan S (2011) Direct axillary shoot regeneration from the mature seed explant of the hairy vetch (Vicia villosa Roth). Biol Sci, Belgrade 63(3):757–762

Abbas A, Ali A, Hussain A, Ali A, Alrefaei AF, Naqvi SAH et al (2023) Assessment of genetic variability and evolutionary relationships of rhizoctonia solani inherent in legume crops. Plan Theory 12(13):2515

Abraham EM, Ganopoulos I, Madesis P, Mavromatis A, Mylona P, Nianiou-Obeidat I, Parissi Z, Polidoros A, Tani E, Vlachostergios D (2019) The use of lupin as a source of protein in animal feeding: Genomic tools and breeding approaches. Int J Mol Sci 20(4):851. https://doi.org/10.3390/ijms20040851

AghaKouchak A, Cheng L, Mazdiyasni O, Farahmand A (2014) Global warming and changes in risk of concurrent climate extremes: insights from 2014 California drought. Geophys Res Lett 41:8847–8852. https://doi.org/10.1002/2014GL062308

Alghamdi SS (2009) Chemical composition of faba bean (Vicia faba L.) genotypes under various water regimes. Pak J Nutr 8:477–482. https://doi.org/10.3923/pjn.2009.477.482

Ali et al (2022) Recent advancement in OMICS approaches to enhance abiotic stress tolerance in legumes Frontiers in Plant Science https://doi.org/1310.3389/fpls.2022.952759

Ambachew D, Mekbib F, Asfaw A, Beebe SE, Blair MW (2015) Trait associations in common bean genotypes grown under drought stress and field infestation by BSM Bean fly. Crop J 3:305–316. https://doi.org/10.1016/j.cj.2015.01.006

Amede T, Schubert S (2023) Mechanisms of drought resistance in grain legumes I: osmotic adjustment. Ethiop J Sci 26:37–46. https://doi.org/10.4314/sinet.v26i1.18198

Anderson W, Johansen C, Siddique KH (2016) Addressing the yield gap in rainfed crops: a review. Agron Sustain Dev 36:18. https://doi.org/10.1007/s13593-015-0341-y

Anwar R, Ali S, Zeshan MA, Ghani MU, Ali A, Iftikhar Y, Bakhsh A (2022) Effect of Fusarium wilt disease on varied Chickpea germplasm and its management. J Pure Appl Agric 7(1):1–9

Arrese-Igor C, Gonzalez EM, Marino D, Ladrera R, Larrainzar E, Gil-Quintana E (2011) Physiological responses of legume nodules to drought. Plant Stress 5:24–31

Asati R, Tripathi MK, Tiwari S, Yadav RK, Tripathi N (2022, 1846) Molecular breeding and drought tolerance in chickpea. Life 12(11)

Awasthi R, Kaushal N, Vadez V, Turner NC, Berger J, Siddique KN, Nayyar H (2014) Individual and combined effects of transient drought and heat stress on carbon assimilation and seed filling in chickpea. Funct Plant Biol 41:1148–1167. https://doi.org/10.1071/FP13340

Bagheri M, Santos CS, Rubiales D, Vasconcelos MW (2023) Challenges in pea breeding for tolerance to drought: status and prospects. Ann Appl Biol 183(2):108–120. https://doi.org/10.1111/aab.12840

Barnett TP, Adam JC, Lettenmaier DP (2005) Potential impacts of a warming climate on water availability in snow-dominated regions. Nature 438:303–309. https://doi.org/10.1038/nature04141

Belchew KY, Soderholm-Emas A, Topp CFE, Watson CA, Stoddard FI (2023) Management factors affecting faba bean yield. Legume. Perspectives 24

Bénézit M, Biarnès V, Jeuffroy MH (2017) Impact of climate and diseases on pea yields: what perspectives with climate change? Oléagineux Corps Gras Lipides 24(1):np

Benjamin JG, Nielsen DC (2006) Water deficit effects on root distribution of soybean, field pea and chickpea. Field Crop Res 97(2–3):248–253. https://doi.org/10.1016/j.fcr.2005.10.005

Bokan N, Dugalić G, Tomić D, Vasiljević S, Karagić Đ, Milić D, Milošević B, Katanski S (2016) Značaj leguminoza za organsku poljoprivredu. XXI savetovanje o biotehnologiji sa međunarodnim učešćem. Zbornik radova. Čačak, 11. – 12. Mart, 123–129

Brunner M, van Loon AF, Stahl K (2022) Moderate and severe hydrological droughts in Europe differ in their hydrometeorological drivers. Water Resour Res 58:e2022WR032871. https://doi.org/10.1029/2022WR032871

Cernay C, Ben-Ari T, Pelzer E, Meynard J-M, Makowski D (2015) Estimating variability in grain legume yields across Europe and the Americas. Sci Rep 5:11171. https://doi.org/10.1038/srep11171

Collier R, Tegeder M (2012) Soybean ureide transporters play a critical role in nodule development, function and nitrogen export. Plant J 72:355–367. https://doi.org/10.1111/j.1365-313X.2012.05086.x

Condon AG, Reynolds MP, Rebetzke GJ, Ginkel MVAN, Richards RA, Farquhar GD (2007) Using stomatal aperture related traits to select for high yield potential in bread wheat. In: Buck HT, Nisi JE, Salomón N (eds) Wheat production in stressed environments, Developments in plant breeding, vol 12. Sprigner, Dordrecht

Dai A (2012) Increasing drought under global warming in observations and models. Nat Clim Change 3:52–58. https://doi.org/10.1038/nclimate1633

Daryanto S, Wang L, Jacinthe PA (2015) Global synthesis of drought effects on food legume production. PLoS One 10(6):e0127401. https://doi.org/10.1371/journal.pone.0127401

De Notaris C, Enggrob EE, Olsen JE, Sorensen P, Rasmussen J (2023) Faba bean productivity, yield stability and N2-fixation in long-term organic and conventional crop rotations. Field Crop Res 295:108894. https://doi.org/10.1016/j.fcr.2023.108894

Deskoy ESM, Mansour E, Yasmin MA, El Sobky ESE, Rady MM (2020) Improvement of drought tolerance in five different cultivars of Vicia faba with foliar application of ascorbic acid or silicon. Span J Agric Res 18:e0802. https://doi.org/10.5424/sjar/2020182-16122

Dornbos DL, Mullen RE, Shibles RM (1989) Drought stress effects during seed fill on soybean seed germination and vigor. Crop Sci 29:476–480. https://doi.org/10.2135/cropsci1989.0011183X002900020047x

Dozet G, Cvıjanovıc G, Djukıc V, Cvıjanovıc D, Kostadinovic LJ (2014) Effect of microbial fertilizer on soybean yield in organic and conventional production. Türk Tarım Ve Doğa Bilimleri Dergisi 1(Özel Sayı-1):1333–1339

Drummond EA, Rabb JL, Melville DR (1983) Effect of irrigation on soybean seed quality. La Agriculturae 26:9–16

Đukić V, Miladinov Z, Dozet G, Cvijanović G, Miladinović J, Ranđelović P, Kandelinskaja OL (2020) The impact of a pulsed electromagnetic field on the seed protein content of soybean. J Agric Sci (Belgrade) 65(4):311–320

Dybing CD, Ghiasi H, Paech C (1986) Biochemical characterization of soybean ovary growth from anthesis to abscission of aborting ovaries. Plant Physiol 81:1069–1074

Etienne P, Bruno K, Wolfgang S, Frédéric M, Marcus M (2021) Pea and rapeseed acreage and land use for plant-based meat alternatives in the EU. OCL 28:54. https://doi.org/10.1051/ocl/2021037

EUROSTAT (2022) Eurostat Database, Crops Production., *European Union*, https://ec.europa.eu/eurostat/web/main/data/database

Fahad S, Hussain S, Matloob A, Khan FA, Khaliq A, Saud S, Hassan S, Shan D, Khan F, Ullah N, Faiq M, Khan MR, Tareen AK, Khan A, Ullah A, Ullah N, Huang J (2015) Phytohormones and plant responses to salinity stress. A review. Plant Growth Regul 75:391–404. https://doi.org/10.1007/s10725-014-0013-y

Fang X, Turner NC, Yan G, Li F, Siddique KH (2010) Flower numbers, pod production, pollen viability, and pistil function are reduced and flower and pod abortion increased in chickpea (Cicer arietinum L.) under terminal drought. J Exp Bot 61:335–345. https://doi.org/10.1093/jxb/erp307

Farooq M, Aziz T, Basra SMA, Cheema MA, Rehman H (2008) Chilling tolerance in hybrid maize induced by seed priming with salicylic acid. J Agron Crop Sci 194:161–168. https://doi.org/10.1111/j.1439-037X.2008.00300.x

Farooq MA, Wahid N, Kobayashi DF, Basra SMA (2009a) Plant drought stress: effects, mechanisms and management. Agron Sustain Dev 29:185–212

Farooq M, Wahid A, Kobayashi NSMA, Fujita DBSMA, Basra SMA (2009b) Plant drought stress: Effects, mechanisms and management. Sustain Agric 153-188. https://doi.org/10.1051/agro:2008021

Farooq M, Gogoi N, Barthakur S, Baroowa B, Bharadwaj N, Alghamdi SS, Siddique KHM (2017) Drought stress in grain legumes during reproduction and grain filling. J Agron Crop Sci 203:81–102. https://doi.org/10.1111/jac.12169

Farooq M, Hussain M, Usman M, Farooq S, Alghamdi S, Siddique K (2018) Impact of abiotic stresses on grain composition and quality in food legumes. J Agric Food Chem 66(34):8887–8897. https://doi.org/10.1021/acs.jafc.8b02924

Farooq et al (2021) Morphological physiological and biochemical aspects of zinc seed priming-induced drought tolerance in faba bean Scientia Horticulturae https://281109894-10.1016/j.scienta.2021.109894

Fathi A, Tari DB (2016) Effects of drought stress and its mechanisms in plants. Int J Life Sci 10:1–6

FAOSTAT (2024) Statistical database, *Food and Agriculture Organization of the United Nations. (2024)*, https://www.fao.org/faostat/en/#home

Franca MGC, Thi ATP, Pimentel C, Rossiello ROP, Zuiliy-Fodil Y, Laffray D (2000) Differences in growth and water relations among Phaseolus vulgaris cultivars in response to induced drought stress. Environ Exp Bot 43:227–237. https://doi.org/10.1016/S0098-8472(99)00060-X

Francis C, Clegg M (2020) Crop rotations in sustainable production systems. In: Sustainable agricultural systems. CRC Press, pp 107–122

Garg BK (2003) Nutrient uptake and management under drought: Nutrient-moisture interaction. Cur Agric 27:1–8

Giller KE, Cadisch G (1995) Future benefits from biological nitrogen fixation: an ecological approach to agriculture. Plant Soil (Historical Archive) 174:255–277

Gliessman SR (2000) Agroecology, Ecological processes in sustainable agriculture. Lewis Publisher, London/New York/Washington, D.C., p 330

Gselman A, Kramberger B (2008) Benefits of winter legume cover crops require early seeding. Aust J Agric Res 59:1156–1163. https://doi.org/10.1071/AR08015

Guo YY, Liu J, Zhu YL, Bo YW, Li HJ, Xue JQ, Zhang RH (2018) Response of photosynthesis and antioxidant enzyme activities in maize leaves to drought stress. J Plant Physiol. 54:98–105. https://doi.org/10.13592/j.cnki.ppj.2018.0437

Haffani S, Mezni M, Chaïbi W (2014) Agronomic performances of three vetch species growing under different drought levels. Chil J Agric Res 74(3):263–272. https://doi.org/10.4067/S0718-58392014000300003

Hamann K (2020) Facilitating the EU market demand for legume-grain and -fodder as feeds. Prepared in Transition paths to sustainable legume-based systems in Europe H2020 project

Hanly JA, Gregg PEH (2004) Green-manure impact on nitrogen availability to organic sweetcorn (Zea mays). N Z J Crop Hortic Sci 32:295–307. https://doi.org/10.1080/01140671.2004.9514308

Harris D, Pathan AK, Gothkar P, Joshi A, Chivasa W, Nyamudeza P (2001) On farm seed priming: using participatory method to revive and refine a key technology. Agric Syst 69:151–164. https://doi.org/10.1016/S0308-521X(01)00023-3

Henn K, Olsen SB, Goddyn H, Bredie WLP (2022) Willingness to replace animal-based products with pulses among consumers in different European countries. Food Res Int 157:111403. https://doi.org/10.1016/j.foodres.2022.111403

Hosseinzadeh-Mahootchi A, Ghassemi-Golezani K (2012) The impact of seed priming and aging on physiological performance of chickpea under different irrigation treatments. Plant Breed Seed Sci 67:13–26

Hrustić M, Miladinović J, Đukić V, Tatić M, Balešević Tubić S (2008) Privredni značaj, osobine i tehnologija proizvodnje soje. Biljni lekar 36(3–4):171–186

Ilić A, Zorić M, Živanov D, Medić-Pap S, Vasić M (2023) Productivity assessment of common bean (Phaseolus vulgaris) germplasm from Serbia. Crop Pasture Sci 74(5):470–482. https://doi.org/10.1071/CP22275

Imran M, Latif Khan A, Shahzad R, Aaqil Khan M, Bilal S, Khan A, Kang SM, Lee IJ (2021) Exogenous melatonin induces drought stress tolerance by promoting plant growth and antioxidant defence system of soybean plants. AoB PLANTS 13(4):plab026. https://doi.org/10.1093/aobpla/plab026

Jarak MN, Đukić D, Đurić S, Stevović V, Đalović I (2004) Aktiviranje mikrobioloških procesa zemljišta s ciljem povećanja prinosa krmnih leguminoza. Acta agriculturae Serbica 9:221–228

Jiang XY, Dou JX, Wang ZQ (2001) Comparison of the effects of NaCl on photosynthesis and osmotic adjustment ability of corn and cotton (briefing). Plant Physiol Commun 37:303–305

Karges K, Bellingrath-Kimura SD, Watson CA, Stoddard FI, Halwani M, Reckling M (2022) Agro-economic prospects for expanding soybean production beyond its current northerly limit in Europe. Eur J Agron 133:126415

Kasim WAEA, Nessem AA, Gaber A (2017) Alleviation of drought stress in Vicia faba by seed priming with ascorbic acid or extracts of garlic and carrot. Egypt J Bot 57:45–45. https://doi.org/10.21608/ejbo.2017.831.1057

Kellogg WW, Schware R (2019) Climate change and society: consequences of increasing atmospheric carbon dioxide. Routledge, p 192. https://doi.org/10.4324/9780429048739

Kendir H, Sahin-Demirbag N, Aasim M, Khawar KM (2009) In vitro plant regeneration from Turkish Narbon Vetch (Vicia narbonensis L. var. narbonensis L.). Afr J Biotechnol 8(4):614–618

Khalil SE, Ismail EG (2010) Growth, yield and seed quality of Lupinus termisas affected by different soil moisture levels and different ways of yeast application. J Amer Sci 6:141–153

Khatun M, Sarkar S, Era FM, Islam AKMM, Parvez Anwar M, Fahad S, Datta R, Aminul Islam AKM (2021) Drought stress in grain legumes: effects, tolerance, mechanisms and management. Agronomy 11:2374. https://doi.org/10.3390/agronomy11122374

King CA, Purcell LC (2001) Soybean nodule size and relationship to nitrogen fixation response to water deficit. Crop Sci 41:1099–1107. https://doi.org/10.2135/cropsci2001.4141099x

Lahuta LB, Szablinska-Piernik J, Horbowicz M (2022) Changes in metabolic profiles of pea (Pisum sativum L.) as a result of repeated short-term soil drought and subsequent re-watering. Int J Mol Sci 23(3):1704. https://doi.org/10.3390/ijms23031704

Lawlor D, Cornic G (2002) Photosynthetic carbon assimilation and associated metabolism in relation to water deficit in higher plants. Plant Cell Environ 25:275–294. https://doi.org/10.1046/j.0016-8025.2001.00814.x

Leng G, Hall J (2019) Crop yield sensitivity of global major agricultural countries to droughts and the projected changes in the future. Sci Total Environ 654:811–821. https://doi.org/10.1016/j.scitotenv.2018.10.434

Leonardis AMD, Petrarulo M, Vita PD, Mastrangelo AM (2012) Genetic and molecular aspects of plant response to drought in annual crop species. In: Montanaro G (ed) Advances in selected plant physiology aspects. Rijeka, InTech

Lioi L, Piergiovanni A (2013) European common bean. In: Singh M, Upadhyaya HD, Singh Bisht I (eds) . Elsevier, Genetic and genomic resources of grain legume improvement, pp 11–40. https://doi.org/10.1016/B978-0-12-397935-3.00002-5

Liu Y, Gai JY, Lu HN, Wang YJ, Chen SY (2005) Identification of drought tolerant germplasm and inheritance and QTL mapping of related root traits in soybean (Glycine max (L.) Merr.). Acta Genet Sin 32(8):855–863

López-Galiano MJ, García-Robles I, González-Hernández AI, Camañes G, Vicedo B, Real MD, Rausell C (2019) Expression of miR159 is altered in tomato plants undergoing drought stress. Plan Theory 8:201. https://doi.org/10.3390/plants8070201

Losa A, Vorster J, Cominelli E, Sparvoli F, Paolo D, Sala T, Ferrari M, Carbonaro M, Marconi S, Camilli E, Reboul E, Waswa B, Ekesa B, Aragao F, Kunert K (2021) Drought and heat affect common bean minerals and human diet – what we know and where we go. Food Energy Secur 11:e351. https://doi.org/10.1002/fes3.351

Luan X, Bommarco R, Scaini A, Vico G (2021) Combined heat and drought suppress rainfed maize and soybean yields and modify irrigation benefits in the USA. Environ Res Lett 16:064023. https://doi.org/10.1088/1748-9326/abfc76

Lucas MM, Stoddard FL, Annicchiarico P, Frías J, Martínez-Villaluenga C, Sussmann D, Duranti M, Seger A, Zander PM, Pueyo JJ (2015) The future of lupin as a protein crop in Europe. Front Plant Sci 6:160197. https://doi.org/10.3389/fpls.2015.00705

Mamlić Z, Uhlarik A, Đukić V, Vasiljević S, Katanski S, Dozet G (2021) Influence of growth regulators on soybean morphology depending on weather conditions during the vegetation period. J Agron Technol Eng Manag 4(4):619–624

Mamlić Z, Đukić V, Dozet G, Abdurhman A, Dolapčev-Rakić A, Sinjušin A, Đurić N (2023) Značaj gajenja leguminoza u biljnoj proizvodnji. Zbornik radova nacionalnog naučnog skupa sa međunarodnim učešćem „Biotehnologija i savremeni pristup u gajenju i oplemenjivanju bilja, Smederevska Palanka, Srbija, 02. Novembar, str. 216–225

Manjeru P, Madanzi T, Makeredza B, Nciizah A, Sithole M (2007) Effects of water stress at different growth stages on components and grain yield of common bean (Phaseolus vulgaris L.). In Proceedings of the 8th African Crop Science Society Conference, El-Minia, Egypt, 27–31 October 2007, pp 299–303

Market Information – The European market potential for fresh peas. https://www.cbi.eu/market-information/fresh-fruit-vegetables/peas/market-potential

Marković JP, Milenković JR, Petrović MP, Zornić VG, Racić NL, Stepić MD, Blagojević MB (2023) Nutritive value of field pea-oat and common vetch-oat bicrops. J Agric Sci (Belgrade) 68(1):1–18

Markonis Y, Kumar R, Hanel M, Rakovec O, Máca P, AghaKouchak A (2021) The rise of compound warm-season droughts in Europe Science Advances 7(6) https://doi.org/10.1126/sciadv.abb9668

Martin N (2014) What is the way forward for protein supply? The European perspective OCL 21, D403

Masclaux C, Quilleré I, Gallais A, Hirel B (2001) The challenge of remobilisation in plant nitrogen economy. A survey of physio-agronomic and molecular approaches. Ann Appl Biol 138:68–81. https://doi.org/10.1111/j.1744-7348.2001.tb00086.x

Miladinov Z (2020) Predsetveni tretmani semena soje (Glycine max L.) u cilju poboljšanja kliavosti i uticaj zemljišne suše na kvalitet formiranog semena. Doktorska disertacija, Poljoprivredni fakultet, Univrezitet u Novom Sadu. 133 str

Miladinov Z., Đukić V., Dozet G., Ćeran M., Petrović K., Ranđelović P., Cvijanović G (2019) Sadržaj ulja i proteina u NS sortama soje. Zbornik radova 60. Savetovanje Proizvodnja i prerada uljarica, 16–21. Jun 2019. Herceg Novi, str. 63–69

Miladinov Z, Maksimovic I, Balesevic-Tubic S, Miladinovic J, Djordjevic V, Vasiljevic M, Radic V (2020a) The impact of water deficit on the soybean (Glycine max L.). Legume Res 43(5):693–697. https://doi.org/10.18805/LR-517

Miladinov Z, Maksimović I, Balešević-Tubić S, Đukić V, Nikolić Z, Milošević B, Katanski S (2020b) Priming seeds-method for increasing the germination of soybean seeds under drought stress conditions. Acta Agric Serb 25(50):105–111. https://doi.org/10.5937/AASer2050105M

Mirić M, Brkić M (2002) Dorada semena. Društvo selekcionera i semenara Republike Srbije, Beograd

Mladenov P, Aziz S, Topalova E, Renaut J, Planchon S, Raina A, Tomlekova N (2023) Physiological responses of common bean genotypes to drought stress. Agronomy 13:1022. https://doi.org/10.3390/agronomy13041022

Muktadir MA, Adhikari KN, Merchant A, Belachew KY, Vandenberg A, Stoddard FL, Khazaei H (2020) Physiological and biochemical basis of faba bean breeding for drought adaptation—a review. Agronomy 10:1345. https://doi.org/10.3390/agronomy10091345

Nadeem M, Li J, Yahya M, Sher A, Ma C, Wang X, Qiu L (2019) Research progress and perspective on drought stress in legumes: a review. Int J Mol Sci 20:2541. https://doi.org/10.3390/ijms20102541

Nadeem MA, Baloch FS, Fiaz S, Aasim M, Habyarimana E, Sönmez O, Nusret Z (2023) Legumes biofortification. Springer International Publishing, Berlin. https://doi.org/10.1007/978-3-031-33957-8

Nendel C, Reckling M, Debaeke P, Schulz S, Berg-Mohnicke M, Constantin J, Fronzek S, Hoffmann M, Jakšić S, Kersebaum KC (2023) Future area expansion outweighs increasing drought risk for soybean in Europe. Glob Chang Biol 29:1340–1358. https://doi.org/10.1111/gcb.16562

Neugschwandtner RW, Wagentristl H, Kaul HP (2015) Nitrogen yield and nitrogen use of chickpea compared to pea, barley and oat in Central Europe. Int J Plant Prod 9(2):291–304. https://doi.org/10.22069/IJPP.2015.2049

Nguyen V, Riley S, Nagel S, Fisk I, Searle IR (2020) Common vetch: a drought tolerant, high protein neglected leguminous crop with potential as a sustainable food source. Front Plant Sci 11:521832. https://doi.org/10.3389/fpls.2020.00818

Notz I, Reckling M (2022) Why farmers grow lupin. Insights from a survey of German farmers. Legume Hub

Notz I, Reckling M, Schuler J (2022) The role of legume production and use in European agri-food systems. Legumes Translated Practice Guide 2

Oljača S, Dolijanović Ž (2013) Ekologija i agrothnika združenih useva. Monografija, Poljoprivredni fakultet, Beograd, str. 1–182

Omomowo OI, Babalola OO (2021) Constraints and prospects of improving cowpea productivity to ensure food, nutritional security and environmental sustainability. Fron Plant Sci 12:751731. https://doi.org/10.3389/fpls.2021.751731

Our World In Data – Peas, dry. https://ourworldindata.org/grapher/pea-yields

Ozturk A, Aydin F (2004) Effect of water stress at various stages on some quality parameters of winter wheat. Crop Sci 190:93–99. https://doi.org/10.1046/j.1439-037X.2003.00080.x

Palta JA, Berger JD, Bramley H (2012) Physiology of the yield under drought: lessons from studies with lupin. In: Plant responses to drought stress: from morphological to molecular features. Springer, Berlin, Heidelberg, pp 417–440

Papathanasiou F, Ninou E, Mylonas I, Baxevanos D, Papadopoilou F, Avdikos I, Sistanis I, Koskosidis A, Vlachostergios D, Stefanou S, Tigka E, Kargiotidou A (2022) The evaluation of common bean (Phaseolus vulgaris L.) genotypes under water sytess based on physiological and agronomic parameters. Plan Theory 11:2432. https://doi.org/10.3390/plants11182432

Parvin S, Uddin S, Tausz-Posch S, Fitzgerald G, Armstrong R, Tausz M (2019) Elevated CO2 imroves yield and N fixation but not grain N concentration of faba bean (Vicia faba L.) subjected to terminal drought. Environ Exp Bot 165:161–173. https://doi.org/10.1016/j.envexpbot.2019.06.003

Pecetti L, Annicchiarico P, Crosta M, Notario T, Ferrari B, Nazzicari N (2023) White lupin drought tolerance: genetic variation, trait genetic architecture, and genome-enabled prediction. Int J Mol Sci 24:2351. https://doi.org/10.3390/ijms24032351

Petrychenko VF (2009) Production and use of soybean in Ukraine. Agronomists 3:79–81

Philandras C, Nastos P, Kapsomenakis J, Douvis K, Tselioudis G, Zerefos C (2011) Long term precipitation trends ana variability within the Mediterranean region. Nat Hazards Earth Syst Sci 11:3235–3250. https://doi.org/10.5194/nhess-11-3235-2011

Priyanka JP, Goral RT, Rupal KS, Saraf M (2019) Rhizosperic microflora: a natural alleviator of drought stress in agricultural crops. In: Sayyed R, Arora N, Reddy M (eds) Plant growth promoting rhizobacteria for sustainable stress management. Microorganisms for sustainability, vol 12. Springer, Singapore. https://doi.org/10.1007/978-981-13-6536-2_6

Rahman M, Mostofa MG, Keya SS, Rahman A, Das AK, Islam R, Abdelrahman M, Bhuiyan SU, Naznin T, Ansary MU, Tran LS (2021) Acetic acid improves drought acclimation in soybean: an integrative response of photosynthesis, osmoregulation, mineral uptake and antioxidant defence. Physiol Plant 172(2):334–350. https://doi.org/10.1111/ppl.13191

Rainey KM, Griffiths PD (2005) Differential response of common bean genotypes to high temperature. J Am Soc Hortic Sci 130:18–23. https://doi.org/10.21273/JASHS.130.1.18

Ramírez-Parra E, De la Rosa L (2023) Designing novel strategies for improving old legumes: an overview from common vetch. Plan Theory 12(6):1275. https://doi.org/10.3390/plants12061275

Redden R (2013) New approaches for crop genetic adaptation to the abiotic stresses predicted with climate change. Agronomy 3:419–432. https://doi.org/10.3390/agronomy3020419

Reddy AR, Chaitanya K, Jutur P, Sumithra K (2004) Differential antioxidative responses to water stress among five mulberry (Morus alba L.) cultivars. Environ Exp Bot 52:33–42. https://doi.org/10.1016/j.envexpbot.2004.01.002

Rodrigues FA, Marcolino-Gomes J, de Fátima Corrêa Carvalho J, do Nascimento LC, Neumaier N, Farias JR, Carazzolle MF, Marcelino FC, Nepomuceno AL (2012) Subtractive libraries for prospecting differentially expressed genes in the soybean under water deficit. Genet Mol Biol 35:304–314. https://doi.org/10.1590/S1415-47572012000200011

Rui PH, Han KL, Wang CJ, Li WY (2018) Effects of high temperature at grain filling stage on antioxidant enzyme activities and osmotic adjustment substances in maize leaves. Jiangsu Agric Sci 46:82–84

Sabry GE (2018) The importance of using high quality seeds in agriculture systems. Agric Res Technol 15(4)

Saget S, Costa M, Barilli E, de Vasconcelos MW, Santos CS, Styles D, Williams M (2020) Substituting wheat with chickpea flour in pasta production delivers more nutrition at a lower environmental cost. Sustain Prod Consump 24:26–38. https://doi.org/10.1016/j.spc.2020.06.012

Saglam S, Sibel DAY, Gamze KAYA, Gurbuz A (2010) Hydropriming increases germination of lentil (Lens culinaris Medil.) under water stress. Not Sci Biol 2:103–106. https://doi.org/10.15835/nsb.2.2.4602

Sarkar S, Khatun M, Era FM, Islam AKMM, Anwar MP, Danish S, Datta R, Islam AKMA (2021) Abiotic stresses: alteration of composition and grain quality in food legumes. Agronomy 11:2238. https://doi.org/10.3390/agronomy11112238

Savić A, Zorić M, Brdar-Jokanović M, Zdravković M, Dimitrijević M, Petrović S, Živanov D, Vasić M (2020) Origin and diversity study of local common bean (Phaseolus vulgaris L.) germplasm from Serbia: phaseolin and phenotyping approach. Genet Resour Crop Evol 67(8):2195–2212. https://doi.org/10.1007/s10722-020-00974-9

Savić A, Pipan B, Vasić M, Meglič V (2021) Genetic diversity of common bean (Phaseolus vulgaris L.) germplasm from Serbia, as revealed by single sequence repeats (SSR). Sci Hortic 288:110405. https://doi.org/10.1016/j.scienta.2021.110405

Sepngang BK, Stauss W, Stute I, Mergenthaler M (2018) The market of grain legumes in Germany. Results of the EU – project LegValue project, H2020

Sepngang KB, Muel F, Smadja T, Stauss W, Stute I, Simmen M, Mergenthaler M (2020) Report on legume markets in the EU. Results of the EU-project LegValue project, H2020

Siddiqui MH, Al-Khaishany MY, Al-Qutami MA, Al-Whaibi MH, Grover A, Ali HM, Al-Wahibi MS, Bukhari NA (2015) Response of different genotypes of faba bean plant to drought stress. Int J Mol Sci 16:10214–10227. https://doi.org/10.3390/ijms160510214

Simiciklas KD, Mullen RE, Carlson RE, Knapp AD (1989) Drought induced stress effects on soybean seed calcium and seed quality. Crop Sci 29:1519–1523. https://doi.org/10.2135/cropsci1989.0011183X002900060041x

Simova-Stoilova L, Demirevska K, Petrova T, Tsenov N, Feller U (2009) Antioxidative protection and proteolytic activity in tolerant and sensitive wheat (Triticum aestivum L.) varieties subjected to long-term field drought. Plant Growth Regul 58:107–117. https://doi.org/10.1007/s10725-008-9356-6

Sinclair TR, Purcell LC, King A, Sneller CH, Chen P, Vadez V (2007) Drought tolerance and yield increase of soybean resulting from improved symbiotic N2 fixation. Field Crop Res 101:68–71. https://doi.org/10.1016/j.fcr.2006.09.010

Singh SP (2007) Drought resistance in the race Durango dry bean landraces and cultivars. Agron J 99:1219–1225. https://doi.org/10.2134/agronj2006.0301
Sionit N, Strain BR, Flint EP (1987) Interaction of temperature and CO2 enrichment on soybean: photosynthesis and seed yield. Can J Plant Sci 67:629–636
Skovbjerg CK, Knudsen JN, Fuchtbauer W, Stougaard J, Stoddard FL, Janss L, Andersen SU (2020) Evaluation of yield, yield stability and yield-protein relationship in 17 commercial faba bean cultivars. Leg Sci 2:e39
Sorensen JN, Edelenbos M, Wienberg L (2003) Drought effects on green pea texture and related physical-chemical properties at comparable maturity. J Amer Soc Hort Sci 128(1):128–135. https://doi.org/10.21273/JASHS.128.1.0128
Soussa P, Trigo R, Aizpurua P, Nieto R, Gimeno L, Garcia-Herrera R (2011) Trends and extremes of drought indices throughout zhe 20th century in the Mediterranean. Nat Hazarsd Earth Syst Sci 11:33–51. https://doi.org/10.5194/nhess-11-33-2011
Stoddard FI, Balko C, Erskine W, Khan H, Link W, Sarker A (2006) Screening techniques and sources of resistance to abiotic stresses in cool-season food legumes. Euphytica 147:167–186. https://doi.org/10.1007/s10681-006-4723-8
Stolf-Moreira R, Medri ME, Neumaier N, Lemos NG, Pimenta JA, Tobita S, Brogin RL, Marcelino-Guimarães FC, Oliveira MCN, Farias JR, Abdelnoor RV, Nepomuceno AL (2010) Soybean physiology and gene expression during drought. Genet Mol Res 9:1946–1956
Stone L, Goodrum RDE, Jafar MN, Khan AH (2001) Rooting front and water depletion Depths in grain sorghum and sunflower. Agron J 1:1105–1110. https://doi.org/10.2134/agronj2001.9351105x
Stričević RJ, Stojaković N, Vujadinović-Mandić M, Todorović M (2018) Impact of climate change on yield, irrigation requirements and water productivity of maize cultivated under the moderate continental climate of Bosnia and Herzegovina. J Agric Sci 156(5):618–627. https://doi.org/10.1017/S0021859617000557
Talebi R, Esnafi MH, Baghebani N, Karami E, Mohammadi K (2013) Physiological responses of chickpea (Cicer arietinum) genotypes to drought stress. Environ Exp Biol 11:9–15
Tassoni A, Tedeschi T, Zurlini C, Cigognini IM, Petrusan JI, Rodriguez O, Neri S, Celli A, Sisti L, Cinelli P, Signori F, Tsatsos G, Bondi M, Verstringe S, Bruggerman G, Corvini PFX (2020) State-of-the-art production chains for peas, beans and chickpeas—valorization of agro-industrial residues and applications of derived extracts. Molecules 25:1383. https://doi.org/10.3390/molecules25061383
Tekle AT, Alemu MA (2016) Drought tolerance mechanisms in field crops. World J Biol Med Sci 3:15–39
Tetsuji O, Nepomuceno AL, Neumaier N, Farias JRB, Tobita S, Ito O (2004) Drought tolerance characteristics of Brazilian soybean cultivars-evaluation and characterization of drought tolerance of various Brazilian soybean cultivars in the field. Plant Prod Sci 7(2):129–137. https://doi.org/10.1626/pps.7.129
Tiwari P, Srivastava D, Chauhan AS, Indoliya Y, Singh PK, Tiwari S, Fatima T, Mishra SK, Dwivedi S, Agarwal L et al (2020) Root system architecture, physiological analysis and dynamic transcriptomics unravel the drought-responsive traits in rice genotypes. Exotoxicol Environ Safety 207:111252. https://doi.org/10.1016/j.ecoenv.2020.111252
Tonitto C, David MB, Drinkwater LE (2006) Replacing bare fallows with cover crops in fertilizer-intensive cropping systems: a meta-analysis of crop yield and N dynamics. Agric Ecosyst Environ 112:58–72. https://doi.org/10.1016/j.agee.2005.07.003
Tzenova V, Kirkova Y, Stoimenov G (2008) Methods for plant water stress evaluation of soybean canopy. In Balwois 2008 – Water Observation and Information System for Decision Support 2008: Ohrid, Republic of Macedonia
Tzortzakis N, Chrysargyris A, Aziz A (2020) Adaptive response of a native Mediterranean grapevine cultivar upon short-term exposure to drought and heat stress in the context of climate Chang. Agronomy 10:249. https://doi.org/10.3390/agronomy10020249

Uddin S, Ullah S, Nafees M (2021) Effect of seed priming on growth and performance of Vigna radiata L. under induced drought stress. J Agric Food Res 4:100140. https://doi.org/10.1016/j.jafr.2021.100140

Uhlarik A, Ćeran M, Živanov D, Grumeza R, Skøt L, Sizer-Coverdale E, Lloyd D (2022) Phenotypic and genotypic characterization and correlation analysis of pea (Pisum sativum L.). Divers Panel Plants 11:1321. https://doi.org/10.3390/plants11101321

Uhlarik A, Ćeran M, Mamlić Z, Vasiljević S, Katanski S, Rakić AD, Đukić V (2024) comparative analysis of agronomic traits, seed color variability and protein content in Pisum sativum L. Cultivars across diverse european environments. Legum Res. https://doi.org/10.18805/LRF-803

Ullah A, Farooq M (2022) The challenge of drought stress for grain legumes and options for improvement. Arch Agron Soil Sci 68(11):1601–1618. https://doi.org/10.1080/03650340.2021.1906413

Van Loon A, Tijdeman E, Wanders N, van Lanen HJ, Teuling A, Uijlenhoet R (2014) How climate seasonality modifies drought duration and deficit. J Geophys Res Atmos 119:4640–4656. https://doi.org/10.1002/2013JD020383

Van Loon AF, Gleeson T, Clark J, van Dijk AI, Stahl K, Hannaford J, Baldassarre G Di, Teuling AJ, Tallaksen LM, Uijenhoet R, Hannah DM, Sheffield DM, Svoboda M, Verbeiren B, Wagener T, Rangecroft S, Wanders N, van Lanen HAJ (2016) Drought in the antropocene. Nat Geosci 9:89–91. https://doi.org/10.1038/ngeo2646

van Loon M, Alimagham S, Pronk A, Fodor N, Ion V, Kryvoshein O, Kryvobok O, Marrou H, Mihail R, Minguez MI, Pulina A, Reckling M, Rittler L, Roggero PP, Stoddard F, Top C, van der Wel J, Watson C, van Ittersum MK (2023) Grain legume production in Europe for food, feed and meat-substitution. Glob Food Secur 39:100723. https://doi.org/10.1016/j.gfs.2023.100723

Vargas-Blandino D, Cárdenas-Travieso RM (2021) Chickpea cultivation, a possible solution to climate change. Cultivos Tropicales 42(1):e09

Vidić M, Hrustić M, Miladinović J, Đukić V, Đorđević V (2009) Sortni ogledi soje u 2008. godini. Zbornik radova Instituta za ratarstvo i povrtarstvo, Novi Sad 46:261–270

Vieira RD, Tekrony DM, Egli DB (1992) Effect of drought and defoliation stress in the filed onsoybean seed germination and vigor. Crop Sci 32:471–475. https://doi.org/10.2135/cropsci1992.0011183X003200020037x

Vujaković M, Balesević-Tubić S, Jovicić D, Taski-Ajduković K, Petrović D, Nikolić Z, Kostić M (2011) Klijavost i vigor semena soje proizvedenog u različitim agrometeorološkim uslovima. J Process Energy Agric 15(3):157–159

Walker J, Hertel K, Parker P, Edwards J (2011) Lupin growth & development. Procrop, Industry & Investment, NSW Government

Wang R, Gao M, Ji S, Wang S, Meng Y, Zhou Z (2016) Carbon allocation, osmotic adjustment, antioxidant capacity and growth in cotton under long-term soil drought during flowering and boll-forming period. J Plant Biochem Physiol 107:137–146. https://doi.org/10.1016/j.plaphy.2016.05.035

Watson CA, Reckling M, Preissel S, Bachinger J, Bergkvist G, Khulman T, Lindström AK, Nemecek T, Topp CFE, Vanhatalo AO, Zander P, Murphy-Bokern D, Stoddard FL (2017) Grain legume production and use in European agricultural systems. Adv Agron 144:235–303. https://doi.org/10.1016/bs.agron.2017.03.003

Wu Z, Lei T, Fan X, Zhang H, Du Z, Yu L (2020) Recent changes in the drought of China from 1960 to 2014. Int J Climatol 40(7):3281–3296

Xu W, Cui K, Xu A, Nie L, Huang J, Peng S (2015) Drought stress condition increases root to shoot ratio via alteration of carbohydrate partitioning and enzymatic activity in rice seedlings. Acta Physiol Plant 37:9. https://doi.org/10.1007/s11738-014-1760-0

Ye H, Roorkiwal M, Valliyodan B, Zhou L, Chen P, Varshney RK, Nguyen HT (2018) Genetic diversity of root system architecture in response to drought stress in grain legumes. J Exp Bot 69:3267–3277. https://doi.org/10.1093/jxb/ery082

Zhang C, Shi S, Liu Z, Yang F, Yin G (2019) Drought tolerance in alfalfa (Medicago sativa L.) varieties is associated with enhanced antioxidative protection and declined lipid peroxidation. J Plant Physiol 232:226–240. https://doi.org/10.1016/j.jplph.2018.10.023

Živanov D, Tančić Živanov S, Savić A, Uhlarik A, Miladinov Z, Medic Pap S, Nagl N (2021) First report of Fusarium oxysporum f. sp. ciceris on chickpea (Cicer arietinum L.) in Serbia. Plant Dis., ISSN: 0191-2917. https://doi.org/10.1094/PDIS-09-21-1998-PDN

Legume Breeding for Drought Tolerance Experience from the Past Using Classical Breeding Approaches

5

Shumaila Ijaz, Javed Iqbal, Banzeer Ahsan Abbasi, Zakir Ullah, Akhtar Munir, Tabassum Yaseen, Muhammad Majeed, Noshin Ilyas, Amin Fathi, Sobia Kanwal, and Tariq Mahmood

Abstract

The primary environmental stressor that affects agriculture globally is drought, and one of the main goals of plant breeding is to increase production in such conditions. Legume crops like peas and chickpeas have benefited greatly from improved drought resistance thanks to classical breeding techniques. In order to create cultivars that are resistant to drought, these methods include: screening, mutation, hybridization, and marker-assisted selection. Morphological parameters are important in drought resistance breeding; in legumes for example, root characteristics like biomass and depth are critical in controlling terminal drought tolerance. When it comes to creating drought-tolerant legume varieties, the practice of genomics-assisted breeding techniques like marker-aided selection has greatly increased breeding efficiency and precision. Rice cultivars with improved drought resistance are the result of classical breeding, which focuses on biometric factors such as root depth, avoidance, escape, and tolerance. The quick pro-

Shumaila Ijaz and Javed Iqbal contributed equally with all other contributors.

S. Ijaz
School of Biomedical Engineering, Shenzhen University Medical School, Shenzhen University, Shenzhen, China

J. Iqbal (✉) · T. Yaseen
Department of Botany, Bacha Khan University, Charsadda, Khyber Pakhtunkhwa, Pakistan

B. A. Abbasi
Department of Botany, Rawalpindi Women University, Rawalpindi, Pakistan

Z. Ullah · T. Mahmood
Department of Plant Sciences, Faculty of Biological Sciences, Quaid-i-Azam University, Islamabad, Pakistan

© The Author(s), under exclusive license to Springer Nature Singapore Pte Ltd. 2025

M. A. Nadeem et al. (eds.), *Marker-Assisted Breeding in Legumes for Drought Tolerance*, https://doi.org/10.1007/978-981-96-4112-3_5

duction of high-yielding drought-tolerant legume varieties is facilitated by the accelerated genetic gain for drought tolerance in legumes that results from the merging of traditional breeding techniques with contemporary molecular biology tools and biometrical approaches. These methods are essential for tackling the problems caused by drought. Legumes like chickpea have benefited from enhanced drought tolerance through classical breeding techniques like hybridization and mutation, which select for desired features. Legumes' resilience to drought has been greatly increased via classical breeding, which makes use of genetic variety and identifies important characteristics linked to stress tolerance. In order to facilitate the transfer of desirable traits to elite cultivars, this technique entails the identification of possible donors, quantitative trait loci (QTLs)/ genes, and candidate genes connected to traits relevant to drought tolerance. The effective integration of drought-tolerant characteristics into legume cultivars has increased their resilience to water shortages and climate change issues due to the prudent use of molecular markers such as single nucleotide polymorphism (SNP), Simple Sequence Repeat (SSR), and Insertion–deletion mutations (INDELs). We can make a substantial contribution to sustainable agriculture by preparing legumes for drought conditions, which will lessen the effects of drought and increase global food security.

Keywords

Legume crops · Classical breeding techniques · Drought · Marker-assisted selection · Morphological parameters · Genomics-assisted breeding · Accelerated genetic gain

5.1 Introduction

With the frequency of droughts rising owing to climate change, legume breeding for drought resistance is an important topic of research. Numerous processes and genetic variables contributing to drought tolerance in leguminous crops have been

A. Munir
Department of Chemistry, Quaid-I-Azam University, Islamabad, Pakistan

Department of Chemistry, University of Sialkot, Sialkot, Pakistan

M. Majeed
Department of Botany, University of Gujrat, Punjab, Pakistan

N. Ilyas
Department of Botany, PMAS Arid Agriculture University, Rawalpindi, Pakistan

A. Fathi
Department of Agronomy, Ayatollah Amoli Branch, Islamic Azad University, Amol, Iran

S. Kanwal
Department of Biology and Environmental Sciences, Allama Iqbal Open University, Islamabad, Pakistan

emphasized by studies on lentils (El Haddad et al. 2023), peas (Bagheri et al. 2023), soybeans (Saleem et al. 2022, and chickpeas (Asati et al. 2022; Ullah et al. 2023a). These comprise increased levels of enzyme functions such as superoxide dismutase, catalase, and ascorbate peroxidase, in addition to the build-up of osmoprotectants such proline, flavonoids, and phenolics. Candidate genes linked to abscisic acid (ABA) signaling, stomatal conductance, and root system properties that are essential for drought stress responses have been found through genomic investigations. The development of drought-tolerant legume cultivars has showed promise thanks to marker-assisted breeding and omics approaches, highlighting the significance of combining molecular tools with conventional breeding methods to increase agricultural resistance to water shortage (Fig. 5.1) (Savitri 2023).

Heat, salt, and drought are three environmental stress variables that have an impact on nearly every part of the plant, from germination to maturity (Negrao et al. 2017). According to Hussain et al. (2018), drought poses a serious risk and is the most unexpected restriction, having a negative impact on agricultural output globally. According to Chowdhury et al. (2016), drought has a catastrophic effect on plants because it disrupts a number of their processes, including the rate at which carbon is assimilated, oxidative damage, turgor, and alterations in leaf gas exchange. Consequently, yield is decreased. The complex phenomena of plant susceptibility to drought are influenced by a multitude of factors, including the plant's developmental stage, genetic potential, and the duration and severity of the stress (Zhu 2002). In addition, drought impacts ion balance, enzyme activity, and leaf growth, which eventually results in decreased yield (Todaka et al. 2017). Since crops were first domesticated, the selection of desirable features identified at the phenotypic level

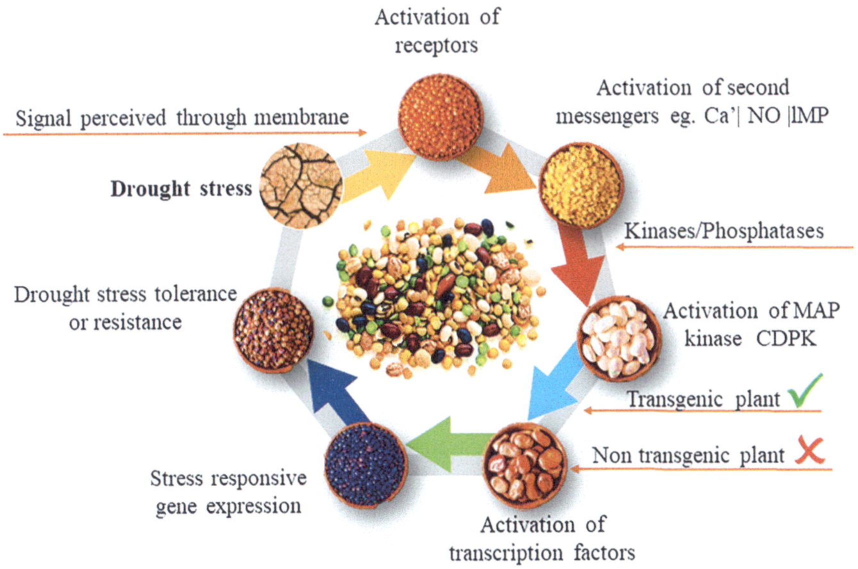

Fig. 5.1 The transgenic approach's fundamental working mechanism under drought stress

has propelled their evolution (Piepho 2000). The delayed development in yield enhancement in drought-prone regions can be attributed to the intricacy of drought tolerance systems. Recent advances in drought tolerance brought about by advances in agricultural physiology and genetics have given breeders new information and resources for plant development (Tuberosa and Salvi 2006). The purpose of this chapter is to give a summary of the accomplishments made in drought tolerance breeding and to highlight potential future directions for plant breeding that might arise from combining the most current developments in physiology and genetics.

Because of their capacity to fix nitrogen in the atmosphere, legumes provide a wholesome and affordable source of protein (Rubiales and Mikic 2015). Their adaptability to situations with nitrogen shortage is increased by these unique qualities (Graham and Vance 2003). Abiotic hazards can affect legume crops, and one of the primary factors restricting crop productivity has been identified as drought (Farooq et al. 2016). Legume crops are often farmed in rain-fed areas, and the concern of water shortage has been highlighted by many models (such as the Global Climate Model) that project a rise in the frequency and intensity of drought (Mittal et al. 2014). Any stage of water shortage can impact plant growth by decreasing crop yield, especially during the reproductive stage and grain filling (Pushpavalli et al. 2015). Grain yield, related legume components, and plant biomass are all impacted by the frequency and severity of droughts (Ghassemi-golezani et al. 2013). The amount and crop developmental stage, genotypic diversity, and duration of drought stress (DS), all influence how much production is reduced. To minimize production losses in water-scarce conditions, innovative methods for improving legumes' resistance to drought are therefore essential. The development of cultivars that are more water-efficient and drought-tolerant might boost crop yield in dry areas (Ulemale et al. 2013).

The main factors that lead to improved legume production in severe situations are the significant development and integration of cutting-edge techniques for arid environments. Promising strategies to lessen the severe concerns of drought include the creation of different characteristics for drought tolerance, creative breeding, and water-efficient techniques like mulching and drip irrigation (Miyauchi et al. 2012). There has been a prior analysis of the negative effects of drought on a number of different crops (Shavrukov et al. 2017: Kamanga et al. 2018).

5.2 Drought Stress's Effect on Legumes

Although legumes react and are sensitive to droughts differently, the end yield is always dramatically reduced. Numerous variables are associated with this, such as carbon fixation and reduced assimilate translocation (Mondal et al. 2011; Zlatev and Lidon 2012), fewer pods and lower grain set, decreased sink activity, and suppressed blooming time and its influence on reproductive organs (Andersen et al. 2015; Samarah et al. 2009). Drought affects legume growth and development in several ways, such as reproductive stage, germination, photosynthesis, root, and shoot development.

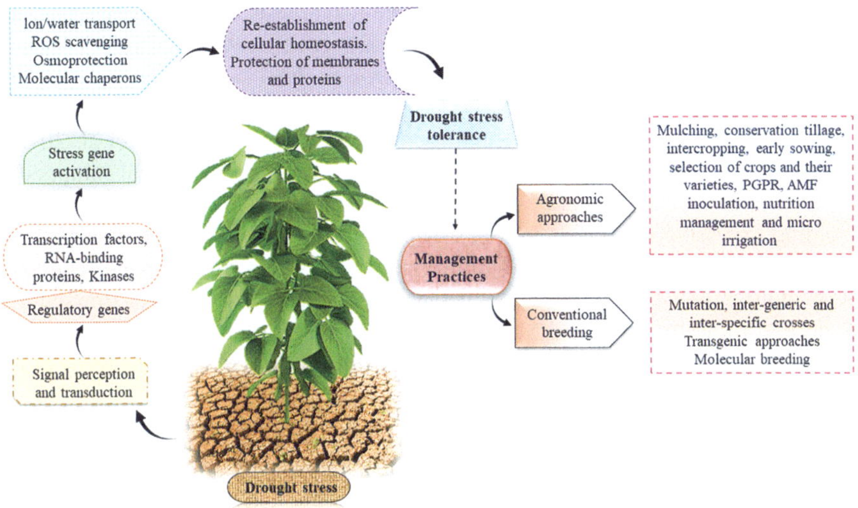

Fig. 5.2 Progress and aspects of the drought stress in legumes

Drought is becoming one of the most unpredictable and uncontrollable factors restricting agricultural productivity and negatively affecting legume crops as an end result of global climate change (Fig. 5.2). Research has indicated that extreme drought circumstances impact the morphology, physiology, and growth period of plants. Conversely, moisture content is crucial for the activation of enzymes during germination, which may provide insight into the susceptibility of plants to drought at this stage of development. The phases of germination and reproduction are extremely susceptible to a lack of water. Soybean germination rate was much lower under DS (Draught Stress). Awari and Mate observed that when there is a water deficit, the germination rate of chickpeas decreases. According to Li et al. (2018), DS frequently manifests itself in the seedling stage and drastically lowers faba bean production (Li et al. 2018; Awari and Mate 2015).

Numerous environmental stressors have an impact on photosynthesis, a vital function in plants that drives growth and development (Andrianasolo et al. 2016). The kind, intensity, and duration of the stress will determine how a plant responds to a water shortage. By interfering with the enzymatic activity of many enzymes, including PEPCase, PPDK, Rubisco, and EBPase, DS impairs carbon fixation (Farooq et al. 2009). Depending on the severity and length of the stress, DS affects the photosynthetic system and ultimately lowers the yield of legumes. During DS, there was a significant reduction in the rate of photosynthetic rate, transpiration rate, and stomatal conductance (Ohashi et al. 2006).

Similar to this, a recent study found that drought-stricken soybeans had less net photosynthesis, which in turn reduced the amount of dry matter accumulated and the pace at which podding occurred, hence lowering yield (Wang et al. 2018a, b). According to Hao et al. (2013), drought-stressed soybean plants have 31% less chlorophyll than non-stressed plants (Hao et al. 2013). According to Siddiqui et al.

(2015), DS dramatically reduced the rate of photosynthesis and chlorophyll content in faba beans, which hampered plant growth and productivity. The activities of antioxidant enzymes and chlorophyll fluorescence in faba beans are impacted by drought (Abid et al. 2016). Similarly, DS influences photosynthesis, chlorophyll fluorescence, and chlorophyll concentration in chickpeas (Mafakheri et al. 2010). Additionally, it is thought that stomatal regulation is a key physiological component for maximizing water consumption during drought stress (DS) (Makbul et al. 2011), avoiding excessive water loss in times of prolonged drought. For example, under DS, stomatal conductance in soybeans dropped by 60% in comparison to the control (Mak et al. 2014). The stomatal conductance decreased as the drought stressed the plant, according to Abdel and Stutzel (2016), confirming the discovery that under extreme stress as opposed to medium stress, the impacts of drought were more noticeable (Atti et al. 2004).

They clarified that, in contrast to what had previously been reported, there was a stronger correlation between the stomatal conductance and the transpiration rate (De Souza et al. 1997). Drought's complicated effects on several processes, including gametogenesis, fertilization, embryogenesis, and grain production, have been documented (Farooq et al. 2014). Water shortage greatly affects the blossoming and reproductive stages of a plant's life cycle. By reducing the formation of pollen tubes and pollen grain germination, DS shortens the blooming period, alters flower development, and results in sterility in the pollen grains (Phillips et al. 2018; Gusmao et al. 2012). Drought significantly reduces a plant's capacity to develop new blooms, pods, and seeds, which eventually diminishes the output. For example, several studies observed that when there is a water deficit, the number of pods per plant decreases (Desclaux and Roumet 2000; Li et al. 2013a, b). The combined consequences of a decrease in young pod abortion and pod induction (Kokubun et al. 2001) and the reduction in flower number and pod enlargement (He et al. 2016) resulted in a reduction in pod number of 92.7% for DS at the start of pod development and 81.6% for controls during pod lengthening. Less seeds per plant under DS were seen in earlier investigations. The blooming stage saw the largest decrease in the amount of seeds produced by each plant (Maleki et al. 2013).

According to José et al. (2018), the incidence of dryness during flower creation resulted in a shorter flowering time, which in turn generated fewer flowers, fewer pods, and fewer seeds per plant. Nevertheless, it was previously determined that drought stress (DS) lowers seed quantity during the seed set stage, which in turn lowers eventual yield (Meckel et al. 1984; Rotundo and Westgate 2009). Soybean seed production under DS significantly decreased (73–82%) per plant (Wei et al. 2018). Drought reduces output by 27–87% and is detrimental to several developmental stages, such as the formation and function of reproductive organs. For example, a deficiency of water drastically lowers the production of chickpea (Samarah et al. 2009), soybean (Kobraee et al. 2011), and mash bean (Baroowa and Gogoi 2014).

5.3 Formation of Legumes Resistant to DDT (dichlorodiphenyltrichloroethane) Through Molecular and Biotechnological Methods

It is strategically crucial to develop drought-resistant genotypes and effective water management techniques at the same time in order to maximize legume output. Therefore, combining contemporary techniques with traditional breeding methods may have a big impact. The breeding, molecular, and transgenic strategies used to increase legume resistance to drought stress are covered in this section.

5.3.1 Breeding Approach

Using conventional breeding to increase DS tolerance in plants is a useful technique and an important crop improvement approach. To achieve DS tolerance in legume crops, breeding and selection methods require significant heritable variation (Beebe et al. 2008). However, the quantitative genetic foundation of characteristics and our limited understanding of the physiological basis of crop production response to drought frequently impede the advancement of breeding (Torres et al. 2010). Furthermore, strong genotype and environment interactions, as well as differences in the timing and quantity of precipitation established under dry circumstances, frequently result in poor heritability. Identifying features that offer yield potential and/or stability under DS is important, regardless of this.

Moreover, enhancing the environment's characterization is necessary to increase the efficacy of target qualities (Mir et al. 2012). To acquire the necessary phenotypic features based on the qualities connected with the yield, screening and mass selection could be helpful. In legumes including cowpea, common bean, soybean, and chickpea, a few root characteristics, such as root length, density, rooting depth, and fibrous root system, may be important in determining the avoidance of DS. They might also be helpful in DS tolerance genotype screening. Traits that provide an escape route, for example early blooming, podding, and maturity, can be used for mass screening. During droughts, cooler canopies and greater stomatal conductance have frequently been linked to increased grain yields; these characteristics may function as indirect selection effects (Duc et al. 2015). Chapman (2008) state that canopy spectral reflectance is a useful noninvasive high-throughput phenotyping tool that allows the rapid and simple assessment of a range of dynamic, complicated features, such as biomass accumulation, carbon absorption, and plant canopy growth. Therefore, canopy spectral reflectance might be applied as a mass screening method for genotypes of legumes that are drought tolerant. Thermal infrared imaging, sometimes called infrared thermography, can be used to screen grain legumes for drought resistance by taking a measurement of the temperature of the leaves or canopy (Martynenko et al. 2016). Plant water status may be inferred from plant canopy temperature, a commonly observed characteristic that is strongly correlated with canopy conductance during the vegetative stage (Hou et al. 2018). Throughout time, the entire plant or the canopy level may be assessed for conductivity using thermal infrared imaging.

In breeding programs aiming to promote drought tolerance, evaluation of delayed senescence may be useful for indirect selection of grain and biomass yields (Muchero et al. 2013). Crucial physiological characteristics such as leaf temperature, root growth, water consumption efficiency, carbon isotope discrimination (Δ13C), and leaf temperature may aid in the screening of legume genotypes for drought resistance (Ribeiro et al. 2019). Improvements in root architecture may be made using chickpea genotypes with significant genetic variety for carbon isotope discrimination (Krishnamurthy et al. 2013). This is an indirect indication of chickpea genetic diversity. Breeding techniques such as wide hybridization are used to produce desired features within or between species. Grain legumes have been the subject of several interspecific crosses, with varying degrees of success (Singh et al. 2018).

According to this viewpoint, *Phaseolus vulgaris* can be crossed with its wild plant *Phaseolus acutifolius*, which requires interspecific hybridization to be transferred to cultivated beans since it adjusts to osmotic pressure more than the previous one did. However, depending on the plant's physiological stage or stress level, osmotic adjustment may have less stability (Basu et al. 2007). A number of legume crops with drought-associated origins have produced promising germplasm accessions; some of these have been discovered in chickpea (ICC 4958, for example), and they are connected to canopy temperature, root depth, root length density, and terminal drought. According to reports, wild genes from *C. reticulatum* were also reportedly incorporated into farmed chickpea, producing nine varieties that were very drought-tolerant (Kashiwagi et al. 2005).

5.3.2 Quantitative Trait Loci (QTL) for Drought Tolerance

Identifying favored alleles through genome-based techniques is advantageous as distinct QTLs could impact intended results. Physiological and anatomical traits that influence the drought tolerance mechanism are inherited quantitatively. Determining QTLs linked to drought resilience is, thus, among the most promising uses of marker-assisted selection (MAS). Additionally, a range of breeding strategies have been used to improve legumes depending on the drought resilience of MAS (Farooq et al. 2014).

There are five QTLs associated with fibrous roots in soybeans: Gm01, Gm02, Gm03, Gm04, and Gm20 (Abdel-Haleem et al. 2011). SNPs obtained from *Medicago truncatula* L. were utilized by Khazaei et al. (2014) to find QTLs linked to stomatal features in faba beans. Similar to this, Mukeshimana et al. (2014) identified 14 QTLs for characteristics linked to drought tolerance in common beans using SNPs from the BARCBean6K_3 Beadchip. Moreover, recent studies (Mukeshimana et al. 2014; Briñez et al. 2017) reported using a cross between AND-277xSEA-5 to find QTLs associated with stress tolerance and assess the factors affecting the common bean's drought response. QTLs for leaf biomass dry weight, leaf temperature, number of seeds per plant, total yield, number of pods per plant (22), leaf and stem fresh biomass, days to flowering, chlorophyll, and leaf biomass dry weight were found in both drought and well-watered growth conditions.

5.3.3 Functional Genomics and Biotechnology

With the creation of crop transgenic tools, gene-based technology has become a crucial means of doing in-depth research on the complex mechanisms behind drought resistance. It is also believed to be an additional tactic for introducing desired genetic changes into plants. Recent developments in biotechnology have made it possible for us to identify genes that are resistant to abiotic stress from other organisms or even from different species. This has allowed us to modify the genetic makeup of grain and legume crops to safeguard against devastating drought conditions. There are several methods for transforming legumes into transgenic plants, including agrobacterium- or biolistic-mediated transformation. Targeted resistance genes have also been shown in earlier research to enhance plant performance in drought conditions without having an adverse effect on plant production when integrated into different genomes. While there are several initiatives in progress, certain studies have already produced successful transgenes in various legume crops by the use of distinct genes taken from other species as well as the genome.

Legume plants were genetically modified via single-gene transformation to encode late embryogenesis proteins, osmoprotectant-producing enzymes, and membrane lipid modifications (Kishor et al. 1995). Numerous factors, such as the severity of the drought, the capacity to spread resistance mechanisms, and the effect on biomass and plant production, influence the entire process. After the osmoregulatory P5CSF129A gene was inserted into a genotype of chickpea, there was a decrease in malonaldehyde and free radical levels and an increase in proline synthesis, but no appreciable increase in biomass building (Bhatnagar-Mathur et al. 2009).

Numerous genes, including DRED transcription factors, are part of the AP2/ERF gene family. These transcription factors are important for plant growth and development as well as for adapting to a variety of stresses in the environment. Overexpressing DREB1A transgenic chickpea plants driven by the Arabidopsis rd29A promoter demonstrated an increase in DREB1A gene expression prior to 50% soil dryness conditions (Anbazhagan et al. 2014).

Prior studies have shown that rd29A: DREB1A was beneficial in improving the mechanisms underlying transpiration efficiency, rooting architecture, water absorption, and stomatal responsiveness in arid environments in comparison to controls. Li et al. (2013a, b) reported that overexpressing LOSS/ABA3 increased ABA accumulation, which activates stress-upregulated gene expression and initiates a sequence of physiological and biochemical resistance processes, thereby improving drought resistance in soybeans. VuNCED1 is necessary for cowpea ABA synthesis in dry environments.

5.3.4 OMICS-Based Approaches

The precise function of the target characteristic genes has been identified through the application of OMICS-based technologies. Utilizing data from the transcriptome, microme, genome, metabolome, and proteome (Fig. 5.3), this novel method helps with QTL mapping by identifying potential genes (Ali et al. 2022a, b). More recently, study series and scientific studies have been made accessible to clarify the roles of proteins, metabolites, and genes in legume DS sensitivity. In the OMICS technique, phenomics is an additional method for trait discovery. Next-generation sequencing (NGS) has ushered in a new age for transcriptomics-based sequencing of legumes. sRNA genome-scale surveys have been conducted using a variety of NGS techniques (Song et al. 2011). For example, the development of a transcriptome atlas for soybeans has made it possible to use the NGS technology to perform RNA sequencing of samples from 14 different drought-stressed environments (Joshi et al. 2010). Wang et al. (2018a, b) have demonstrated recently that RNA-sequencing may be used to identify the soybean transcriptional response to DS. Comparative transcriptome analysis clarifies the transcriptional changes in drought-tolerant and drought-sensitive soybean cultivars during drought stress in a different research (Prince et al. 2015). A transcriptome analysis of the genes in chickpeas that produce oxylipin revealed that early jasmonate induction in the roots under a water deficit (De Domenico et al. 2012). A total of 35 proteins were found in the roots under DS, according to proteomics research on soybeans. In roots experiencing drought, ferritin-type proteins that offer a protective barrier against oxidative stress expressed upregulation as opposed to the corresponding controls (Yamaguchi et al. 2010). Studies on drought-related topics were also conducted using chickpeas; SUPERSAGE analysis discovered 106 SSR and expressed sequence tags

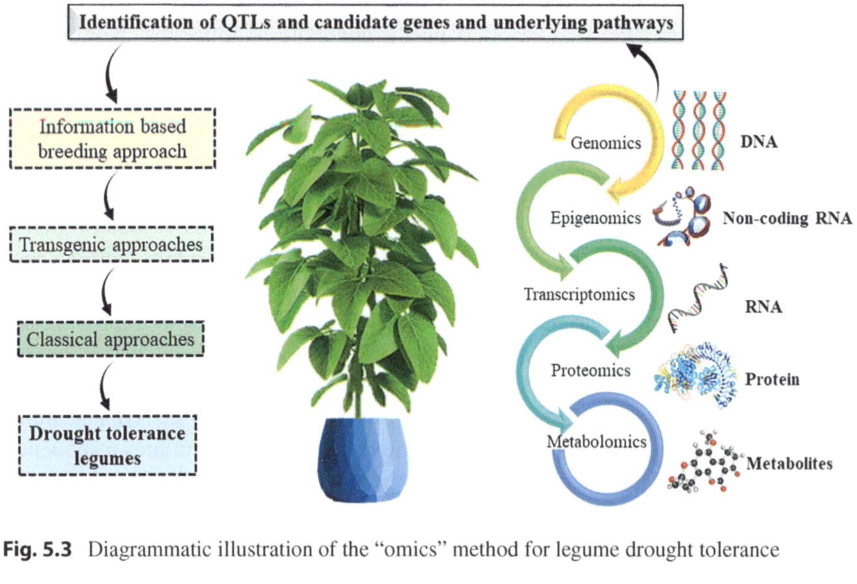

Fig. 5.3 Diagrammatic illustration of the "omics" method for legume drought tolerance

(EST)-based markers, as well as root features. ESTs offer high-quality transcripts for the identification of genes and the development of functional markers connected to drought tolerance, which may be useful in the breeding of drought-tolerant legumes (Varshney and Dubey 2009). Pandey et al. (2008) identified dehydration-responsive proteins in chickpea, which are critical for signal transmission and cell wall remodeling during dehydration. They revealed that 205 differently regulated protein locations and 147 differentially expressed proteins were involved in cell signaling, nucleocytoplasmic transport, ROS pathway, chromatin remodeling, molecular chaperones, replication and transcription of genes, and chromatin remodeling. In a similar vein, Pandey et al. (2008) discovered that under stress, certain LEA proteins known as dehydrins (CaN-600) were generated, scavenging ROS to protect enzyme function. Furthermore, proteomes are related in many biochemical processes and will synthesize several metabolic products in the event of a drought. In a recent study on soybeans, Das et al. (2017) demonstrated that in situations of water deprivation, phytochemical metabolism, sugar and nitrogen metabolism, and metabolomic profiling have primary relevance. In conclusion, the combination of these "omics" techniques would result in legumes that are resistant to drought.

5.3.5 CRISPR/Cas9

CRISPR/Cas9 is the most effective and precise genome editing (GE) method currently on the market. Under unpredictable environmental conditions, scientists, breeders, and policymakers must priorities sustainable crop production to ensure food security in the face of a fast-growing human population. Nevertheless, crop enhancement via random mutagenesis or genetic recombination is very time-consuming and cannot keep up with the growing need for food. With the use of CRISPR/Cas9, it is now possible to more effectively modify any genomic sequence while retaining any desired target gene. With the use of CRISPR/Cas9, desirable features in non-genetically modified plants may be produced, which can help increase agricultural yields under abiotic stress situations (Fig. 5.4). CRISPR/Cas9 application has been documented in a number of crops in recent years, including maize (Agarwal et al. 2018), potatoes (Nakayasu et al. 2018; Ahmad et al. 2022, 2024a), barley (Kapusi et al. 2017), wheat (Sanchez-Leon et al. 2018; Ahmad et al. 2022, 2024b), and rice (Huang et al. 2018). According to recent assessments, CRISPR/Cas9-mediated GE is crucial for producing crops that are more resilient to abiotic stressors (Abdelrahman et al. 2018). Even though CRISPR/Cas9 has only been used in a small number of studies to alter genes associated to drought tolerance in legumes, it is essential for future use in molecular breeding to improve drought tolerance.

The first effective CRISPR/Cas9-mediated GE in soybean was accomplished by Cai and colleagues. Six sgRNAs that targeted different locations of two genes (GmFE12 and GmSHR) in soybeans were employed, and a single sgRNA for transgene (bar) was used to investigate the effectiveness of sgRNAs in hairy roots. The soybean gene (GmFT2) was shown to be linked to flowering time (Cai et al. 2018).

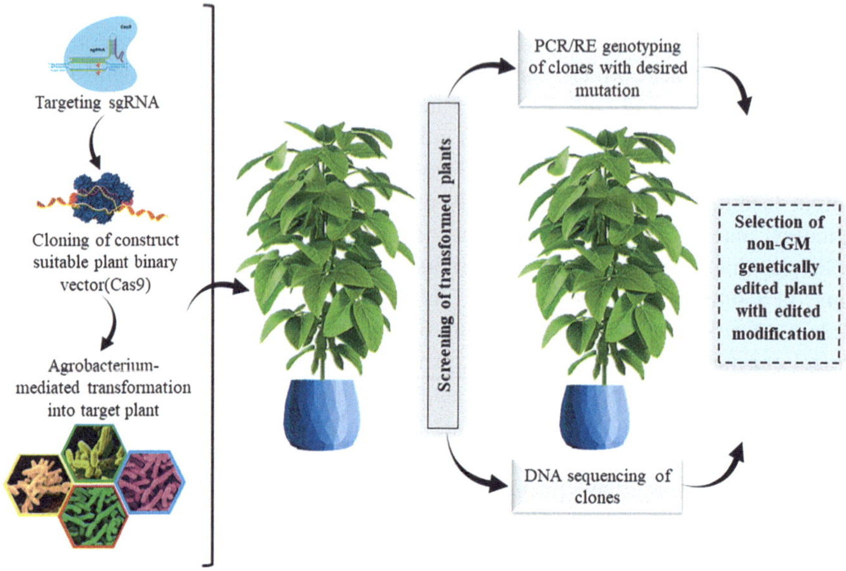

Fig. 5.4 Diagram illustrating the Cas9/sgRNA method of genome editing (GE). The target gene is first chosen, and then internet tools are used to create and synthesize sgRNAs. Target sgRNA cloning using Cas9 (or its variation) often involves the use of a plant binary vector in order to facilitate Agrobacterium-mediated transformation into the desired plant species. Following PCR/RE genotyping, transformed plants are then chosen based on the presence of Cas9 and sgRNA. Lastly, the plants with the appropriate editing or mutation are chosen using DNA sequencing

GmFT2 mutants showed late blooming in both short-day and long-day environments. Therefore, the focused and precise mutagenesis of CRISPR/Cas9 GE offers great promise for creating top cultivars of legumes with improved and long-lasting climate resistance. In conclusion, CRISPR/Cas9 will be the most effective technique for ensuring food security and reducing world hunger since it can target gene variation of complex quantitative features in crops.

5.4 Recent Advances in Legume Breeding

The goal of recent developments in legume breeding has been to increase crop yield and tolerance to biotic and abiotic challenges through the merging of genomics-based and classical genetics techniques. Legume breeding programs have been transformed by the use of varied plant genetic resources, high-quality genome sequences, and omics technologies, which have made it possible to identify important genes linked to stress tolerance and yield potential (Udawat 2023; Ali et al. 2022a, b). Through marker-assisted selection, genomic selection, and pyramiding of resistance genes, these developments have made it easier to develop high-yielding, climate-resilient legume cultivars. In the end, this has helped to address issues with scarce agricultural resources and unpredictable stresses brought on by climate change, thereby promoting global food security. Modern molecular technologies combined with conventional breeding techniques have sped up the genetic advancements in legume crops, opening the door for more sustainable farming practices and increased food legume output globally.

5.5 Molecular Techniques Revolutionized Legume Breeding

Legume breeding has undergone a revolution because of molecular technology, which offers instruments for precise genetic improvement and improved crop growth. Breeding programs have been accelerated through the usage of Marker-Assisted Selection (MAS), which has been made possible by the development of linkage and QTL maps, germplasm characterization, and the identification of genes linked to abiotic and biotic stresses. These techniques include DNA-based molecular markers like Random Amplified Polymorphic DNA (RAPD), Amplified fragment length polymorphism (AFLP), Simple sequence repeats (SSRs), and Single nucleotide polymorphisms (SNPs) (Nadeem et al. 2018). Furthermore, improvements in high-throughput genotyping (HTG), high-throughput phenotyping (HTP), and next-generation sequencing (NGS) technology have allowed legume breeders to rearrange gene combinations, optimize genetic potential, and create improved haplotypes for customized crop varieties (Fig. 5.5). Legume breeding programs are now much more efficient and precise because of the integration of several molecular markers, MAS, genome editing technologies like CRISPR/Cas9, and other omics

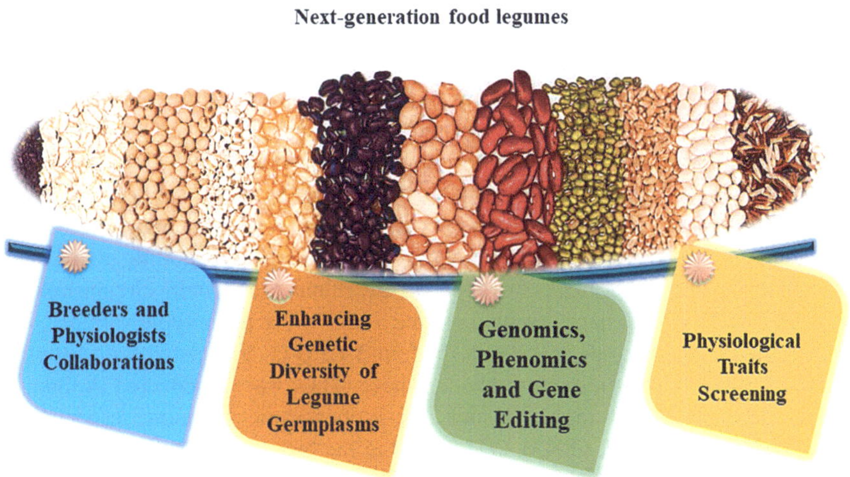

Fig. 5.5 A theoretical foundation for creating drought-tolerant legumes of the future. Breeders and physiologists can work together to generate drought-tolerant legumes by utilizing developments in genomics, phenomics, and gene editing

methods. This has boosted the productivity and quality of legumes for global food security (Maurya et al. 2023).

Several molecular approaches are frequently used in legume breeding to improve crop growth and solve issues pertaining to biotic and abiotic stressors. As mentioned in Maurya et al. (2023), these methods include the use of DNA-based molecular markers for linkage and QTL mapping, such as RAPD, AFLP, SSRs, SNPs, and DArTs. Breeding programs are accelerated by the use of Marker-Assisted Selection (MAS), which links DNA markers to certain features. Furthermore, by utilizing single nucleotide polymorphism (SNP) markers, next-generation molecular breeding techniques identify and exploit resistance genes from landraces and wild relatives (Pakeerathan 2023; Ijaz et al. 2024). Legume breeding efforts are being advanced and stress-related losses are being mitigated by the integration of many molecular tools, including genome-editing technologies like CRISPR/Cas9, with omics techniques (Maurya et al. 2023).

5.6 Methods of Breeding Legumes to Increase Their Resistance to Drought

By modifying traditional breeding procedures, efforts have been undertaken to improve and increase drought tolerance; nevertheless, this is a long process that involves complicated indicators (Chamarthi et al. 2019). Because of the properties of molecular markers, the use of Marker-Assisted Selection (MAS) to enhance drought tolerance in cowpea and many other leguminous crops has gained popularity. Several molecular markers are used to identify and categorize certain genes and genomic regions that influence and promote the development of the complex drought tolerance traits of cowpea (Boukar et al. 2016; Ullah et al. 2023b). Cowpea's drought resistance is elaborated through a variety of biochemical and physiological processes that are well understood by molecular markers (Nadeem et al. 2019). According to Gupta et al. (2017), quantitative trait loci (QTL) map for a number of morpho-physiological variables under drought stress help find loci for markers-assisted breeding. Using SSR markers, the cowpea QTLs were evaluated for agronomic variables associated with grain bulk and pod breaking (Andargie et al. 2011; Ijaz et al. 2024). QTL pertaining to seed size were identified, exhibiting phenotypic differences within an 8.9–19.1% range. In greenhouse and field circumstances, QTL analyzed for leaf shape revealed around 71.5–74.7% phenotypic variance.

In cowpea, Muchero et al. (2009a, b) found 12 QTL associated with drought tolerance in seedlings and mature plants. These QTLs were discovered in linkage groups 1 and 10, exhibiting 4.7% and 24.2% phenotypic variation, respectively. According to Lucas et al. (2013), QTL found in the bigger seed genotypes showed improved growth and production. For cowpea QTL analysis at the seedling stage, the assessment of drought tolerance in morpho-physiological parameters and heritable attributes is useful. The most stable yield QTLs were discovered on Pv08 and Pv01 in a marker-assisted selection of common beans, which made them appropriate candidate genes for marker development (Diaz et al. 2018; Ullah et al. 2023 a, b).

By combining QTL identification with conventional breeding, better germplasms were able to be sealed sooner (Bernardo 2008).

Thanks to molecular techniques, superior genes and genetics are currently accessible, which will further improve the breeding program (Andleeb et al. 2020). Understanding the second phase is aided by the assignment of candidates to clone genes, which alters certain features or genes referred by closed linkage map to loci and controlling attributes (Pflieger et al. 2001). Physiological analysis and genetic transmission are then used to establish the final validity of candidate genes linked to drought QTLs that are involved in ABA, dehydrin synthesis, invertase activity, and transcription factors (Latif et al. 2020). A number of potential genes linked to certain QTLs were examined and mapped for drought-responsive characteristics in various crops (Mir et al. 2012; Gudys et al. 2018). Carreno-Quintero et al. (2013) reported that potential phosphatidate phosphatases, ascorbate peroxidase, glutathione reductase, transferase, and ABA biosynthesis genes were cloned from cowpea under water-stressed conditions. More focus is being paid to cowpea ESTs and gene space sequences, which may accelerate the advancement of possible orthologues in other crop species (Abdou 2012).

Comparative genomics, which performs extensive genome conservation by comparing the genomes of mungbean, cowpea, pea, and lentil, is the third method. Genomes of tropical and temperate legumes are supported by a genetic link between model crops and legumes that was discovered by comparative mapping beforehand. Comparative genomics is a potentially useful and illuminating method that plays a major genetic role in a number of physiological functions. Since cowpea comparative genomics has published data that is freely accessible, it may be used for other leguminous crops such as soybean and common beans (Blair et al. 2002). This will allow the legume species that are important for saturation mapping for the drought QTLs to be identified. Additionally, because cowpea and model legumes share comparable micro- and macro-synthetic processes, it will be easier to find appropriate markers for map-based gene isolation and marker-assisted attribute selection, two processes that are crucial for cowpea development (Timko et al. 2008).

Techniques for assessing the drought resilience of legumes Legume crops are gaining popularity globally owing to their drought-tolerant, nutritious, and economical characteristics (Ravelombola et al. 2017). However, significant progress has not yet been made in breeding to produce genotypes that are both highly productive and drought-tolerant (Nadeem et al. 2019; Haider et al. 2021a, b). The lack of straightforward, affordable, and trustworthy screening methods for choosing plants and progenies resistant to drought from segregating populations may be the cause of this discrepancy. Two strategies for drought tolerance have been developed by scientists to screen and breed agricultural plants. The first method uses criteria and factual data that are based on crop yield (Rauf et al. 2016). The second strategy is physiological or analytical, in which particular morphological or physiological characteristics expressively influence yield and growth under conditions of water deficiency (Gonçalves et al. 2019). Yield-based line selection throughout various locales and periods has resulted in great success in cowpea drought-tolerant genotypes (Zhang

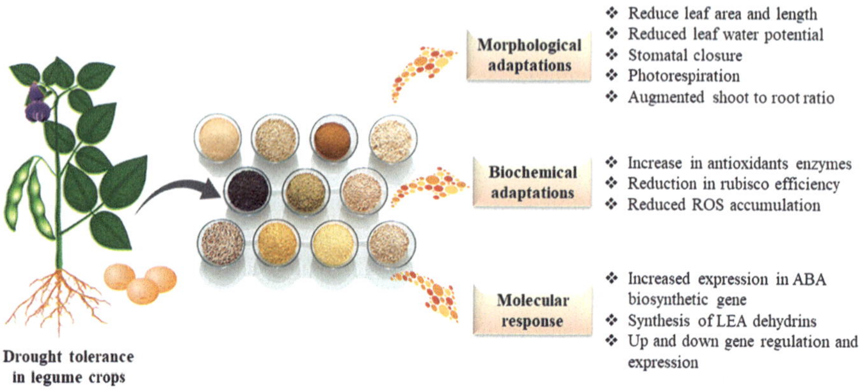

Fig. 5.6 Mechanisms of legume adaptation that contribute to drought tolerance

QingWei and Bartels 2018). Because the analytical technique requires less time and money to analyze, it has superseded the empirical approach. Numerous lines' yields from various periods and places were assessed for significant environmental variations and impacts, as well as genotype-environment interactions.

Additionally, it can be assumed that the strategy of expressing physiological indices in conjunction with selection for potential yield under favorable conditions is drought tolerant; on the other hand, repeated stress conditions might be considered the quick, most effective, and cost-effective in the given situation (Fig. 5.6). The development of genotypes resistant to drought may be accomplished quickly and easily by combining physiological traits with actual environmental circumstances; nevertheless, association with yield remains a difficulty (Gonçalves et al. 2019). A number of drought-tolerant genotypes in cowpea have been generated through the screening process that falls under the purview of breeding Munjonji et al. 2018; Ali et al. 2022a, b). Other leguminous crops that were eliminated in favor of drought-tolerant qualities included the larger root system of chickpeas, increased physiological indices in soybean genotypes, high-yielding characteristics in groundnuts, and improved agro-morphological and physiological traits in common beans (Kashiwagi et al. 2008; Koolachart et al. 2013; Hossain et al. 2014; Arruda et al. 2019).

5.7 Classical Breeding Methods in Legume Crops for Drought Tolerance

In order to increase drought tolerance and discover genetic variety, traditional breeding techniques are essential for legume crops. Because of the advantages legume-based cropping systems provide for the environment, there is a growing need for targeted breeding efforts to keep up with the needs of a changing climate (Rubiales 2023). Molecular markers are crucial instruments for evaluating genetic diversity, characterizing banks of germplasm, and choosing parents from a variety of backgrounds for breeding initiatives (Bunjkar et al. 2024). According to Vergata et al. (2024), certain accessions of legumes, such chickpeas and lentils, exhibit encouraging characteristics when exposed to unfavorable circumstances, like heat stress and water stress. Furthermore, it is crucial to take the plant microbiome into account when utilizing microbiome engineering techniques to improve drought tolerance in legumes (Vergata et al. 2024; Haider et al. 2022a). Legume cultivars with enhanced genetic diversity and drought tolerance may be generated by combining traditional breeding techniques, molecular markers, and microbiome research, guaranteeing sustainable agricultural production in the face of climate change problems.

Through the use of genetic variety, traditional breeding techniques significantly contribute to the improvement of drought tolerance in legume crops by creating cultivars that are resistant to water shortage. Classical breeding seeks to find and include advantageous features for drought resistance in legumes using techniques including screening, breeding, and marker-assisted selection (Bagheri et al. 2023). Furthermore, the use of genomics-assisted breeding methods, such as fine mapping and association mapping, aids in identifying certain genes and molecular markers linked to drought tolerance in crops like chickpea (Thakro et al. 2023). In the end, these techniques help create climate-resilient legume varieties that can flourish in water-limited environments by facilitating the selection and development of cultivars with enhanced root and shoot biomass, photosynthetic efficiency, and yield under drought conditions (Thakro et al. 2023; Thudi et al. 2024).

5.8 Role of Genetic Markers in Enhancing Drought Tolerance in Legume Crops

Genetic markers are essential for improving drought tolerance in crops such as wheat, rice, and chickpeas because they facilitate the identification of important genes linked to drought resistance. Research on rice has demonstrated that by introducing drought-effect QTLs using marker-assisted backcross breeding, better lines with higher yields under water-limited circumstances were produced. Similar findings were made on genes regulating drought regulation and plant development when researchers studied chickpeas and found SNP markers linked to a variety of drought-linked phenotypes (Seeli et al. 2024). Additionally, GWAS investigation in wheat revealed important SNPs regulating physiological and morphological characteristics linked to drought tolerance at the seedling stage, offering useful data for breeding initiatives focused on creating genotypes resistant to drought (Istanbuli et al. 2024; Haider et al. 2022b). All things considered, genetic markers are effective instruments for deciphering the genetic basis of drought tolerance and aiding in the creation of robust crop types.

Genetic markers are essential for identifying drought-tolerant plants because they make it possible to identify certain genes linked to drought tolerance. Single nucleotide polymorphism (SNP) markers associated with drought tolerance traits in wheat seedlings have been identified by studies using genome-wide association studies (GWAS). This has revealed significant SNPs controlling both morphological and physiological traits under drought stress (Istanbuli et al. 2024). Likewise, Liu et al. underline the significance of genetic variants found by linkage and association mapping in figuring out the genetic foundation of crop resistance to drought, opening the door for genome editing and genetic modification to create cultivars resistant to stress (Liu et al. 2023). In addition, Abegunde et al. (2022) emphasize the use of SSR markers to evaluate genetic diversity in genotypes of drought-tolerant maize, assisting in the breeding programs' selection of a variety of parent materials (Abegunde et al. 2022). These markers aid in the identification of genes resistant to drought, and they also make marker-assisted selection easier, allowing for the development of plants that are more resilient to drought stress.

5.9 Conclusion and Future Perspective

In conclusion, the production of drought-tolerant legume varieties has advanced significantly thanks to classical breeding techniques, proving the effectiveness of these strategies in the face of growing climatic unpredictability. The historical background of drought-tolerant legume breeding demonstrates a distinct trajectory of using genetic variation to promote physiological features linked to drought resistance, such as better root structure, increased water efficiency, and enhanced osmotic adjustment capacity. Adaptability and yield stability in low-water circumstances are demonstrated by cultivars such as cowpea and mungbean, which have become prime examples of resilience. The potential to further improve drought tolerance in legumes has greatly expanded as researchers use discoveries from both conventional breeding and contemporary biotechnological technologies. It is anticipated that genetic engineering methods, such as marker-assisted selection, combined with physiological understanding would completely transform the breeding environment. There are still difficulties, though, especially when it comes to using these qualities in a variety of agro-ecological settings in a sustainable manner. In addition to production, future breeding programs should prioritize resilience qualities that guarantee food security in the face of changing climate circumstances. The adoption of drought-tolerant legume varieties is contingent upon the active participation of local farmers in participatory breeding programs, which guarantees that the generated cultivars are tailored to the unique requirements of agricultural communities. Lessons from classical breeding experiences in drought tolerance will continue to teach and inspire future agricultural innovations, eventually boosting the sustainability of legume production systems globally as efforts to tackle food insecurity escalate on a global scale. Recurrent selection, crossbreeding, and phenotypic selection are examples of classical breeding techniques that have helped breeders uncover and add essential drought resistance characteristics to legume cultivars. Thus, effective cultivators have produced legume varieties with increased resistance to water shortage, strong root systems, and enhanced water-use efficiency.

The identification of important genetic variables linked to drought tolerance, the careful selection and assessment of germplasm, and the use of successful breeding techniques are responsible for the success of classical breeding techniques. The cooperation of molecular scientists, agronomists, and breeders has accelerated progress by combining cutting-edge instruments and methods to speed up the breeding process. It is important to recognize the constraints associated with traditional breeding methodologies. To get desirable features, the procedure is laborious and involves crossover and selection over many generations. It is also difficult to find new sources of drought tolerance in legume germplasm due to the very low genetic diversity in this population. Future breeding programs for legumes should use cutting-edge technologies like genomic and marker-assisted selection to circumvent these constraints, speeding up the breeding process and improving the accuracy of trait selection. Furthermore, undiscovered sources of drought tolerance in legumes can be unlocked by using the rich genetic variety found in exotic germplasm and wild cousins. Our capacity to generate robust legume varieties that can tolerate water shortage has greatly progressed thanks to the use of traditional breeding techniques in legume breeding for drought tolerance. Future breeding efforts will be guided by the information gathered from past experiences, which will allow for the development of legume cultivars that both overcome the problems posed by climate change and contribute to food security. Legume breeding for drought resistance has enormous potential to provide sustainable agriculture, improve farmer livelihoods, and lessen the effects of climate change on legume output with sustained innovation and cooperation. As we advance, the secret to realizing legumes' full potential for flourishing and adapting to water-limited conditions will be to combine traditional breeding methods with state-of-the-art technology.

Author Contributions S.I., J.I., B.A.A., and Z.K. designed and conceived the chapter idea. S.I., J.I., B.A.A., and Z.K., A.M., T.Y., M.M., N.I. wrote the original draft. N.I., A.F., S.K., and T.M. reviewed and edited the manuscript and provided valuable insights. All authors have made valuable revisions and edited the manuscript and approved the last version.

References

Abdel CG, Stutzel H (2016) Effects of water stress on Epidermis and stomata population of sixteen water stressed and irrigated barley (*Hordeum vulgare* L.) genotypes. Euphrates J Agric Sci 8:48–62

Abdel-Haleem H, Lee GJ, Boerma RH (2011) Identification of QTL for increased fibrous roots in soybean. Theor Appl Genet 122:935–946

Abdelrahman M, Al-Sadi AM, Pour-Aboughadareh A, Burritt DJ, Tran LSP (2018) Genome editing using *CRISPR/Cas9*–targeted mutagenesis: an opportunity for yield improvements of crop plants grown under environmental stresses. Plant Physiol Biochem 131:31–36

Abdou RIB (2012) Using morphological and physiological factors to evaluate six cowpea varieties for drought tolerance (Doctoral dissertation)

Abegunde OI, Idehen EO, Oduwaye OA, Fabunmi TO, Kehinde OB (2022) Genetic characterization of drought-tolerant maize genotypes based on SSR markers. Nigerian J Biotechnol 39(2):97–105

Abid G, Mahmoud M, Mingeot D, Aouida M (2016) Effect of drought stress on chlorophyll fluorescence, antioxidant enzyme activities and gene expression patterns in faba bean (*Vicia faba* L.). Arch Agron Soil Sci 63:536–552

Agarwal A, Yadava P, Kumar K, Singh I (2018) Insights into maize genome editing via *CRISPR/Cas9*. Physiol Mol Biol Plants 24:175–183. [Google Scholar] [CrossRef]

Ahmad Y, Haider S, Iqbal J, Abbasi BA, Yaseen T, Mahmood T (2022) The mechanisms of genome editing technologies in crop plants. In: Principles and practices of OMICS and genome editing for crop improvement. Springer International Publishing, Cham, pp 295–313

Ahmad Y, Haider S, Iqbal J, Naseer S, Attia KA, Mohammed AA, Mahmood T (2024a) In-silico analysis and transformation of OsMYB48 transcription factor driven by CaMV35S promoter in model plant–Nicotiana tabacum L. conferring abiotic stress tolerance. GM Crops Food 15(1):130–149

Ahmad Y, Haakim Z, Iqbal J, Abbasi BA, Mahmood T, Kazi M (2024b) Technological innovations for abiotic stress resistance in horticultural crops. In: OMICs-based techniques for global food security. Wiley, pp 233–244

Ali A, Altaf MT, Nadeem MA, Karaköy T, Shah AN, Azeem H, Baloch FS, Baran N, Hussain T, Duangpan S, Aasim M (2022a) Recent advancement in OMICS approaches to enhance abiotic stress tolerance in legumes. Front Plant Sci 13:952759

Ali I, Salah KBH, Sher H, Ali H, Ullah Z, Ali A, Mahmood T (2022b) Drought stress enhances the efficiency of floral dip method of agrobacterium-mediated transformation in Arabidopsis thaliana. Braz J Biol 84:e259326

Anbazhagan K, Bhatnagar-Mathur P, Vadez V, Dumbala SR, Kishor PK, Sharma KK (2014) DREB1A overexpression in transgenic chickpea alters key traits influencing plant water budget across water regimes. Plant Cell Rep 34:199–210

Andargie M, Pasquet RS, Gowda BS, Muluvi GM, Timko MP (2011) Construction of a SSR-based genetic map and identification of QTL for domestication traits using recombinant inbred lines from a cross between wild and cultivated cowpea (V. unguiculata (L.) Walp.). Mol Breed 28:413–420

Andersen MN, Asch F, Wu Y, Jensen CR, Næsted H, Mogensen VO, Koch KE (2015) Soluble invertase expression is an early target of drought stress during the critical, abortion-sensitive phase of young ovary development in maize. Plant Physiol 130:591–604

Andleeb T, Shah T, Nawaz R, Munir I, Munsif F, Jalal A (2020) QTL mapping for drought stress tolerance in plants. In: Salt and drought stress tolerance in plants: signaling networks and adaptive mechanisms. Springer, Cham, pp 383–403

Andrianasolo FN, Debaeke P, Champolivier L, Maury P (2016) Analysis and modelling of the factors controlling seed oil concentration in sunflower: a review. OCL 23:D206

Arruda IM, Moda-Cirino V, Koltun A, Zeffa DM, Nagashima GT, Gonçalves LSA (2019) Combining ability for agromorphological and physiological traits in different gene pools of common bean subjected to water deficit. Agronomy 9(7):371

Asati R, Tripathi MK, Tiwari S, Yadav RK, Tripathi N (2022) Molecular breeding and drought tolerance in chickpea. Life 12(11):1846

Atti S, Bonnell R, Smith D, Prasher S (2004) Response of an indeterminate soybean {*Glycine Max* (L.) Merr} to chronic water deficit during reproductive development under greenhouse conditions. Can Water Resour J 29:209–222

Awari VR, Mate SN (2015) Effect of drought stress on early seedling growth of chickpea (*Cicer arietinum* L.) genotypes. Life Sci Int Res J 2:356–361

Bagheri M, Santos CS, Rubiales D, Vasconcelos MW (2023) Challenges in pea breeding for tolerance to drought: status and prospects. Ann Appl Biol 183(2):108–120

Baroowa B, Gogoi N (2014) Biochemical changes in black gram and green gram genotypes after imposition of drought stress. J Food Legum 27:350–353

Basu PS, Berger JD, Turner NC, Chaturvedi SK, Ali M, Siddique KHM (2007) Osmotic adjustment of chickpea (*Cicer arietinum* L.) is not associated with changes in carbohydrate composition or leaf gas exchange under drought. Ann Appl Biol 150:217–225

Beebe SE, Rao IM, Cajiao C, Grajales M (2008) Selection for drought resistance in common bean also improves yield in phosphorus limited and favorable environments. Crop Sci 48:582–592

Bernardo R (2008) Molecular markers and selection for complex traits in plants: learning from the last 20 years. Crop Sci 48(5):1649–1664

Bhatnagar-Mathur P, Vadez V, Jyostna Devi M, Lavanya M, Vani G, Sharma KK (2009) Genetic engineering of chickpea (*Cicer arietinum* L.) with the *P5CSF129A* gene for osmoregulation with implications on drought tolerance. Mol Breed 23:591–606

Blair MW, Munoz MC, Beebe S (2002) 1.2. 1 QTL analysis of drought and abiotic stress tolerance in common bean RIL populations. PROJECT SB-2 ASSESSING AND UTILIZING AGROBIODIVERSITY THROUGH BIOTECHNOLOGY, 2

Boukar O, Fatokun CA, Huynh BL, Roberts PA, Close TJ (2016) Genomic tools in cowpea breeding programs: status and perspectives. Front Plant Sci 7:757

Briñez B, Morini J, Cardoso K, Rosa JS, Bassi D, Gonçalves GR, Almeida C, Fausto J, Paulino DC, Blair MW et al (2017) Mapping QTLs for drought tolerance in a SEA 5 × AND 277 common bean cross with SSRs and SNP markers. Genet Mol Biol 823:813–823

Bunjkar A, Walia P, Sandal SS (2024) Unlocking genetic diversity and germplasm characterization with molecular markers: strategies for crop improvement. J Adv Biol Biotechnol 27(6):160–173

Cai Y, Chen L, Liu X, Guo C, Sun S, Wu C, Jiang B, Han T, Hou W (2018) *CRISPR/Cas9*-mediated targeted mutagenesis of *GmFT2a* delays flowering time in soybean. Plant Biotechnol J 16:176–185

Carreno-Quintero N, Bouwmeester HJ, Keurentjes JJ (2013) Genetic analysis of metabolome–phenotype interactions: from model to crop species. Trends Genet 29(1):41–50

Chamarthi SK, Belko N, Togola A, Fatokun CA, Boukar O (2019) Genomics-assisted breeding for drought tolerance in cowpea. In: Genomics assisted breeding of crops for abiotic stress tolerance, Vol. II. Springer, Cham, pp 187–209

Chapman SC (2008) Use of crop models to understand genotype by environment interactions for drought in real-world and simulated plant breeding trials. Euphytica 161:195–208

Chowdhury JA, Karim MA, Khaliq QA, Ahmed AU, Khan MSA (2016) Effect of drought stress on gas exchange characteristics of four soybean genotypes. Bangladesh J Agr Res 41:195–205

Das A, Rushton P, Rohila J (2017) Metabolomic profiling of soybeans (*Glycine max* L.) reveals the importance of sugar and nitrogen metabolism under drought and heat stress. Plan Theory 6:21

De Domenico S, Bonsegna S, Horres R, Pastor V, Taurino M, Poltronieri P, Imtiaz M, Kahl G, Flors V, Winter P et al (2012) Transcriptomic analysis of oxylipin biosynthesis genes and chemical profiling reveal an early induction of jasmonates in chickpea roots under drought stress. Plant Physiol Biochem 61:115–122

De Souza PI, Egli DB, Bruening WP (1997) Water stress during seed filling and leaf senescence in soybean. Agron J 89:807–812

Desclaux D, Roumet P (2000) Identification of soybean plant characteristics that indicate the timing of drought stress. Crop Sci 40:716–722

Diaz LM, Ricaurte J, Tovar E, Cajiao C, Teran H, Grajales M et al (2018) QTL analyses for tolerance to abiotic stresses in a common bean (Phaseolus vulgaris L.) population. PLoS One 13(8):e0202342

Duc G, Agrama H, Bao S, Berger J, Bourion V, De Ron AM, Gowda CLL, Mikic A, Millot D, Singh KB et al (2015) Breeding annual grain legumes for sustainable agriculture: new methods to approach complex traits and target new cultivar ideotypes. CRC Crit Rev Plant Sci 34:381–411

El Haddad N, En-Nahli Y, Choukri H, Aloui K, Mentag R, El-Baouchi A et al (2023) Metabolic mechanisms underlying heat and drought tolerance in lentil accessions: implications for stress tolerance breeding. Plan Theory 12(23):3962

Farooq M, Wahid A, Kobayashi N, Fujita D, Basra SMA (2009) Plant drought stress: effects, mechanisms and management. Agron Sustain Dev 29:185–212

Farooq M, Farooq M, Hussain M, Siddique KHM (2014) Drought stress in wheat during flowering and grain- filling periods. Crit Rev Plant Sci 33:331–349. https://doi.org/10.1080/07352689.2014.875291

Farooq M, Gogoi N, Barthakur S, Baroowa B, Bharadwaj N, Alghamdi SS (2016) Drought stress in grain legumes during reproduction and grain filling. J Agron Crop Sci 203:81–102

Ghassemi-golezani K, Ghassemi S, Bandehhagh A (2013) Effects of water supply on field performance of chickpea (*Cicer arietinum* L.) cultivars. Int J Agron Plant Prod 4:94–97

Gonçalves JGR, Andrade ERD, Silva DAD, Esteves JADF, Chiorato AF, Carbonell SAM (2019) Drought tolerance evaluated in common bean genotypes. Ciência e Agrotecnologia 43:e001719

Graham PH, Vance CP (2003) Update on legume utilization legumes: importance and constraints to greater use. Plant Physiol 131:872–877

Gudys K, Guzy-Wrobelska J, Janiak A, Dziurka MA, Ostrowska A, Hura K et al (2018) Prioritization of candidate genes in QTL regions for physiological and biochemical traits underlying drought response in barley (Hordeum vulgare L.). Front Plant Sci 9:769

Gupta PK, Balyan HS, Gahlaut V (2017) QTL analysis for drought tolerance in wheat: present status and future possibilities. Agronomy 7(1):5

Gusmao M, Siddique KHM, Flower K, Nesbitt H, Veneklaas EJ (2012) Water deficit during the reproductive period of grass Pea (*Lathyrus sativus* L.) reduced grain yield but maintained seed size. J Agron Crop Sci 198:430–441

Haider S, Iqbal J, Naseer S, Yaseen T, Shaukat M, Bibi H, Mahmood T (2021a) Molecular mechanisms of plant tolerance to heat stress: current landscape and future perspectives. Plant Cell Rep 40:1–25

Haider S, Iqbal J, Shaukat M, Naseer S, Mahmood T (2021b) The epigenetic chromatin-based regulation of somatic heat stress memory in plants. Plant Gene 27:100318

Haider S, Raza A, Iqbal J, Shaukat M, Mahmood T (2022a) Analyzing the regulatory role of heat shock transcription factors in plant heat stress tolerance: a brief appraisal. Mol Biol Rep 49(6):5771–5785

Haider S, Iqbal J, Naseer S, Shaukat M, Abbasi BA, Yaseen T, Mahmood T (2022b) Unfolding molecular switches in plant heat stress resistance: a comprehensive review. Plant Cell Rep 41(3):775–798

Hao X, Li P, Feng Y, Han X, Gao J, Lin E, Han Y (2013) Effects of fully open-air [CO_2] elevation on leaf photosynthesis and ultrastructure of *Isatis indigotica* Fort. PLoS One 8:e74600

He J, Du Y, Wang T, Turner NC, Yang R, Jin Y, Xi Y, Zhang C, Cui T, Fang X et al (2016) Conserved water use improves the yield performance of soybean. Agric Water Manag 179:236

Hossain MM, Liu X, Qi X, Lam HM, Zhang J (2014) Differences between soybean genotypes in physiological response to sequential soil drying and rewetting. Crop J 2(6):366–380

Hou M, Tian F, Zhang L, Li S, Du T, Huang M, Yuan Y (2018) Estimating crop transpiration of soybean under different irrigation treatments using thermal infrared remote sensing imagery. Agronomy 9:8

Huang L, Zhang R, Huang G, Li Y, Melaku G, Zhang S, Chen H, Zhao Y, Zhang J, Zhang Y et al (2018) Developing superior alleles of yield genes in rice by artificial mutagenesis using the *CRISPR/Cas9* system. Crop J 6:475–481

Hussain M, Farooq S, Hasan W, Ul-allah S, Tanveer M (2018) Drought stress in sunflower: physiological effects and its management through breeding and agronomic alternatives. Agric Water Manag J 201:152–166

Ijaz S, Iqbal J, Abbasi BA, Yaseen T, Rehman S, Kazi M, Mahmood T (2024) Role of OMICS-based technologies in plant sciences. In: OMICs-based techniques for global food security. Wiley, pp 45–66

Istanbuli T, Nassar AE, Abd El-Maksoud MM, Tawkaz S, Alsamman AM, Hamwieh A (2024) Genome-wide association study reveals SNP markers controlling drought tolerance and related agronomic traits in chickpea across multiple environments. Front Plant Sci 15:1260690

José A, Rodrigues J, Filho M, Rodrigues C, Sales G, Célia R, Pires DM, Machado EC (2018) Source-sink relationships in two soybean cultivars with indeterminate growth under water deficit. Bragantia 77:23–35

Joshi T, Yan Z, Libault M, Jeong D, Park S, Green PJ, Sherrier DJ, Farmer A, May G, Meyers BC et al (2010) Prediction of novel *miRNAs* and associated target genes in *Glycine max* L. BMC Bioinform 11:S14

Kamanga RM, Mbega E, Ndakidemi P (2018) Drought tolerance mechanisms in plants: physiological responses associated with water deficit stress in *Solanum lycopersicum*. Adv Crop Sci Technol 6:1–8

Kapusi E, Corcuera-Gómez M, Melnik S, Stoger E (2017) Heritable genomic fragment deletions and small indels in the putative *ENGase* gene induced by *CRISPR/Cas9* in barley. Front Plant Sci 8:540

Kashiwagi J, Krishnamurthy L, Upadhyaya HD, Krishna H, Chandra S, Vadez V, Serraj R (2005) Genetic variability of drought-avoidance root traits in the mini-core germplasm collection of chickpea (*Cicer arietinum* L.). Euphytica 146:213–222

Kashiwagi J, Krishnamurthy L, Gaur PM, Chandra S, Upadhyaya HD (2008) Estimation of gene effects of the drought avoidance root characteristics in chickpea (C. arietinum L.). Field Crop Res 105(1–2):64–69

Khazaei H, Sullivan DMO, Sillanpää MJ, Stoddard FL (2014) Use of synteny to identify candidate genes underlying QTL controlling stomatal traits in faba bean (*Vicia faba* L.). Theor Appl Genet 127:2371–2385

Kishor PBK, Hong Z, Miao C-H, Hu C-AA, Verma DPS (1995) Overexpression of A1-Pyrroline-5-Carboxylate synthetase increases proline production and confers osmotolerance in transgenic plants. Plant Physiol 108:1387–1394

Kobraee S, Shamsi K, Rasekhi B (2011) Soybean production under water deficit conditions. Sch Res Libr 2:423–434

Kokubun M, Shimada S, Takahashi M (2001) Flower abortion caused by preanthesis water deficit is not attributed to impairment of pollen in soybean. Crop Sci 41:1517–1521

Koolachart R, Jogloy S, Vorasoot N, Wongkaew S, Holbrook CC, Jongrungklang N et al (2013) Rooting traits of peanut genotypes with different yield responses to terminal drought. Field Crop Res 149:366–378

Krishnamurthy L, Kashiwagi J, Tobita S, Ito O, Upadhyaya HD, Gowda CLL, Gaur PM, Sheshshayee MS, Singh S, Vadez V et al (2013) Variation in carbon isotope discrimination and its relationship with harvest index in the reference collection of chickpea germplasm. Funct Plant Biol 40:1350

Latif S, Shah T, Munsif F, D'Amato R (2020) Genetic manipulation of drought stress signaling pathways in plants. In: Salt and drought stress tolerance in plants: signaling networks and adaptive mechanisms. Springer, Cham, pp 367–382

Li D, Liu H, Qiao Y, Wang Y, Cai Z, Dong B, Shi C, Liu Y, Li X, Liu M (2013a) Effects of elevated CO_2 on the growth, seed yield, and water use efficiency of soybean (*Glycine max*.L.) under drought stress. Agric Water Manag 129:105–112

Li Y, Zhang J, Zhang J, Hao L, Hua J, Duan L, Zhang M, Li Z (2013b) Expression of an *Arabidopsis* molybdenum cofactor sulphurase gene in soybean enhances drought tolerance and increases yield under field conditions. Plant Biotechnol J 11:747–758

Li P, Zhang Y, Wu X, Liu Y (2018) Drought stress impact on leaf proteome variations of faba bean (*Vicia faba* L.) in the Qinghai–Tibet Plateau of China. 3 Biotech 8:110

Liu F, Andersen MN, Jacobsen S, Jensen CR (2005) Stomatal control and water use efficiency of soybean (*Glycine max* L. Merr.) during progressive soil drying. Environ Exp Bot 54:33–40

Liu S, Wang H, Qin F (2023) Genetic dissection of drought resistance for trait improvement in crops. Crop J 11(4):975–985

Lucas MR, Ehlers JD, Huynh BL, Diop NN, Roberts PA, Close TJ (2013) Markers for breeding heat-tolerant cowpea. Mol Breed 31:529–536

Mafakheri A, Siosemardeh A, Bahramnejad B, Struik PC, Sohrabi Y (2010) Effect of drought stress on yield, proline and chlorophyll contents in three chickpea cultivars. Aust J Crop Sci 4:580–585

Mak M, Babla M, Xu S, Carrigan AO, Liu X, Gong Y, Holford P, Chen Z (2014) Leaf mesophyll K^+, H^+ and Ca^{2+} fluxes are involved in drought-induced decrease in photosynthesis and stomatal closure in soybean. Environ Exp Bot 98:1–12

Makbul S, Güler NS, Durmuş N, Güven S (2011) Changes in anatomical and physiological parameters of soybean under drought stress. Turk J Botany 35:369–377

Maleki A, Naderi A, Naseri R, Fathi A, Bahamin S, Maleki R (2013) Physiological performance of soybean cultivars under drought stress. Bull Environ Pharmacol Life Sci 2:38–44

Martynenko A, Shotton K, Astatkie T, Petrash G, Fowler C, Neily W, Critchley AT (2016) Thermal imaging of soybean response to drought stress: the effect of Ascophyllum nodosum seaweed extract. Springerplus 5:1393

Maurya V, Singh N, Sharma A, Kumar R (2023) DNA markers-assisted crop improvement for biotic and abiotic stresses in legumes. In: Molecular marker techniques: a potential approach of crop improvement. Springer Nature, Singapore, pp 161–197

Meckel L, Egli DB, Phillips RE, Radcliffe D, Leggett JE (1984) Effect of moisture stress on seed growth in soybeans. Agron J 76:647–650

Mir RR, Zaman-Allah M, Sreenivasulu N, Trethowan R, Varshney RK (2012) Integrated genomics, physiology and breeding approaches for improving drought tolerance in crops. Theor Appl Genet 125:625–645

Mittal N, Mishra A, Singh R, Kumar P (2014) Assessing future changes in seasonal climatic extremes in the Ganges river basin using an ensemble of regional climate models. Clim Chang 123:273–286

Miyauchi Y, Isoda A, Li Z, Wang P (2012) Soybean cultivation on desert sand using drip irrigation with mulch. Plant Prod Sci 15:310–316

Mondal MMA, Fakir SA, Juraimi AS, Hakim MA, Islam MM, Shamsuddoha ATM (2011) Effects of flowering behavior and pod maturity synchrony on yield of mungbean [*Vigna radiata* (L.) Wilczek]. Aust J Crop Sci 5:945–953

Muchero W, Diop NN, Bhat PR, Fenton RD, Wanamaker S, Pottorff M et al (2009a) A consensus genetic map of cowpea [Vigna unguiculata (L) Walp.] and synteny based on EST-derived SNPs. Proc Natl Acad Sci 106(43):18159–18164

Muchero W, Je V, Close TJ, Roberts PA (2009b) Mapping QTL for drought stress-induced premature senescence and maturity in cowpea (*Vigna unguiculata* L.). Theor Appl Genet 118:849–863

Muchero W, Roberts PA, Diop NN, Drabo I, Cisse N, Close TJ, Muranaka S, Boukar O, Ehlers JD (2013) Genetic architecture of delayed senescence, biomass, and grain yield under drought stress in cowpea. PLoS One 8:e70041

Mukeshimana G, Butare L, Cregan PB, Blair MW, Kelly JD (2014) Quantitative trait loci associated with drought tolerance in common bean. Crop Sci 54:923–938

Munjonji L, Ayisi KK, Haesaert G, Boeckx P (2018) Screening cowpea genotypes for high biological nitrogen fixation and grain yield under drought conditions. Agron J 110(5):1925–1935

Nadeem MA, Nawaz MA, Shahid MQ, Doğan Y, Comertpay G, Yıldız M et al (2018) DNA molecular markers in plant breeding: current status and recent advancements in genomic selection and genome editing. Biotechnol Biotechnol Equip 32(2):261–285

Nadeem M, Li J, Yahya M, Sher A, Ma C, Wang X, Qiu L (2019) Research progress and perspective on drought stress in legumes: a review. Int J Mol Sci 20(10):2541

Nakayasu M, Akiyama R, Lee HJ, Osakabe K, Osakabe Y, Watanabe B, Sugimoto Y, Umemoto N, Saito K, Muranaka T et al (2018) Generation of α-solanine-free hairy roots of potato by *CRISPR/Cas9* mediated genome editing of the *St16DOX* gene. Plant Physiol Biochem 131:70–77

Negrao S, Schmöckel SM, Tester M (2017) Evaluating physiological responses of plants to salinity stress. Ann Bot 119:1–11

Ohashi Y, Nakayama N, Saneoka H, Fujita K (2006) Effects of drought stress on photosynthetic gas exchange, chlorophyll fluorescence and stem diameter of soybean plants. Biol Plant 50:138–141

Pakeerathan K (2023) Molecular characterization and mapping of stress resistance genes using SNP platform in legumes. In: Legumes: physiology and molecular biology of abiotic stress tolerance. Springer Nature, Singapore, pp 149–188

Pandey A, Chakraborty S, Datta A, Chakraborty N (2008) Proteomics approach to identify dehydration responsive nuclear proteins from chickpea (*Cicer arietinum* L.). Mol Cell Proteomics 7:88–107

Pflieger S, Lefebvre V, Causse M (2001) The candidate gene approach in plant genetics: a review. Mol Breed 7(4):275–291

Phillips BB, Shaw RF, Holland MJ, Fry EL, Bardgett RD, Bullock JM, Osborne JL (2018) Drought reduces floral resources for pollinators. Glob Chang Biol 24:3226–3235

Piepho HP (2000) A mixed-model approach to mapping quantitative trait loci in barley on the basis of multiple environment data. Genetics 156(4):2043–2050

Prince SJ, Joshi T, Mutava RN, Syed N, Vitor MDSJ, Patil G, Song L, Wang JJ, Lin L, Chen W et al (2015) Comparative analysis of the drought-responsive transcriptome in soybean lines contrasting for canopy wilting. Plant Sci 240:65–78

Pushpavalli R, Zaman-allah M, Turner NC, Baddam R, Rao MV, Vadez V (2015) Higher flower and seed number leads to higher yield under water stress conditions imposed during reproduction in chickpea. Funct Plant Biol 42:162–174

Rauf S, Al-Khayri JM, Zaharieva M, Monneveux P, Khalil F (2016) Breeding strategies to enhance drought tolerance in crops. In: Advances in plant breeding strategies: agronomic, abiotic and biotic stress traits. Springer, Cham, pp 397–445

Ravelombola WS, Shi A, Weng Y, Clark J, Motes D, Chen P, Srivastava V (2017) Evaluation of salt tolerance at germination stage in cowpea [Vigna unguiculata (L.) Walp]. HortScience 52(9):1168–1176

Ribeiro T, da Silva DA, Esteves JADF, Azevedo CVG, Gonçalves JGR, Carbonell SAM, Chiorato AF (2019) Evaluation of common bean genotypes for drought tolerance. Bragantia 78:1–11

Rotundo L, Westgate ME (2009) Meta-analysis of environmental effects on soybean seed composition. F Crop Res 110:147–156

Rubiales D (2023) Plant breeding is needed to meet agroecological requirements: legume crops as a case study. Outlook Agric 52(3):294–302

Rubiales D, Mikic A (2015) Introduction: Legumes in sustainable agriculture. CRC Crit Rev Plant Sci 34:2–3

Saleem A, Roldán-Ruiz I, Aper J, Muylle H (2022) Genetic control of tolerance to drought stress in soybean. BMC Plant Biol 22(1):615

Samarah NH, Haddad N, Alqudah AM (2009) Yield potential evaluation in chickpea genotypes under late terminal drought in relation to the length of reproductive stage. Ital J Agron 3:111–117

Sanchez-Leon S, Gil-humanes J, Ozuna CV, Sousa C, Voytas DF, Barro F (2018) Low-gluten, nontransgenic wheat engineered with *CRISPR/Cas9*. Plant Biotechnol J 16:902–910

Savitri ES (2023) Molecular breeding for the development of drought stress tolerance in soybean. In: QTL mapping in crop improvement. Academic Press, pp 311–323

Seeli FP, Manoharan M, Ayyenar B, Kambale R, Mohanavel V, Rajagopalan VR, Swaminathan M (2024) Genetic improvement of drought tolerance in a mega-rice variety improved white Ponni through marker-assisted backcross breeding. Agriculture 14(3):431

Shavrukov Y, Kurishbayev A, Jatayev S, Shvidchenko V, Zotova L, Koekemoer F, De Groot S, Soole K, Langridge P (2017) Early flowering as a drought escape mechanism in plants: how can it aid wheat production? Front Plant Sci 8:1950

Siddiqui MH, Al-khaishany MY, Al-qutami MA (2015) Response of different genotypes of faba bean plant to drought stress. Int J Mol Sci 16:10214–10227

Singh M, Rani S, Malhotra N, Katna G, Sarker A (2018) Transgressive segregations for agronomic improvement using interspecific crosses between *c. Arietinum* L. × *c. Reticulatum* Ladiz. And *c. Arietinum* L. × *c. Echinospermum* davis species. PLoS One 13:e0203082

Song Q, Liu Y, Hu X, Zhang W, Ma B, Chen S, Zhang J (2011) Identification of *miRNAs* and their target genes in developing soybean seeds by deep sequencing. BMC Plant Biol 11:5

Thakro V, Malik N, Basu U, Srivastava R, Narnoliya L, Daware A et al (2023) A superior gene allele involved in abscisic acid signaling enhances drought tolerance and yield in chickpea. Plant Physiol 191(3):1884–1912

Thudi M, Samineni S, Li W, Boer MP, Roorkiwal M, Yang Z et al (2024) Whole genome resequencing and phenotyping of MAGIC population for high resolution mapping of drought tolerance in chickpea. Plant Genome 17(1):e20333

Timko MP, Rushton PJ, Laudeman TW, Bokowiec MT, Chipumuro E, Cheung F, Chen X (2008) Sequencing and analysis of the gene-rich space of cowpea. BMC Genomics 9:1–20

Todaka D, Zhao Y, Yoshida T, Kudo M, Kidokoro S, Mizoi J, Kodaira K (2017) Temporal and spatial changes in gene expression, metabolite accumulation and phytohormone content in rice seedlings grown under drought stress conditions. Plant J 90:61–78

Torres AM, Avila CM, Gutierrez N, Palomino C, Moreno MT, Cubero JI (2010) Marker-assisted selection in faba bean (*Vicia faba* L.). F Crop Res 115:243–252

Tuberosa R, Salvi S (2006) Genomics-based approaches to improve drought tolerance of crops. Trends Plant Sci 11(8):405–412

Udawat P (2023) Recent advancements in legumes: next generation sequencing and omics approaches. Indian J Agric Sci 93(5):467–474

Ulemale CS, Mate SN, Deshmukh DV (2013) Physiological indices for drought tolerance in chickpea (*Cicer arietinum* L.). World J Agric Sci 9:123–131

Ullah Z, Iqbal J, Abbasi BA, Ijaz S, Sharifi-Rad J, Yaseen T, Mahmood T (2023a) How genome editing can be helpful in the biofortification of legumes. In: Legumes biofortification. Springer International Publishing, Cham, pp 207–232

Ullah Z, Iqbal J, Abbasi BA, Akhtar W, Kanwal S, Ali I, Mahmood T (2023b) Assessment of Gus expression induced by anti-sense Os PPO gene promoter and antioxidant enzymatic assays in response to drought and heavy metal stress in transgenic Arabidopsis thaliana. Sustain For 15(17):12783

Vadez V, Berger JD, Warkentin T, Asseng S, Ratnakumar P, Rao KPC, Gaur PM (2012) Adaptation of grain legumes to climate change: a review. Agron Sustain Dev 32:31–44

Vergata C, Karalija E, Caleri F, Calvani M, Piergiovanni AR, Martinelli F (2024) Genotype diversity as a source of tolerance to drought stress in legumes. Genet Appl 8(1)

Wang L, Dong S, Liu L, Ma Y, Li S, Zu W (2018a) Transcriptome profiling reveals PEG-simulated drought, heat and combined stress response mechanisms in soybean. Comput Biol Chem 77:413

Wang W, Wang C, Pan D, Zhang Y, Luo B, Ji J (2018b) Effects of drought stress on photosynthesis and chlorophyll fluorescence images of soybean (*Glycine max* L.) seedlings. Int J Agric Biol Eng 11:196–201

Wei Y, Jin J, Jiang S, Ning S, Liu L (2018) Quantitative response of soybean development and yield to drought stress during different growth stages in the Huaibei Plain, China. Agronomy 8:97

Yamaguchi M, Valliyodan B, Zhang J, Lenoble ME, Yu O, Rogers EE, Nguyen HT, Sharp RE (2010) Regulation of growth response to water stress in the soybean primary root. I. Proteomic analysis reveals region-specific regulation of phenylpropanoid metabolism and control of free iron in the elongation zone. Plant Cell Environ 33:223–243

Zhang QingWei ZQ, Bartels D (2018) Molecular responses to dehydration and desiccation in desiccation-tolerant angiosperm plants. J Exp Bot 69(13):3211

Zhu JK (2002) Salt and drought stress signal transduction in plants. Annu Rev Plant Biol 53:247–273

Zlatev Z, Lidon FC (2012) An overview on drought induced changes in plant growth, water relations and photosynthesis. Emirates J Food Agric 24:57–72

6 Genome-Wide Association Studies (GWAS) in Pulses: A Genomic Perspective to Drought

Hafiz Muhammad Ahmad, Muhammad Farhan Khan Pasha, Shazia Arif, Saira Ishaq, Syed Hamza Mahfooz, Sumaira Hafeez, and Sadaf Zahra

Abstract

Pulses are high-nutrient foods that are good sources of protein and nutrients such as phosphorus, zinc, selenium, iron, fiber, complex carbohydrates, protein, and folates. Identification of appropriate genes responsible for a specific plant trait has always been a challenging task for plant breeders. However, genome-wide association study (GWAS) has been successfully implemented in pluses for the identification of marker trait associations (MTAs), quantitative trait loci (QTL), and marker-assisted selection (MAS) for drought-related traits. Though improvement has been achieved in genomic technology using computational and statistical approaches, the challenges lie in the detection of noble variants having minimal adverse effects. This chapter sheds light on recent trends and future prospects of GWAS for the identification and mitigation of drought stress tolerance in pulses. Further we have highlighted the various drought-related traits that have been identified and improved in pluses using GWAS.

H. M. Ahmad (✉) · S. Zahra
Department of Bioinformatics and Biotechnology, Government College University, Faisalabad, Pakistan

M. F. K. Pasha
Pakistan Oilseed Department, Ministry of National Food Security and Research, Islamabad, Pakistan

S. Arif · S. Hafeez
Department of Plant Breeding and Molecular Genetics, University of Poonch, Rawalakot, Pakistan

S. Ishaq
Department of Food Science and Technology, University of Poonch, Rawalakot, Pakistan

S. H. Mahfooz
Barani Agriculture Research Institute, Chakwal, Pakistan

© The Author(s), under exclusive license to Springer Nature Singapore Pte Ltd. 2025
M. A. Nadeem et al. (eds.), *Marker-Assisted Breeding in Legumes for Drought Tolerance*, https://doi.org/10.1007/978-981-96-4112-3_6

Keywords

Drought stress · Food legumes · Speed breeding · Computational biology

6.1 Introduction to Pulses

Pulses belong to the family (*Leguminosae or Fabaceae*), and this family has about 750 genera and more than 20,000 species (Sahruzaini et al. 2020). Commercially grown pulses include Chickpea (*Cicer arietinum*), lentils (*Lens Culinaris*), soybean (*Glycine max*), faba beans (*Vicia faba*), lupin (*Lupinus* spp.), grass peas (*Lathyrus sativus*), common beans (*Pisum sativum*), common vetch (*Vicia sativa*), cowpea (*Vigna unguiculata*), mash bean (*Vigna mungo*), and mung bean (*Vigna radiate*) (Ahmad et al. 2023; Arif et al. 2024; Kumari et al. 2021). Pulses possess a good quality of protein, carbohydrate, micronutrients, and low contents of fats. The fatty acids found in pulses are linked to lowest risk of obesity, cholesterol, and heart diseases in humans (Jamil et al. 2022). Pulses are well documented for their impact in maintaining sustainable agricultural systems along with their health and nutritional benefits (Ahmad et al. 2023). In addition to human and animal consumption pulses also improve the soil and environment health (Ali et al. 2022; Ahmad et al. 2012). Since pulses are typically produced in rainfed or marginal regions, they have developed defenses against abiotic challenges during the cropping season, including salinity, waterlogging, cold, drought, and heat (Raina and Khan 2023). Thus, the development of climate-resilient pulse types is of vital importance. While conventional breeding methods have proven effective in increasing yield, cutting-edge molecular methods could be employed to increase precision and resistance to abiotic stress. In areas where availability to animal protein is restricted, pulses offer a cost-effective supply of plant-based proteins (Agarwal and Fulgoni 2023). The nutritional content, environmental sustainability, and economic advantages of pulses make them essential for maintaining food security (Ahmad et al. 2023). A more resilient and sustainable food system is facilitated by their supply of vital nutrients, enhancement of soil health, and support of farmers' livelihoods. To ensure that pulses are grown, consumed, and included in agricultural and nutrition policy worldwide, it is imperative that their importance in addressing food security be acknowledged (Ahmad et al. 2023). Pulses can be grown for food for the smallholder farming community, either in tandem with other crops or as cash crops, which means they are sold in markets. One of the primary issues with food security is food waste. According to an estimate one-third of the food produced for human use is thought to be lost or wasted globally (Blakeney 2019). There are losses and wastes across the whole agricultural supply chain. Most food losses in developing nations happen during manufacturing or transit, whereas in developed nations, a significant amount of food is lost during consumption (Xue and Liu 2019). Pulses offer significant financial potential for farmers of all sizes to diversify their crop and revenue stream portfolios, increase income, and lower risk. In addition to their positive effects on the environment, pulse production has social and economic benefits as well. It reduces soil erosion, increases soil fertility, and promotes food production and consumption diversity. Pulse growers and other stakeholders need to be aware of the effects that increased production and consumption of pulses would have on livelihood and development (Rajendra Prasad et al. 2016). Changing climatic conditions particularly drought stress has proven the serious threat for the production of pulses globally. Present chapter shed light on mechanism and tools used for GWAS in pulses improvement against drought stress. Further we have highlighted the recent progress achieved in pulses genetics using GWAS techniques to eliminate the effects of drought stress.

6.2 Drought and Its Impact on Pulse Production

Drought is the major abiotic stress that eliminates pulse production throughout the globe (Ahmad et al. 2022). Major environmental factors of drought stress may include soil water holding capacity, improper rainfall pattern, and evapotranspiration rate. Drought is a complex trait that is linked with various morphological and physiological traits. Drought stress may cause reduced germination and lowered photosynthetic activity, decreased assimilate translocation and carbon fixation, suppressed flowering time and an influence on reproductive organs, pollen grain sterility, fewer pods and lower grain set, and decreased sink activity (Mahdi Dar et al. 2018). The genetic basis of drought tolerance in cereals involves a complex interplay of genes and regulatory mechanisms that control various physiological and biochemical processes. Important genetic components include proteins, signaling molecules, and transcription factors that are essential for signal transduction, stress detection, and the activation of stress-responsive pathways (Zahid et al. 2023). Morphologically drought affects the plant height, leaf angle, wilting, shoot biomass, and root attributes (Javed et al. 2016). Physiologically drought imparts an impact on relative water contents, chlorophyll contents, antioxidant activities, and membrane stability index (Ahmad et al. 2021b). It was observed that reduced leaf area, nutrient uptake, CO_2 influx, shoot and root growth, chlorophyll content, stomatal conductance, epicuticular wax contents, and water-use efficiency (WUE) eventually affect pulses production (Khatun et al. 2021). Various direct and indirect mechanisms of drought stress on plant life have been shown in Fig. 6.1. However, it has been proved that the seedling stage of pulses is more sensitive to drought stress. Pulses are usually grown in arid, semiarid, or dryland environments, and water deficiency affects the yield due to terminal drought stress (Kushwah et al. 2022). Drought stress reduces 73–82% the yield in *G. max* (Wei et al. 2018), 40–45% in *C. arietinum* (Rani et al. 2020), 40% in *V. faba* (Daryanto et al. 2015), and 29% in *V. unguiculata* (Kyei-Boahen et al. 2017). So it is necessary to screen the pulse genotypes that respond well to drought stress, improve the soil texture, and select appropriate agroclimatic regions to grow the pulse crops. Further genetic modification of pulses against drought stress can be achieved by identification and incorporation of drought resistance genes and transcription factors.

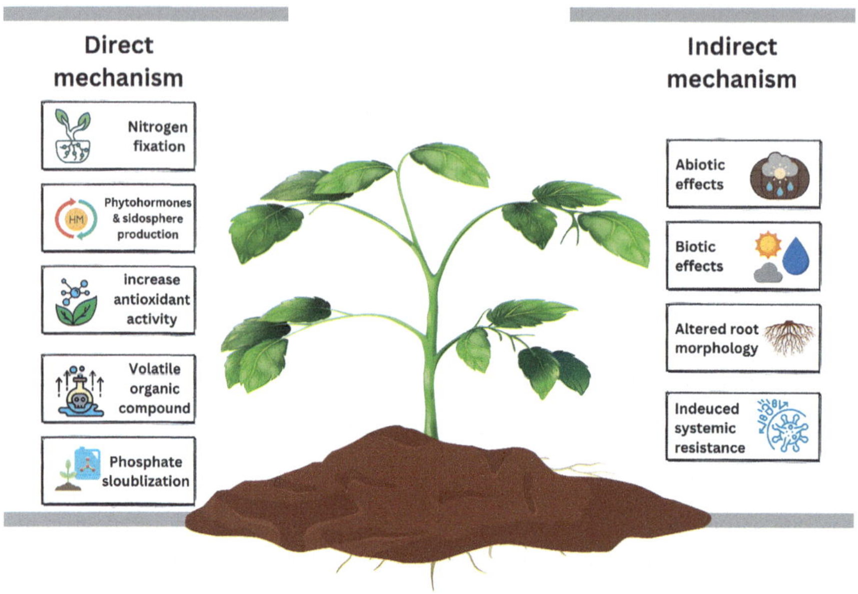

Fig. 6.1 Physiological, morphological, and molecular response of drought to plant life

6.3 Genome-Wide Association Study (GWAS) and Its Mechanism

The first ever genome of *Arabidopsis thaliana* was sequenced in 2000, and recently about 4604 plant genomes have been sequenced from 1482 species (Bernal-Gallardo and de Folter 2024). The GWAS analysis has gained popularity in recent years due to the capability to detect genomic locations of genes and SNP in genomes. GWAS have identified genomic regions associated with a number of agronomic traits, physiological, biochemical, cellular, and fitness traits such as plant height, flowering time, number of kernels, abiotic and biotic stress tolerance, and grain yield (Sahito et al. 2024; Karaköy et al. 2024). GWAS investigates a multitude of genetic variations across various genomes to discover those associated statistically with a certain trait or common disease (Bush 2018). GWAS is a complex and ongoing research process, and collaborating with experts in the field is highly recommended because it also requires bioinformatics expertise, computational resources, and access to genetic and phenotypic data (Ali et al. 2024). Various types of marker systems have been used for the genotyping of pulse genomes and have become important components of pulse breeding programs (Arif et al. 2024). GWAS involves numerous molecular and computational techniques along with statistical tools to compare the genome as shown in Fig. 6.2. However, some important steps are as follows.

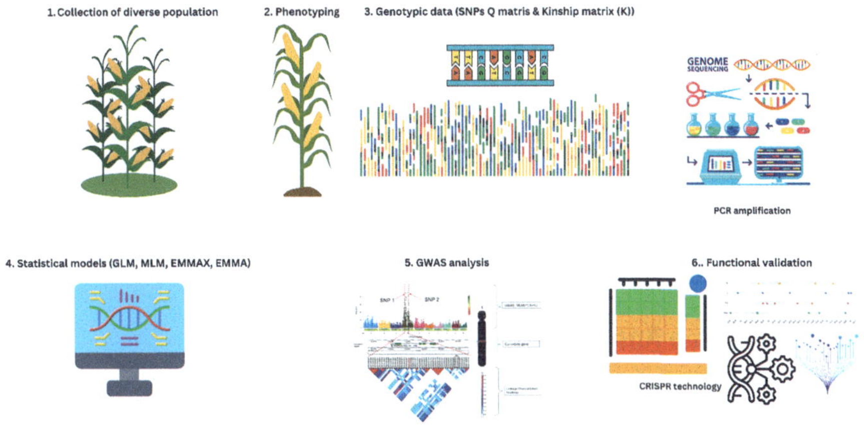

Fig. 6.2 Functional mechanism of genome-wide association study in pulses

6.3.1 Collection of Diverse Germplasm

Germplasm collection is the first step to carry the GWAS in crop plants In order to identify novel genotypes with respect to genes of desirable traits, plant breeders always urge to identify and characterize the targeted germplasm and use it efficiently in plant improvement programs.

6.3.2 Genotyping

Genotyping is the process of determining the specific DNA sequences and their position within an individual genome. The well-known techniques of microarrays and methods of next-generation sequencing including whole exome sequencing (WES) or whole genome sequencing (WGS) are being used for genotyping of the population. High-throughput genotyping focuses on the detection of a sizeable number of single-nucleotide polymorphisms (SNPs).

6.3.3 High-Throughput Techniques for Phenotyping in GWAS

The techniques of phenotyping can detect the genetic variants associated with agronomic, biochemical, or molecular phenotypes of crop plants. These methods will probably enhance the usefulness of GWAS methods and their imputation in plant sciences. To acquire information about the phenotypic traits of interest for each sample, such as yield, disease susceptibility, or nutritional composition physiological and stress-related traits, is a prerequisite. This can be achieved through field trials, laboratory analysis, or accessing public databases.

6.3.4 Genotype-Phenotype Association Analysis (GPA Analysis)

Statistical methods may be diploid to detect associations between genetic markers (SNPs) and phenotypic traits. This may involve performing regression analysis, controlling for population structure, and other confounding factors that can influence the trait variation. GPA analysis has made the GWAS an effective tool for identifying the association between single-nucleotide polymorphisms (*SNPs*) and phenotypic traits.

6.3.5 Genome Annotation

Genome annotation is a mechanism used to predict the structural and functional information of a gene or protein from raw data through bioinformatics tools. Genome annotation may be divided into nucleotide, protein, and process-level annotation (Vuruputoor et al. 2023).

6.3.6 Functional Validation and Fine-Mapping of Candidate Gene

The next step is to validate the discovered genes and QTLs in a sample or population. The complex nature of plant traits and the involvement of multiplex genes to get the desired results make functional validation challenging. However, fine-mapping successfully narrows down the genomic regions containing the causal variants or undesired traits. The approaches like heuristic models, penalized regression models, and Bayesian methods play a substantial role in fine-mapping (Schaid et al. 2018).

6.4 Advantages of GWAS Over Traditional Breeding Methods

The advancement of sequencing technologies and the combined efforts of the scientific community have allowed GWAS to grow steadily over the last few decades. GWAS has evolved as an emerging tool that helps to identify the genotypic and phenotypic variation within a plant species to find the promising alleles linked with desirable traits (Ahmad et al. 2020). This technique has also been implemented for the identification of genotype-phenotype association and evaluation of genetic variation across the genome of various plant species. GWAS helps in mapping the desired economical and physiological traits among the diverse range of genetic populations (Alsamman et al. 2024). GWAS has successfully been implemented in various pulses such as *C. arietinum*, peas, lentils, and soybeans for biotic and abiotic stress tolerance (Srungarapu et al. 2022). Additionally, GWAS has helped to segregate traits such as disease resistance, nutritional contents, and high-yielding components in pluses. Potential applications of GWAS in genetic analysis include the identification of candidate genes for early-generation selections in the breeding pipeline in pulses and validation in a variety of populations (Sinha et al. 2023). Genetic variations, most commonly single nucleotide polymorphisms (SNPs), are statistically associated with specific diseases or traits. By avoiding the need for population construction, GWAS uses a natural population as compared to linkage analysis and thus saves a significant time span (Alqudah et al. 2020). Furthermore, compared to bi-parental QTL mapping, GWAS can uncover more loci responsible for the traits by using high-resolution markers and naturally existing populations with substantial variation Istanbuli et al. (2024). Due to the extensive amount of information accumulated on historical recombinations made feasible by the various and unstructured natural populations employed in GWAS, a rather high mapping resolution has been achieved (Alqudah et al. 2020). The genetic analysis of drought tolerance in a variety of crops, including *Triticum aestivum* (Javed et al. 2016), *A. thaliana* (Oranab et al. 2023), *Helianthus annuus* (Ahmad et al. 2021a), and *V. faba* (Gutiérrez et al. 2023), has made considerable use of association studies in recent years. However, the genetic dissection of drought tolerance pathways in pulses using GWAS has been limited. Furthermore, due to the developments in HTP technology, GWAS assists in measuring a wide range of plant properties that are frequently studied using GWAS techniques (Demirjian et al. 2023). In recent years, GWAS approaches have addressed computation-complexity problems or improved statistical power. These methods can detect genetic differences associated with biochemical, agronomic, or molecular traits (Baloglu et al. 2022). This will enhance the use of GWAS approaches in the future and its implications for plant research (Uffelmann et al. 2021). Major legume crops have shown success using GWAS to analyze or pinpoint the genetic underpinnings of a variety of agronomic variables, quality traits, and biotic stress (Faridi et al. 2021; Shaibu et al. 2019; Zhang et al. 2016). Numerous GWASs have been carried out in some of the major and minor legume crops utilizing various marker types (SSR, SNPs, etc.) (Faridi et al. 2021). GWAS has advantages over conventional breeding techniques as it makes it possible to identify genetic variants associated with traits without a great deal of phenotyping. Further it can pinpoint specific genetic variations associated with traits, enabling more accurate breeding individual selection. Its ability to be applied to a wide range of species and characteristics makes it a versatile tool for genetic research and breeding programs (Korte and Farlow 2013).

6.5 Achievements of GWAS for Drought Tolerance in Pulses

Drought is one of the most severe biotic and abiotic stressors, decreasing crop yield and endangering global food security in arid and semi-arid areas (Istanbuli et al. 2024). Water scarcity can significantly impair the production and grain quality of pulses, particularly during the flowering and grain development stages (Khatun et al. 2021). Pulses exhibit a range of morphological, physiological, biochemical, and molecular reactions in response to drought stress (Khatun et al. 2021). To increase food legume adaptability to drought-prone areas, altered gene expression and understanding of these mechanisms are prerequisites (Kouighat et al. 2024). However, the understanding of drought tolerance at the physiological and molecular levels is extremely complex due to the polygenic character of stress tolerance, which is controlled by multiple genes with tiny effects and high genotype by environment interactions (Javed et al. 2016). Using GWAS, genetic loci linked to pulses' ability to withstand drought have been found. Researchers have successfully discovered candidate genes and genetic variations connected to numerous drought-related traits by examining vast collections of varied bean accessions. Numerous GWAS investigations have identified the genetic areas linked to characteristics that help pulses withstand drought stress (Istanbuli et al. 2024).

6.5.1 QTLs Mapping for Identification of Drought Tolerance Genes in Pulses

Drought stress, a complex quantitative trait regulated by numerous small-effect genes, has a significant impact on pulse production. Finding the genomic areas linked to drought tolerance component traits could help design drought-tolerant cultivars and advance our knowledge of drought tolerance mechanisms (Ahmad et al. 2023). In a pulse breeding effort, identifying QTLs for drought tolerance component characteristics can be a wise strategy given the difficulties in breeding drought-tolerant types. For example, QTLs have aided in analyzing the genetic basis of *C. arietinum* features associated to drought tolerance (Sivasakthi et al. 2018). QTLs for drought and its related traits such as secondary metabolite biosynthesis, relative water contents, root length, angle, and surface have been successfully identified in *C. arietinum* (Jaganathan et al. 2015; Yadava et al. 2023), *G. max* (Chen et al. 2021), *V. unguiculata* (Burridge et al. 2017; Prince et al. 2015), and *P. sativum* (Mukeshimana et al. 2014). Research conducted on genetic mapping for drought tolerance in *C. arietinum* helped to identify four QTLs that effect drought-related traits such as relative water contents, membrane stability index, seed weight, and yield under drought (Yadava et al. 2023). It was observed that "QTL-hotspot" regions contain QTLs for a number of drought-related characteristics in the *C. arietinum* (Kushwah et al. 2022). Twelve drought-related QTLs (NCPGR-50, TR-50, SCEA19, TAA-58, H6C-07, H5E-02, H5G-01, H6C-07, H1B-04, TA-113, H6C-07, H1F-21) were identified in *C. arietinum* that were associated to drought tolerance at seedling stage (Hamwieh et al. 2013). Recently, researchers evaluated the root-related traits in desi genotypes of *C. arietinum* under drought stress (Asati et al. 2022). In horsegram (*Macrotyloma uniflorum*) QTLs associated with relative water contents, root volume, and root length were found to tolerate the drought stress in this crop (Choudhary et al. 2022). Water deficit–related QTLs (*qSV_Gm03*, *qSV_Gm05*, *qSV_Gm10*, *and qSV_Gm12*) were identified in *G. max* (Carpentieri-Pipolo et al. 2012). QTLs identified in various pulses have been listed in Table 6.1. It was also revealed that enhanced lines with different genetic origins that were subjected to drought stress outperformed their respective recurrent parents (Bharadwaj et al. 2021).

Table. 6.1 QTLs discovered in various pulses for drought and its related traits

Name of pulses	QTL identified	Drought trait	References
C. arietinum	RLD4, RSA6, RDWR4	Root length, surface area, root dry weight ratio	Jaganathan et al. (2015)
C. arietinum	qYLD7.1	Secondary metabolite biosynthesis	Yadava et al. (2023)
C. arietinum	qRWC1.1	Relative water contents	Yadava et al. (2023)
C. arietinum	LG3 and LG4	Drought resistance score	Hamwieh et al. (2013)
G. max	TRL_Gm06, TRL_Gm08	Root length	Manavalan et al. (2015)
G. max	LRN_Gm08	Lateral root number	Manavalan et al. (2015)
G. max	FR_Gm01, FR_Gm03, FR_Gm04	Fibrous rooting	Abdel-Haleem et al. (2011)
G. max	*qRL16.1*	Root length	Chen et al. (2021)
V. unguiculata	Dro-1, Dro-2	Drought induced senescence	Muchero et al. (2009)
V. unguiculata	RDT2_Gm08	Root density and thickness	Prince et al. (2015)
V. unguiculata	Brg10, RD1–1	Basal root angle, root diameter	Burridge et al. (2017)
P. vulgaris	Brg1.1, Brg5.1, Brg5.2	Basal root angle	Liao et al. (2004)
P. sativum	rl1, rl2, rl3	Root length	Fondevilla et al. (2011)
P. sativum	Pv03	Drought tolerance	Mukeshimana et al. (2014)

6.5.2 Identification of Marker Trait Association and Single Nucleotide Polymorphisms (SNP) for Drought Tolerance in Pulses

Drought stress is a severe threat to world food security, particularly for vital pulses grown in dry and semi-arid environments. As climate change affects the frequency and severity of droughts, there is an urgent need to improve the drought-tolerant resilience of these important crops (Kouighat et al. 2024). Using molecular markers, breeders can select target phenotypes based on genotypes and environmental interactions for crop genetic improvement and selection (Sinha et al. 2023). Finding the molecular markers and genetic regions linked to the desired qualities is essential in order to carry the MAS (Nadeem et al. 2018; Istanbuli et al. 2024; Baloch et al. 2024). Therefore, high-density QTL mapping populations with target trait characteristics are needed to conduct effective MAS investigations in various crops. However, breeding materials and research methods that can replace it are needed, as it takes a lot of time and effort to develop such a QTL mapping population. Recently, it has proven helpful to demonstrate associations between markers and attributes using association mapping techniques (Salgotra and Chauhan 2023). Specifically, marker-trait associations (MTAs) for MAS can be found via association mapping analysis, which has several benefits over quantitative trait loci (QTL) mapping, including shorter experimental times and lower costs (Srungarapu et al. 2022). Furthermore, three benefits of association mapping are larger allele numbers, shorter research times, and improved mapping resolution (Sinha et al. 2023). Thus, because of improvements in high throughput genomic technology and better alleles, association mapping has continued to gain favor in genetic research (Giordani et al. 2022). DNA markers have been widely used subsequently for genomic research and crop breeding, as well as for QTL mapping and association studies (Kushwah et al. 2022). Especially because SSR markers are highly polymorphic, typically codominant, repeatable, and widely distributed throughout the plant genome, they offer important information regarding genetic diversity, genetic linkages, and population structure in genetic material of pulses (Ahmadi and Fotokian 2011). According to Ahmadi and Fotokian (2011), SSR markers have been effectively used for crop characterization, genetic diversity research, and targeted gene association analysis. The identification of molecular markers linked to drought tolerance in pulses helps in the selection of inbred lines and cultivars, hence supporting breeding initiatives for pulses that aim to improve productivity and yield while simultaneously addressing drought resistance. Simple sequence repeat markers were successfully implemented to explore the genetic diversity in *P. sativum* landraces (Arif et al. 2024). In another research Srungarapu et al. (2022) found that trait-linked markers may be utilized for agronomic and grain nutrient traits in diverse sets of chickpea germplasm. In addition to biotic and abiotic stresses some markers also have been identified in *P. vulgaris* for grain mineral contents (Delfini et al. 2021).

GWAS has been implemented to perform SNP for the identification and annotation of drought-related genes in various pulses. Whole-genome resequencing WGRS has provoked practical application in plant breeding by identifying the presence or absence of variation for genome-wide association analysis. The study of resequencing the entire genome (whole genome sequencing) in bean genotypes has become potentially important with the aim to generate an extensive database system for the detection of single nucleotide polymorphisms (de Souza et al. 2023). Furthermore, by employing the targeted sequence capture (CaptureSeq) technology to search for SNPs in certain genomic regions, hundreds of SNPs in interest regions were instantly identified, enabling a refinement of genotyping. Genetic improvement through the identification and utilization of single nucleotide polymorphisms (SNPs) associated with drought tolerance represents a promising approach in breeding programs. SNPs, as genetic variations at single nucleotide positions in the DNA sequence, can influence the expression and function of these genes, thereby affecting the plant's ability to withstand drought stress. QTL mapping and GWAS are frequently used to identify SNPs associated with drought tolerance. These approaches enable researchers to correlate specific genetic markers with phenotypic traits related to drought tolerance, such as water use efficiency, leaf wilting, and yield under water-limited conditions Jaganathan et al. (2015).

Estimation of genetic variants improves the plant's capacity to maintain cellular homeostasis during drought stress by changing the transcriptional control of downstream genes involved in osmotic adjustment and antioxidant defense (Sobia et al. 2014). By using marker-assisted selection to include the SNPs in breeding programs, it is possible to create pulse cultivars that are resistant to drought and can sustain stable yields even in the face of unfavorable weather (Mutari et al. 2023). In chickpea, SNPs associated with drought tolerance have been identified in genes encoding pivotal transcription factors and proteins involved in stress response pathways (Harish et al. 2024). In cowpea, it was discovered that SNPs may be utilized to identify the complex drought traits such as stomatal conductance and their opening and closure (Wu et al. 2021). In *V. faba* SNPs linked to stomatal conductance, density, length, and canopy temperature were identified that were located on chromosome II (Khazaei et al. 2014). It was revealed that plant cuticular waxes also effects drought tolerance, so keeping this in view some SNPs controlling petiole and lamina wax were discovered in common bean (Tafesse et al. 2021). A list of drought-related SNPs discovered in pulses is listed in Table 6.2. Food security is ensured in areas vulnerable to drought-induced yield losses by the incorporation of these genetic markers into breeding programs, which aid in the production of pulses with enhanced drought resistance and yield stability. In conclusion, SNPs offer significant potential for enhancing drought tolerance in pulses, albeit with challenges that must be addressed for effective implementation in breeding programs (Deokar et al. 2014). SNPs provide precise genetic markers that enable targeted selection of favorable alleles associated with drought resilience, thereby accelerating the development of climate-resilient crop varieties (Acuña-Galindo et al. 2015). However,

Table 6.2 Various SNPs identified for drought tolerance traits in pulses

Name of crop	SNPs/MTAs discovered	Genes identified	Traits effected	References
G. max	26	41	Drought tolerance	Zhao et al. (2022)
G. max	47	N/A	Water use efficiency	Kaler et al. (2017)
P. sativum	3166	489	Drought stress in root and leaves	Pereira et al. (2020)
P. sativum	89	39	Aquaporin	Wu et al. (2022)
P. sativum	7	16	Petiole and lamina wax	Tafesse et al. (2021)
V. faba	188	N/A	Stomatal characteristics	Khazaei et al. (2014)
P. vulgaris	68	22	Drought tolerance	Mutari et al. (2023)
C. arietinum	11	28	Drought link traits	Istanbuli et al. (2024)
C. arietinum	52	5	Hydrogen peroxide-resistant, drought resistance	Harish et al. (2024)
V. unguiculata	30	28	Stomatal characteristics	Wu et al. (2021)

challenges such as the complex genetic architecture of drought tolerance, variability in SNP effects across genetic backgrounds, and the need for robust functional validation underscore the necessity for continued research and technological innovation (Moore et al. 2010). Moreover, ethical considerations and regulatory frameworks must be carefully navigated to ensure equitable and responsible deployment of SNP-based breeding technologies. Overall, SNP markers represent valuable tools for sustainable agriculture, contributing to global food security by enhancing the resilience of pulses to drought stress and other environmental pressures in a changing climate (Yadava et al. 2023).

6.6 Conclusions and Future Prospects

To precisely incorporate undiscovered adaptive features from diverse pulses gene pools into elite genotypes, pre-breeding activities and advanced breeding techniques are essential. The growing problems of drought stress in pulses could be addressed in a number of ways. In pulses drought-tolerant mechanisms have been thoroughly investigated; however, further investigation is required to comprehend the molecular underpinnings of drought stress tolerance. There are a number of QTLs for drought tolerance in grain legumes that have been identified and confirmed, but the relationships between these QTLs and drought tolerance are still unknown. Thus, more field research is needed to determine how these QTLs affect yield stability in drought-prone environments. In agricultural genetics, conventional GWAS methods have emerged for investigating such intricate pulse characteristics. With the current advances in machine learning, biotechnology, deep understanding of machines, and various bioinformatics tools, it is now possible to explore such massive datasets and assess linkage disequilibrium (LD) based associations between genome-wide assembly of single-nucleotide polymorphisms (SNPs) and preferred phenotypic feature.

References

Abdel-Haleem H, Lee G-J, Boerma RH (2011) Identification of QTL for increased fibrous roots in soybean. Theor Appl Genet 122(5):935–946. https://doi.org/10.1007/s00122-010-1500-9

Acuña-Galindo MA, Mason RE, Subramanian NK, Hays DB (2015) Meta-analysis of wheat QTL regions associated with adaptation to drought and heat stress. Crop Sci 55:477. https://doi.org/10.2135/cropsci2013.11.0793

Agarwal S, Fulgoni VL (2023) Effect of adding pulses to replace protein foods and refined grains in healthy dietary patterns. Nutrients 15(20):1–12. https://doi.org/10.3390/nu15204355

Ahmad HM, Ahsan M, Ali Q, Javed I (2012) Genetic variability, heritability and correlation studies of various quantitative traits of mungbean (Vigna radiate L.) at different radiation levels. Int Res J Microbiol 3(11):352–362

Ahmad HM, Mahmood-ur-Rahman, Azeem F, Shaheen T, Irshad MA (2020) Genome-wide analysis of long chain Acyl-CoA synthetase (LACS) genes in sunflower (Helianthus annuus) suggests their role in drought stress. Int J Agric Biol 24(4):863–870. https://doi.org/10.17957/IJAB/15.1510

Ahmad HM, Wang X, Fiaz S, Mahmood-Ur-Rahman, Azhar Nadeem M, Aslam Khan S et al (2021a) Comprehensive genomics and expression analysis of eceriferum (CER) genes in sunflower (Helianthus annuus). Saudi J Biol Sci 28:6884. https://doi.org/10.1016/j.sjbs.2021.07.077

Ahmad HM, Wang X, Mahmood-Ur-Rahman, Fiaz S, Azeem F, Shaheen T (2021b) Morphological and physiological response of Helianthus annuus L. to drought stress and correlation of wax contents for drought tolerance traits. Arab J Sci Eng 47(6):6747–6761. https://doi.org/10.1007/s13369-021-06098-1

Ahmad HM, Wang X, Ijaz M, Mahmood-Ur-Rahman, Oranab S, Ali MA, Fiaz S (2022) Molecular aspects of MicroRNAs and phytohormonal signaling in response to drought stress: A review. Curr Issues Mol Biol 44(8):3695–3710. https://doi.org/10.3390/cimb44080253

Ahmad HM, Zahra S, Oranab S, Arif S, Zakia S, Gul-E-Zahra et al (2023) Contribution of conventional breeding approaches in legumes biofortification. In: Legumes biofortification. Springer International Publishing, Cham. https://doi.org/10.1007/978-3-031-33957-8_5

Ahmadi J, Fotokian M (2011) Identification and mapping of quantitative trait loci asso- ciated with salinity tolerance in rice (Oryza Sativa) using SSR markers. Iran J Biotechnol 9(1):21–30

Ali A, Altaf MT, Nadeem MA, Karaköy T, Shah AN, Azeem H et al (2022) Recent advancement in OMICS approaches to enhance abiotic stress tolerance in legumes. Front Plant Sci 13:952759

Ali MA, Zahoor A, Niaz Z, Jabran M, Anas M, Shafique I et al (2024) Bioinformatics and computational biology. In: Trends in plant biotechnology. Springer Nature, Singapore, pp 281–334. https://doi.org/10.1007/978-981-97-0814-7_10

Alqudah AM, Sallam A, Stephen Baenziger P, Börner A (2020) GWAS: fast-forwarding gene identification and characterization in temperate Cereals: lessons from Barley – A review. J Adv Res 22:119–135. https://doi.org/10.1016/j.jare.2019.10.013

Alsamman AM, Mousa HK, Istanbuli T, Abd El-Maksoud MM, Tawkaz S, Hamwieh A (2024) Unveiling the genetic basis of Fusarium wilt resistance in chickpea using GWAS analysis and characterization of candidate genes. Front Genet 14. https://doi.org/10.3389/fgene.2023.1292009

Arif U, Ali S, Alharthi B, Pasha MFK, Ahmad HM, Ahmed MJ, Arif AA (2024) Genetic diversity analysis of pea (Pisum sativum l.) landraces by utilizing simple sequence repeat markers. Genet Resour Crop Evol 71:4255. https://doi.org/10.1007/s10722-023-01796-1

Asati V, Deepa PR, Sharma PK (2022) Desert legume Prosopis cineraria as a novel source of antioxidant flavonoids/isoflavonoids: biochemical characterization of edible pods for potential functional food development. Biochem Biophys Rep 29:101210. https://doi.org/10.1016/j.bbrep.2022.101210

Baloch FS, Ali A, Tajibayev D, Nadeem MA, Ölmez F, Aktaş H et al (2024) Stripe rust resistance gene Yr15 in Turkish and Kazakhstan wheat germplasms and the potential of Turkish wild emmer for stripe rust breeding. Genet Resour Crop Evol 71(6):2699–2719

Baloglu MC, Celik Altunoglu Y, Baloglu P, Yildiz AB, Türkölmez N, Özden Çiftçi Y (2022) Gene-editing technologies and applications in legumes: progress, evolution, and future prospects. Front Genet 13. https://doi.org/10.3389/fgene.2022.859437

Bernal-Gallardo JJ, de Folter S (2024) Plant genome information facilitates plant functional genomics. Planta 259(5):117. https://doi.org/10.1007/s00425-024-04397-z

Bharadwaj C, Tripathi S, Soren KR, Thudi M, Singh RK, Sheoran S et al (2021) Introgression of “ QTL-hotspot ” region enhances drought tolerance and grain yield in three elite chickpea cultivars. Plant Genome 14(1):e20076. https://doi.org/10.1002/tpg2.20076

Blakeney M (2019) Food loss and food waste: causes and solutions. In: Food loss and food waste: causes and solutions. https://doi.org/10.4337/9781788975391

Burridge JD, Schneider HM, Huynh B-L, Roberts PA, Bucksch A, Lynch JP (2017) Genome-wide association mapping and agronomic impact of cowpea root architecture. Theor Appl Genet 130(2):419–431. https://doi.org/10.1007/s00122-016-2823-y

Bush WS (2018) Genome-wide association studies. In: Encyclopedia of bioinformatics and computational biology: ABC of bioinformatics, vol 1–3, pp 235–241. https://doi.org/10.1016/B978-0-12-809633-8.20232-X

Carpentieri-Pipolo V, Pipolo AE, Abdel-Haleem H, Boerma HR, Sinclair TR (2012) Identification of QTLs associated with limited leaf hydraulic conductance in soybean. Euphytica 186(3):679–686. https://doi.org/10.1007/s10681-011-0535-6

Chen H, Kumawat G, Yan Y, Fan B, Xu D (2021) Mapping and validation of a major QTL for primary root length of soybean seedlings grown in hydroponic conditions. BMC Genomics 22(1):132. https://doi.org/10.1186/s12864-021-07445-0

Choudhary S, Isobe S, Chahota RK (2022) Elucidation of drought tolerance potential of horsegram (Macrotyloma uniflorum Var.) germplasm using genome wide association studies. Gene 819:146241. https://doi.org/10.1016/j.gene.2022.146241

Daryanto S, Wang L, Jacinthe P-A (2015) Global synthesis of drought effects on food legume production. PLoS One 10(6):e0127401. https://doi.org/10.1371/journal.pone.0127401

de Souza IP, de Azevedo BR, Coelho ASG, de Souza TLPO, Valdisser PAMR, Gomes-Messias LM et al (2023) Whole-genome resequencing of common bean elite breeding lines. Sci Rep 13(1):12721. https://doi.org/10.1038/s41598-023-39399-6

Delfini J, Moda-Cirino V, dos Santos Neto J, Zeffa DM, Nogueira AF, Ribeiro LAB et al (2021) Genome-wide association study for grain mineral content in a Brazilian common bean diversity panel. Theor Appl Genet 134(9):2795–2811. https://doi.org/10.1007/S00122-021-03859-2/TABLES/3

Demirjian C, Vailleau F, Berthomé R, Roux F (2023) Genome-wide association studies in plant pathosystems: success or failure? Trends Plant Sci 28(4):471–485. https://doi.org/10.1016/j.tplants.2022.11.006

Deokar AA, Ramsay L, Sharpe AG, Diapari M, Sindhu A, Bett K et al (2014) Genome wide SNP identification in chickpea for use in development of a high density genetic map and improvement of chickpea reference genome assembly. BMC Genomics 15(1):1–19. https://doi.org/10.1186/1471-2164-15-708

Faridi R, Koopman B, Schierholt A, Ali MB, Apel S, Link W (2021) Genetic study of the resistance of faba bean (Vicia faba) against the fungus Ascochyta fabae through a genome-wide association analysis. Plant Breed 140(3):442–452. https://doi.org/10.1111/pbr.12918

Fondevilla S, Almeida NF, Satovic Z, Rubiales D, Vaz Patto MC, Cubero JI, Torres AM (2011) Identification of common genomic regions controlling resistance to Mycosphaerella pinodes, earliness and architectural traits in different pea genetic backgrounds. Euphytica 182(1):43–52. https://doi.org/10.1007/s10681-011-0460-8

Giordani W, Gama HC, Chiorato AF, Garcia AAF, Vieira MLC (2022) Genome-wide association studies dissect the genetic architecture of seed shape and size in common bean. G3 Genes|Genomes|Genetics 12(4). https://doi.org/10.1093/g3journal/jkac048

Gutiérrez N, Pégard M, Balko C, Torres AM (2023) Genome-wide association analysis for drought tolerance and associated traits in faba bean (Vicia faba L.). Front Plant Sci 14. https://doi.org/10.3389/fpls.2023.1091875

Hamwieh A, Imtiaz M, Malhotra RS (2013) Multi-environment QTL analyses for drought-related traits in a recombinant inbred population of chickpea (Cicer arientinum L.). Theor Appl Genet 126(4):1025–1038. https://doi.org/10.1007/s00122-012-2034-0

Harish D, Pappula Reddy SP, Kumar N, Bharadwaj C, Kumar T, Parida S et al (2024) Integrating multilocus genome-wide association studies in chickpea landraces to discern the genetics of drought tolerance. Front Sustain Food Syst 8. https://doi.org/10.3389/fsufs.2024.1389970

Istanbuli T, Nassar AE, Abd El-Maksoud MM, Tawkaz S, Alsamman AM, Hamwieh A (2024) Genome-wide association study reveals SNP markers controlling drought tolerance and related agronomic traits in chickpea across multiple environments. Front Plant Sci 15. https://doi.org/10.3389/fpls.2024.1260690

Jaganathan D, Thudi M, Kale S, Azam S, Roorkiwal M, Gaur PM et al (2015) Genotyping-by-sequencing based intra-specific genetic map refines a "QTL-hotspot" region for drought tolerance in chickpea. Mol Gen Genomics 290(2):559–571. https://doi.org/10.1007/s00438-014-0932-3

Jamil S, Ilyas M, Khan MZ, Awan SI, Rehman S, Shafique S et al (2022) Heritability, genetic gain, and path coefficient analyses in black gram at Poonch Rawalakot, Azad Jammu and Kashmir. Sabrao J Breed Genet 54(3):537–548. https://doi.org/10.54910/SABRAO2022.54.3.7

Javed I, Awan S, Ahmad H, Rao A (2016) Assesment of genetic diversity in wheat synthetic double haploids for yield and drought related traits through factor and cluster analyses. Plant Gene Trait. https://doi.org/10.5376/pgt.2016.07.0003

Kaler AS, Dhanapal AP, Ray JD, King CA, Fritschi FB, Purcell LC (2017) Genome-wide association mapping of carbon isotope and oxygen isotope ratios in diverse soybean genotypes. Crop Sci 57(6):3085–3100. https://doi.org/10.2135/cropsci2017.03.0160

Karaköy T, Toklu F, Karagöl ET, Uncuer D, Çilesiz Y, Ali A et al (2024) Genome-wide association studies revealed DArTseq loci associated with agronomic traits in Turkish faba bean germplasm. Genet Resour Crop Evol 71(1):181–198

Khatun M, Sarkar S, Era FM, Islam AKMM, Anwar MP, Fahad S et al (2021) Drought stress in grain legumes: effects, tolerance mechanisms and management. Agronomy 11(12):2374. https://doi.org/10.3390/agronomy11122374

Khazaei H, O'Sullivan DM, Sillanpää MJ, Stoddard FL (2014) Use of synteny to identify candidate genes underlying QTL controlling stomatal traits in faba bean (Vicia faba L.). Theor Appl Genet 127(11):2371–2385. https://doi.org/10.1007/s00122-014-2383-y

Korte A, Farlow A (2013) The advantages and limitations of trait analysis with GWAS : a review self-fertilisation makes Arabidopsis particularly well suited to GWAS. Plant Methods 9(1):29

Kouighat M, Kettani R, El Fechtali M, Nabloussi A (2024) Exploring mechanisms of drought-tolerance and adaptation of selected sesame mutant lines. J Agric Food Res 15:100911. https://doi.org/10.1016/j.jafr.2023.100911

Kumari VV, Roy A, Vijayan R, Banerjee P, Verma VC, Nalia A et al (2021) Drought and heat stress in cool-season food legumes in sub-tropical regions: consequences, adaptation, and mitigation strategies. Plan Theory 10(6):1038. https://doi.org/10.3390/plants10061038

Kushwah A, Bhatia D, Barmukh R, Singh I, Singh G, Bindra S et al (2022) Genetic mapping of QTLs for drought tolerance in chickpea (Cicer arietinum L.). Front Genet 13. https://doi.org/10.3389/fgene.2022.953898

Kyei-Boahen S, Savala CEN, Chikoye D, Abaidoo R (2017) Growth and yield responses of cowpea to inoculation and phosphorus fertilization in different environments. Front Plant Sci 8. https://doi.org/10.3389/fpls.2017.00646

Liao H, Yan X, Rubio G, Beebe SE, Blair MW, Lynch JP (2004) Genetic mapping of basal root gravitropism and phosphorus acquisition efficiency in common bean. Funct Plant Biol 31(10):959. https://doi.org/10.1071/FP03255

Mahdi Dar Z, Masood A, Hussain Mughal A, Asif M, Ahamd Malik M (2018) Review on drought tolerance in plants induced by plant growth promoting Rhizobacteria. Int J Curr Microbiol App Sci 7(05):412–422. https://doi.org/10.20546/ijcmas.2018.705.053

Manavalan LP, Prince SJ, Musket TA, Chaky J, Deshmukh R, Vuong TD et al (2015) Identification of novel QTL governing root architectural traits in an interspecific soybean population. PLoS One 10(3):e0120490. https://doi.org/10.1371/journal.pone.0120490

Moore JH, Asselbergs FW, Williams SM (2010) Bioinformatics challenges for genome-wide association studies. Bioinformatics 26(4):445–455. https://doi.org/10.1093/bioinformatics/btp713

Muchero W, Ehlers JD, Close TJ, Roberts PA (2009) Mapping QTL for drought stress-induced premature senescence and maturity in cowpea [Vigna unguiculata (L.) Walp.]. Theor Appl Genet 118(5):849–863. https://doi.org/10.1007/s00122-008-0944-7

Mukeshimana G, Butare L, Cregan PB, Blair MW, Kelly JD (2014) Quantitative trait loci associated with drought tolerance in common bean. Crop Sci 54(3):923–938. https://doi.org/10.2135/cropsci2013.06.0427

Mutari B, Sibiya J, Shayanowako A, Chidzanga C, Matova PM, Gasura E (2023) Genome-wide association mapping for component traits of drought tolerance in dry beans (Phaseolus vulgaris L.). PLoS One 18(5):e0278500. https://doi.org/10.1371/JOURNAL.PONE.0278500

Nadeem MA, Nawaz MA, Shahid MQ, Doğan Y, Comertpay G, Yıldız M et al (2018) DNA molecular markers in plant breeding: current status and recent advancements in genomic selection and genome editing. Biotechnol Biotechnol Equip 32(2):261–285

Oranab S, Ghaffar A, Ahmad A, Pasha M, Munir B, Arif S et al (2023) Genome-wide analysis of cyclic nucleotide-gated ion channels (Cngcs) of Arabidopsis Thaliana under abiotic stresses. SABRAO J Breed Genet 55(1):38–49. https://doi.org/10.54910/sabrao2023.55.1.4

Pereira WJ, de Melo ATO, Coelho ASG, Rodrigues FA, Mamidi S, de Alencar SA et al (2020) Genome-wide analysis of the transcriptional response to drought stress in root and leaf of common bean. Genet Mol Biol 43(1):e20180259. https://doi.org/10.1590/1678-4685-gmb-2018-0259

Prince SJ, Song L, Qiu D, Maldonado dos Santos JV, Chai C, Joshi T et al (2015) Genetic variants in root architecture-related genes in a Glycine soja accession, a potential resource to improve cultivated soybean. BMC Genomics 16(1):132. https://doi.org/10.1186/s12864-015-1334-6

Raina A, Khan S (2023) Field assessment of yield and its contributing traits in cowpea treated with lower, intermediate, and higher doses of gamma rays and sodium azide. Front Plant Sci 14. https://doi.org/10.3389/fpls.2023.1188077

Rajendra Prasad B, Singh P, Shahi B (2016) Munich Personal RePEc archive pulses for sustainable livelihood and food security pulses for sustainable livelihood and food security, (80269)

Rani A, Devi P, Jha UC, Sharma KD, Siddique KHM, Nayyar H (2020) Developing climate-resilient Chickpea involving physiological and molecular approaches with a focus on temperature and drought stresses. Front Plant Sci 10. https://doi.org/10.3389/fpls.2019.01759

Sahito JH, Zhang H, Gishkori ZGN, Ma C, Wang Z, Ding D et al (2024) Advancements and prospects of Genome-Wide Association Studies (GWAS) in maize. Int J Mol Sci 25(3):1918. https://doi.org/10.3390/ijms25031918

Sahruzaini NA, Rejab NA, Harikrishna JA, Khairul Ikram NK, Ismail I, Kugan HM, Cheng A (2020) Pulse crop genetics for a sustainable future: where we are now and where we should be heading. Front Plant Sci 11(April):1–9. https://doi.org/10.3389/fpls.2020.00531

Salgotra RK, Chauhan BS (2023) Genetic diversity, conservation, and utilization of plant genetic resources. Genes 14(1). https://doi.org/10.3390/genes14010174

Schaid DJ, Chen W, Larson NB (2018) From genome-wide associations to candidate causal variants by statistical fine-mapping. Nat Rev Genet 19(8):491–504. https://doi.org/10.1038/s41576-018-0016-z

Shaibu AS, Motagi B, Ayittah P (2019) Genome wide association studies for fatty acids, mineral and proximate compositions in groundnut (arachis hypogaea L.) seeds. Pertanika J Trop Agric Sci 42(3):939–955

Sinha MK, Aski MS, Mishra GP, Kumar MBA, Yadav PS, Tokas JP et al (2023) Genome wide association analysis for grain micronutrients and anti-nutritional traits in mungbean [Vigna radiata (L.) R. Wilczek] using SNP markers. Front Nutr 10. https://doi.org/10.3389/fnut.2023.1099004

Sivasakthi K, Thudi M, Tharanya M, Kale SM, Kholová J, Halime MH et al (2018) Plant vigour QTLs co-map with an earlier reported QTL hotspot for drought tolerance while water saving QTLs map in other regions of the chickpea genome. BMC Plant Biol 18(1):29. https://doi.org/10.1186/s12870-018-1245-1

Sobia A, Ahmad HM, Awan SI, Kang SA, Iqbal MS, Ali MA (2014) Estimation of genetic variability, heritibility and correlation for some morphological traits in spring wheat. J Biol Agric Healthc 4(5):10–16

Srungarapu R, Mahendrakar MD, Mohammad LA, Chand U, Jagarlamudi VR, Kondamudi KP et al (2022) Genome-wide association analysis reveals trait-linked markers for grain nutrient and agronomic traits in diverse set of Chickpea Germplasm. Cells 11(15):2457. https://doi.org/10.3390/cells11152457

Tafesse EG, Gali KK, Lachagari VBR, Bueckert R, Warkentin TD (2021) Genome-Wide Association mapping for heat and drought adaptive traits in pea. Genes 12(12):1897. https://doi.org/10.3390/genes12121897

Uffelmann E, Huang QQ, Munung NS, de Vries J, Okada Y, Martin AR et al (2021) Genome-wide association studies. Nat Rev Methods Primers 1(1). https://doi.org/10.1038/s43586-021-00056-9

Vuruputoor VS, Monyak D, Fetter KC, Webster C, Bhattarai A, Shrestha B et al (2023) Welcome to the big leaves: best practices for improving genome annotation in non-model plant genomes. Appl Plant Sci 11(4):e11533. https://doi.org/10.1002/aps3.11533

Wei Y, Jin J, Jiang S, Ning S, Liu L (2018) Quantitative response of soybean development and yield to drought stress during different growth stages in the Huaibei Plain, China. Agronomy 8(7):97. https://doi.org/10.3390/agronomy8070097

Wu X, Sun T, Xu W, Sun Y, Wang B, Wang Y et al (2021) Unraveling the genetic architecture of two complex, stomata-related drought-responsive traits by high-throughput physiological phenotyping and GWAS in Cowpea (Vigna. Unguiculata L. Walp). Front Genet 12. https://doi.org/10.3389/fgene.2021.743758

Wu L, Chang Y, Wang L, Wang S, Wu J (2022) The aquaporin gene PvXIP1;2 conferring drought resistance identified by GWAS at seedling stage in common bean. Theor Appl Genet 135(2):485–500. https://doi.org/10.1007/s00122-021-03978-w

Xue L, Liu G (2019) Introduction to global food losses and food waste. In: Saving food: production, supply chain, food waste and food consumption, vol 3. Elsevier Inc. https://doi.org/10.1016/B978-0-12-815357-4.00001-8

Yadava YK, Chaudhary P, Yadav S, Rizvi AH, Kumar T, Srivastava R et al (2023) Genetic mapping of quantitative trait loci associated with drought tolerance in chickpea (Cicer arietinum L.). Sci Rep 13(1):17623. https://doi.org/10.1038/s41598-023-44990-y

Zahid G, Iftikhar S, Shimira F, Ahmad HM, Aka Kaçar Y (2023) An overview and recent progress of plant growth regulators (PGRs) in the mitigation of abiotic stresses in fruits: A review. Sci Hortic. Elsevier B.V 309:111621. https://doi.org/10.1016/j.scienta.2022.111621

Zhang J, Song Q, Cregan PB, Jiang GL (2016) Genome-wide association study, genomic prediction and marker-assisted selection for seed weight in soybean (Glycine max). Theor Appl Genet 129(1):117–130. https://doi.org/10.1007/s00122-015-2614-x

Zhao X, Liu Z, Li H, Zhang Y, Yu L, Qi X et al (2022) Identification of drought-tolerance genes in the germination stage of soybean. Biology 11(12):1812. https://doi.org/10.3390/biology11121812

Marker-Assisted Breeding in Legumes for Drought Tolerance

7

Aiman Hina, Ahtsham Ul Hassan, Muhammad Tajammal Khan, Adeela Haroon, Asim Abbasi, Awais Hanif, Hayssam M. Ali, and Nader R. Abdelsalam

Abstract

Grain legumes, such as chickpeas, lentils, and beans, are indispensable for international food security, offering a sustainable source of protein and enhancing soil fertility through nitrogen fixation. However, their productivity is significantly hindered by abiotic stresses, particularly drought, which is exacerbated by cli-

A. Hina (✉)
Ministry of Agriculture (MOA) Key Laboratory of Biology and Genetic Improvement of Soybean (General), Soybean Research Institute, National Centre for Soybean Improvement, State Key Laboratory for Crop Genetics and Germplasm Enhancement, Nanjing Agricultural University, Nanjing, China

A. U. Hassan
Centre of Agricultural Biochemistry and Biotechnology (CABB), University of Agriculture, Faisalabad, Pakistan

M. T. Khan
Division of Science and Technology, Department of Botany, University of Education, Lahore, Pakistan

A. Haroon
Department of Botany, The Women University Multan, Multan, Pakistan

A. Abbasi
Department of Entomology, University of Agriculture, Faisalabad, Pakistan

A. Hanif
Department of Plant Breeding and Genetics, University of Agriculture, Faisalabad, Pakistan

H. M. Ali
Department of Botany and Microbiology, College of Science, King Saud University, Riyadh, Saudi Arabia

N. R. Abdelsalam
Agricultural Botany Department, Faculty of Agriculture (Saba Basha), Alexandria University, Alexandria, Egypt

© The Author(s), under exclusive license to Springer Nature Singapore Pte Ltd. 2025

M. A. Nadeem et al. (eds.), *Marker-Assisted Breeding in Legumes for Drought Tolerance*, https://doi.org/10.1007/978-981-96-4112-3_7

mate change. This chapter covers a comprehensive overview of the methodologies and applications of marker-assisted breeding (MAB) in breeding drought-tolerant grain legumes. It emphasizes the critical role of MAB in overcoming the challenges posed by abiotic stresses and highlights the potential of these technologies to enhance the resilience and sustainability of legume cultivation in a changing climate. MAB leverages molecular markers, such as quantitative trait loci (QTLs), to identify and select for drought-tolerant traits with precision. The integration of backcross breeding and recurrent selection methods within MAB programs facilitates the introgression of these traits into elite cultivars, accelerating the breeding process and improving genetic gain. By focusing on traits such as deep root systems, osmotic adjustment, and efficient water use, MAB enables the advancement of drought-resistant legume varieties. This chapter highlights the identification and utilization of QTLs linked to drought tolerance, the advantages of combining multiple traits for improved stress resilience, and the potential of MAB to address the challenges posed by climate change, ultimately underscoring the importance of MAB in creating drought-tolerant legume varieties that contribute to sustainable agriculture and food security in the face of increasing environmental stresses.

Keywords

Grain legumes · Marker-Assisted Breeding · Backcross breeding · Quantitative trait loci · Molecular markers · Abiotic stresses · Drought tolerance · Recurrent selection · Crop resilience

7.1 Introduction

Legumes stood second to cereals in terms of global food production. Humans rely heavily on legumes for their food (27%) and protein (33%) requirements (Pandey et al. 2016; Khatun et al. 2021; Abbasi et al. 2023). The leguminous crops can be grown successfully on diverse soil types under varying climatic conditions. Besides providing nutrition, these crops are ideal candidates for inclusion within resource-efficient farming methods owing to their adaptability to mitigate emission of certain greenhouse gases and fix atmospheric nitrogen (Hina et al. 2023; Ali et al. 2022). Moreover, their addition in different cropping systems not only diversifies the agro-ecosystem settings but also helps in breaking the cycles of notorious pathogens and insect pests (Voisin et al. 2014; Pradhan et al. 2019; Shahid et al. 2023). The cultivation of leguminous crops provides a net benefit of 31 billion USD to almost 700 million farmers belonging to different developing countries (Pradhan et al. 2022). Among these, the majority of economic returns from export comes from four major crops, with soybean contributing 83.8%, common bean 8.8, 4.9 and chickpea 2.4 per cent (Abate et al. 2012).

The grain legume production has been compromised due to a number of ecological factors (Razzaq et al. 2023). Among these, water scarcity has emerged as a major concern limiting the growth and yield of legume crops (Fang and Xiong 2015; Foyer et al. 2016). Grain legumes are usually grown in rain-fed agroecosystems due to which they are more prone to drought (Farooq et al. 2016; Lu et al. 2021). Water-deficit condition interrupts stomatal oscillations, plant nutrient uptake, and water use efficiency of crop plants (Ahmed et al. 2023). Moreover, water scarcity is also associated with cell dehydration which ultimately affect stem elongation, leaf size, root proliferation, and cell division (Kaushal and Wani 2016; Gillani et al. 2022). The severe and more pronounced effects of drought were evident during the reproductive and grain-filling stage of crop plants (Pushpavalli et al. 2015; Tiwari et al. 2016; Tomer and Singh 2020; Khan et al. 2022). Nevertheless, the extent of reduction in crop yield largely depends on crop stage, genetic variability, and duration and intensity of drought condition (Baroowa and Gogoi 2013; Ghassemi-golezani et al. 2013; Zhang et al. 2019). Therefore, the introduction of new cost-effective approaches is critical for enhancing grain yields of legume crops particularly under drought stress (Ulemale et al. 2013; Chaudhry et al. 2023).

Until now, limited success regarding drought stress mitigation strategies has been achieved possibly due to complex genetic structure of drought stress-tolerant quantitative trait loci (QTLs) (Fleury et al. 2010; Deshmukh et al. 2014). Improvements in drought tolerance mechanisms can be made possible through harnessing genetic variations from available germplasm possessing elite agronomic characters and adaptive mechanisms to cope with adverse environmental conditions (Valliyodan et al. 2017). It had been reported that legume crops possess ample genetic variability regarding drought tolerance traits (Agbicodo et al. 2009; Krishnamurthy et al. 2010). The drought-tolerant genotypes in legume crops can be better assembled by evaluating certain morphological, physiological, and yield-related plant traits associated with drought tolerance (Valliyodan et al. 2017).

Several drought-resilient legume varieties have been developed in recent past using conventional breeding approaches. However, these approaches could not bear satisfactory results regarding gains in genetic diversity and crop yield (Foyer et al. 2016; Varshney et al. 2018). This seeks incessant progression in pulse breeding programs with focus on assimilating conventional breeding programs with molecular markers for better exploitation of genetic resources (Baloch et al. 2024). In the past decade, a variety of modest-scale genomic resources have been developed in these crops to accelerate breeding cycles (Varshney et al. 2013; Pandey et al. 2016). Various molecular markers, along with hybrid and natural populations, were utilized for the analysis and enhancement of traits (Thudi et al. 2014). Rapid enhancement of drought-relevant traits, supported by marker technologies like marker-assisted backcrossing (MABC) and marker-assisted recurrent selection (MARS), has yielded promising outcomes in major legumes like chickpea and groundnut (Pandey et al. 2016; Collard and MacKill 2008). In grain legumes, MAS/MABC has proven effective for transferring a small number of QTLs that greatly influence the phenotype (Varshney et al. 2019). Despite its advantages, MABC encounters significant challenges when used to improve complex traits like yield under drought stress, which are regulated by many small-effect QTLs (Collins et al.

2008). To tackle complex quantitative traits like yield, scientists are progressively relying on genomic selection (GS) (Singh et al. 2015). Ensuring sustainable agriculture and food security depends on breeding legumes for stable yields and improved drought tolerance (Kunert and Vorster 2020). Despite the slow response to selection observed for various drought tolerance traits, which has hindered the development of tolerant cultivars in grain legumes, recent advancements in legume genetics and genomics have enhanced our ability to address drought challenges more effectively (Jha et al. 2020). This chapter explores recent advancements in marker-assisted breeding for legumes, concentrating on strategies for improving drought resilience.

7.2 Drought Effects in Grain Legumes

Generally, legume crops depend on rainfall for irrigation and are susceptible to drought stress particularly during reproductive and vegetative phases (Savita et al. 2020). However, plants have developed numerous complex mechanisms at physiological, morphological, and biochemical levels to cope with drought stress (Khoyerdi et al. 2016). The negative impacts of drought stress on crop plants have been mentioned below in detail (Fig. 7.1).

7.2.1 Drought-Driven Morphological Changes

7.2.1.1 Plant Vegetative Development

The growth and development of a crop plant are predominantly governed by cell expansion, cell division, and cell differentiation, which are interconnected with certain morphological and physiological plant events. Water scarcity is responsible for reduced turgor pressure which ultimately affects cell development (Subbaramamma et al. 2017; Khatun et al. 2021). Similarly, the interruption in xylem-mediated water flow to the rest of the plant cells hinders cell elongation, leaf area, plant height, and crop growth (Farooq et al. 2008; Vessal et al. 2020). Water scarcity also limits plant's ability to uptake nutrients from soil which ultimately affects crop growth and development (Fathi and Tari 2016). Yet, particular drought-adapted lines in grain legumes are capable of growing deeper roots much more than lines that are sensitive to drought (Ye et al. 2018).

7.2.1.2 Foliage Coverage

Drought stress also affects the magnitude and number of plant leaves to a great extent depending on its duration and severity (Fathi and Tari 2016). The internal leaf mesophyll cells become desiccated when subjected to water stress. Similarly, an inhibited cell proliferation followed by reduction in leaf area development can be observed under drought-stressed conditions (Fathi and Tari 2016). In several grain legume crops, reduction in soil moisture contents had a more pronounced effect on leaf area development rather than leaf abscission (Rozita et al. 2012). However, a strong relationship exists between drought-induced leaf senescence and reductions

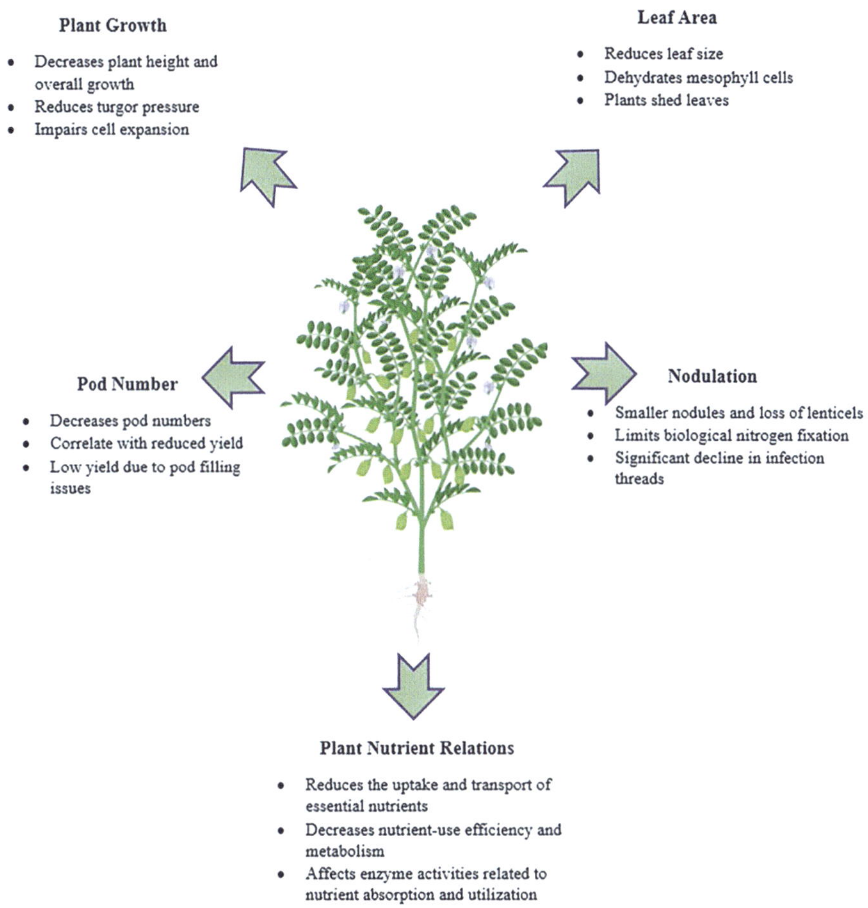

Fig. 7.1 Impact of drought on plant structure, function, and biochemistry

in leaf-area development (Sarkar et al. 2021). The rapid increase in leaf senescence during drought stress mainly occurs due to reduction in nitrogen uptake which regulates radiation efficiency and chlorophyll production (Khatun et al. 2021; Lu et al. 2021).

7.2.1.3 Number of Pods

Drought stress also had a negative effect on number of pods and pod yield of legume crops. Mafakheri et al. (2010) reported that a chickpea cultivar named Pirouz had a greater number of pods even under drought stress but lowest yield mainly due to significant reductions in pod filling and grain weight. Similarly, Sarkar et al. (2021) reported that drought have a negative effect on the number of pods per plant which ultimately affected crop yield (Fig. 7.1).

7.2.2 Drought-Driven Physiological Changes

7.2.2.1 Leaf Surface Temperature

Leaf temperature of legume crops is also affected by drought stress. Plants subjected to drought stress had higher leaf temperature than well-saturated plants (Fahad et al. 2017). However, variations among leaf temperature exist between crop species or even between the varieties of same species (Chowdhury et al. 2017). However, stomatal closure facilitates enhanced heat dissipation in crop leaves (Fahad et al. 2017).

7.2.2.2 Chlorophyll Measurement

The chlorophyll depletion associated with drought stress has long been considered to be a symptom of either chlorophyll degradation or pigment photo-oxidation (Jan et al. 2024; Anjum et al. 2011). Drought stress is responsible for diminishing the chlorophyll contents of crop plants which largely depends on extent and length of drought (Fathi and Tari 2016). Plants encountering drought stress during their vegetative stages usually exhibit reduced chlorophyll a content. Moreover, reduction in chlorophyll b and total chlorophyll content were evident when drought encounters crop plants at vegetative and flowering stages (Mafakheri et al. 2010). It had also been reported that chlorophyll b is more prone to adverse effects of drought stress than chlorophyll a (Jain et al. 2010).

7.2.2.3 Photosynthesis

The photosynthetic efficiency of crop plants is also affected by drought stress which had a direct impact on carbon reduction cycle, electron transport, and stomatal CO_2 regulation (Awasthi et al. 2014). The total chlorophyll contents of crop plants were also affected under drought stress, inferring a reduction in light harvesting ability and thus reduced photosynthesis (Fathi and Tari 2016). Similarly, variations in leaf photosynthesis during drought stress also happen due to mesophyll cell collapse and partial stomatal closure owing to loss of turgor pressure (Farooq et al. 2009; Sohrawardy and Hossain 2014; Zhu et al. 2021).

7.2.2.4 Nutrient Dynamics in Plants

Water scarcity also affects the nutrients uptake from soil and their transfer from roots to shoots and other parts of plants (Farooq et al. 2009; Nazir et al. 2021). The uptake of many vital elements such as Si, N, Ca, and Mg depends on adequate water availability as dryness obstructs their transportation from roots to upper plant parts (Farooq et al. 2009). Drought also limits the nitrogen availability by interfering with root nitrate levels and leaf nitrate reductase activity (Ullah and Farooq 2022). The nitrogen-use efficiency of legume crops is also affected by drought stress as it hinders sucrose synthase, nitrate reductase, and legume-rhizobium symbiotic activities (Ullah and Farooq 2022).

7.2.3 Drought-Driven Biochemical Changes

Drought stress also upsets the production of reactive oxygen species (ROS) and compromises the antioxidant defense mechanisms of crop plants by exposing them to oxidative stresses (Farooq et al. 2009). Plant defense system largely relies on production of carotenes or isoprenoid molecules, but their function also destabilizes under oxidative damage (Savita et al. 2020). Dry circumstances often lead to rapid salt accumulation in the surface soil and plant roots which later becomes a major reason for ion toxicity and osmotic stress (Fathi and Tari 2016). Similarly, water stress also degrades β-carotene which is a vital element of PSI (Photosystem I) and PSII (Photosystem II) core complexes. This degradation is mainly attributed to a rapid rise in ROS concentrations in the thylakoids which cause oxidative damage to different cell components including lipids, proteins, and genetic blueprint (Savita et al. 2020).

7.3 Types of Marker-Assisted Breeding

Cutting-edge breeding techniques and genetic innovations, including marker-assisted selection (MAS), genomic selection (GS), and targeted gene editing, would expedite the development and commercialization of legume varieties (Hasan et al. 2021). Identifying genetic markers linked to drought tolerance and key economic characteristics such as high seed yield, oil content, and oleic acid content, as well as disease and insect resistance, is essential for developing new varieties with these important characteristics (Shaibu et al. 2020; Devate et al. 2022).

7.3.1 Marker-Assisted Selection (MAS)

During the first decade of the twenty-first century, there was a notable rise in the development and utilization of genetic resources for research on marker-trait associations in pulse crops (Maurya et al. 2023). Modern breeding techniques, powered by genomics, offer tools to tackle challenges related to increasing food yield, quality, and production stability (Ali et al. 2024). The use of crossbreeding has proven the value of employing DNA markers to support marker-assisted selection in plant breeding (Manickam et al. 2023). MAS enables the breeding of plants for specific traits by using markers within the plant's genotype, eliminating the need to observe the associated phenotype (Table 7.1) (Habib et al. 2024). This approach holds a momentous ability for enhancing the efficacy and efficiency of selecting plant genotypes with desired traits (Varshney et al. 2009). This method relies on establishing a strong link between a molecular marker and the chromosome region containing the genetic markers for the desired trait in a specific environment. Once this correlation is made, selection can occur in the lab without the need for the phenotype to be expressed (Maurya et al. 2023). In early generations, molecular markers tightly linked with the desired trait are identified, allowing for the selection of plants for

Table 7.1 Drought resistance genotypes of grain legumes developed through Marker-Assisted Breeding

Crop	Genotypes	Linked marker	Improved traits	Breeding technique	References
Cicer arietinum	ICCV 92944 (Chania Desi II)	SSR and SNP markers	Root traits	Marker-assisted backcrossing (MABC)	Kosgei et al. (2022)
	ICCV 00108 (LDT 068)				
Cicer arietinum	JG 11	QTL-hotspot	Root and drought tolerance traits	Marker-assisted backcrossing (MABC)	Varshney et al. (2013)
Cicer arietinum	Pusa 372	SSR markers in QTL-hotspot region	Drought tolerance and yield gain	Assisted backcrossing (MABC)	Bharadwaj et al. (2021)
Phaseolus vulgaris	Sierra/AC1028	RAPD markers	Yield under drought stress	Marker-assisted selection (MAS)	Schneider et al. (1997)
	Sierra/Lef-2RB				
Vigna unguiculata	M499_BC3F3_48P85	EST-derived SNP markers	Drought tolerance	Marker-assisted selection (MAS)	Batieno et al. (2016)
	M499_BC3F3_4P67				
	M499_BC3F3_48P90				
Arachis hypogaea	JL 24	Not specified	High yield, early bunch, and drought resistance	Conventional breeding	Patil et al. (1981) and Upadhyaya et al. (2005)

further breeding. These traits are typically selected indirectly, which is particularly useful when the traits are challenging to express phenotypically. This approach enables plant breeders to bypass the costly process of directly evaluating certain traits (Nadeem et al. 2018). To fully leverage the potential of MAS, it is essential to verify the impact of the target and accurately determine its chromosomal location before using the marker to identify and select the most desirable plants (Liu et al. 2003). For optimal effectiveness in MAS, a characteristic should have a heritability of 0.2 or higher; anything below this threshold diminishes the efficiency of the selection process (Moreau et al. 1998). It is suggested to incorporate only those QTLs that demonstrate a substantial and consistent effect across multiple environments and avoid those that show inconsistency (Liu et al. 2007). Phenotypic selection helps preserve minor QTL and identifies loci involved in epistatic interactions. When combined with MAS, this approach enhances the effectiveness of selecting complex traits (Kumar et al. 2014).

7.3.2 Marker-Assisted Backcrossing (MABC)

In the back-crossing technique, specific desirable genes are transferred from a donor genotype to a recipient genotype. The donor, also known as the non-recurrent parent, is used only once for crossing, while the recipient, or recurrent parent, is repeatedly used in each generation to ensure that most of its genome is retained in the progeny. Incorporating markers into this breeding process improves precision and accelerates the achievement of desired outcomes. MABC breeding has been effectively exploited to improve both biotic and abiotic resistance in legumes, particularly in chickpeas (Varshney et al. 2013; Bohra et al. 2020; Roorkiwal et al. 2020).

7.3.3 Marker-Assisted Recurrent Selection (MARS)

Using marker-assisted recurrent selection, plant breeders can opt for particular trait like crop production under drought. Nonetheless, it is important to also consider other factors that might be important, such as harvest in ideal circumstances, and resistance against disease (Suvarna et al. 2023). Various strategies exist for MARS, and choosing the right one depends on several factors. While theoretical comparisons can offer guidance, their practical implementation often requires costly experimental validation. Research into the genetic features associated with various marker-assisted recurrent selection methods bridges this gap by demonstrating their effectiveness and real-world applicability. Few studies compared MARS with other techniques, including marker-aided mass selection and phenotypic recurrent selection, and explored strategies such as MS, half-sib families, full-sib families, St families, test cross progeny, etc., (Xie and Xu 1998; Van Berloo and Stam 2001; Moreau et al. 2004). Other studies have utilized MARS to enhance drought resistance by selecting genetic markers linked to improved performance under water stress. For instance, Varshney et al. (2013) conducted a study to develop drought-tolerant chickpea varieties using MARS. They identified and selected specific genetic markers associated with drought resilience, leading to the development of chickpea lines that exhibit better yield and adaptability in drought-prone environments. This approach has significantly contributed to the improvement of chickpea varieties and can be seen as a model for similar applications in other legumes.

7.3.4 Genomic Selection (GS)

Recently, the use of genomic selection has gained popularity in breeding schemes for incorporating poly genes with small additive effects. The concept of GS was first introduced by Meuwissen et al. (2001) in the field of animal breeding. GS accelerates genetic progress by enabling the identification of superior individuals from a breeding pool, even without phenological data. This approach enhances genetic gains per unit of time by leveraging genetic information to guide selection (Crossa et al. 2017). To enhance quantitative trait variation, there has been a move toward new molecular breeding techniques like GS, which differ from traditional methods that depend on identifying specific significant markers. Instead, GS utilizes comprehensive genome-wide genetic marker information to leverage the small-effect QTLs distributed across the legume genome, thus capturing the quantitative variations effectively (Table 7.2). Predictive models in GS are created using genotypic and phenotypic data from a training set. These models generate Genomic Estimated Breeding Values (GEBVs), which are then employed to choose breeding candidates, eliminating the need for repeated phenotypic evaluations. Recent applications in chickpea have demonstrated encouraging results, with various models employed (Roorkiwal et al. 2016). The advancements in affordable genotyping technologies have paved the way for the introduction of GS in legume crops. Recent reviews have thoroughly examined GS in plant breeding, focusing on the effectiveness of various methods and algorithms for predicting accuracy (Heslot et al. 2014; Crossa et al.

Table 7.2 QTLs with their corresponding characteristics found in different grain legume crops

Crop	QTLs identified	Linked traits	References
Cicer arietinum	qrsr-01, qrsr-02, qrsr-03, qrsr-04, qrld-01, qrdw-01 and qrdwtdw-01	Root traits	Kushwah et al. (2022)
Cicer arietinum	qRWC1.1, qRWC5.1, qMSI2.1, qMSI4.1, qMSI7.1 and qMSI7.2	Drought tolerance	Yadava et al. (2023)
Phaseolus vulgaris	DPM8.1 and DF1.1/DPM1.1	Root traits	Nakedde et al. (2016) and Singh et al. (2019)
Phaseolus vulgaris	Wp1.1 and Wp5.1	Water potential	Sedlar et al. (2020)
Vigna unguiculata	Stg 2_25850, Stg 1_0274, Stg 2_07162 and Dro-7 1_0067	Drought resistance traits	Wu et al. (2021)
Vicia faba	qCT-2012, qCT-2013, qCT-drought-2013, qSD-2013-1, SL-2013, qgs2013, qSD-2014 and qCT-2014	Drought adaptation traits	Khazaei (2014)
Lens culinaris	QRSratioIX-2.30, QLRNIII-98.64 and QSRLIV-61.63	Root traits	Idrissi et al. (2016)
Glycine max	qPH2, qPH6, qPH7, qPH17, qPH19-1, qPH19-2, qPH19-3, qSWPP2, qSWPP6, qSWPP13, qSWPP17 and qSWPP19	Drought tolerance traits	Ren et al. (2020)
Arachis hypogaea	qHW-A01.1, qHW-A05.4, qHW-B01.4, qHW-B09.1, qPW-A03.1, qPW-A02.1, qPW-A03.2, qSW-A03.1, qDW-A05.3, qDW-A05.2, qSLA-A03.4, qSLA-A04.5, qISC-A04.1, qTR-A09.1, qNB-A07.1, qSCMRd-A04.4, qHW-A03.1, qHW-A03.2, qTE-A03.1, qTE-A03.2, qDW-A05.1, qHW-A05.1, qTE-A07.1, qDW-B01.1, qDW-B01.2, qPW-B01.1, qPW-B01.2, qSW-B01.1, qSW-B01.2, qTR-B01.1, qHW-B09.1 and qHW-B09.2	Drought tolerance traits	Pandey et al. (2020)

2017). The application of GS for improving crops faces numerous obstacles, such as the relationship between phenotyped and non-phenotyped plants, the scale of training datasets, and the selection and quantity of DNA markers. Additionally, addressing interactions between genotype × environment and marker × environment is crucial for effective integration (Crossa et al. 2017). As shown in chickpea, GS can simultaneously account for multiple variables, which enhances prediction accuracy in breeding programs by incorporating G × E effects (Roorkiwal et al. 2020).

7.4 Marker-Assisted Breeding for Enhancing Drought Tolerance in Major Legume

As water resources become scarcer and competition for them intensifies, irrigation is no longer a practical solution for alleviating drought. Therefore, adopting genetic or agronomic drought management practices is essential for optimising soil moisture extraction and improving crop yield. Strategies based on genetic and agronomic solutions without using external irrigation can help mitigate the impacts of drought; however, crop yield will still be lower than what can be achieved with irrigation. Evidence shows reduced gains from traditional breeding methods, highlighting the need to explore more precise approaches for genetically enhancing tolerance toward drought. The advancement of genomic technologies has led to the growing popularity of marker-assisted breeding as a method to improve drought resilience in key legumes.

7.4.1 Soybean (*Glycine max*)

Soybean, native to China and cultivated for more than three millennial spans (Sedivy et al. 2017; Hina et al. 2024a), has transformed into a worldwide staple due to its significant composition. Soybean is a key provider of plant-based protein and oil for human use and also serves as animal feed (Zhao et al. 2021; Hina et al. 2024b). Moreover, its distinctive nodule structure makes it an efficient organic nitrogen fertilizer, adding to its overall value (Sun et al. 2023).

Soybean is particularly susceptible to drought, especially during the germination stage, compared to other crops (Zhao et al. 2022). During drought stress, soybean yields can fall by more than 50%, emphasizing the urgent need to improve drought tolerance in soybeans to ensure food sustainability (Arya et al. 2021; Cotrim et al. 2021). Since seed germination initiates plant growth, identifying drought-resistant genotypes at this stage is essential for successful plant breeding (Waterworth et al. 2015).

The swift advancement in high-throughput sequencing technology has led to the widespread use of genome-wide association studies for analyzing plant genetics. This approach has significantly revealed several SNPs and QTLs associated with drought tolerance in soybean (Hacisalihoglu et al. 2018). Lately, Zhang et al. (2024) reported major drought-related QTLs and discovered the gene *GmPrx16* with underlying drought tolerance. In a study, 359 soybean accessions were subjected to short and lengthy drought trials during early growth, leading to the discovery of major SNPs for the short-duration test and 22 for the long-duration test (Saleem et al. 2022). In another study, 26 SNPs were identified on multiple chromosomes, linked with drought tolerance (Zhao et al. 2022). Owing to the variation in genetic background and environmental factors, it becomes difficult to consistently detect stable QTLs across multiple environments, which also hinders their functional validation (Zhang et al. 2024; Elattar et al. 2021; Hina et al. 2020).

7.4.2 Chickpea (*Cicer arietinum*)

Chickpea, being a diploid annual and highly self-pollinated crop, globally ranks as the third foremost legume grain, following field peas and faba bean (Singh et al. 2018; Madurapperumage et al. 2021; Razzaq et al. 2022). Chickpea commonly described as a budget-friendly meat due to its importance as a key protein source (Shabbir et al. 2014; Grewal et al. 2020; Sahu et al. 2020; Singh et al. 2021a). This crop is primarily grown in the post-rainy season, utilizing the remains of soil moisture from the prior rain. Consequently, chickpea often endures extreme temperature and water-deficit conditions, which lead to yield decline of approximately 50% due to impaired photosynthesis and membrane damage (Upadhyaya et al. 2012; Arif et al. 2021; Choudhary et al. 2021; Korbu et al. 2022). In chickpea prolonged drought also influences water retention, chlorophyll index, membrane stability index, and biomass (Sapes and Sala 2021; Raddi et al. 2022). Studies reported that microsatellite markers and sequence-tagged microsatellite site markers are considered more advantageous (Palit et al. 2020). In 2019, the drought-resistant, high-performance chickpeas of Ethiopian origin were developed via marker-assisted back-crossing under multiple environments. This germplasm was created by introgressing the QTL hotspot associated with drought tolerance from ICC4958 into the Indian chickpea (gldc.cgiar.org). Similarly, various drought tolerance traits were incorporated into three superior chickpea varieties from India from ICC 4958. Lately, Kenyan chickpea varieties have been enhanced with drought-resistant root traits using MABC method (Kosgei et al. 2022). A drought-tolerant chickpea variety, Pusa 372 was originally introduced for commercial cultivation in various zones of India. However, its performance under drought conditions has declined recently. To address this, the MABC method was used to incorporate the “QTL hotspot” from ICC 4958 into Pusa 372, leading to the development of Pusa 10216, which offers improved drought resistance (Asati et al. 2022).

7.4.3 Groundnut (*Arachis hypogaea*)

Groundnut, commonly known as peanuts, is a crucial oilseed and protein crop cultivated globally, known for its high-quality oil and nutritional benefits for humans (Moharana et al. 2020). Global production of peanuts is about 50 million metric tons, with the United States ranking fourth, following China, India, and Nigeria (Dang et al. 2024). Groundnut is more drought-tolerant in comparison to other traditional leguminous species. However, drought stress can still impact the quantity and excellence of its grains, seed oil, and fodder values (Wan et al. 2014; Abady et al. 2021). Research indicates that extreme water scarcity during the reproductive growth period can result in up to 33% decrease in yield (de Lima Pereira et al. 2016; Carvalho et al. 2017).

Drought is a damaging abiotic stress that impacts all key stages of peanut growth, such as seedling stage, flowering, fruit setting, and seed maturation (Dang et al. 2024). The release of drought-tolerant peanut varieties is a sustainable strategy for ensuring long-term production, but breeding for drought resilience is challenging due to the significant influence of complex gene interactions and environmental factors (Puppala et al. 2023). In 2009, the first genetic map for domesticated groundnut was developed after screening a vast number of SSR markers, both from public sources and unpublished collections (Varshney et al. 2009). In one study, Ravi et al. (2011) developed a thorough and precise genetic map with 191 SSR loci to identify candidate QTLs for drought tolerance. This map was based on a single mapping population that segregated for drought and associated traits. They extensively analyzed genotyping and phenotyping measurement for more than ten drought-related attributes over two to three seasons to uncover both main effect QTLs and epistatic interactions. In a separate study, Gautami et al. (2012) employed two novel recombinant inbred line (RIL) mapping populations to explore the genetic foundations and identify QTLs for drought tolerance. They screened 3215 SSR markers across the parental genotypes of these populations, resulting in the development of two new genetic maps with 119 and 82 SSR loci, respectively. There is limited understanding of the number of genetic markers and their associations with drought tolerance across broad genetic variability in groundnut. Understanding these marker-trait associations is essential for implementing marker-assisted selection, integrating traits, and advancing precision breeding in groundnut.

7.4.4 Cow Pea (*Vigna unguiculata*)

Cowpea is primarily used as either a dry grain or a fresh vegetable. It is rich in protein, carbohydrates, vitamins, and fiber, making it crucial for addressing today's global food security challenges (Hall 2012; Singh et al. 2023). The crop is typically grown for its fresh and dry grains, as well as for its fresh leaves and fodder, across Africa and in other regions including South America, Asia, and the United States (Boukar et al. 2019; Attamah et al. 2024). Recent progress in genomics, along with advanced phenotyping and genotyping technologies, has been leveraged to discover genes and QTLs linked to traits relevant to climate change in cowpea (Sahay et al. 2022). According to Agbicodo et al. (2009), there are two primary approaches for identifying drought-tolerant genes in cowpea. One approach involves identifying candidate genes related to drought tolerance from existing research on cowpea and similar crops. The other, more frequently used, method involves analyzing variations in mRNA expression between drought-stressed plants and control group plants. Drought-tolerant genes in cowpea were also explored (Horn and Shimelis 2020). Several studies have successfully constructed genetic maps for cowpea using SNPs and other markers. Muchero et al. (2009) created a genetic map to establish a genomic framework for identifying quantitative trait loci, performing map-based cloning, evaluating genetic variation, and conducting association mapping. This framework was used to enhance marker-assisted selection in seven populations of cowpea RILs using SNPs. By utilizing a consensus genetic map and the F2:3 population along with composite interval mapping (CIM), 11 QTLs related to yield traits were identified across 7 chromosomes in cowpea. This characterization of molecular markers is highly beneficial for breeders engaged in marker-assisted breeding and enhancement of cowpea (Nkomo et al. 2021).

7.4.5 Pea (*Pisum sativum* L.)

Peas are grown globally for various purposes, including food, fodder, and feed, and are regarded as a significant legume for the winter season (Parihar et al. 2020; Ramzan et al. 2014). Developing resistant pea cultivars to abiotic stresses like drought is a key strategy for improving productivity, and Kreplak et al. (2019) provided a whole genome sequence of pea at the chromosome level to support this effort. The availability of the pea reference genome makes it easier to pinpoint desirable alleles affecting important agronomic traits, thereby enhancing these traits via genomic technologies such as MAS and GS (Alemu et al. 2022; Pandey et al. 2023). A comprehensive study on diversity analysis and genotyping of the genus *Pisum* has been carried out utilizing various molecular markers. Jing et al. (2010) conducted the most extensive research so far, analyzing 3020 *Pisum* accessions with retrotransposon-based insertion polymorphism (RBIP) markers. Additionally, other molecular markers, such as SNPs, can also be used (Powers et al. 2021). The extensive variation in potential traits among wild and domesticated peas offers significant potential for strategically hybridizing matching genotypes. Leveraging the genetic diversity within *Pisum* allows for the enhancement of key traits, including agronomic, biochemical, and physiological attributes, which can significantly improve drought resistance and increase yield potential in peas (Yang et al. 2022; Rispail et al. 2023).

7.4.6 Common Bean (*Phaseolus vulgaris* L.)

The common bean is a vital legume for human dietary use. It is extensively cultivated across the globe, particularly in the tropical and subtropical regions of Africa and the Americas (Joshi and Rao 2017). Bean production faces significant challenges from various abiotic and biotic factors. It is estimated that around 60% of common bean yields are impacted by terminal or intermittent drought conditions (Habib 2021; Hamabwe 2023). If current climate change trends persist, much of the southeastern African regions where common beans are currently grown may become unsuitable for cultivation by 2050. This could lead to reduced yields for existing bean varieties and potentially impact the crop's nutritional quality (Hummel et al. 2018).

Drought poses a significant risk for bean farmers in tropical regions, making the development of improved, climate-resilient varieties crucial for future generations. As a result, enhancing drought resistance is a primary breeding objective for common beans. Identifying molecular markers is essential to advancing breeding efforts for drought tolerance (Wu et al. 2024). Genetic mapping identifies genomic regions with variations that influence phenotypic trait expression. This data is then applied in breeding through MAS, enabling the identification of breeding lines with desirable traits via genetic screening (Diaz et al. 2020). Ideally, genetic mapping results in the discovery of genes with functional alleles that drive the observed phenotypic variation. This process is most advanced in model crops such as rice, where over 2000 genes associated with key agronomic traits have been successfully cloned (Li et al. 2018; Liu and Yan 2019). In common bean, only a few genes have been cloned. Among them is the bc-3 gene, which provides antiviral resistance to bean common mosaic virus (BCMV) and has been identified as the eIF4E gene. Another example is the Mrp1 gene, responsible for the lpa1 mutant, which regulates low phytic acid concentrations in bean seeds (Naderpour et al. 2010; Panzeri et al. 2011). In other research, candidate genes linked to agronomic traits in common bean have been identified by examining annotations near markers with significant associations (Katuuramu et al. 2018). The advancement of research on candidate genes is constrained by the limited genetic resources and the scarcity of genetic transformation protocols available for common bean (Hnatuszko-Konka et al. 2014; Diaz et al. 2020).

7.4.7 Faba Bean (*Vicia faba*)

The faba bean (*Vicia faba* L.) is a cold-season pulse cultivated globally for human consumption and livestock feed. Renowned for its high protein content, encompassing 25–37%, it is a valuable crop for both nutritional and agricultural purposes (Warsame et al. 2020). The faba bean is highly sensitive to drought stress throughout its growth, from seedling to maturity, which hinders its ability to achieve full performance potential (Khan et al. 2010). Over a span of 34 years, a meta-analysis of global field studies and drought experiments revealed that when water availability dropped by 65%, faba bean yields declined by 40%. This represents the most substantial yield reduction due to drought among the legume crops studied (Daryanto et al. 2015). In faba bean, to counteract the severe effects of drought, it is crucial to identify drought-tolerant genotypes and develop effective molecular breeding strategies. Faba bean accessions have shown a wide range of genetic variation in traits associated with drought resilience (Muktadir et al. 2020). Despite extensive research on the faba bean's reaction to water deficit, only a few molecular techniques have been applied to boost drought tolerance in this crop (Ammar et al. 2015; Siddiqui et al. 2015). In 2014, Khazaei et al. (2014) were the first to report QTLs linked to stomatal traits located on chromosome II. They leveraged the genetic similarities with *Medicago truncatula* to pinpoint potential candidate genes associated with QTLs. Later, Abid et al. (2015) used suppression subtractive hybridization to identify six potential drought stress-responsive genes in faba beans. Similarly, in 2016, Ali et al. (2016) performed the inaugural genome-wide association study on drought in faba beans, evaluating 189 German winter faba bean lines that are derived from 11 original parental lines. Their study evaluated various physiological traits associated with dehydration tolerance, based on findings from earlier research (Alderfasi and Alghamdi 2010; Ammar et al. 2015).

7.4.8 Lentil (*Lens culinaris*)

Drought stress offers a substantial obstacle to lentil cultivation in drought-prone and semi-drought-prone areas (El Haddad et al. 2020). Drought stress in lentils reduces germination, decreases leaf area, stunts shoot and root growth, and lowers relative water content. It also disrupts membrane stability, hinders photosynthesis, and induces oxidative stress, ultimately leading to a major drop in organic matter production and severe yield decline (Gupta et al. 2020; Saha et al. 2020; Akter et al. 2021; Zafar-ul-Hye et al. 2021; Venugopalan et al. 2022). Moreover, drought often coincides with other stressors, exacerbating the severity of losses, particularly during the reproductive phase (Choukri et al. 2020, 2022; El Haddad et al. 2021).

As a nutritious and affordable crop rotation option, lentils serve a pivotal role in international food and nutrition safety. Therefore, developing climate-adapted lentil varieties to increase productivity in harsh environments is an essential task that could help counteract the effects of climate change. Improving lentil yields in such conditions necessitates the creation of climate-resilient varieties (Kumari et al. 2021; Ullah and Farooq 2022). Achieving this goal requires a thorough understanding of how lentils respond to and tolerate drought stress. Successful crop breeding and genetic improvement rely on having a diverse gene pool, identifying and characterizing novel alleles, and obtaining comprehensive crossbreeding information to select appropriate parent taxa (Rajpal et al. 2023). Multiple studies have shown that domesticated lentils exhibit a significantly narrower genetic range than their wild counterparts, *L. culinaris* ssp. *Orientalis* (Alo et al. 2011). Research has consistently shown that wild *Lens* taxa possess resistance to a range of abiotic stress environment (Gupta and Sharma 2006). The application of genomics-assisted and molecular breeding techniques, combined with traditional breeding methods, has been employed to uncover the hidden variability in wild relatives of lentil crops. Research reveals that *L. nigricans* exhibits the highest drought tolerance, while *L. orientalis* may offer valuable genetic resources for improving drought resistance in low-rainfall regions of Africa (Gupta and Sharma 2006). Furthermore, testing of wild *Lens* genetic resources has revealed that *L. odemensis*, *L. ervoides*, *L. lamottei*, *L. tomentosus*, and *L. nigricans* exhibit resistance to drought (Gorim and Vandenberg 2017). In response to drought, several accessions of *L. odemensis*, *L. ervoides*, and

L. orientalis developed enhanced deep root development, with some also showing delayed flowering. Meanwhile, *L. tomentosus* exhibited drought tolerance by lowering its transpiration rates (Gorim and Vandenberg 2017). In their study, Omar et al. (2019) examined drought resilience in superior lentil varieties hybridized with wild relatives. They discovered that drought tolerance was associated with enhanced cell membrane firmness, a higher root-to-shoot ratio, fuzzy leaves, increased leaf moisture content, and reduced evaporation and withering. Sanderson et al. (2019) conducted an analysis of RILs to study drought tolerance, using crosses between lentil cultivars and wild relatives. The broad range of drought responses observed in lentils highlights the importance of wild relatives for future lentil development. This progress hinges on successfully crossing wild and cultivated species to produce viable hybrids (Rajpal et al. 2023).

Numerous lentil mapping populations have been created from intra- and interspecific crosses between drought-sensitive and drought-resistant varieties to investigate drought tolerance inheritance and identify linked genetic markers. In total, 7 out of 51 SSR markers, related to drought resilience were identified using Bulk segregant analysis. These markers were then screened and mapped in an F2 population comprising 101 plants, covering a distance of 133.2 cM (Singh et al. 2016). SNP markers have been widely used in lentils, making substantial contributions to linkage mapping, the study of genetic diversity, and the investigation of trait associations (García-García et al. 2019; Wang et al. 2020). Many studies have used SNP markers to identify genetic factors associated with drought tolerance and to generate detailed genetic insights. Approximately 377 SNPs from TOG sequences in *L. ervoides* were identified and utilized to develop a map with seven separate linkage clusters (Gujaria-Verma et al. 2014). Genetic linkage maps, draft genome assemblies, and high-throughput markers have advanced lentil breeding, enabling the development of climate-resilient varieties that can handle multiple stresses, such as drought.

7.4.9 Mung Bean (*Vigna radiata*)

Mung bean ranks as the third key pulse crop, following chickpea and pigeon pea. It is primarily cultivated in Asia and has also spread to parts of Africa, Australia, and South America. Nevertheless, mung bean's sensitivity to drought stress at all growth stages significantly impacts its productivity (Pratap et al. 2020). Mung bean is especially sensitive to drought during the reproductive period, which includes the time from anthesis to maturity. Any drought stress during these critical phases can lead to significant yield reductions. Thus, selecting genotypes that excel during these key growth stages is vital for enhancing drought resilience (Singh et al. 2021b). Understanding the genetics behind traits related to abiotic stress adaptation can assist breeders in integrating these traits into enhanced cultivars. A significant breakthrough for mung bean was the completion of the first draft of its whole-genome sequence, which has opened new avenues for genomic research, allowing for detailed exploration of the genome and marker positioning (Kang et al. 2014).

Researchers have identified several QTLs linked to drought adaptation using a mung bean RIL population (Liu et al. 2017). Effective identification of genomic regions related to complex traits in mung bean has been achieved through QTL mapping and marker-assisted techniques. Evaluating large populations phenotypically and conducting QTL mapping are crucial strategies for enhancing the yield potential, dietary quality, and stress endurance of mung bean (Huppertz et al. 2023). It is crucial to explore novel adaptive traits for drought stress from diverse gene pools and accurately introgress these traits into promising genotypes using marker-assisted and advanced breeding techniques.

7.5 Conclusion

Legumes are a significant source of protein and are cultivated globally. They provide a substantial portion of the daily protein requirements for both humans and animals. Common examples include mung beans, kidney beans, cowpeas, soybeans, peas, peanuts, lentils, chickpeas, and horse gram. This family of plants is recognized for its rich nutritional content and beneficial phytochemicals. Besides being rich in protein, legume crops also provide significant amounts of carbohydrates and oils. They are packed with fiber, vitamins, probiotics, polyphenols, antioxidants, minerals, and notable phenolic acids. However, abiotic factors such as arid conditions, high salt concentration, metal contaminants, UV radiation, extreme temperatures, and flooding can adversely impact legume crops, particularly in hot and arid regions, leading to reduced growth and yield. A key challenge in pulse cultivation is the significant genotype-environment interactions affecting important quantitative traits. This leads to slow progress in genetic improvement and stability of pulse yields. To enhance traditional agricultural growth objectives, there is an urgent need to integrate accelerated molecular breeding strategies with DNA marker technologies. This approach will help identify desirable genotypes and improve the effectiveness of conventional breeding methods. Pulses exhibit significant phenotypic variability, making traditional improvement methods challenging. Utilizing molecular markers can facilitate rapid and substantial progress by enabling the swift identification and development of high-yielding genotypes.

References

Abady S, Shimelis H, Janila P, Yaduru S, Shayanowako AI, Deshmukh D, Manohar SS (2021) Assessment of the genetic diversity and population structure of groundnut germplasm collections using phenotypic traits and SNP markers: implications for drought tolerance breeding. PLoS One 16:e0259883

Abate T, Alene AD, Bergvinson D, Shiferaw B, Silim S, Or A, Asfaw S (2012) Tropical grain legumes in Africa and South Asia: knowledge and opportunities. International Crops Research Institute for the Semi-Arid Tropics, Patancheruvu

Abbasi A, Hina A, Shahid S, Imtiaz S, Intisar A, Khan MT (2023) Micronutrients enrichments in legumes through agronomic and cultural practices. In: Nadeem MA et al (eds) Legumes biofortification. Springer, Cham. https://doi.org/10.1007/978-3-031-33957-8_3

Abid G, Muhovski Y, Mingeot D, Watillon B, Toussaint A, Mergeai G, Jebara M (2015) Identification and characterization of drought stress responsive genes in faba bean (*Vicia faba* L.) by suppression subtractive hybridization. Plant Cell Tissue Organ Cult 121:367–379

Agbicodo EM, Fatokun CA, Muranaka S, Visser RGF, van der Linden CG (2009) Breeding drought tolerant cowpea: constraints, accomplishments, and future prospects. Euphytica 167:353–370

Ahmed S, Khan MT, Abbasi A, Haq IU, Hina A, Mohiuddin M, Tariq MAUR, Afzal MZ, Ng AWM, Li Y (2023) Characterizing stomatal attributes and photosynthetic induction in relation to biochemical changes in *Coriandrum sativum* by foliar applied zinc oxide nanoparticles under drought conditions. Front Plant Sci 13:1079283

Akter S, Jahan I, Hossain MA, Hossain MA (2021) Laboratory-and field-phenotyping for drought stress tolerance and diversity study in lentil (*Lens culinaris* Medik.). Phyton 90:949

Alderfasi AA, Alghamdi SS (2010) Integrated water supply with nutrient requirements on growth, photosynthesis productivity, chemical status and seed yield of faba bean. Am Eur J Agro 3:08–17

Alemu A, Brantestam AK, Chawade A (2022) Unraveling the genetic basis of key agronomic traits of wrinkled vining pea (*Pisum sativum* L.) for sustainable production. Front Plant Sci 13:844450

Ali MB, Welna GC, Sallam A, Martsch R, Balko C, Gebser B, Link W (2016) Association analyses to genetically improve drought and freezing tolerance of faba bean (*Vicia faba* L.). Crop Sci 56:1036–1048

Ali A, Altaf MT, Nadeem MA, Karaköy T, Shah AN, Azeem H et al (2022) Recent advancement in OMICS approaches to enhance abiotic stress tolerance in legumes. Front Plant Sci 13:952759

Ali A, Altaf MT, Bedir M, Liaqat W, Jamil A, Nadeem MA, Baloch FS (2024) Biotechnological advancements toward sugarcane crop improvement. In: Industrial crop plants. Springer, Singapore, pp 333–352

Alo F, Furman BJ, Akhunov E, Dvorak J, Gepts P (2011) Leveraging genomic resources of model species for the assessment of diversity and phylogeny in wild and domesticated lentil. J Hered 102:315–329

Ammar MH, Anwar F, El-Harty EH, Migdadi HM, Abdel-Khalik SM, Al-Faifi SA, Alghamdi SS (2015) Physiological and yield responses of faba bean (*Vicia faba* L.) to drought stress in managed and open field environments. J Agron Crop Sci 201:280–287

Anjum SA, Xie XY, Wang LC, Saleem MF, Man C, Lei W (2011) Morphological, physiological and biochemical responses of plants to drought stress. Afr J Agric Res 6:2026–2032

Arif A, Parveen N, Waheed MQ, Atif RM, Waqar I, Shah TM (2021) A comparative study for assessing the drought-tolerance of chickpea under varying natural growth environments. Front Plant Sci 11:607869

Arya H, Singh MB, Bhalla PL (2021) Towards developing drought-smart soybeans. Front Plant Sci 12:750664

Asati R, Tripathi MK, Tiwari S, Yadav RK, Tripathi N (2022) Molecular breeding and drought tolerance in chickpea. Life 12:1846

Attamah P, Kusi F, Kena AW, Awuku FJ, Lamini S, Mensah G, Akromah R (2024) Pyramiding aphid resistance genes into the elite cowpea variety, Zaayura, using marker-assisted backcrossing. Heliyon 10:e31976. https://doi.org/10.1016/j.heliyon.2024.e31976

Awasthi R, Kaushal N, Vadez V, Turner NC, Berger J, Siddique KH, Nayyar H (2014) Individual and combined effects of transient drought and heat stress on carbon assimilation and seed filling in chickpea. Funct Plant Biol 41:1148–1167

Baloch FS, Ali A, Tajibayev D, Nadeem MA, Ölmez F, Aktaş H et al (2024) Stripe rust resistance gene Yr15 in Turkish and Kazakhstan wheat germplasms and the potential of Turkish wild emmer for stripe rust breeding. Genet Resour Crop Evol 71(6):2699–2719

Baroowa B, Gogoi N (2013) Biochemical changes in two Vigna spp. during drought and subsequent recovery. Ind J Plant Physiol 18:319–325

Batieno BJ, Danquah E, Tignegre JB, Huynh BL, Drabo I, Close TJ, Ouedraogo TJ (2016) Application of marker-assisted backcrossing to improve cowpea (*Vigna unguiculata* L. Walp) for drought tolerance. J Plant Breed Crop Sci 8:273–286

Bharadwaj C, Tripathi S, Soren KR, Thudi M, Singh RK, Sheoran S, Varshney RK (2021) Introgression of "QTL-hotspot" region enhances drought tolerance and grain yield in three elite chickpea cultivars. Plant Genom 14:e20076

Bohra A, Saxena KB, Varshney RK, Saxena RK (2020) Genomics-assisted breeding for pigeonpea improvement. Theor Appl Genet 133:1721–1737

Boukar O, Belko N, Chamarthi S, Togola A, Batieno J, Owusu E, Fatokun C (2019) Cowpea (*Vigna unguiculata*): genetics, genomics and breeding. Plant Breed 138:415–424

Carvalho MJ, Vorasoot N, Puppala N, Muitia A, Jogloy S (2017) Effects of terminal drought on growth, yield and yield components in Valencia peanut genotypes. SABRAO J Breed Genet 49:270–279

Chaudhry A, Hassan AU, Khan SH, Abbasi A, Hina A, Khan MT, Abdelsalam NR (2023) The changing landscape of agriculture: role of precision breeding in developing smart crops. Funct Integr Genom 23:167

Choudhary ML, Tripathi MK, Gupta N, Tiwari S, Tripathi N, Parihar P, Pandya RK (2021) Screening of pearl millet [*Pennisetum glaucum* [L.] R. Br.] germplasm lines against drought tolerance based on biochemical traits. Curr Appl Sci Technol 40:1–12

Choukri H, Hejjaoui K, El-Baouchi A, El Haddad N, Smouni A, Maalouf F, Kumar S (2020) Heat and drought stress impact on phenology, grain yield, and nutritional quality of lentil (*Lens culinaris* Medikus). Front Nutr 7:596307

Choukri H, El Haddad N, Aloui K, Hejjaoui K, El-Baouchi A, Smouni A, Kumar S (2022) Effect of high temperature stress during the reproductive stage on grain yield and nutritional quality of lentil (*Lens culinaris* Medikus). Front Nutr 9:857469

Chowdhury JA, Karim MA, Khaliq QA, Ahmed AU, Mondol AM (2017) Effect of drought stress on water relation traits of four soybean genotypes. SAARC J Agric 15:163–175

Collard BC, MacKill DJ (2008) Marker-assisted selection: an approach for precision plant breeding in the twenty-first century. Philos Trans R Soc B Biol Sci 363:557–572. https://doi.org/10.1098/rstb.2007.2170

Collins NC, Tardieu F, Tuberosa R (2008) Quantitative trait loci and crop performance under abiotic stress: where do we stand? Plant Physiol 147:469–486

Cotrim MF, Gava R, Campos CNS, de David CHO, Reis IDA, Teodoro LPR, Teodoro PE (2021) Physiological performance of soybean genotypes grown under irrigated and rainfed conditions. J Agro Crop Sci 207:34–43

Crossa J, Pérez-Rodríguez P, Cuevas J, Montesinos-López O, Jarquín D, de Los Campos G, Burgueño J, González-Camacho JM, Pérez-Elizalde S, Beyene Y, Dreisigacker S, Singh R, Zhang X, Gowda M, Roorkiwal M, Rutkoski J, Varshney RK (2017) Genomic selection in plant breeding: methods, models, and perspectives. Trends Plant Sci 22:961–975

Dang P, Patel J, Sorensen R, Lamb M, Chen CY (2024) Genome-wide association analysis identified quantitative trait loci (QTLs) underlying drought-related traits in cultivated peanut (*Arachis hypogaea* L.). Genes 15:868

Daryanto S, Wang L, Jacinthe PA (2015) Global synthesis of drought effects on food legume production. PLoS One 10:e0127401

de Lima Pereira JW, Albuquerque MB, Melo Filho PA, Nogueira RJMC, de Lima LM, Santos RC (2016) Assessment of drought tolerance of peanut cultivars based on physiological and yield traits in a semiarid environment. Agric Water Manag 166:70–76

Deshmukh R, Sonah H, Patil G, Chen W, Prince S, Mutava R, Nguyen HT (2014) Integrating omic approaches for abiotic stress tolerance in soybean. Front Plant Sci 5:244

Devate NB, Krishna H, Parmeshwarappa SKV, Manjunath KK, Chauhan D, Singh S, Singh PK (2022) Genome-wide association mapping for component traits of drought and heat tolerance in wheat. Front Plant Sci 13:943033

Diaz S, Ariza-Suarez D, Izquierdo P, Lobaton JD, de La Hoz JF, Acevedo F, Raatz B (2020) Genetic mapping for agronomic traits in a MAGIC population of common bean (Phaseolus vulgaris L.) under drought conditions. BMC Genomics 21:1–20

El Haddad N, Rajendran K, Smouni A, Es-Safi·NE, Benbrahim N, Mentag R, Kumar S (2020) Screening the FIGS set of lentil (*Lens culinaris* Medikus) germplasm for tolerance to terminal heat and combined drought-heat stress. Agronomy 10:1036

El Haddad N, Choukri H, Ghanem ME, Smouni.A, Mentag R, Rajendran K, Kumar S (2021) High-temperature and drought stress effects on growth, yield and nutritional quality with transpiration response to vapor pressure deficit in lentil. Plan Theory 11:95

Elattar MA, Karikari B, Li S, Song S, Ce CY, Aslam M, Hina A, Abou-Elwafa SF, Zhao T (2021) Identification and validation of major QTLs, epistatic interactions and candidate genes for soybean seed shape and weight using two related RIL populations. Fron Genet 12:666440

Fahad S, Bajwa AA, Nazir U, Anjum SA, Farooq A, Zohaib A, Huang J (2017) Crop production under drought and heat stress: plant responses and management options. Front Plant Sci 8:1147

Fang Y, Xiong L (2015) General mechanisms of drought response and their application in drought resistance improvement in plants. CMLS 72:673–689

Farooq M, Aziz T, Basra SMA, Cheema MA, Rehman H (2008) Chilling tolerance in hybrid maize induced by seed priming with salicylic acid. J Agron Crop Sci 194:161–168

Farooq M, Wahid A, Kobayashi NSMA, Fujita DBSMA, Basra SM (2009) Plant drought stress: effects, mechanisms and management. Agron Sustain Dev 29:185–212

Farooq M, Gogoi N, Barthakur S, Baroowa B, Bharadwaj N, Alghamdi SS (2016) Drought stress in grain legumes during reproduction and grain filling. J Agron Crop Sci 203:81–102

Fathi A, Tari DB (2016) Effect of drought stress and its mechanism in plants. Int J Life Sci 10:1

Fleury D, Jefferies S, Kuchel H, Langridge P (2010) Genetic and genomic tools to improve drought tolerance in wheat. J Exp Bot 61:3211–3222

Foyer CH, Lam HM, Nguyen HT, Siddique KH, Varshney RK, Colmer TD, Considine MJ (2016) Neglecting legumes has compromised human health and sustainable food production. Nat Plants 2:1–10

García-García P, Vaquero F, Vences FJ, de Miera LES, Polanco C, González AI, de La Vega MP (2019) Transcriptome profiling of lentil in response to *Ascochyta lentis* infection. Span J Agric Res 17:e0703

Gautami B, Pandey MK, Vadez V, Nigam SN, Ratnakumar P, Krishnamurthy L, Varshney RK (2012) Quantitative trait locus analysis and construction of consensus genetic map for drought tolerance traits based on three recombinant inbred line populations in cultivated groundnut (*Arachis hypogaea* L.). Mol Breed 30:757–772

Ghassemi-golezani K, Ghassemi S, Bandehhagh A (2013) Effects of water supply on field performance of chickpea (*Cicer arietinum* L.) cultivars. Int J Agron Plant Prod 4:94–97

Gillani SFA, Rasheed A, Abbasi A, Majeed Y, Abbas M, Hassan MU, Qari SH, Amin NBN, Kashgry TA, Tahir MM, Yunling P (2022) Comparative gene enrichment analysis for drought tolerance in contrasting maize genotype. Genes 14:31

Gorim LY, Vandenberg A (2017) Evaluation of wild lentil species as genetic resources to improve drought tolerance in cultivated lentil. Front Plant Sci 8:1129

Grewal SK, Sharma KP, Bharadwaj RD, Hegde V, Tripathi S, Singh S, Mondal B (2020) Understanding genotypic variation and identification of promising genotypes for iron and zinc content in chickpea (*Cicer arietinum* L.). J Food Compos Anal 88:103458

Gujaria-Verma N, Vail SL, Carrasquilla-Garcia N, Penmetsa RV, Cook DR, Farmer AD, Bett KE (2014) Genetic mapping of legume orthologs reveals high conservation of synteny between lentil species and the sequenced genomes of Medicago and chickpea. Front Plant Sci 5:676

Gupta D, Sharma SK (2006) Evaluation of wild lens taxa for agro-morphological traits, fungal diseases and moisture stress in North Western Indian Hills. Genet Resour Crop Evol 53:1233–1241

Gupta D, Dadu RH, Sambasivam P, Bar I, Azad M, Beera N, Biju S (2020) Conventional and biotechnological approaches for targeted trait improvement in lentil. In: Gosal SS, Wani SH (eds) Accelerated plant breeding, vol 3. Springer, Cham. https://doi.org/10.1007/978-3-030-47306-8_3

Habib RR (2021) Understanding the effect of terminal drought on gene expression in the root of different genotypes of common bean (*Phaseolus Vulgaris*. L). Master's thesis, Delaware State University
Habib M, Hina A, Sajjad M, Ahmad MQ, Azhar MT, Qayyum A (2024) MarkerAssisted selection in wheat. In: Zencirci N, Altay F, Baloch FS, Nadeem MA, Ludidi N (eds) Advances in wheat breeding. Springer, Singapore. https://doi.org/10.1007/978-981-999478-6_15
Hacisalihoglu G, Burton AL, Gustin JL, Eker S, Asikli S, Heybet EH, Settles AM (2018) Quantitative trait loci associated with soybean seed weight and composition under different phosphorus levels. J Integr Plant Biol 60:232–241
Hall AE (2012) Phenotyping cowpeas for adaptation to drought. Front Physiol 3:155
Hamabwe SM (2023) Identification of morpho-physiological traits for drought tolerance and their associated genomic regions in Andean Gene Pool of common bean (*Phaseolus Vulgaris* L.). Doctoral dissertation, University of Nairobi
Hasan N, Choudhary S, Naaz N, Sharma N, Laskar RA (2021) Recent advancements in molecular marker-assisted selection and applications in plant breeding programmes. J Genet Eng Biotechnol 19:128
Heslot N, Akdemir D, Sorrells ME, Jannink JJ (2014) Integrating environmental covariates and crop modeling into the genomic selection framework to predict genotype by environment interactions. Theor Appl Genet 127:463–480
Hina A, Cao Y, Song S, Li S, Sharmin RA, Elattar MA, Bhat JA, Zhao T (2020) High-resolution mapping in two RIL populations refines major "QTL hotspot" regions for seed size and shape in soybean (*Glycine max* L.). Int J Mol Sci 21:1040
Hina A, Abbasi A, Imtiaz S, Haq SIU, Hassan AU, Chaudhry (2023) General aspects of genetic improvement (traditional and transgenic methods) aiming to food biofortification. In: Nadeem MA, et al. (Eds). Legumes biofortification. Springer, Cham. https://doi.org/10.1007/978-3-031-33957-8_4
Hina A, Razzaq MK, Abbasi A, Shehzad MB, Arshad M, Sanaullah T, Arshad K, Raza G, Ali HM, Hayat F, Akhtar N, Abdelsalam NR (2024a) Genomic blueprints of soybean pathogen resistance: unveiling key genes for sustainable agriculture. Funct Plant Biol 51. https://doi.org/10.1071/FP23295
Hina A, Khan N, Kong K, Lv W, Karikari B, Abbasi A, Zhao T (2024b) Exploring the role of FBXL gene family in soybean: implications for plant height and seed size regulation. Physiol Plant 176:14191
Hnatuszko-Konka K, Kowalczyk T, Gerszberg A, Wiktorek-Smagur A, Kononowicz AK (2014) Phaseolus vulgaris – recalcitrant potential. Biotechnol Adv 32:1205–1215
Horn LN, Shimelis H (2020) Production constraints and breeding approaches for cowpea improvement for drought prone agro-ecologies in Sub-Saharan Africa. Annal Agric Sci 65:83–91
Hummel M, Hallahan BF, Brychkova G, Ramirez-Villegas J, Guwela V, Chataika B, Spillane C (2018) Reduction in nutritional quality and growing area suitability of common bean under climate change induced drought stress in Africa. Sci Rep 8:16187
Huppertz M, Kachhap D, Dalai A, Yadav N, Baby D, Khan MA, Panigrahi KC (2023) Exploring the potential of mung bean: from domestication and traditional selection to modern genetic and genomic technologies in a changing world. J Agric Res 14:100786
Idrissi O, Udupa SM, De Keyser E, McGee RJ, Coyne CJ, Saha GC, De Riek J (2016) Identification of quantitative trait loci controlling root and shoot traits associated with drought tolerance in a lentil (*Lens culinaris* Medik.) recombinant inbred line population. Front Plant Sci 7:1174
Jain M, Tiwary S, Gadre R (2010) Sorbitol-induced changes in various growth and biochemical parameters in maize. Plant Soil Environ 56:263–267
Jan MF, Li M, Liaqat W, Altaf MT, Liu C, Ahmad H et al (2024) Chlorophyll fluorescence: a smart tool for maize improvement. Cereal Res Commun:1–32
Jha UC, Bohra A, Nayyar H (2020) Advances in "omics" approaches to tackle drought stress in grain legumes. Plant Breed 139:1–27
Jing R, Vershinin A, Grzebyta J, Shaw P, Smýkal P, Marshall D, Ellis THN, Flavell AJ (2010) The genetic diversity and evolution of field pea (Pisum) studied by high throughput retrotransposon-based insertion polymorphism (RBIP) marker analysis. BMC Evol Biol 10:44

Joshi PK, Rao PP (2017) Global pulses scenario: status and outlook. Ann N Y Acad Sci 1392:6–17

Kang YJ, Kim SK, Kim MY, Lestari P, Kim KH, Ha BK, Lee SH (2014) Genome sequence of mungbean and insights into evolution within Vigna species. Nat Commun 5:5443

Katuuramu DN, Hart JP, Porch TG, Grusak MA, Glahn RP, Cichy KA (2018) Genome-wide association analysis of nutritional composition-related traits and iron bioavailability in cooked dry beans (*Phaseolus vulgaris* L.). Mol Breed 38:1–18

Kaushal M, Wani SP (2016) Rhizobacterial-plant interactions: strategies ensuring plant growth promotion under drought and salinity stress. Agric Ecosyst Environ 231:68–78

Khan HR, Paull JG, Siddique KHM, Stoddard FL (2010) Faba bean breeding for drought-affected environments: a physiological and agronomic perspective. Field Crop Res 115:279–286

Khan MT, Ahmed S, Sardar R, Shareef M, Abbasi A, Mohiuddin M, Ercisli S, Fiaz S, Marc RA, Attia K, Khan N, Golokhvast KS (2022) Impression of foliar applied folic acid on coriander (*Coriandrum sativum* L.) to regulate aerial growth, biochemical activity, and essential oil profiling under drought stress. Front. Plant Sci 13:1005710

Khatun M, Sarkar S, Era FM, Islam AM, Anwar MP, Fahad S, Islam AA (2021) Drought stress in grain legumes: effects, tolerance mechanisms and management. Agronomy 11:2374

Khazaei H (2014) Leaf traits associated with drought adaptation in faba bean (*Vicia faba* L.). University of Helsinki, Finland 15

Khazaei H, O'Sullivan DM, Sillanpää MJ, Stoddard FL (2014) Use of synteny to identify candidate genes underlying QTL controlling stomatal traits in faba bean (*Vicia faba* L.). Theor Appl Genet 127:2371–2385

Khoyerdi FF, Shamshiri MH, Estaji A (2016) Changes in some physiological and osmotic parameters of several pistachio genotypes under drought stress. Sci Hortic 198:44–51

Korbu L, Fikre A, Tesfaye K, Funga A, Bekele D, Ojiewo CO (2022) Response of chickpea to varying moisture stress conditions in Ethiopia. Agrosyst Geosci Environ 5:e20234

Kosgei AJ, Kimurto PK, Gaur PM, Yeboah MA, Offei SK, Danquah EY (2022) Introgression of drought tolerance root traits into Kenyan commercial chickpea varieties using marker assisted backcrossing. Afr Crop Sci J 30:31–50

Kreplak J, Madoui MA, Cápal P, Novák P, Labadie K, Aubert G, Burstin J (2019) A reference genome for pea provides insight into legume genome evolution. Nat Genet 51:1411–1422

Krishnamurthy L, Kashiwagi J, Gaur PM, Upadhyaya HD, Vadez V (2010) Sources of tolerance to terminal drought in the chickpea (*Cicer arietinum* L.) minicore germplasm. Field Crop Res 119:322–330

Kumar R, Sharma H, Singh S, Sud RK, Gulati A, Ahuja PS, Sharma RK (2014) Molecular markersbased approach for genetic improvement of tea (*Camellia sinensis*). New Biotechnol 31:207

Kumari VV, Roy A, Vijayan R, Banerjee P, Verma VC, Nalia A, Hossain A (2021) Drought and heat stress in cool-season food legumes in sub-tropical regions: consequences, adaptation, and mitigation strategies. Plan Theory 10:1038

Kunert K, Vorster BJ (2020) In search for drought-tolerant soybean: is the slow-wilting phenotype more than just a curiosity? J Exp Bot 71:457–460

Kushwah A, Bhatia D, Barmukh R, Singh I, Singh G, Bindra S, Singh S (2022) Genetic mapping of QTLs for drought tolerance in chickpea (*Cicer arietinum* L.). Front Genet 13:953898

Li Y, Xiao J, Chen L, Huang X, Cheng Z, Han B, Wu C (2018) Rice functional genomics research: past decade and future. Mol Plant 11:359–380

Liu HJ, Yan J (2019) Crop genome-wide association study: a harvest of biological relevance. Plant J 97:8–18

Liu P, Zhu J, Lou X, Lu Y (2003) A method for marker-assisted selection based on QTLs with epistatic effects. Genetica 119:75–86

Liu S, Banik M, Yu K, Park SJ, Poysa V, Guan Y (2007) Marker-assisted selection (MAS) in major cereal and legume crop breeding: current progress and future directions. Int J Plant Breed 1:75–78

Liu C, Wu J, Wang L, Fan B, Cao Z, Su Q, Wang S (2017) Quantitative trait locus mapping under irrigated and drought treatments based on a novel genetic linkage map in mungbean (Vigna radiata L.). Theor Appl Genet 130:2375–2393

Lu H, Tabassum A, Zhou G (2021) Plant hydraulic conductivity determines photosynthesis in rice under PEG-induced drought stress. Pak J Bot 53:409–417

Madurapperumage A, Tang L, Thavarajah P, Bridges W, Shipe E, Vandemark G, Thavarajah D (2021) Chickpea (*Cicer arietinum* L.) as a source of essential fatty acids – a biofortification approach. Front Plant Sci 12:734980

Mafakheri A, Siosemardeh AF, Bahramnejad B, Struik PC, Sohrabi Y (2010) Effect of drought stress on yield, proline and chlorophyll contents in three chickpea cultivars. Aust J Crop Sci 4:580–585

Manickam S, Rajagopalan VR, Manoharan B, Natesan S, Muthurajan R (2023) Genomics-assisted breeding for climate-resilient crops. In: Kumar N (ed) Molecular marker techniques. Springer, Singapore. https://doi.org/10.1007/978-981-99-1612-2_17

Maurya V, Singh N, Sharma A, Kumar R (2023) DNA markers-assisted crop improvement for biotic and abiotic stresses in legumes. In: Kumar N (ed) Molecular marker techniques. Springer, Singapore. https://doi.org/10.1007/978-981-99-1612-2_10

Meuwissen TH, Hayes BJ, Goddard M (2001) Prediction of total genetic value using genome-wide dense marker maps. genetics. 157(4):1819–1829

Moharana A, Lenka B, Singh AP, Kumar NK, Nagaraju B, Das SR (2020) Peanut as a food source: a review. J Pharmacogn Phytother 9:225–232

Moreau L, Charcosset A, Hospital F, Gallais A (1998) Marker-assisted selection efficiency in populations of finite size. Genetics 148:1353–1365

Moreau L, Charcosset A, Gallais A (2004) Experimental evaluation of several cycles of markerassisted selection in maize. Euphytica 137:111–118

Muchero W, Ehlers JD, Close TJ, Roberts PA (2009) Mapping QTL for drought stress-induced premature senescence and maturity in cowpea [*Vigna* unguiculata (L.) Walp.]. Theor Appl Genet 118:849–863

Muktadir MA, Adhikari KN, Merchant A, Belachew KY, Vandenberg A, Stoddard FL, Khazaei H (2020) Physiological and biochemical basis of faba bean breeding for drought adaptation – a review. Agronomy 10:1345

Nadeem MA, Nawaz MA, Shahid MQ, Doğan Y, Comertpay G, Yıldız M, Hatipoğlu R, Ahmad F, Alsaleh A, Labhane N, Özkan H (2018) DNA molecular markers in plant breeding: current status and recent advancements in genomic selection and genome editing. Biotechnol Equip 32:261–285

Naderpour M, Lund OS, Larsen R, Johansen E (2010) Potyviral resistance derived from cultivars of Phaseolus vulgaris carrying bc-3 is associated with the homozygotic presence of a mutated eIF4E allele. Mol Plant Pathol 11:255–263

Nakedde T, Ibarra-Perez FJ, Mukankusi C, Waines JG, Kelly JD (2016) Mapping of QTL associated with Fusarium root rot resistance and root architecture traits in black beans. Euphytica 212:51–63

Nazir S, Zaman Q, Abbasi A, Komal N, Riaz U, Ashraf K, Ahmad N, Agarwal S, Nasir R, Chen Y (2021) Bioresource nutrient recycling in the rice–wheat cropping system: cornerstone of organic agriculture. Plan Theory 10:2323

Nkomo GV, Sedibe MM, Mofokeng MA (2021) Production constraints and improvement strategies of cowpea (*Vigna unguiculata* L. Walp.) genotypes for drought tolerance. Int. J Agron 2021:5536417

Omar GI, Saqer MM, Adwan GM (2019) Phylogenetic relationship among some species of the genera *Lens*, *Vicia*, *Lathyrus* and *Pisum* (Leguminosae) in Palestine. Jordan J Biol Sci 12:296

Palit P, Ghosh R, Tolani P, Tarafdar A, Chitikineni A, Bajaj P, Varshney RK (2020) Molecular and physiological alterations in chickpea under elevated CO2 concentrations. Plant Cell Physiol 61:1449–1463

Pandey MK, Roorkiwal M, Singh VK, Ramalingam A, Kudapa H, Thudi M, Chitikineni A, Rathore A, Varshney RK (2016) Emerging genomic tools for legume breeding: current status and future prospects. Front Plant Sci 7:455

Pandey MK, Gangurde SS, Sharma V, Pattanashetti SK, Naidu GK, Faye I, Varshney RK (2020) Improved genetic map identified major QTLs for drought tolerance-and iron deficiency tolerance-related traits in groundnut. Genes 12:37

Pandey J, Devadasu E, Saini D, Dhokne K, Marriboina S, Raghavendra AS, Subramanyam R (2023) Reversible changes in structure and function of photosynthetic apparatus of pea (*Pisum sativum*) leaves under drought stress. Plant J 113:60–74

Panzeri D, Cassani E, Doria E, Tagliabue G, Forti L, Campion B, Sparvoli F (2011) A defective ABC transporter of the MRP family, responsible for the bean lpa1 mutation, affects the regulation of the phytic acid pathway, reduces seed myo-inositol and alters ABA sensitivity. New Phytol 191:70–83

Parihar AK, Dixit GP, Bohra A, Sen Gupta D, Singh AK, Kumar N, Singh NP (2020) Genetic advancement in dry pea (*Pisum sativum* L.): retrospect and prospect. Accelerated plant breeding. Food Legumes 3:283–341

Patil GD, Desale SC, Patil PS, Patil SS (1981) 'Phule-Pragati', a high yielding early bunch groundnut for Maharashtra

Powers S, Boatwright JL, Thavarajah D (2021) Genome-wide association studies of mineral and phytic acid concentrations in pea (*Pisum sativum* L.) to evaluate biofortification potential. G3 11:227

Pratap A, Gupta S, Basu PS, Tomar R, Dubey S, Rathore M, Prajapati US, Singh P, Kumari G (2020) Towards development of climate smart mungbean: challenges and opportunities. Genomic designing of climate-smart pulse crops. 2019:235–264

Pradhan J, Katiyar D, Hemantaranjan A (2019) Drought mitigation strategies in pulses. Pharm Innov J 8:567–576

Pradhan G, Meena RS, Kumar S, Jhariya MK, Khan N, Shukla UN, Singh AK, Sheoran S (2022) Legumes for eco-friendly weed management in agroecosystem. InAdvances in legumes for sustainable intensification (pp. 133–154). Academic Press

Puppala N, Nayak SN, Sanz-Saez A, Chen C, Devi MJ, Nivedita N, Sharma V (2023) Sustaining yield and nutritional quality of peanuts in harsh environments: physiological and molecular basis of drought and heat stress tolerance. Front Genet 14:1121462

Pushpavalli R, Zaman-allah M, Turner NC, Baddam R, Rao MV, Vadez V (2015) Higher flower and seed number leads to higher yield under water stress conditions imposed during reproduction in chickpea. Funct Plant Biol 42:162–174

Raddi S, Giannetti F, Martini S, Farinella F, Chirici G, Tani A, Mariotti B (2022) Monitoring drought response and chlorophyll content in Quercus by consumer-grade, near-infrared (NIR) camera: a comparison with reflectance spectroscopy. New For 53:241–265

Rajpal VR, Singh A, Kathpalia R, Thakur RK, Khan MK, Pandey A, Raina SN (2023) The prospects of gene introgression from crop wild relatives into cultivated lentil for climate change mitigation. Front Plant Sci 14:1127239

Ramzan A, Noor T, Khan TN, Hina A (2014) Correlation, cluster and regression analysis of seed yield and its contributing traits in pea (*Pisum sativum* L.). J Agric Res 52:481–488

Ravi K, Vadez V, Isobe S, Mir RR, Guo Y, Nigam SN, Varshney RK (2011) Identification of several small main-effect QTLs and a large number of epistatic QTLs for drought tolerance related traits in groundnut (*Arachis hypogaea* L.). Theor Appl Genet 122:1119–1132

Razzaq MK, Akhter M, Ahmad RM, Cheema KL, Hina A, Karikari B (2022) CRISPR-Cas9 based stress tolerance: new hope for abiotic stress tolerance in chickpea (*Cicer arietinum*). Mol Biol Rep 49:8977–8985

Razzaq MK, Hina A, Abbasi A, Karikari B, Ashraf HJ, Mohiuddin M, Maqsood S, Maqsood A, Haq IU, Xing G, Raza G, Bhat JA (2023) Molecular and genetic insights into secondary metabolic regulation underlying insect-pest resistance in legumes. Funct Integr Genom 23:217

Ren H, Han J, Wang X, Zhang B, Yu L, Gao H, Qiu LJ (2020) QTL mapping of drought tolerance traits in soybean with SLAF sequencing. Crop J 8:977–989

Rispail N, Wohor OZ, Osuna-Caballero S, Barilli E, Rubiales D (2023) Genetic diversity and population structure of a wide Pisum spp. core collection. Int J Mol Sci 24:2470

Roorkiwal M, Rathore A, Das RR, Singh MK, Jain A, Srinivasan S, Gaur PM, Chellapilla B, Tripathi S, Li Y, Hickey JM, Lorenz A, Sutton T, Crossa J, Jannink JL, Varshney RK (2016) Genome-enabled prediction models for yield related traits in chickpea. Front Plant Sci 7:1666

Roorkiwal M, Bharadwaj C, Barmukh R, Dixit GP, Thudi M, Gaur PM, Varshney RK (2020) Integrating genomics for chickpea improvement: achievements and opportunities. Theor Appl Genet 133:1703–1720

Rozita K, Hassan F, Fatemeh N (2012) Salicylic acid ameliorates the effects of oxidative stress induced by water deficit in hydroponic culture of Nigella sativa. J Stress Physiol Biochem 8:13–22

Saha M, Bandyopadhyay PK, Sarkar A, Nandi R, Singh KC, Sanyal D (2020) Understanding the impacts of sowing time and tillage in optimizing the micro-environment for rainfed lentil (*Lens culinaris* Medik) production in the lower Indo-Gangetic Plain. JPNSS 20:2536–2551

Sahay G, Mehta BK, Dikshit N, Shashikumara P, Bhargavi HA, Priyadarshini P (2022) Developing climate-resilient cowpea (*Vigna unguiculata* [L.] Walp.) through genomics-assisted breeding approaches. In: Developing climate resilient grain and forage legumes. Springer, Singapore, pp 77–103

Sahu VK, Tiwari S, Tripathi MK, Gupta N, Tomar RS, Yasin M (2020) Morpho-physiological and biochemical traits analysis for Fusarium wilt disease using gene-based markers in desi and Kabuli genotypes of chickpea (*Cicer arietinum* L.). Indian J Genet 80:163–172

Saleem A, Roldán-Ruiz I, Aper J, Muylle H (2022) Genetic control of tolerance to drought stress in soybean. BMC Plant Biol 22:615

Sanderson LA, Caron CT, Tan R, Shen Y, Liu R, Bett KE (2019) KnowPulse: a web-resource focused on diversity data for pulse crop improvement. Front Plant Sci 10:965

Sapes G, Sala A (2021) Relative water content consistently predicts drought mortality risk in seedling populations with different morphology, physiology and times to death. Plant Cell Environ 44:3322–3335

Sarkar S, Khatun M, Era FM, Islam AM, Anwar MP, Danish S, Islam AA (2021) Abiotic stresses: alteration of composition and grain quality in food legumes. Agronomy 11:2238

Savita, Tomer A, Singh SK (2020) Drought stress tolerance in legume crops. In: Agronomic crops: volume 3: stress responses and tolerance, pp 149–155

Schneider KA, Rosales-Serna R, Ibarra-Perez F, Cazares-Enriquez B, Acosta-Gallegos JA, Ramirez-Vallejo P, Kelly JD (1997) Improving common bean performance under drought stress. Crop Sci 37:43–50

Sedivy EJ, Wu F, Hanzawa Y (2017) Soybean domestication: the origin, genetic architecture and molecular bases. New Phytol 214:539–553

Sedlar A, Zupin M, Maras M, Razinger J, Šuštar-Vozlič J, Pipan B, Meglič V (2020) QTL mapping for drought-responsive agronomic traits associated with physiology, phenology, and yield in an Andean intra-gene pool common bean population. Agronomy 10:225

Shabbir MZ, Arshad M, Hussain B, Nadeem I, Ali S, Abbasi A, Ali Q (2014) Genotypic response of chickpea (*Cicer arietinum* L.) for resistance against gram pod borer (*Helicoverpa armigera*). Adv Life Sci 2:23–30

Shahid S, Hina A, Abbasi A, Maqsood S, Ansari M, Haq IU Zaman QU (2023) Organic approaches toward iron and zinc biofortification in legume crops. In: Nadeem MA et al. (Eds). Legumes biofortification. Springer, Cham. https://doi.org/10.1007/978-3-031-33957-8_15

Shaibu AS, Sneller C, Motagi BN, Chepkoech J, Chepngetich M, Miko ZL, Mohammed SG (2020) Genome-wide detection of SNP markers associated with four physiological traits in groundnut (*Arachis hypogaea* L.) mini core collection. Agronomy 10:192

Siddiqui MH, Al-Khaishany MY, Al-Qutami MA, Al-Whaibi MH, Grover A, Ali HM, Bukhari NA (2015) Response of different genotypes of faba bean plant to drought stress. Int J Mol.Sci 16:10214–10227

Singh B, Bohra A, Mishra S, Joshi R, Pandey S (2015) Embracing new-generation 'omics' tools to improve drought tolerance in cereal and food-legume crops. Biol Plant 59:413–428

Singh A, Ganapathysubramanian B, Singh AK, Sarkar S (2016) Machine learning for high-throughput stress phenotyping in plants. Trends Plant Sci 21:110–124

Singh A, Nath O, Singh S, Kumar S, Singh IK (2018) Genome-wide identification of the MAPK gene family in chickpea and expression analysis during development and stress response. Plant Gene 13:25–35

Singh J, Gezan SA, Vallejos CE (2019) Developmental pleiotropy shaped the roots of the domesticated common bean (*Phaseolus vulgaris*). Plant Physiol 180:1467–1479

Singh S, Babu KS, Arora A, Panwar RK, Verma SK (2021a) Genetic studies for biofortification traits in chickpea. J Food Legume 34:17–20

Singh CM, Singh P, Tiwari C, Purwar S, Kumar M, Pratap A, Mishra AK (2021b) Improving drought tolerance in mungbean (*Vigna radiata* L. Wilczek): morpho-physiological, biochemical and molecular perspectives. Agronomy 11:1534

Singh A, Mamo T, Singh A, Mahama AA (2023) Cowpea breeding. In: Suza WP, Lamkey KR (eds) Crop improvement. Iowa State University Digital Press

Sohrawardy H, Hossain ML (2014) Response of short duration tropical legumes and maize to water stress: a glasshouse study. Adv Agric 2014:641319

Subbaramamma P, Sangamitra M, Manjusha D (2017) Mitigation of drought stress in production of pulses. Int J Multidiscip Adv Res Trends 4:41–62

Sun B, Wang Y, Yang Q, Gao H, Niu H, Li Y, Ren B (2023) A high-resolution transcriptomic atlas depicting nitrogen fixation and nodule development in soybean. J Integr Plant Biol 65:1536–1552

Suvarna, Ashwini K, Yashaswini R (2023) Marker assisted recurrent selection for crop improvement. In: Kumar N (ed) Molecular marker techniques. Springer, Singapore. https://doi.org/10.1007/978-981-99-1612-2_3

Thudi M, Gaur PM, Krishnamurthy L, Mir RR, Kudapa H, Fikre A, Varshney RK (2014) Genomics-assisted breeding for drought tolerance in chickpea. Funct Plant Biol 41:1178–1190

Tiwari S, Lata C, Chauhan PS, Nautiyal CS (2016) Pseudomonas putida attunes morphophysiological, biochemical and molecular responses in *Cicer arietinum* L. during drought stress and recovery. Plant Physiol Biochem 99:108–117

Tomer A, Singh SK (2020) Drought stress tolerance in legume crops. In: Agronomic crops. Springer, Singapore, pp 149–155

Ulemale CS, Mate SN, Deshmukh DV (2013) Physiological indices for drought tolerance in chickpea (*Cicer arietinum* L.). World J Agric Sci 9:123–131

Ullah A, Farooq M (2022) The challenge of drought stress for grain legumes and options for improvement. Arch Agron Soil Sci 68:1601–1618

Upadhyaya HD, Hildebrand GL, Nigam SN, Yellaiah N (2005) Groundnut cultivar Nyanda (ICGV 93437) released in Zimbabwe. Int Arachis Newslett 25:22–24

Upadhyaya HD, Kashiwagi J, Varshney RK, Gaur PM, Saxena KB, Krishnamurthy L, Singh IP (2012) Phenotyping chickpeas and pigeonpeas for adaptation to drought. Front Physiol 3:179

Valliyodan B, Ye H, Song L, Murphy M, Shannon JG, Nguyen HT (2017) Genetic diversity and genomic strategies for improving drought and water logging tolerance in soybeans. J Exp Bot 68:1835–1849

Van Berloo R, Stam P (2001) Simultaneous marker-assisted selection for multiple traits in autogamous crops. Theor Appl Genet 102:1107–1112

Varshney RK, Bertioli DJ, Moretzsohn MDC, Vadez V, Krishnamurthy L, Aruna R, Hoisington DA (2009) The first SSR-based genetic linkage map for cultivated groundnut (*Arachis hypogaea* L.). Theor Appl Genet 118:729–739

Varshney RK, Gaur PM, Chamarthi SK, Krishnamurthy L, Tripathi S, Kashiwagi J, Samineni S, Singh VK, Thudi M, Jaganathan D (2013) Fast-track introgression of 'QTL-hotspot' for root traits and other drought tolerance traits in JG 11, an elite and leading variety of chickpea. Plant Genome 6:1–9

Varshney RK, Thudi M, Pandey MK, Tardieu F, Ojiewo C, Vadez V, Whitbread AM, Siddique KHM, Nguyen HT, Carberry PS, Bergvinson D (2018) Accelerating genetic gains in legumes for the development of prosperous smallholder agriculture: integrating genomics, phenotyping, systems modelling and agronomy. J Exp Bot 69:3293–3312

Varshney RK, Pandey MK, Bohra A, Singh VK, Thudi M, Saxena RK (2019) Towards sequence-based breeding in legumes in the post-genome sequencing era. Theor Appl Genet 132:797–816
Venugopalan VK, Nath R, Sengupta K, Pal AK, Banerjee S, Banerjee P, Siddique KH (2022) Foliar spray of micronutrients alleviates heat and moisture stress in lentil (*Lens culinaris* Medik) grown under rainfed field conditions. Front Plant Sci 13:847743
Vessal S, Arefian M, Siddique KH (2020) Proteomic responses to progressive dehydration stress in leaves of chickpea seedlings. BMC Genomics 21:1–15
Voisin AS, Guéguen J, Huyghe C, Jeuffroy MH, Magrini MB, Meynard JM (2014) Legumes for feed, food, biomaterials and bioenergy in Europe: a review. Agron Sustain Dev 34:361–380
Wan L, Wu Y, Huang J, Dai X, Lei Y, Yan L, Liao B (2014) Identification of ERF genes in peanuts and functional analysis of AhERF008 and AhERF019 in abiotic stress response. Funct Integr Genom 14:467–477
Wang D, Yang T, Liu R, Li N, Wang X, Sarker A, Ding H (2020) RNA-Seq analysis and development of SSR and KASP markers in lentil (*Lens culinaris* Medikus subsp. *culinaris*). Crop J 8:953–965
Warsame AO, Michael N, O'Sullivan DM, Tosi P (2020) Identification and quantification of major faba bean seed proteins. J Agric Food Chem 68:8535–8544
Waterworth WM, Bray CM, West CE (2015) The importance of safeguarding genome integrity in germination and seed longevity. J Exp Bot 66:3549–3558
Wu L, Chang Y, Wang L, Wang S, Wu J (2024) Genome-wide association study dissecting drought resistance-associated loci based on physiological traits in common bean. J Integr Agric 23:3657. https://doi.org/10.1016/j.jia.2024.03.079
Xie C, Xu S (1998) Strategies of marker-assisted selection. Crop Sci 38:1526–1535
Yadava YK, Chaudhary P, Yadav S, Rizvi AH, Kumar T, Srivastava R, Jain PK (2023) Genetic mapping of quantitative trait loci associated with drought tolerance in chickpea (*Cicer arietinum* L.). Sci Rep 13:17623
Yang T, Liu R, Luo Y, Hu S, Wang D, Wang C, Zong X (2022) Improved pea reference genome and pan-genome highlight genomic features and evolutionary characteristics. Nat Genet 54:1553–1563
Ye H, Roorkiwal M, Valliyodan B, Zhou L, Chen P, Varshney RK, Nguyen HT (2018) Genetic diversity of root system architecture in response to drought stress in grain legumes. J Exp Bot 69:3267–3277
Zafar-ul-Hye M, Akbar MN, Iftikhar Y, Abbas M, Zahid A, Fahad S, Danish S (2021) Rhizobacteria inoculation and caffeic acid alleviated drought stress in lentil plants. Sustain For 13:9603
Zhang X, Xu M, Hina A, Kong J, Gai J, He X, Zhao T (2019) Seed storability of summer-planting soybeans under natural and artificial aging conditions. Legum Res 42:250–259
Zhang Z, Ma J, Yang X, Liu Z, Liu Y, Liu X, Tian Z (2024) Natural allelic diversities of GmPrx16 confer drought tolerance in soybean. Plant Biotechnol J 22:535
Zhao Y, Cao P, Cui Y, Liu D, Li J, Zhao Y, Chen Q (2021) Enhanced production of seed oil with improved fatty acid composition by overexpressing NAD+-dependent glycerol-3-phosphate dehydrogenase in soybean. J Integr Plant Biol 63:1036–1053
Zhao X, Liu Z, Li H, Zhang Y, Yu L, Qi X, Qiu L (2022) Identification of drought-tolerance genes in the germination stage of soybean. Biol 11:1812
Zhu G, Gu L, Shi Y, Chen H, Liu Y, Lu F, Zhou G (2021) Plant hydraulic conductivity determines photosynthesis in rice under PEG-induced drought stress. Pak J Bot 53:409–417

8 Marker-Assisted Breeding in Common Bean for Drought Tolerance

Xubo Ke, Jia Yao, Xinyue Gu, Zhihan Jiang, and Pei Xu

Abstract

The common bean (*Phaseolus vulgaris* L.), a globally cultivated crop valued for its taste and nutritional benefits, is rich in protein, dietary fiber, vitamins, minerals, and other essential nutrients. However, escalating global climate change has made drought a significant challenge for common bean cultivation, emphasizing the need for improved drought resistance through breeding. While traditional genetic breeding has been crucial in developing common bean varieties, the long breeding cycles and the complexity of stabilizing genetic traits have slowed progress. To address these challenges, researchers are integrating molecular biotechnology into breeding programs to accelerate the development of drought-resistant varieties. Molecular marker-assisted breeding has revolutionized this process by increasing the efficiency and precision of breeding programs, enabling the rapid identification and selection of varieties with superior drought resistance. This document reviews recent advancements in marker-assisted breeding for drought resistance in common beans, covering the current state of molecular breeding, genetic diversity, mechanisms of drought resistance, and milestones in genomic research. It highlights the critical role of marker-assisted breeding in enhancing drought resistance and explores future directions and potential of this approach.

Keywords

Phaseolus vulgaris · Genomics · Drought · Breeding · MAS

Xubo Ke and Jia Yao contributed equally with all other contributors.

X. Ke · J. Yao · X. Gu · Z. Jiang · P. Xu (✉)
Key Laboratory of Specialty Agri-product Quality and Hazard Controlling Technology, College of Life Sciences, China Jiliang University, Hangzhou, People's Republic of China
e-mail: peixu@cjlu.edu.cn

© The Author(s), under exclusive license to Springer Nature Singapore Pte Ltd. 2025

M. A. Nadeem et al. (eds.), *Marker-Assisted Breeding in Legumes for Drought Tolerance*, https://doi.org/10.1007/978-981-96-4112-3_8

8.1 Introduction

The common bean (*Phaseolus vulgaris* L.) is of considerable social and economic importance, serving as a key source of protein and essential nutrients globally (Karavidas et al. 2023). This versatile crop is among the most consumed legumes worldwide, with certain regions in Africa and the Americas heavily relying on it for their daily caloric needs (Aasim et al. 2022). In these areas, common beans contribute significantly to the diet, providing 36% of protein intake. In sub-Saharan Africa, where over 200 million people depend on common beans as a staple, their role is especially pronounced (Schmutz et al. 2014).

Common beans are not only a dietary staple but also a rich source of protein, fats, carbohydrates, dietary fiber, and a range of vitamins, including A, C, and K, as well as minerals, such as calcium, phosphorus, iron, and magnesium (Nadeem and Baloch 2023; Wiesinger et al. 2016). These nutrients are essential for maintaining normal physiological functions and metabolic processes. The low-calorie and low-fat content of common beans make them an ideal choice for weight management, while their dietary fiber content promotes satiety, improves intestinal movement, and supports digestive health, helping to prevent constipation and other intestinal disorders. For instance, *Phaseolus vulgaris* extracts have been explored as potential weight loss supplements (Watanabe et al. 2020).

In summary, the common bean is more than just a dietary staple; it is a valuable asset for health and well-being. Given its nutritional benefits, further research into the full spectrum of advantages provided by common beans is essential. Additionally, understanding their genetic composition and breeding practices is crucial for enhancing their nutritional value, disease resistance, and adaptability to varying environmental conditions (Wu et al. 2020). A comprehensive approach to studying and improving common beans will not only optimize their cultivation and consumption but also significantly contribute to global food security and public health.

8.1.1 Impacts of Drought on Common Bean

In the natural world, plants face a multitude of environmental constraints throughout their life cycle, with drought stress emerging as one of the most critical non-biological challenges (Munné-Bosch and Villadangos 2023). This stress is a leading cause of crop yield loss, and common beans are particularly vulnerable to such conditions in agricultural settings. The cultivation environment for common beans is complex and subject to various biotic and abiotic factors, with drought being especially detrimental. Approximately 60% of the global common bean harvest is lost annually due to water scarcity, and this figure can rise to 80% in regions experiencing sudden droughts (Nguyen et al. 2023). Regions such as the arid northwest of Latin America, the highlands of central and eastern Mexico, Kenya, and South Africa frequently experience droughts that severely limit common bean cultivation. Additionally, areas like northern Tanzania, Uganda, and Ethiopia suffer from annual droughts, leading to significant declines in crop yields and, in cases of abrupt droughts, even complete crop failures (Nadeem et al. 2019).

As global climate patterns shift, drought conditions in common bean-producing regions are expected to worsen. Research indicates that physiological indicators of common beans—such as stomatal conductance, osmotic adjustment substance levels, water use efficiency, and antioxidant enzyme activities—vary considerably among different drought-resistant strains. Compared to related species like cowpea and adzuki bean, common beans exhibit relatively weaker drought tolerance (Meza et al. 2022). To enhance yields in arid regions, two primary strategies are considered: optimizing the external water environment to meet the plants' physiological needs and improving the plants' drought resilience through breeding programs that develop drought-resistant varieties. The former approach faces challenges, including high costs, instability, and resource wastage. Consequently, breeding for improved drought resistance is viewed as the most effective strategy for increasing productivity in dry regions (Zhang et al. 2024). As a result, research into the drought resistance of common beans has become a major focus for scholars worldwide and a key objective in breeding initiatives.

8.1.2 Concepts and Importance of Marker-Assisted Breeding

In the next three decades, the global population is expected to increase by 25%. Although traditional agricultural practices have been effective in cultivating high-yield, nutrient-rich crops to meet the dietary needs of this expanding population, the current rate of yield improvement for staple crops like wheat, rice, and corn is insufficient to meet the projected demand (Martre et al. 2024). This challenge places breeders and plant scientists in a difficult position, requiring them to both enhance existing crop varieties and innovate new ones that offer higher yields, better nutritional profiles, robust pest resistance, and greater adaptability to climate variability.

Genetic variation is crucial for the enhancement and innovation of crop varieties (Salgotra and Stewart 2020; Iqbal et al. 2023). This variation arises from the complex interplay of genetic inheritance and environmental factors, a process that is inherently unpredictable. With the integration of high-throughput sequencing and other advanced technologies into breeding practices, the use of molecular markers in plant breeding has greatly improved (Wang et al. 2020). These advancements accelerate the breeding process, facilitating the development of crops that can better withstand water scarcity, a key factor in adapting to changing environmental conditions. Identifying molecular markers associated with drought tolerance is essential for enhancing the drought resilience of crops such as common beans. Breeding for drought-resistant varieties is not only a scientific endeavor but also a strategic necessity for food security. Consequently, incorporating marker-assisted breeding for drought-resistant crops represents a proactive approach, ensuring that agricultural practices are both responsive to current needs and prepared for future uncertainties. This strategy is vital for maintaining agricultural productivity and ensuring global food security amid a growing population and a changing climate.

8.2 Genetic Diversity and Mechanisms of Drought Tolerance in Common Bean

8.2.1 Genetic Diversity of Common Bean

Genetic diversity is crucial for the adaptability, evolution, and survival of species (Yıldız and Arbizu 2022; Yıldız et al. 2016). The common bean is categorized into two main gene pools: Mesoamerican and Andean, each further divided into distinct races based on geographical origin and genetic lineage (Nadeem et al. 2018a; Mensack et al. 2010). The Mesoamerican gene pool includes races such as Durango, Jalisco, Central America, and Guatemala, while the Andean gene pool encompasses races from Chile, New Granada, and Peru (Long et al. 2020). These classifications reflect the crop's evolutionary history and the impact of human selection. The spread of the common bean to other regions is a complex process involving introductions from various parts of the Americas and exchanges with Mediterranean and European countries. Today, the common bean is cultivated globally, either as pure gene pools or hybrids between the two gene pools, for its edible dry seeds or immature pods. Domestication and spread have led to changes in the bean's morphology, including larger seeds and leaves, altered growth habits, and variations in seed color and markings (Gaut 2014; Nadeem et al. 2018a; Wu et al. 2020; Cortinovis et al. 2024).

Researchers have used various "omics" techniques, such as transcriptomics, proteomics, and metabolomics, to study the genetic diversity of common beans. These approaches have provided deep insights into genetic variations and their effects on agronomic traits (Nadeem et al. 2021). For example, local germplasm is vital for conserving genetic diversity and developing cultivars suited to specific environmental conditions. Studies of Chinese local varieties, such as those from Chongqing, have highlighted how regional environmental factors and human preferences shape the evolution of the common bean (Long et al. 2020). Recent advancements in genomics, including large-scale resequencing of 683 common bean genotypes, have identified numerous loci associated with yield traits (Wu et al. 2020). These studies have not only uncovered the genetic basis of phenotypic variations but also provided valuable information for molecular breeding and crop improvement.

8.2.2 Impacts of Drought on Physiological Characters of Common Bean

In response to drought stress, the common bean employs a sophisticated array of physiological and molecular mechanisms, including morphological adjustments, stomatal regulation, osmotic adjustment, and modulation of antioxidant capacity. These processes work together to help the plant adapt to arid conditions, ensuring its survival and productivity even when water is limited (Lanna et al. 2018; Abd El Mageed et al. 2023; Geleta et al. 2024; Martínez-Barradas et al. 2024). One critical defense mechanism during drought is stomatal closure, a rapid response aimed at reducing transpiration and conserving water. Under drought stress, the common bean adjusts stomatal conductance to minimize water loss while improving water use efficiency (WUE) to sustain photosynthetic activity (Jiang et al. 2024).

Morphological adaptations are also significant. As soil moisture declines, common beans may slow their growth or shed older leaves to reduce water demand—a strategic move to conserve resources. Simultaneously, the plants often increase the length and density of their root systems to enhance water uptake from the soil (Alghamdi et al. 2022). Drought-tolerant cultivars, in particular, exhibit a more extensive root architecture that enables them to reach deeper water reserves and nutrients, supporting continued growth and development under drought conditions (Wu et al. 2021).

Drought-tolerant varieties also upregulate genes associated with drought resistance, including those encoding enzymes involved in osmotic regulation and antioxidant defense (Ponce et al. 2024). Osmotic regulation is crucial for mitigating drought effects. The accumulation of inorganic ions like potassium (K^+) and chloride (Cl^-) in vacuoles helps maintain the plant's internal osmotic balance (Razi and Muneer 2021). This adjustment restores cell osmotic potential and maintains turgor pressure, thereby preserving the plant's structural integrity. Proline and soluble sugars, as primary osmolytes, play a key role in maintaining leaf turgor and water balance.

In addition to osmotic adjustments, common beans utilize both enzymatic and non-enzymatic antioxidant systems to preserve redox homeostasis and protect against oxidative damage. These systems include enzymes such as superoxide dismutase (SOD), peroxidase (POD), catalase (CAT), ascorbate peroxidase (APX), and glutathione reductase (GR), as well as non-enzymatic antioxidants like ascorbic acid and glutathione. These antioxidants effectively scavenge reactive oxygen species and protect cell membranes. For instance, a study by Yin et al. (2021) demonstrated that under varying levels of drought stress, the activities of APX, SOD, CAT, and POD were significantly higher in common beans compared to those under normal watering conditions. Antioxidant enzymes also work in tandem with photorespiration to reduce lipid peroxidation and minimize oxidative damage to leaf tissue (Table 8.1).

8.2.3 Molecular Mechanisms of Drought Tolerance in Common Bean

Plants respond to drought stress through a complex and multifaceted array of biological processes. Central to this response is the activation of drought-tolerant genes, which are crucial for the plant's survival under water-deficient conditions. Jorge et al. (2020) identified a total of 1005 genes involved in the drought response of common beans, with 645 genes upregulated and 360 genes downregulated under drought stress. The upregulated genes were predominantly associated with carbohydrate metabolism, particularly polysaccharide metabolism, while the downregulated genes were mainly related to protein folding, chloroplast functions, and REDOX processes.

Table 8.1 Drought resistance physiological trait in common bean

Biochemical indices	References
Shoot biomass at harvest and flowering; lodging score; hundred-seed weight; leaf elongation rate; wilting score	Hoyos-Villegas et al. (2017)
Internode length; plant height; number of flowers; number of nodes; chlorophyll content; protein content; ascorbate peroxidase activity; catalase activity	Saeidi et al. (2019)
Leaf relative water content; biomass of roots and shoots; total chlorophyll contents; catalase activities; superoxide dismutase activities	López et al. (2020)
Water use efficiency; transpiration efficiency; rate of photosynthesis; photosynthetic efficiency; stomatal density; stomatal index; stomatal size; threshold for transpiration decline	Polania et al. (2022)
Grain yield; pod number	Labastida et al. (2023)
Harvest index; pod number per plant; seed number per pod; seed number per plant; plant height; pod harvest index; pod length; hundred-seed weight; seed yield; seed yield per plant	Lei Wu et al. (2023)
Net photosynthesis rate; stomatal conductance; transpiration; intercellular CO_2 concentration; total chlorophyll concentration; sucrose concentration; fructose concentration; glucose concentration; starch concentration	González-Lemes et al. (2023)
Shoot total nitrogen content; shoot dry weight; nodules' dry weights; leaf relative water content; nitrogenase activity	Da Silva et al. (2024)

Transcription factors play a key role in the plant's response to drought. Several families of transcription factors, including MYB, bZIP, bHLH, WRKY, NAC, ERF, and DREB, are essential for regulating signal transduction pathways involved in drought tolerance. Laskar et al. (2021) reported that the *Phvul.010G072900* gene, encoding a member of the GRAS transcription factor family, was significantly upregulated under low-temperature stress, suggesting its involvement in the common bean's stress response. Additionally, genes encoding members of the bHLH and MYB transcription factor families are crucial in enhancing the plant's ability to scavenge reactive oxygen species (ROS), thereby contributing significantly to drought stress tolerance (Liang et al. 2022; Yao et al. 2022).

Recent research has also highlighted the role of small RNAs in drought response. For example, Cowpea mild mottle virus (CPMMV) infection in common bean induces the production of a 21 nt vsiRNA6163. This small RNA targets the 5′-UTR of the common bean transcriptional repressor gene *PvTCP2*, leading to the activation of the autophagy-related gene PvATG8c and an increase in cellular autophagy levels. This process enhances the degradation of the stomatal opening regulator PvERD15, reducing stomatal conductance and water loss. As a result, common beans exhibit improved drought resistance under combined stress conditions of virus infection and drought (Wu et al. 2024a, b). Taken together drought tolerance in common bean is a complex trait involving interactions at physiological, biochemical, and molecular levels. Thorough research into these mechanisms provides a scientific foundation for developing effective drought-resistant breeding strategies and improving common bean production, thereby enhancing their resilience to environmental challenges.

8.3 Marker-Assisted Breeding Facilitates Drought Tolerance Improvement in Common Bean

8.3.1 Types of Molecular Markers

As molecular biotechnology advances, molecular markers have become an essential tool in the field of breeding (Nadeem et al. 2018b). Characterized by their stability, reliability, and excellent repeatability, these markers are invaluable for a variety of applications, including genetic breeding, gene mapping, gene cloning, construction of genetic maps, systematic taxonomy, research on genetic germplasm resources, and assisted selective breeding. Molecular markers fall into four primary categories:

- *Hybridization-Based Markers*: These include restriction fragment length polymorphism (RFLP), which uses DNA hybridization techniques to detect variations in DNA sequences.
- *Markers Based on Restriction Enzymes and PCR*: Amplified fragment length polymorphism (AFLP), which identifies DNA polymorphisms through the selective amplification of DNA fragments.
- *PCR-Based Markers*: Simple sequence repeat (SSR) focuses on the polymorphism of repeated DNA sequences. Random amplified polymorphic DNA (RAPD) utilizes random primers to amplify and detect DNA polymorphisms. Sequence-related amplified polymorphism (SRAP) detects polymorphisms based on the sequence composition of DNA.
- *Next-Generation Markers*: Single nucleotide polymorphism (SNP) detects the smallest DNA sequence variations, offering high precision in genetic studies. Kompetitive allele-specific PCR (KASP) enables precise the identification of SNPs and insertion-deletion polymorphisms (InDels) in genomic DNA samples.

These markers provide a sophisticated means to dissect the genetic architecture of complex traits, enabling breeders to make informed decisions and accelerate the development of improved crop varieties. The ongoing evolution of molecular marker technology is crucial for enhancing our understanding of genetic diversity and advancing innovation in plant breeding (Hasan et al. 2021).

8.3.2 Types and Principles of Marker-Assisted Breeding

Molecular breeding techniques have revolutionized plant genetics, encompassing a range of strategies known collectively as marker-assisted breeding (MAB). Key methods include:

- *Marker-Assisted Back-Crossing (MABC)*: MABC aims to transfer a major gene from a donor cultivar into an elite line. This strategy is highly effective for integrating desirable traits into well-adapted commercial cultivars, making the genetic enhancement process more efficient (Steele et al. 2013).

- *Marker-Assisted Recurrent Selection (MARS)*: MARS involves assembling and incorporating favorable alleles from diverse sources to express quantitative traits. This method is particularly useful for identifying multiple genomic regions and detecting both minor and major quantitative trait loci (QTLs). MARS has the potential to achieve greater genetic gains compared to MABC due to its comprehensive approach to allele selection (Abdulmalik et al. 2017).
- *Genomic Selection (GS)*: GS is an advanced approach that relies on marker-based selection, which can be performed without extensive testing or prior knowledge of marker × trait associations. Unlike MABC and MARS, GS directly identifies superior genotypes through predicted breeding values (BV), using a genome-wide distribution of markers. This method provides a holistic view of an individual's genetic potential, leading to more precise and efficient breeding outcomes (Keller et al. 2020).

These techniques represent the forefront of molecular breeding, equipping scientists and breeders with powerful tools to enhance crop performance and resilience, and to address the challenges of modern agriculture (Fig. 8.1).

8.3.3 Marker Resources for Improving Drought Tolerance in Common Bean

Common beans require significant amounts of water for both vegetative and reproductive growth, making them particularly vulnerable to drought stress compared to other grain legumes. Traditional methods for identifying drought resistance in common bean germplasm are limited by screening efficiency. The advent of molecular marker-assisted selection (MAS) has markedly improved the efficiency of screening for drought-resistant germplasm, thereby accelerating the breeding process for drought tolerance. Identifying drought tolerance traits is crucial for developing molecular tools that enhance breeding programs, particularly through marker-assisted selection. Developing drought-resistant (DR) varieties is crucial for mitigating the impacts of climate change and can be significantly expedited through marker-assisted selection. Identifying quantitative trait loci (QTLs) associated with drought resistance is essential for improving the reproductive capacity of common beans.

Historically, the lack of robust linkage maps led researchers to use various mapping populations and molecular markers for genetic analysis of drought resistance in common beans. For example, Schneider et al. (1997) used crosses between Sierra/AC1028 and Sierra/Lef2RB to produce isolated populations, identifying nine RAPD (random amplified polymorphic DNA) markers linked to drought resistance based on yield traits. However, these markers were not integrated into the linkage groups of the RIL populations derived from the BAT477 and DOR364 cross under drought conditions (Blair et al. 2012). By combining 186 AFLP (amplified fragment length polymorphism), RAPD, and SSR (simple sequence repeat) markers, a genetic map spanning 1087.5 cM was constructed. Phenotypic evaluations conducted in both

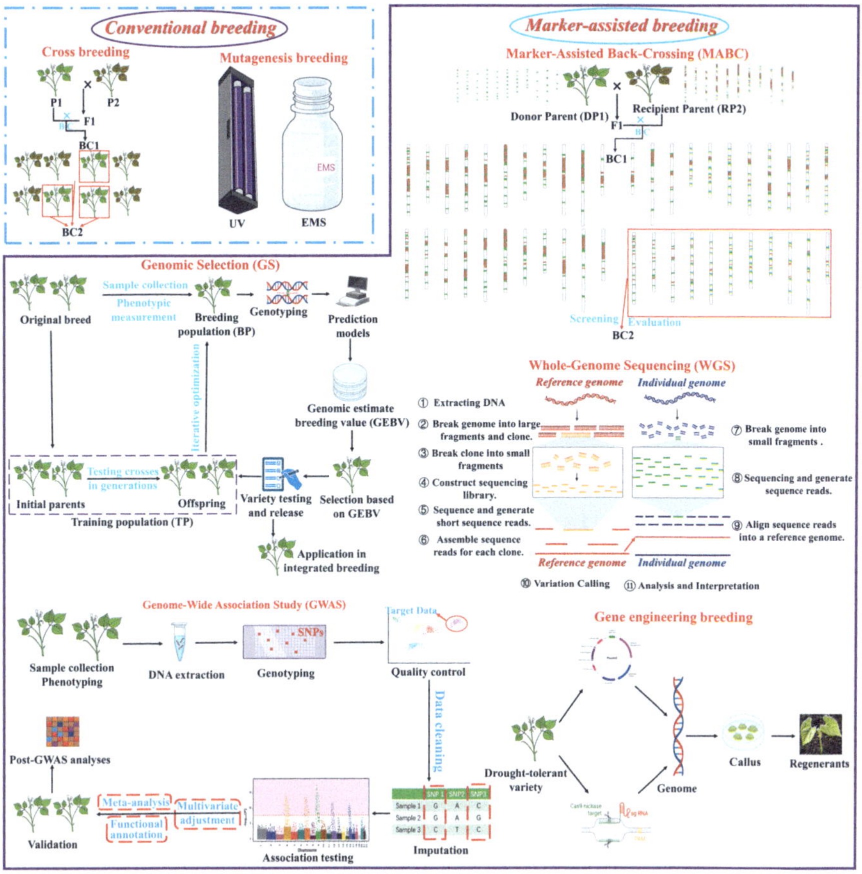

Fig. 8.1 Overview of breeding methods in common bean marker-assisted breeding. This figure illustrates the various breeding methods employed in common bean marker-assisted breeding. It depicts cross-breeding, molecular marker-assisted breeding, genome-wide breeding, and genetic engineering techniques used to enhance common bean varieties. Each method is shown to highlight its role in the breeding process

field and greenhouse settings, under well-watered and drought conditions, led to the detection of five yield QTLs on linkage groups Pv04, Pv06, Pv08, and Pv10 under drought stress. Additionally, an analysis of the genetic basis of common bean roots resulted in the identification of nine root QTLs, six of which were located on chromosome 11. QTLs related to total root length, root volume, and root dry weight were all found on chromosome 11, with phenotypic interpretation rates of 33.2%, 37.7%, and 40%, respectively (Asfaw and Blair 2012). In Ethiopia, Colombia, and Malawi, the same population was used to map yield QTLs on chromosome Pv08, with additional QTLs related to canopy biomass, pod harvest index, and chlorophyll content also identified (Asfaw et al. 2012). Another study investigated the genetic correlation between root traits and drought resistance, identifying eight root QTLs under drought conditions using the PVE tube method to assess root characteristics in a greenhouse (Asfaw and Blair 2012). Mukeshimana et al. (2014) constructed a genetic map with 2122 SNPs using an RIL population from the SEA5/CAL-96 cross, identifying 14 QTLs related to drought resistance, many of which were consistently detected across multiple years. Trapp et al. (2015) used RIL populations from the Buster/Roza cross to construct SNP genetic maps, repeatedly detecting a major drought-resistance-related QTL in linkage groups Pv01 and Pv02, with a phenotypic interpretation rate of 37%. Despite these efforts, relatively few genetic studies on common bean drought resistance have been conducted, and further research is needed to identify major effect QTLs that are consistently expressed across diverse drought environments.

With advancements in sequencing technology, single nucleotide polymorphisms (SNPs) have become a prominent class of molecular markers. For example, a previous study constructed a genetic map featuring 2122 SNPs using a recombinant inbred line (RIL) population from the SEA5/CAL-96 cross. This cross included common bean materials from various gene banks and led to the identification of 14 quantitative trait loci (QTLs) associated with drought resistance, many of which were consistently observed across multiple years of experimentation (Mukeshimana et al. 2014). The co-localization of phenological traits and yield QTLs highlights the importance of phenological plasticity in drought adaptation. Further research used the SEA5/AND277 cross along with a genetic map created using SSR and SNP markers to explore QTLs related to drought resistance (Briñez et al. 2017). All detected QTLs under drought conditions showed a positive effect from the SEA5 alleles, reinforcing SEA5 as a valuable resource for drought resistance in common bean (Table 8.2).

8.3.4 Mining Drought Tolerance Genes in Common Bean Based on Genomic Resources and Technologies

Whole-genome (re)sequencing provides a comprehensive view of an organism's genetic makeup, enabling the identification of single nucleotide polymorphisms (SNPs), insertions and deletions (InDels), and structural variations (SVs) that are crucial for traits like drought tolerance. This advanced technique involves sequencing the entire genome to reveal genetic variations between drought-tolerant and susceptible varieties. Following the completion of the common bean genome sequencing and assembly, research into drought response genes has accelerated, allowing for the identification of genes related to drought tolerance at the whole-genome level (Schmutz et al. 2014). For example, Laskar et al. (2021) identified 55 *PvuGRAS* genes across the genome, and real-time quantitative PCR revealed that at least nine of these genes (including *PvuGRAS25*, *PvuGRAS17*, and *PvuGRAS29*) are responsive to drought stress. Wu et al. (2022) identified 12 loci significantly associated with drought stress through genome-wide association analysis, highlighting a stable locus on chromosome 11 that harbors genes involved in late embryonic development, such as those encoding LEA (late embryogenesis abundant) proteins, aquaporins (AQP), and proline-rich proteins (PRR).

Table 8.2 List of DNA markers linked with drought tolerance in common bean

Breeding technique	Chromosome	Gene	References
RFL, RAPD	Pv03	–	Nodari et al. (1993) and Park et al. (2001)
RAPD	–	$OA08_{780}$, $OA04_{560}$, $OX11_{680}$, $OZO8_{750}$, $OXI8_{980}$	Schneider et al. (1997)
SNP	Pv01	–	Asfaw et al. (2012)
SSR, AFLP, RAPD	4, 6, 8, 10	YId4.1, YId6.1, YId8.1, YId10.1	Blair et al. (2012)
SSR, AFLP, RAPD	3, 5, 6, 8, 9	Cbm3.1I, Ppi3.1, Hri3.1, Stc5.1, Stc6.1, Scr6.1, YId8.1, Sbr9.1	Asfaw et al. (2012)
PILs, SNP	Pv03, Pv08	–	Mukeshimana (2013)
SSR, AFLP, RAPD	1, 3, 7, 9	PHII.I, NP3.1, SW3.1, SW7.2, SY9.2	Mukeshimana et al. (2014)
SNP, SSR	Pv03, Pv09	SEA	Mukeshimana et al. (2014)
SSR, AFLP, RAPD	1, 10, 11	RId1.1, Srl10.1, Rd11.1, R111.1, Fr11.1, Tt11.1, Rv11.1, Rbm11.1, Rsr11.1	Asfaw and Blair (2014)
PILs, SNP	Pv01	Phvul3G081100	Mukeshimana et al. (2014)
SSR, AFLP, RAPD	1, 2	SY1.1 BR, SY2.1 BR	Trapp et al. (2015)
SNP	Pv02	Phvul1G169300	Diaz et al. (2020)

SNPs are distributed across all chromosomes, though their distribution is uneven. Of the 19 bioclimatic variables studied, 17 showed significant associations with SNPs on chromosomes Pv01, Pv02, Pv03, Pv04, Pv06, Pv09, Pv10, and Pv11 of common bean (Elias et al. 2021). Notably, two SNPs on chromosome Pv09, within a 4.8 Mb region between positions 30,671,474 bp and 35,475,381 bp, were identified. One of these SNPs is near the gene model Phvul.009G207800, which encodes leucine-rich repeat (LRR) proteins, some with an F-box domain. Another SNP on chromosome Pv09, S09_35475381, was significantly associated with nine of the nineteen bioclimatic variables, explaining 11.9% to 36.6% of the phenotypic variation. This SNP is close to the candidate gene Phvul.009G241300, which encodes a PLATZ transcription factor protein family, known to play a crucial role in drought stress responses (Elias et al. 2021). Furthermore, by resequencing 683 common bean accessions from various geographical locations, a haplotype map with 4.8 million SNPs was developed. Concurrently, drought resistance tests were conducted at the bud and seedling stages using these natural populations. These tests evaluated germination rate, germination potential, taproot length, root dry weight, and survival rate under repeated drought stress at the seedling stage, as well as under flood and drought conditions.

Genome-wide association studies (GWAS) have been conducted to dissect the loci associated with drought resistance based on physiological traits, identifying 32 quantitative trait loci associated with drought resistance and revealing candidate genes involved in drought stress responses, such as those encoding MYB, bZIP, bHLH, ERF, and protein kinases (Wu et al. 2024a, b). Furthermore, the construction of a *Phaseolus vulgaris* pan-genome has provided a comprehensive resource for studying the genetic diversity of this crop. The pan-genome, which includes both wild and domesticated genotypes, offers insights into the presence/absence variants (PAVs) and their implications for adaptive traits such as drought resistance (Cortinovis et al. 2024). In conclusion, the integration of genomic resources, such as the pan-genome, with advanced genotyping techniques has accelerated the identification of genes and mechanisms related to drought resistance in common beans.

8.4 Technical Challenges and Future Directions

Despite giant technical leaps, there are still several challenges facing researcher. The phenotypic expression of drought resistance in plants is influenced by environmental factors and is a quantitative trait controlled by multiple genes. This trait involves complex gene-environment interactions, which pose challenges for marker-assisted selection. The genetic architecture of common beans is complex, with issues such as linkage disequilibrium and gene interactions potentially impacting the effectiveness of marker-assisted breeding. Balancing drought resistance with other traits, such as yield and quality, is also critical. Furthermore, developing stable and reliable molecular markers closely associated with drought resistance requires substantial technical and economic investment, which can limit the application of molecular marker-assisted breeding for large-scale development of drought-resistant common bean varieties.

As genomic resources for common beans continue to expand, the potential to enhance drought resistance through molecular-assisted breeding (MAB) becomes increasingly promising. Future research should focus on integrating various omics data—encompassing genomics, transcriptomics, proteomics, and metabolomics—to gain a comprehensive understanding of drought resistance mechanisms. Additionally, the creation of predictive models that combine genomic and environmental data, along with the application of artificial intelligence and machine learning algorithms, could further optimize breeding decision-making and improve efficiency. While marker-assisted breeding for drought resistance in common beans faces several challenges, it also holds significant potential. By intensifying foundational research, refining marker selection, increasing technical efficiency, reducing costs, and fostering interdisciplinary collaboration and industrial application, the field can make substantial progress. Advances in marker-assisted breeding technology for drought resistance are poised to contribute significantly to the sustainable development of agricultural production.

References

Aasim M, Katirci R, Baloch FS, Mustafa Z, Bakhsh A, Nadeem MA et al (2022) Innovation in the breeding of common bean through a combined approach of in vitro regeneration and machine learning algorithms. Front Genet 13:897696

Abd El Mageed TA, Semida W et al (2023) Glutathione-mediated changes in productivity, photosynthetic efficiency, osmolytes, and antioxidant capacity of common beans (*Phaseolus vulgaris*) grown under water deficit. PeerJ 11:e15343

Abdulmalik RO, Menkir A, Meseka SK et al (2017) Genetic gains in grain yield of a maize population improved through marker assisted recurrent selection under stress and non-stress conditions in West Africa. Front Plant Sci 8:841

Alghamdi SA, Alharby HF, Bamagoos AA et al (2022) Rebalancing nutrients, reinforcing antioxidant and osmoregulatory capacity, and improving yield quality in drought-stressed *Phaseolus vulgaris* by foliar application of a bee-honey solution. Plants-Basel 12(1):63

Ali A, Altaf MT, Bedir M, Liaqat W, Jamil A, Nadeem MA, Baloch FS (2024) Biotechnological advancements toward sugarcane crop improvement. In: Industrial crop plants. Singapore, Singapore, pp 333–352

Asfaw A, Blair MW (2012) Quantitative trait loci for rooting pattern traits of common beans grown under drought stress versus non-stress conditions. Mol Breed 30(2):681–695

Asfaw A, Blair MW (2014) Quantification of Drought Tolerance in Ethiopian Common Bean Varieties. Agric Sci 5:124–139

Asfaw A, Blair MW, Struik PC (2012) Multienvironment quantitative trait loci analysis for photosynthate acquisition, accumulation, and remobilization traits in common bean under drought stress. G3-Genes Genom Genet 2(5):579–595

Blair MW, Galeano CH, Tovar E et al (2012) Development of a Mesoamerican intra-genepool genetic map for quantitative trait loci detection in a drought tolerant × susceptible common bean (*Phaseolus vulgaris* L.) cross. Mol Breed 29(1):71–88

Briñez B, Perseguini JMKC, Rosa JS et al (2017) Mapping QTLs for drought tolerance in a SEA 5 x AND 277 common bean cross with SSRs and SNP markers. Genet Mol Biol 40(4):813–823

Cortinovis G, Vincenzi L, Anderson R et al (2024) Adaptive gene loss in the common bean pan-genome during range expansion and domestication. Nat Commun 15(1):6698

da Silva HAP, Caetano VS, Pessôa DDV et al (2024) Unraveling the drought-responsive transcriptomes in nodules of two common bean genotypes during biological nitrogen fixation. Front Plant Sci 15:1345379

Diaz S, Ariza-Suarez D, Izquierdo P et al (2020) Genetic mapping for agronomic traits in a MAGIC population of common bean (*Phaseolus vulgaris* L.) under drought conditions. BMC Genomics 21(799):799

Elias JCF, Gonçalves-Vidigal MC, Ariani A et al (2021) Genome-environment association analysis for bio-climatic variables in common bean (*Phaseolus vulgaris* L.) from Brazil. Plan Theory 10(8):1572

Gaut BS (2014) The complex domestication history of the common bean. Nat Genet 46(7):663–664

Geleta RJ, Roro AG, Terfa MT (2024) Phenotypic and yield responses of common bean (*Phaseolus vulgaris* L.) varieties to different soil moisture levels. BMC Plant Biol 24(1):242

González-Lemes I, Acosta-Maspons A, Cetz-Chel JE et al (2023) Carbon-concentrating mechanisms in pods are key elements for terminal drought resistance in *Phaseolus vulgaris*. J Exp Bot 74(5):1642–1658

Hasan N, Choudhary S, Naaz N, Sharma N, Laskar RA (2021) Recent advancements in molecular marker-assisted selection and applications in plant breeding programmes. J Genet Eng Biotechnol 19(1):128

Hoyos-Villegas V, Song Q, Kelly JD (2017) Genome-wide association analysis for drought tolerance and associated traits in common bean. Plant Genome-Us 10(1):10.3835

Iqbal J, Altaf MT, Jan MF, Raza W, Liaqat W, Haq I et al (2023) Exploring genetic diversity in cotton genotypes using EST-SSR and ISSR markers: a comparative study. Sarhad J Agric 39(4):800–814

Jiang R, Sun T, Shi Z, Moshelion M, Xu P (2024) Leveraging 'golden-hour' WUE for developing superior vegetable varieties with optimal water-saving and growth traits. Veg Res 4:e002

Jorge JG, Villalobos-Lopez MA, Chavarria-Alvarado KL et al (2020) Genome-wide transcriptional changes triggered by water deficit on a drought-tolerant common bean cultivar. BMC Plant Biol 20(1):525

Karavidas I, Ntatsi G, Marka S et al (2023) Fresh pod yield, physical and nutritional quality attributes of common bean as influenced by conventional or organic farming practices. Plan Theory 12:32

Keller B, Ariza-Suarez D, de la Hoz J et al (2020) Genomic prediction of agronomic traits in common bean (*Phaseolus vulgaris* L.) under environmental stress. Front Plant Sci 11:1001

Labastida D, Ingvarsson PK, Rendón-Anaya M (2023) Dissecting the genetic basis of drought responses in common bean using natural variation. Front Plant Sci 14:1143873

Lanna AC, Silva RA, Ferraresi TM et al (2018) Physiological characterization of common bean (*Phaseolus vulgaris* L.) under abiotic stresses for breeding purposes. Environ Sci Pollut Res Int 25(31):31149–31164

Laskar P, Bhattacharya S, Chaudhuri A et al (2021) Exploring the GRAS gene family in common bean (*Phaseolus vulgaris* L.): characterization, evolutionary relationships, and expression analyses in response to abiotic stresses. Planta 254(4):84

Liang BB, Wan SG, Ma QL et al (2022) A novel bHLH transcription factor *PtrbHLH66* from trifoliate orange positively regulates plant drought tolerance by mediating root growth and ROS scavenging. Int J Mol Sci 23(23):15053

Long J, Zhang J, Zhang X (2020) Genetic Diversity of Common bean (*Phaseolus vulgaris* L.) germplasm resources in Chongqing, evidenced by morphological characterization. Front Genet 11:697

López CM, Pineda M, Alamillo JM (2020) Differential regulation of drought responses in two *Phaseolus vulgaris* genotypes. Plants 9(12):1815

Martínez-Barradas V, Galbiati M, Barco-Rubio F, Paolo D, Espinoza C, Cominelli E, Arce-Johnson P (2024) PvMYB60 gene, a candidate for drought tolerance improvement in common bean in a climate change context. Microbiol Res 57(1):52

Martre P, Dueri S, Guarin JR et al (2024) Global needs for nitrogen fertilizer to improve wheat yield under climate change. Nat Plants 10:1081–1090

Mensack MM, Fitzgerald VK, Ryan EP et al (2010) Evaluation of diversity among common beans (*Phaseolus vulgaris* L.) from two centers of domestication using 'omics' technologies. BMC Genomics 11:686

Meza C, Valenzuela F, Echeverría-Vega A et al (2022) Plant-growth-promoting bacteria from rhizosphere of Chilean common bean ecotype (*Phaseolus vulgaris* L.) supporting seed germination and growth against salinity stress. Front Plant Sci 13:1052263

Mukeshimana G (2013) Dissecting the genetic complexity of drought tolerance mechanisms in common bean (*Phaseolus vulgaris* L.). Michigan State University

Mukeshimana G, Butare L, Cregan PB et al (2014) Quantitative trait loci associated with drought tolerance in common bean. Crop Sci 54(3):923–938

Munné-Bosch S, Villadangos S (2023) Cheap, cost-effective, and quick stress biomarkers for drought stress detection and monitoring in plants. Trends Plant Sci 28(5):527–536

Nadeem MA, Baloch FS (2023) Genome-wide association studies revealed DArTseq loci associated with seed traits in Turkish common bean germplasm. Turk J Agric For 47(4):479–496

Nadeem MA, Habyarimana E, Çiftçi V et al (2018a) Characterization of genetic diversity in Turkish common bean gene pool using phenotypic and whole-genome DArTseq-generated silicoDArT marker information. PLoS One 13(10):e0205363

Nadeem MA, Nawaz MA, Shahid MQ, Doğan Y, Comertpay G, Yıldız M et al (2018b) DNA molecular markers in plant breeding: current status and recent advancements in genomic selection and genome editing. Biotechnol Biotechnol Equip 32(2):261–285

Nadeem M, Li J, Yahya M et al (2019) Research Progress and Perspective on Drought Stress in Legumes: A Review. Int J Mol Sci 20(10):2541

Nadeem MA, Habyarimana E, Karaköy T, Baloch FS (2021) Genetic dissection of days to flowering via genome-wide association studies in Turkish common bean germplasm. Physiol Mol Biol Plants 27(7):1609–1622

Nguyen H, Thompson AE, Costello C (2023) Impacts of Historical Droughts on Maize and Soybean Production in the Southeastern United States. Agr Water Manag 281:0378–3774

Nodari RO, Tsai SM, Gilbertson RL, Gepts P (1993) Towards an integrated linkage map of common bean. II. Development of an RFLP-based linkage map. Theor Appl Genet 85:513–520

Park SO, Coyne DP, Steadman JR, Skroch PW (2001) Mapping of QTL for resistance to white mold disease in common bean. Crop Sci 41:1253–1262

Polania JA, Salazar-Chavarría V, Gonzalez-Lemes I et al (2022) Contrasting *Phaseolus* crop water use patterns and stomatal dynamics in response to terminal drought. Front Plant Sci 13:894657

Ponce TP, Bugança MDS, da Silva VS et al (2024) Differential gene expression in contrasting common bean cultivars for drought tolerance during an extended dry period. Genes 15(7):935

Razi K, Muneer S (2021) Drought stress-induced physiological mechanisms, signaling pathways and molecular response of chloroplasts in common vegetable crops. Crit Rev Biotechnol 41(5):669–691

Saeidi K, Zare N, Baghizadeh A, Asghari-Zakaria R (2019) *Phaseolus vulgaris* genome possesses *CAMTA* genes, and *phavuCAMTA1* contributes to the drought tolerance. J Genet 98:31

Salgotra RK, Stewart CN (2020) Functional markers for precision plant breeding. Int J Mol Sci 21(13):4792

Schmutz J, McClean P, Mamidi S et al (2014) A reference genome for common bean and genome-wide analysis of dual domestications. Nat Genet 46:707–713

Schneider KA, Brothers ME, Kelly JD (1997) Marker-assisted selection to improve drought resistance in common bean. Crop Sci 37(1):51–60

Steele KA, Price AH, Witcombe JR et al (2013) QTLs associated with root traits increase yield in upland rice when transferred through marker-assisted selection. Theor Appl Genet 126(1):101–108

Trapp JJ, Urreac A, Cregan PB (2015) Quantitative trait loci for yield under multiple stress and drought conditions in a dry bean population. Crop Sci 55(4):1596–1572

Wang N, Yuan Y, Wang H et al (2020) Applications of genotyping-by-sequencing (GBS) in maize genetics and breeding. Sci Rep 10:16308

Watanabe M, Risi R, Masi D et al (2020) Current evidence to propose different food supplements for weight loss: a comprehensive review. Nutrients 12(9):2873

Wiesinger JA, Cichy KA, Glahn RP et al (2016) Demonstrating a nutritional advantage to the fast-cooking dry bean (*Phaseolus vulgaris* L.). J Agric Food Chem 64(45):8592–8603

Wu J, Wang L, Fu J et al (2020) Resequencing of 683 common bean genotypes identifies yield component trait associations across a north–south cline. Nat Genet 52:118–125

Wu L, Chang Y, Wang L, Wu J, Wang S (2021) Genetic dissection of drought resistance based on root traits at the bud stage in common bean. Theor Appl Genet 134(4):1047–1061

Wu L, Chang YJ, Wang LF et al (2022) The aquaporin gene *PvXIP1;2* conferring drought resistance identified by GWAS at seedling stage in common bean. Theor Appl Genet 135(2):485–500

Wu L, Chang YJ, Wang LF et al (2023) Genetic dissection of yield-related traits in response to drought stress in common bean. Crop J 11(4):1097–1105

Wu L, Chang YJ, Wang LF, Wang SM, Wu J (2024a) Genome-wide association study dissecting drought resistance-associated loci based on physiological traits in common bean. J Integr Agric 23:3657. https://doi.org/10.1016/j.jia.2024.03.079

Wu XY, Chen ST, Zhang ZX et al (2024b) A viral small interfering RNA-host plant mRNA pathway modulates virus-induced drought tolerance by enhancing autophagy. Plant Cell 00:1–18

Yao CY, Li WH, Liang XQ et al (2022) Molecular cloning and characterization of MbMYB108, a Malus baccata MYB transcription factor gene, with functions in tolerance to cold and drought stress in transgenic Arabidopsis thaliana. Int J Mol Sci 23(9):4846

Yildiz M, Arbizu C (2022) Inter-primer binding site (iPBS) retrotransposon markers provide insights into thegenetic diversity and population structure of carrots (Daucus, Apiaceae). Turk J Agric For 46(2):214–223

Yıldız M, Ekbiç E, Düzyaman E, Serçe S, Abak K (2016) Genetic and phenotypic variation of Turkish Okra (Abelmoschus esculentus L. Moench) accessions and their possible relationship with American, Indian and African germplasms. J Plant Biochem Biotechnol 25:234–244

Yin Z, Meng X, Wang Q et al (2021) Effects of drought stress on antioxidant enzymes in common bean (*Phaseolus vulgaris* L.) seedlings. Bangl J Bot 50(5):873–878

Zhang Z, Ma J, Yang X et al (2024) Natural allelic diversities of GmPrx16 confer drought tolerance in soybean. Plant Biotechnol J 22(3):535–537

Marker-Assisted Breeding of Pea for Drought Tolerance

9

Md. Mobinur Rahman, Jaber Bin Azim, Md. Yousuf Ali, Touhidul Islam, Md. Parvez Kabir, and Akbar Hossain

Abstract

Peas are vital legume crops that play crucial roles in global food security, providing rich sources of protein, carbohydrates, and essential vitamins and minerals. However, the production of peas is severely threatened by the increasing impact of climate change, particularly the increasing frequency and intensity of drought events. This chapter commences with an exploration of the adverse effects of drought on the morphology, physiology, and biochemical processes of pea plants. However, plants themselves have several defense mechanisms, such as drought escape, drought avoidance, and drought tolerance. Through the utilization of advanced genomic tools, including molecular markers and high-throughput genotyping technologies, breeders can streamline the process of identifying and selecting superior genotypes with enhanced drought tolerance. Leveraging marker-assisted breeding techniques offers a promising avenue for expediting the selection of drought-tolerant pea genotypes, thereby contributing to sustainable agricultural practices. This chapter also presents successful case studies illustrating the practical application of marker-assisted breeding (MAB) in pea breeding programs worldwide. These examples highlight the efficacy of MABs in accelerating the development of drought-tolerant pea varieties characterized by improved agronomic performance and yield stability. Additionally, the chap-

Md. M. Rahman · J. B. Azim · Md. Y. Ali
Maize Breeding Division, Bangladesh Wheat and Maize Research Institute, Dinajpur 5200, Bangladesh

T. Islam
On-Farm Research Division, Bangladesh Wheat and Maize Research Institute, Dinajpur 5200, Bangladesh

Md. P. Kabir · A. Hossain (✉)
Soil Science Division, Bangladesh Wheat and Maize Research Institute, Dinajpur 5200, Bangladesh

© The Author(s), under exclusive license to Springer Nature Singapore Pte Ltd. 2025

M. A. Nadeem et al. (eds.), *Marker-Assisted Breeding in Legumes for Drought Tolerance*, https://doi.org/10.1007/978-981-96-4112-3_9

ter discusses the challenges and opportunities inherent in deploying MABs, such as marker validation, genotype–environment interactions, and regulatory considerations. In conclusion, marker-assisted breeding has emerged as a potent strategy for addressing the complex trait of drought tolerance in pea, offering a sustainable solution to mitigate the impacts of climate change on agricultural productivity. This comprehensive overview serves as a valuable resource for researchers, breeders, and policymakers alike, elucidating the potential of molecular tools in facilitating crop improvement efforts tailored to environmental resilience.

Keywords

Marker-assisted breeding · QTL · Drought tolerance · Pea · Molecular marker

9.1 Introduction

Pea, scientifically known as *Pisum sativum*, is the fourth leading legume crop in the world (Zhao et al. 2020) and originated approximately 9000 BP (Pavan et al. 2022) from the Middle East, with a dissemination route through southern Asia (Wu et al. 2017), making it one of the earliest crops in agricultural history. Pea is an herbaceous plant belonging to the genus *Pisum* in the legume family and is characterized by dry peas, green peas, and snow peas (Thakur et al. 2018; Sarıkaya et al. 2023). Pea is a valuable crop with significant nutritional and economic importance. According to the FAO's 2023 report, peas are extensively grown globally, with a total production of 14.17 million tons, predominantly in China, which accounts for approximately 10% of the global production share.

Peas are known for their high protein content, establishing them as crucial and cost-effective sources of plant-based proteins (Dhaliwal et al. 2021; Ma et al. 2024) for human nutrition and livestock feeding and as a strategic commodity for global food security (Pavan et al. 2022). In addition to protein, peas are abundant in essential nutrients, including starch, fiber, vitamins, minerals, and phytochemicals (Dahl et al. 2012). Peas are particularly notable for being excellent sources of minerals such as calcium, iron, and zinc, as well as vitamins such as carotenoids and folic acid (Wu et al. 2023). Studies have demonstrated that pea seeds offer additional health benefits beyond their nutritional value, including antiobesity, antifungal, anticancer, antidiabetic, and cardioprotective properties (Beniwal et al. 2024).

In recent years, climate change has adversely affected pea production worldwide, primarily through various abiotic stresses, such as drought. Drought is a natural hazard characterized by an extended period of insufficient precipitation (Nikam et al. 2020). The stress caused by drought poses a significant challenge to global agricultural productivity, and its intensity and severity are anticipated to increase in the coming years (Nadeem et al. 2019). Research has indicated that drought is among the primary factors contributing to reduced crop growth and yield (Noman et al. 2015; Ali et al. 2016). Drought can significantly decrease crop productivity by disrupting essential physiological and biochemical processes such as photosynthesis, water uptake, and nutrient assimilation (Fatima et al. 2024). Legumes exhibit heightened susceptibility to drought stress, particularly in their vegetative and reproductive growth stages (Nikam et al. 2020). Research indicates that legumes such as pea face minimal risk from multiple types of temperature stress because of their short growing cycle. However, drought remains the predominant factor that significantly reduces yields (Potopová et al. 2017). Research has indicated that pea plants exposed to drought experience significant decreases in photosynthesis, transpiration, and glycolate oxidase activity, along with slight decreases in chlorophyll and carotenoid levels and soluble protein content (Moran et al. 1994). Drought stress also reduces both nitrogen accumulation and overall dry weight in every part of the pea plant (Mahieu et al. 2009).

Plants employ various physiological and biochemical adaptations, involving numerous stress-related genes, to cope with drought. While many of these changes are shared among different plant species, some are specific to particular drought responses (Carvalho et al. 2019). The drought tolerance of peas can be increased through strategies such as screening, breeding, and marker-assisted breeding (Bagheri et al. 2023).

Molecular marker-assisted breeding (MAB), also known as molecular-assisted breeding, involves the use of DNA markers, linkage maps, and genomics to improve plant or animal traits on the basis of genotypic assays (Nadeem et al. 2018; Jiang 2013). This approach encompasses various breeding strategies, including marker-assisted selection (MAS), marker-assisted backcrossing (MABC), marker-assisted recurrent selection (MARS), and genome-wide selection (GWS) or genomic selection (GS) (Ribaut et al. 2010). MAB is considered a powerful methodology for the genetic improvement of crop plants and has been widely used across multiple crop species (Jiang 2013; Xu 2010). Thus, marker-assisted breeding is an effective method for developing drought-tolerant pea varieties.

This chapter covers the significant development and deployment of genomic tools for drought-tolerant pea breeding in recent years. Future prospects are discussed, especially in light of current progress toward deciphering the pea genome. The identification and mapping of quantitative trait loci (QTLs) associated with specific components of drought tolerance can help in understanding the genetic and physiological foundations of traits crucial for drought resistance (Serraj et al. 2005).

9.2 Consequences of Drought Stress on Pea Production in a Changing Climate

Drought induces a wide range of adverse morphological, physiological, and biochemical reactions (Table 9.1). Germination disruption is the primary impact of drought stress, as evidenced by various studies (Okçu et al. 2005). Drought accelerates the crop growth cycle, prompting an early shift from vegetative growth to reproduction (Farooq et al. 2012). A reduction in soil moisture availability also diminishes the supply of nutrients for plants, thereby limiting their absorption and

Table 9.1 Impact of drought stress on pea production

Impact category	Consequences	References
Physiological	Reduced photosynthesis and transpiration	Anjum et al. (2017)
	Stomatal closure	Pirasteh-Anosheh et al. (2016)
	Decreased leaf water potential	Flexas et al. (2004)
Morphological	Decreasing flower formation and seed size	Fatima et al. (2024)
	Reduced lateral root growth and an increased root-to-shoot biomass ratio	Shoaib et al. (2022)
	Induced floral abortion and reduced pollen germination, number of fertile pods, and seed dry weight	Kobraee et al. (2011)
Biochemical	Leaf rolling and wilting	Blum (2010)
	Root growth alteration	Ranjan et al. (2022)
	Accumulation of osmolytes (e.g., proline, soluble sugars)	Farooq et al. (2009)
	Increased antioxidant enzyme activities	Gill and Tuteja (2010)
	Elevated levels of reactive oxygen species (ROS)	Impa et al. (2012)
Yield	Decreased pod number and seed size	Arafa et al. (2021)
	Reduced overall yield	Annicchiarico et al. (2020)
	Quality deterioration	Sarkar et al. (2021)

utilization of nutrients (Gregory et al. 2012; Barzana et al. 2021). A decrease in water results in adverse effects on the internal movement of elements within plants due to reduced transpiration, transporter function, and cell membrane permeability (Ahanger et al. 2016). Drought conditions contribute to a decrease in the initiation of nodulation, growth of nodules, and specific activity of nodules, ultimately leading to reduced nitrogen levels in plants (Fig. 9.1) and overall growth (Prudent et al. 2016; Bagheri et al. 2023). Drought stress also causes reactive oxygen species (ROS) production in some cell compartments in pea plants, especially in chloroplasts, peroxisomes, and mitochondria, which can cause lipid peroxidation, membrane damage, protein oxidation, and DNA damage (Bagheri et al. 2023). The generation of ROS by various abiotic and biotic stressors, leading to oxidative stress, ultimately diminishes crop yields (Sharma et al. 2017).

9.3 Advanced Strategies to Mitigate Drought Stress in Pea

Mitigating drought stress in pea cultivation necessitates a comprehensive approach that combines traditional and advanced methods. Peas (*Pisum sativum* L.) are highly vulnerable to water deficit, which adversely affects yield and quality. Advanced strategies include technological and genetic interventions to improve water use efficiency, root architecture, and stress response pathways. Techniques such as marker-assisted selection (MAS), genomic selection (GS), and CRISPR–Cas gene editing

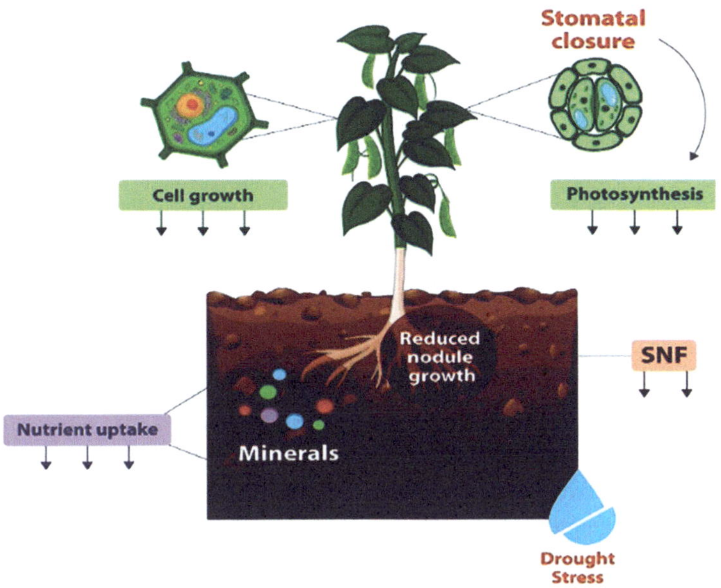

Fig. 9.1 The impact of drought stress (DS) on plants and their potential responses (Bagheri et al. 2023)

are used to develop drought-tolerant cultivars. Furthermore, physiological methods such as proteomics and metabolomics provide insights into the molecular mechanisms of drought tolerance, facilitating targeted enhancements in pea resilience. These integrated strategies offer potential for maintaining pea productivity amid increasingly erratic climatic conditions.

9.3.1 Marker-Assisted Selection (MAS)

The identification of quantitative trait loci (QTLs) is pivotal for marker-assisted selection (MAS) in plant breeding (Nadeem et al. 2018). By associating specific genomic regions with desirable traits such as drought tolerance, QTL mapping enables breeders to use molecular markers for the precise selection of superior plants early in the breeding process. This approach accelerates breeding cycles, reduces resource-intensive field trials, and facilitates the development of cultivars with multiple beneficial traits. The process of utilizing quantitative trait loci (QTLs) in marker-assisted selection (MAS) involves several key steps. Initially, researchers conducted genetic mapping studies to identify genomic regions associated with target traits, such as drought tolerance, through QTL analysis. Once QTLs are identified and validated, molecular markers linked to these regions are developed and used for marker-assisted breeding. Breeders then employ these markers to screen and select plants with desired traits at early stages of development, bypassing lengthy phenotypic evaluations. This accelerates the breeding process, allowing for the rapid development of new crop varieties that are better adapted to challenging environments. Overall, integrating QTLs into MAS enhances breeding precision, efficiency, and the ability to meet agricultural and environmental demands effectively (Nadeem et al. 2018).

9.3.2 Genomic Selection (GS)

Genomic selection (GS) is a modern breeding technique with great potential to increase drought tolerance in pea farming. By analyzing genomic data from various genetic markers spread across the pea genome, GS allows breeders to predict how well pea plants will perform under drought conditions. This method enables early identification of plants with superior drought tolerance traits, expediting the development of new pea varieties that can thrive despite water scarcity. GS is particularly advantageous for complex traits such as drought tolerance, where multiple genes with minor effects collectively contribute to resilience. This approach not only increases breeding efficiency but also supports sustainable agriculture by ensuring that crops can withstand environmental challenges such as drought. Annicchiarico et al. (2020) conducted a study on pea genomic selection tailored for Italian conditions, using genotyping-by-sequencing data to assess its effectiveness across diverse environments. They reported moderate predictive abilities within populations (0.30 for yield) and slightly lower accuracy between populations (0.19 for yield). Compared with traditional phenotypic selection, GS significantly improved breeding efficiency by 80% in terms of yield under various environmental conditions (Maurya et al. 2023). Annicchiarico et al. (2020) further applied GS to increase pea grain yield under severe terminal drought conditions in Mediterranean regions. Their study, which used SNP markers from genotyping-by-sequencing, highlighted significant genotype × environment interactions, especially under severe stress. GS models in controlled stress environments effectively identified top-performing lines similar to traditional phenotypic selection methods. The study underscored the efficiency of GS, showing an 18% improvement over marker-assisted selection (MAS) in enhancing yield under severe drought, emphasizing its cost-effectiveness and robust predictive ability across different pea populations.

9.3.3 Transcriptomics

Transcriptomics plays a vital role in deciphering and mitigating drought stress in pea (*Pisum sativum* L.). By analyzing RNA transcripts produced under drought conditions, scientists can identify crucial genes and regulatory networks involved in stress responses. This information helps pinpoint genes contributing to drought tolerance, facilitating the development of molecular markers for marker-assisted selection (MAS) and genomic selection (GS). Moreover, transcriptomic studies revealed changes in gene expression during drought, shedding light on the physiological and biochemical pathways activated in response. This knowledge supports breeding programs for selecting genotypes with optimal gene expression profiles for stress adaptation, accelerating the creation of drought-tolerant pea varieties, which are essential for sustainable agriculture in arid regions. Zorin et al. (2023) conducted a transcriptomic study comparing the symbiotic responses of the pea breeding line 'Triumph' with those of its parental cultivars 'Vendevil' and 'Classic.' They discovered that 'Triumph' inherited key symbiotic genes from 'Vendevil,' resulting in similar upregulation of genes related to solute transport, hormonal regulation, and flavonoid biosynthesis under combined inoculation. This study identified PsGLP2 as a potential marker for symbiotic responsiveness, improving the understanding of molecular symbiotic traits for future breeding efforts. Liu et al. (2015) used high-throughput RNA-Seq to investigate gene expression changes during seed development in vegetables and grain peas. This study identified 77,273 unigenes and revealed significant differential expression of genes associated with soluble sugar and starch metabolism at different stages of seed maturation in both pea types. These differences impact the metabolic profiles of vegetable and grain pea seeds, providing valuable insights for improving breeding strategies.

9.3.4 Proteomics and Metabolomics

Proteomics and metabolomics play crucial roles in comprehending and addressing drought stress in pea cultivation. Proteomics involves analyzing proteins on a large scale to understand their functions in stress response and cellular maintenance under drought conditions. This helps identify proteins that contribute to stress tolerance. Metabolomics, on the other hand, focuses on profiling metabolites to reveal their biochemical pathways and metabolic adjustments during drought stress. Integrating data from both disciplines helps pinpoint biomarkers and metabolic networks crucial for enhancing drought resilience in pea breeding programs, thereby ensuring sustainable agricultural productivity. Shumilina et al. (2021) studied the effects of moderate drought on the pea root nodule proteome and advanced glycation end product (AGE) patterns. Despite minimal changes in protein expression related to central metabolism and nitrogen fixation in response to osmotic stress, significant glycation was observed in numerous pea and rhizobial proteins. This glycation, particularly carboxymethylation, affects proteins involved in transcription, transport, and metabolic processes, suggesting a regulatory role under drought conditions. Castillejo et al. (2016) conducted a shotgun proteomic study to investigate drought tolerance in three pea genotypes with varying levels of water-deficit tolerance. Their analysis revealed significant protein differences related to primary metabolism and protein regulation. This study suggested that drought tolerance in pea involves maintaining primary metabolic functions and protecting proteins, whereas drought susceptibility may be related to the energy-intensive maintenance of homeostasis.

9.3.5 CRISPR/Cas9 Genome Editing

CRISPR–Cas technology shows immense potential in enhancing drought tolerance in pea (*Pisum sativum*) through precise genetic modifications. By targeting specific genes involved in drought response pathways—such as those regulating water uptake, transport, or stress signalling—CRISPR–Cas enables the development of pea varieties better suited to water-limited environments. Recent advancements have allowed targeted editing of key genes associated with drought stress, potentially improving water use efficiency and overall resilience in pea plants. This approach not only accelerates breeding efforts but also provides a sustainable solution to mitigate the impact of drought on pea crop productivity, addressing challenges posed by climate change.

Genome editing has revolutionized the functional analysis of genes and the introduction of new alleles in commercial crops (Mushtaq et al. 2019). Various genome editing approaches, including CRISPR/Cas9, meganucleases, TALENs, and ZFNs, are extensively used for genetic improvement (Mushtaq et al. 2019). In crop plants, susceptibility (S) or resistance (R) genes are targeted to increase crop protection against diseases or pests (Singh et al. 2016; Ren et al. 2017; Das et al. 2019). Additionally, CRISPR/Cas9 effectively confers disease resistance by targeting conserved regions of viral genomes in different crop species (Iqbal et al. 2016). In peas, transcriptomic analysis can identify genes and pathways related to disease resistance, aiding in the development of resistant cultivars (Barakate and Stephens 2016). Moreover, genome editing can modify epialleles involved in disease or abiotic stress resistance (Latutrie et al. 2019).

9.3.6 Association Mapping and GWAS

Association mapping and genome-wide association studies (GWASs) have become valuable tools for understanding the genetic basis of drought tolerance in pea (*Pisum sativum*). These methods allow researchers to pinpoint quantitative trait loci (QTLs) and candidate genes associated with important traits for drought adaptation, such as water use efficiency, root structure, and stress response pathways. By analyzing genetic diversity across various pea populations, GWASs help identify genomic regions linked to drought tolerance without the need for traditional biparental mapping populations. This approach not only speeds up the development of drought-tolerant pea varieties but also sheds light on the molecular mechanisms governing the drought response in leguminous crops.

In a study by Tafesse et al. (2020), a GWAS revealed that 15 marker trait associations (MTAs) were significantly associated ($p \leq 0.0005$) with stress-adaptive traits across different environments in pea. These traits included characteristics such as leaf and stem wax content, flowering duration, and indices related to pigment and chlorophyll levels. Sixteen candidate genes located near these MTAs were identified, suggesting their potential use in marker-assisted selection for breeding resilient pea varieties against heat and drought stresses.

Additionally, another GWAS conducted by Tafesse et al. (2021) identified 32 MTAs consistently linked to heat-responsive traits across multiple environments in pea. These traits included features such as chlorophyll concentration, canopy temperature, and reproductive characteristics. Forty-eight candidate genes near these MTAs were identified, indicating their promise for marker-assisted selection in the breeding of heat-resistant pea cultivars.

9.3.7 Epigenetic Modifications

Transgenerational epigenetic variation, which persists through successive generations, is emerging as a crucial strategy for breeding climate-resilient crop varieties. These variations alter gene expression through mechanisms such as DNA methylation or histone modification (Kumar et al. 2006). Genome-wide mapping of epigenetic markers can enable breeders to manipulate epigenomic variability, aiding in the development of resilient crop varieties. Such epigenetic variations influence host plant resistance against diverse pathogens, such as fungi, bacteria, viruses, nematodes, oomycetes, and herbivorous insects (Espinas et al. 2016; Ramirez-Prado et al. 2018; Alonso et al. 2019).

For example, in soybeans, methylome studies have identified markers associated with root interactions with cyst nematodes (Rambani et al. 2015). In peas, variations in methylation patterns have been observed among plants propagated over extended periods through in vitro culture (Smýkal et al. 2007). Efforts to identify epigenetic variations responsible for polygenetically inherited disease resistance in peas are crucial, especially since no substantial genetic sources for resistance to many serious diseases have been identified thus far. Novel epigenetic alleles can potentially be generated via methods such as induced gene-specific DNA methylation and epigenome editing.

9.4 Concept and Prospects of Marker-Assisted Breeding for the Development of Pea Cultivars That Are Tolerant to Drought Stress

Marker-assisted breeding plays a crucial role in enhancing the drought tolerance of pea cultivars by identifying genetic markers associated with drought resistance traits. Studies on chickpea and pea have highlighted the importance of genetic markers in the development of drought-tolerant cultivars (Istanbuli et al. 2024). By utilizing high-density genetic maps and genome-wide association studies (GWASs), researchers can pinpoint specific markers linked to traits such as days to maturity, plant height, seed weight, and biomass, which are essential for drought tolerance (Thudi et al. 2021). Marker-assisted selection (MAS) expedites the breeding process by enabling the accurate and efficient selection of desired traits, leading to the development of new pea cultivars with improved drought tolerance (Kumari et al. 2024). The integration of genomics-assisted breeding strategies, such as fine

Table 9.2 QTLs and related markers responsible for drought tolerance in pea

Traits	QTL	Mapping population	Marker linked to QTL	References
Water use efficiency	10	98 F7:8 RILs developed from P665 and cv. Messire	A6, AA175, AC74, AD57, AB141, AB64, Psblox2, PsAAP2_SNP4, and DipeptIV_SNP1	Iglesias-García et al. (2015)

Table 9.3 Markers linked to different drought tolerance traits in leguminous crops

Crops	Traits	Marker type	Marker name/ identifier	References
Chickpea	Associated with root length under drought stress	SSR	AA35	Hamwieh et al. (2013)
	Linked to biomass production in drought conditions	SSR	AB6	Hamwieh et al. (2013)
	Related to drought tolerance traits like water use efficiency	SSR	PSAD27	Jha et al. (2017)
Pea	drought tolerance	SNP_1234	SNP_1234	Sindhu et al. (2014)
	Associated with yield stability under drought conditions	SNP_6789	SNP_6789	Sindhu et al. (2014)
	Related to physiological responses to drought	SNP_5432	SNP_5432	Deulvot et al. (2010)

mapping and transcript profiling, further enhances the identification of key genes, such as CabHLH10 in chickpea, which play a pivotal role in improving drought tolerance in pea cultivars (Thakro et al. 2023).

Advances in genomic technology, along with association mapping studies in pea via morphological and molecular markers, have established a crucial foundation for marker-assisted breeding, which involves the targeting of key agronomic traits and stress resistance (Cheng et al. 2015; Liu et al. 2017). In pea, numerous studies have used QTL mapping with various molecular markers to identify genetic loci associated with key agronomic traits and abiotic stress resistance, such as drought (Rubiales et al. 2015; Jacob et al. 2016). An experiment conducted by Iglesias-García et al. (2015) identified QTLs underlying drought tolerance in pea (Table 9.2).

In the context of *Pisum sativum* (pea), limited research has been conducted on marker-assisted breeding (MAB) for drought tolerance. In leguminous crops such as chickpea (*Cicer arietinum*) and cowpea (*Vigna unguiculata*), significant strides have been made in identifying QTLs linked to drought tolerance. These advancements offer valuable insights and resources that can be leveraged to increase drought tolerance in peas (Table 9.3). The identification and application of quantitative trait loci (QTLs) and associated markers in crop breeding have revolutionized the development of stress-tolerant varieties.

9.5 Challenges in Developing Drought-Tolerant Pea Cultivars Via Marker-Assisted Breeding (MAB)

Pea (*Pisum sativum* L.) is an important cool season legume crop that is valuable for global food security. However, alongside such productivity, it is under the threat of drought due to global climate change (Xu et al. 2019). Marker-assisted breeding (MAB), a potential solution for developing drought-tolerant pea cultivars, involves the use of genetic information with traditional crossbreeding techniques (Assefa et al. 2019). This approach has considerable potential, but several problems need to be solved before we obtain the most out of the MABs.

9.5.1 Complexities of the Drought Tolerance trait

Peas are polygenic for drought tolerance because of the interaction of a number of genetic elements with environmental conditions (Gondalia et al. 2022). In pea, multiple genes are located on multiple chromosomes, which is why it is very difficult to transfer characters in the next generation, resulting in difficulties in QTL mapping. For these reasons, comprehensive genomic studies in combination with deep phenotypic analyses across numerous drought intensities and durations are needed.

Recent studies have identified a number of genes that might be associated with drought tolerance in peas (Bohra et al. 2021), including potential DREB transcription factors. Peas react by activating drought adaptation genes through complicated genetic pathways and associated physiological processes.

9.5.2 Disruptions of the ABA Signaling Pathway

Disruptions in ABA signaling pathways have profound effects on many physiological and biochemical processes in peas (*Pisum sativum* L.). The important plant stress hormone abscisic acid (ABA) can impact drought tolerance by affecting stomatal regulation, osmotic adjustment, gene expression, and overall stress tolerance (Al-Quraan et al. 2021). Ali et al. (2020) also reported that stomatal closure, osmotic adjustment, and gene expression can be affected by disruptions in ABA signaling. Disruptions can make storing osmoprotectants more difficult for plants, reducing their defenses against dehydration and maintaining turgor under drought stress (Rajasheker et al. 2019). Dysregulated gene expression can compromise the ability of plants to trigger stress responses. Understanding these abnormalities via marker-assisted breeding (MAB) can help create peas that are more resilient to water constraints (Gondalia et al. 2022).

9.5.3 Phenotypic and Genetic Variance

The pea germplasm displays high levels of genetic diversity for drought tolerance, and an important resource based on this approach might be useful in breeding programs. However, the proper exploitation of this genetic diversity via MAB demands accurate phenotyping and identification of efficient molecular markers for predicting drought tolerance under a variety of environmental conditions across genetic backgrounds. For example, in a study by Mousavi-Derazmahalleh et al. (2018), genomic analysis of various pea accessions revealed correlations between 935 significant SNPs and yield-related traits mentioned earlier. Mousavi-Derazmahalleh et al. (2019) demonstrated the existence of genetic clusters specific to variable levels of drought resistance. This genetic heterogeneity underlines the importance of selecting appropriate markers and genetic backgrounds tailored to environmental modalities.

9.5.4 Verification and Accessibility of Molecular Markers

The generation of robust molecular markers linked to drought tolerance genes is a prerequisite for the deployment of MABs. The goal is to find and validate markers that can accurately predict phenotypes of drought tolerance in different genetic backgrounds or environments.

For example, Jaganathan et al. (2020) highlighted the importance of SNP markers related to candidate genes for drought tolerance in peas. The validation of crop-derived markers across diverse pea lines and environments has the capacity to fast-track breeding for drought resistance in this legume.

9.5.5 Combining Genetic Information

Optimally, the ATA would involve an integrated platform of genetic data coupled with field-based phenotypic assessments for effective MABs. Collaboration among molecular biologists, breeders, and bioinformaticians is needed to interpret complex genomic data and translate them into practical breeding strategies. For example, by combining genome-wide association study (GWAS) results with physiological and agronomic data, the authors identified candidate genes and markers that should be validated through functional studies to launch breeding programs (Thudi et al. 2021).

9.5.6 Resource and Time Consumption

MAB involves time- and resource-consuming concerns in both the essence of marker discovery and validation for cultivar multiplication and field testing across breeding stages (Bohra et al. 2014). Persistent commitment and interdisciplinary collaboration are needed to overcome these problems to achieve lasting improvements in pea drought resistance.

For example, Bohra et al. (2014) highlighted the iterative cycle of MABs in pulse crops and the importance of continuing with the validation of markers under different agroecological conditions. The use of an iterative method can accelerate the breeding cycle further while still delivering reliable and stable cultivars.

9.5.7 Environment and System Interaction

According to Kant et al. (2017), genotype–environment interactions are typically highly expressed in the drought tolerance traits of peas and, therefore, require extensive testing and validation across agroclimatic range sites. The produced cultivars would only be effectively planted by farmers if they could be stabilized and were able to tolerate a cross range of drought conditions. For example, Kant et al. (2017) conducted long-range sectional learning on numerous websites concerning the function of drought-tolerant pea lines. The evaluation of yield and agronomic traits under varying moisture stress conditions helped to identify good-performing lines in better environments across years to target promising lines with consistent performance. These results highlight the importance of environmental testing for the breeding of tolerant cultivars.

9.5.8 Societal Acceptance and Governance

There must be a balance between meeting legal requirements and winning over the heart, as key stakeholders such as farmers, consumers, and legislators are required for successful commercialization of new cultivars produced by MAB (Wang et al. 2016). Very early in the breeding process, traits that address public expectations and regulatory considerations are needed to simplify the uptake of improved cultivars. For example, Wang et al. (2016) described a body of law that involves adding genetically modified traits to crops and saved seeds, specifically concerning how safety assessments are performed and the role of public input. Proactively addressing these issues by breeders can help speed up licensing and ensure market adoption if drought-tolerant pea cultivars.

9.6 Conclusion

In global agriculture, the importance of developing resilient crop varieties cannot be overemphasized, as climate change continues to pose a major challenge. The combination of molecular tools with traditional breeding approaches will enable scientists and breeders to address present-day challenges, such as drought, in the development of future resilient high-yielding pea varieties. Marker-assisted breeding (MAB) is a revolutionary method of genetic enhancement in pea, specifically for the production of drought-resistant varieties. By using QTL mapping, molecular markers, and advances in genomic technologies, MAB has made it possible to identify specific gene loci for important agronomic traits and stress resistance. The use of MABs in many crop species has potential for coping with climate change and water shortages. In pea, MAB offers a sound position for breeding programs aimed at improving drought tolerance, ensuring food security, and enhancing sustainable agricultural practices. As research progresses, further combining state-of-the-art biotechnologies with traditional breeding methods will accelerate the development of drought-tolerant pea varieties that are capable of thriving in diverse environmental conditions.

References

Ahanger MA, Morad-Talab N, Abd-Allah EF, Ahmad P, Hajiboland R (2016) Plant growth under drought stress: significance of mineral nutrients. In: Water stress and crop plants: a sustainable approach, vol 2, pp 649–668. https://doi.org/10.1002/9781119054450.ch37

Ali Q, Haider MZ, Iftikhar W, Jamil S, Tariq Javed M, Noman A, Perveen R (2016) Drought tolerance potential of Vigna mungo L. lines as deciphered by modulated growth, antioxidant defense, and nutrient acquisition patterns. Rev Bras Bot 39:801–812

Ali S, Hayat K, Iqbal A, Xie L (2020) Implications of abscisic acid in the drought stress tolerance of plants. Agronomy 10(9):1323

Alonso C, Medrano M, Perez R, Canto A, ParraTabla V, Herrera CM (2019) Interspecific variation across angiosperms in global DNA methylation: phylogeny, ecology and plant features in tropical and Mediterranean communities. New Phytol 224:949–960. https://doi.org/10.1111/nph.16046

Al-Quraan NA, Al-Ajlouni ZI, Qawasma NF (2021) Physiological and biochemical characterization of the GABA shunt pathway in pea (*Pisum sativum* L.) seedlings under drought stress. Horticulturae 7(6):125

Anjum SA, Ashraf U, Zohaib A, Tanveer M, Naeem M, Ali I et al (2017) Growth and development responses of crop plants under drought stress: a review. Zemdirbyste 104:267

Annicchiarico P, Nazzicari N, Laouar M, Thami-Alami I, Romani M, Pecetti L (2020) Development and proof-of-concept application of genome-enabled selection for pea grain yield under severe terminal drought. Int J Mol Sci 21(7):2414. https://doi.org/10.3390/ijms21072414

Arafa SA, Attia KA, Niedbała G, Piekutowska M, Alamery S, Abdelaal K et al (2021) Seed priming boost adaptation in pea plants under drought stress. Plan Theory 10(10):2201

Assefa T, Assibi Mahama A, Brown AV, Cannon EK, Rubyogo JC, Rao IM et al (2019) A review of breeding objectives, genomic resources, and marker-assisted methods in common bean (Phaseolus vulgaris L.). Mol Breed 39:1–23

Bagheri M, Santos CS, Rubiales D, Vasconcelos MW (2023) Challenges in pea breeding for tolerance to drought: status and prospects. Ann Appl Biol 183(2):108–120

Barakate A, Stephens J (2016) An overview of CRISPR-based tools and their improvements: new opportunities in understanding plant pathogen interactions for better crop protection. Front Plant Sci 7:765. https://doi.org/10.3389/fpls.2016.00765

Barzana G, Rios JJ, Lopez-Zaplana A, Nicolas-Espinosa J, Yepes-Molina L, Garcia-Ibañez P et al (2021) Interrelations of nutrient and water transporters in plants under abiotic stress. Physiol Plant 171:595–619. https://doi.org/10.1111/ppl.13206

Beniwal A, Savita VS, Punia D (2024) Nutrient-enriched wheat chapatti with fresh pea shells (Pisum sativum l.): a comprehensive quality assessment. Pantnagar J Res 22:1

Blum A (2010) Plant breeding for water-limited environments. Springer Science & Business Media, 255 pages

Bohra A, Pandey MK, Jha UC, Singh B, Singh IP, Datta D, Varshney RK (2014) Genomics-assisted breeding in four major pulse crops of developing countries: present status and prospects. Theor Appl Genet 129(12):2419–2443

Bohra A, Kilian B, Sivasankar S, Caccamo M, Mba C, McCouch SR, Varshney RK (2021) Reap the crop wild relatives for breeding future crops. Trends in Biotechnology 40(4):412–431.

Carvalho M, Castro I, Moutinho-Pereira J, Correia C, Egea-Cortines M, Matos M et al (2019) Evaluating stress responses in cowpea under drought stress. J Plant Physiol 241:153001

Castillejo M-Á, Iglesias-García R, Wienkoop S, Rubiales D (2016) Label-free quantitative proteomic analysis of tolerance to drought in *Pisum sativum*. Proteomics 16:2776–2787. https://doi.org/10.1002/pmic.201600156

Cheng P, Holdsworth W, Ma Y, Coyne CJ, Mazourek M, Grusak MA et al (2015) Association mapping of agronomic and quality traits in USDA pea single-plant collection. Mol Breed 35:1–13

Das A, Parihar AK, Saxena D, Singh D, Singha KD, Kushwaha KPS et al (2019) Deciphering genotype-by-environment interaction for targeting test environments and rust resistant genotypes in field pea (*Pisum sativum* L.). Front Plant Sci 10:825. https://doi.org/10.3389/fpls.2019.00825

Deulvot C, Charrel H, Marty A, Jacquin F, Donnadieu C, Lejeune-Hénaut I et al (2010) Highly-multiplexed SNP genotyping for genetic mapping and germplasm diversity studies in pea. BMC Genomics 11:1–10

Dhaliwal SK, Salaria P, Kaushik P (2021) Pea seed proteins: a nutritional and nutraceutical update. In: Grain and seed proteins functionality, pp 105–114. https://doi.org/10.5772/intechopen.95323

Espinas NA, Saze H, Saijo Y (2016) Epigenetic control of defense signaling and priming in plants. Front Plant Sci 7:1201. https://doi.org/10.3389/fpls.2016.01201

Farooq M, Wahid A, Kobayashi NSMA, Fujita DBSMA, Basra SM (2009) Plant drought stress: effects, mechanisms and management. In: Sustainable agriculture. Springer Netherlands, Dordrecht, pp 153–188

Farooq M, Hussain M, Wahid A, Siddique KHM (2012) Drought stress in plants: an overview. In: Plant responses to drought stress: from morphological to molecular features, pp 1–33. https://doi.org/10.1007/978-3-642-32653-0_1

Fatima S, Aslam R, Shahzadi K, Awan H, Nadeem M, Shabir M, Shafiq Y, Hamid M (2024) Effect of drought stress at growth and development of pea (*Pisum sativum* L.). Asian J Res Crop Sci 9(1):76–80. https://doi.org/10.9734/ajrcs/2024/v9i1247

Flexas J, Bota J, Cifre J, Mariano Escalona J, Galmés J, Gulías J et al (2004) Understanding down-regulation of photosynthesis under water stress: future prospects and searching for physiological tools for irrigation management. Ann Appl Biol 144(3):273–283

Food and Agriculture Organization of the United Nations (2023) With major processing by Our World in Data. Pea production – FAO [dataset]. Food and Agriculture Organization of the United Nations, Production: crops and livestock products [original data]. Retrieved July 16, 2024 from https://ourworldindata.org/grapher/pea-production

Gill SS, Tuteja N (2010) Reactive oxygen species and antioxidant machinery in abiotic stress tolerance in crop plants. Plant Physiol Biochem 48(12):909–930

Gondalia N, Vashi R, Barot V, Sharma F, Anishkumar PK, Chatterjee M et al (2022) Genomic designing for abiotic stress tolerance in pea (Pisum Sativum L.). In: Genomic designing for abiotic stress resistant pulse crops. Springer International Publishing, Cham, pp 45–113

Gregory PJ, Hester R, Harrison R (2012) Soils and food security: challenges and opportunities. In: Soils and food security. RSC Publishing London, Cambridge, pp 1–30

Hamwieh A, Imtiaz M, Malhotra RS (2013) Multi-environment QTL analyses for drought-related traits in a recombinant inbred population of chickpea (Cicer arientinum L.). Theor Appl Genet 126:1025–1038

Iglesias-García R, Prats E, Fondevilla S, Satovic Z, Rubiales D (2015) Quantitative trait loci associated to drought adaptation in pea (Pisum sativum L.). Plant Mol Biol Report 33:1768–1778. https://doi.org/10.1007/s11105-015-0872-z

Impa SM, Nadaradjan S, Jagadish SVK (2012) Drought stress induced reactive oxygen species and anti-oxidants in plants. In: Abiotic stress responses in plants: metabolism, productivity and sustainability, pp 131–147

Iqbal Z, Sattar MN, Shafiq M (2016) CRISPR/Cas9: a tool to circumscribe cotton leaf curl disease. Front Plant Sci 7:475. https://doi.org/10.3389/fpls.2016.00475

Istanbuli T, Nassar AE, Abd El-Maksoud MM, Tawkaz S, Alsamman AM, Hamwieh A (2024) Genome-wide association study reveals SNP markers controlling drought tolerance and related agronomic traits in chickpea across multiple environments. Front Plant Sci 15:1260690

Jacob C, Carrasco B, Schwember AR (2016) Advances in breeding and biotechnology of legume crops. Plant Cell Tissue Organ Cult 127:561–584

Jaganathan D, Bohra A, Thudi M, Varshney RK (2020) Fine mapping and gene cloning in the post-NGS era: advances and prospects. Theor Appl Genet 133(5):1791–1810

Jha UC, Bohra A, Jha R (2017) Breeding approaches and genomics technologies to increase crop yield under low-temperature stress. Plant Cell Rep 36:1–35

Jiang GL (2013) Molecular markers and marker-assisted breeding in plants. In: Plant breeding from laboratories to fields, vol 3, pp 45–83

Kant S, Kant P, Raveh E, Barak S (2017) Evidence for plant-derived xenomiRs based on a large-scale analysis of public small RNA sequencing data from human samples. PLoS One 12(5):e0177448

Kobraee S, Shamsi K (2011) Evaluation of soybean yield under drought stress by path analysis. Australian Journal of Basic and Applied Sciences 5(10):pp.890–895

Kumar S, Srivastava RL, Singh R (2006) Combining ability for yield and its component traits in field pea. Ind J Pulses Res 19:173–175

Kumari M, Dubey AK, Kumar R, Kumar A (2024) Marker-assisted selection in plant breeding for stress tolerance. In: Improving stress resilience in plants. Academic Press, pp 371–387

Latutrie M, Gourcilleau D, Pujol B (2019) Epigenetic variation for agronomic improvement: an opportunity for vegetatively propagated crops. Am J Bot 106:1281. https://doi.org/10.1002/ajb2.1357

Liu N, Zhang G, Xu S, Mao W, Hu Q, Gong Y (2015) Comparative transcriptomic analyses of vegetable and grain pea (Pisum sativum L.) seed development. Front Plant Sci 6:1039. https://doi.org/10.3389/fpls.2015.01039

Liu R, Fang L, Yang T, Zhang X, Hu J, Zhang H et al (2017) Marker-trait association analysis of frost tolerance of 672 worldwide pea (Pisum sativum L.) collections. Sci Rep 7(1):5919

Ma C, Yan J, Li W, Wang Y, McClements DJ, Liu X, Liu F (2024) Enhanced printability of food-grade edible inks: emulsions formulated with modified pea protein and sodium alginate. Food Hydrocoll 152:109946

Mahieu S, Germon F, Aveline A, Hauggaard-Nielsen H, Ambus P, Jensen ES (2009) The influence of water stress on biomass and N accumulation, N partitioning between above and below ground parts and on N rhizodeposition during reproductive growth of pea (*Pisum sativum* L.). Soil Biol Biochem 41(2):380–387

Maurya V, Singh N, Sharma A, Kumar R (2023) DNA Markers-Assisted Crop Improvement for Biotic and Abiotic Stresses in Legumes. In: Kumar, N. (eds) Molecular Marker Techniques. Springer, Singapore. https://doi.org/10.1007/978-981-99-1612-2_10

Moran JF, Becana M, Iturbe-Ormaetxe I, Frechilla S, Klucas RV, Aparicio-Tejo P (1994) Drought induces oxidative stress in pea plants. Planta 194:346–352

Mousavi-Derazmahalleh M, Bayer PE, Nevado B, Hurgobin B, Filatov D, Kilian A et al (2018) Exploring the genetic and adaptive diversity of a pan-Mediterranean crop wild relative: narrow-leafed lupin. Theoretical and Applied Genetics 131:887–901.

Mousavi-Derazmahalleh M, Bayer PE, Hane JK, Valliyodan B, Nguyen HT, Nelson MN, Varshney RK (2019) Adapting legume crops to climate change using genomic approaches. Plant Cell Environ 43(8):1713–1731

Mushtaq M, Sakina A, Wani SH, Shikari AB, Tripathi P, Zaid A (2019) Harnessing genome editing techniques to engineer disease resistance in plants. Front Plant Sci 10:550. https://doi.org/10.3389/fpls.2019.00550

Nadeem MA, Nawaz MA, Shahid MQ, Doğan Y, Comertpay G, Yıldız M et al (2018) DNA molecular markers in plant breeding: current status and recent advancements in genomic selection and genome editing. Biotechnol Biotechnol Equip 32(2):261–285

Nadeem M, Li J, Yahya M, Sher A, Ma C, Wang X, Qiu L (2019) Research progress and perspective on drought stress in legumes: a review. Int J Mol Sci 20(10):2541

Nikam BR, Aggarwal SP, Thakur PK, Garg V, Roy S, Chouksey A, Chauhan P (2020) Assessment of early season agricultural drought using remote sensing. Int Arch Photogramm Remote Sens Spat Inf Sci 43:1691–1695

Noman A, Ali S, Naheed F, Ali Q, Farid M, Rizwan M, Irshad MK (2015) Foliar application of ascorbate enhances the physiological and biochemical attributes of maize (*Zea mays* L.) cultivars under drought stress. Arch Agron Soil Sci 61(12):1659–1672

Okçu G, Kaya MD, Atak M (2005) Effects of salt and drought stresses on germination and seedling growth of pea (*Pisum sativum* L.). Turk J Agric For 29:237–242

Pavan S, Delvento C, Nazzicari N, Ferrari B, D'Agostino N, Taranto F et al (2022) Merging genotyping-by-sequencing data from two ex situ collections provides insights on the pea evolutionary history. Hortic Res 9:uhab062

Pirasteh-Anosheh H, Saed-Moucheshi A, Pakniyat H, Pessarakli M (2016) Stomatal responses to drought stress. In: Water stress and crop plants: a sustainable approach, vol 1. John Wiley & Sons, pp 24–40

Potopová V, Zahradníček P, Štěpánek P, Türkott L, Farda A, Soukup J (2017) The impacts of key adverse weather events on the field-grown vegetable yield variability in The Czech Republic from 1961 to 2014. Int J Climatol 37(3):1648–1664

Prudent M, Vernoud V, Girodet S, Salon C (2016) How nitrogen fixation is modulated in response to different water availability levels and during recovery: a structural and functional study at the whole plant level. Plant Soil 399:1–12. https://doi.org/10.1007/s11104-015-2674-3

Rajasheker G, Jawahar G, Jalaja N, Kumar SA, Kumari PH, Punita DL et al (2019) Role and regulation of osmolytes and ABA interaction in salt and drought stress tolerance. In: Plant signaling molecules. Woodhead Publishing, pp 417–436

Rambani A, Rice JH, Liu J, Lane T, Ranjan P, Mazarei M et al (2015) The methylome of soybean roots during the compatible interaction with the soybean cyst nematode. Plant Physiol 168(4):1364–1377. https://doi.org/10.1104/pp.15.00826

Ramirez-Prado JS, Abulfaraj AA, Rayapuram N, Benhamed M, Hirt H (2018) Plant immunity: from signaling to epigenetic control of defense. Trends Plant Sci 23(9):833–844. https://doi.org/10.1016/j.tplants.2018.06.004

Ranjan A, Sinha R, Singla-Pareek SL, Pareek A, Singh AK (2022) Shaping the root system architecture in plants for adaptation to drought stress. Physiol Plant 174(2):e13651

Ren Y, Singh RP, Basnet BR, Lan CX, Huerta-Espino J, Lagudah ES et al (2017) Identification and mapping of adult plant resistance loci to leaf rust and stripe rust in common wheat cultivar Kundan. Plant Dis 101:456–463. https://doi.org/10.1094/PDIS-06-16-0890-re

Ribaut JM, De Vicente MC, Delannay X (2010) Molecular breeding in developing countries: challenges and perspectives. Curr Opin Plant Biol 13(2):213–218

Rubiales D, Fondevilla S, Chen W, Gentzbittel L, Higgins TJ, Castillejo MA et al (2015) Achievements and challenges in legume breeding for pest and disease resistance. Crit Rev Plant Sci 34(1–3):195–236

Sarıkaya MF, İleri O, Erkovan Ş, Erkovan Hİ, Koç A (2023) Growing forage pea (Pisum arvense L.) for hay: different sowing dates and plant densities in Central Anatolia. Atatürk Üniversitesi Ziraat Fakültesi Dergisi 54(2):75–80

Sarkar S, Khatun M, Era FM, Islam AM, Anwar MP, Danish S et al (2021) Abiotic stresses: alteration of composition and grain quality in food legumes. Agronomy 11(11):2238

Serraj R, Hash CT, Rizvi SMH, Sharma A, Yadav RS, Bidinger FR (2005) Recent advances in marker-assisted selection for drought tolerance in pearl millet. Plant Prod Sci 8(3):334–337

Sharma M, Gupta SK, Deeba F, Pandey V (2017) Effects of reactive oxygen species on crop productivity: an overview. In: Reactive oxygen species in plants: boon or bane-revisiting the role of ROS. John Wiley & Sons, pp 117–136

Shumilina J, Gorbach D, Popova V, Tsarev A, Kuznetsova A, Grashina M, Dorn M, Lukasheva E (2021) Protein glycation and drought response of pea (Pisum sativum L.) root nodule proteome: a proteomics approach. Bio. Comm 66(3):210–224. https://doi.org/10.21638/spbu03.2021.303

Shoaib M, Banerjee BP, Hayden M, Kant S (2022) Roots' drought adaptive traits in crop improvement. Plan Theory 11(17):2256

Sindhu A, Ramsay L, Sanderson LA, Stonehouse R, Li R, Condie J et al (2014) Gene-based SNP discovery and genetic mapping in pea. Theor Appl Genet 127:2225–2241

Singh VK, Khan AW, Saxena RK, Kumar V, Kale SM, Sinha P et al (2016) Next-generation sequencing for identification of candidate genes for *Fusarium* wilt and sterility mosaic disease in pigeonpea (*Cajanu scajan*). Plant Biotechnol J 14:1183–1194. https://doi.org/10.1111/pbi.12470

Smýkal P, Valledor L, Rodriguez R, Griga M (2007) Assessment of genetic and epigenetic stability in long-term in vitro shoot culture of pea (*Pisum sativum* L.). Plant Cell Rep 26:1985–1998. https://doi.org/10.1007/s00299-007-0413-9

Tafesse EG, Gali KK, Lachagari VR, Bueckert R, Warkentin TD (2020) Genome-wide association mapping for heat stress responsive traits in field pea. Int J Mol Sci 21(6):2043

Tafesse EG, Gali KK, Lachagari VR, Bueckert R, Warkentin TD (2021) Genome-wide association mapping for heat and drought adaptive traits in pea. Genes 12(12):1897

Thakro V, Malik N, Basu U, Srivastava R, Narnoliya L, Daware A et al (2023) A superior gene allele involved in abscisic acid signaling enhances drought tolerance and yield in chickpea. Plant Physiol 191(3):1884–1912

Thakur B, Sharma S, Sharma I, Sharma P, Zargar SM (2018) Diversity analysis of pea genotypes using RAPD markers. Legume Res 41(2):196–201

Thudi M, Chitikineni A, Liu X, He W, Roorkiwal M, Yang W, Varshney RK (2021) Recent breeding strategies to enhance drought adaptation in grain legumes. Theor Appl Genet 134(2):609–632

Wang X, Wang H, Liu S, Ferjani A, Li J, Yan J et al (2016) Genetic variation in ZmVPP1 contributes to drought tolerance in maize seedlings. Nat Genet 48(10):1233–1241

Wu X, Li N, Hao J, Hu J, Zhang X, Blair MW (2017) Genetic diversity of Chinese and global pea (Pisum sativum L.) collections. Crop Sci 57(3):1574–1584

Wu DT, Li WX, Wan JJ, Hu YC, Gan RY, Zou L (2023) A comprehensive review of pea (*Pisum sativum* L.): chemical composition, processing, health benefits, and food applications. Food Secur 12(13):2527

Xu Y (2010) Molecular plant breeding. CAB International, Wallingford, Cambridge

Xu C, McDowell NG, Fisher RA, Wei L, Sevanto S, Christoffersen BO et al (2019) Increasing impacts of extreme droughts on vegetation productivity under climate change. Nat Clim Chang 9(12):948–953

Zhao T, Su W, Qin Y, Wang L, Kang Y (2020) Phenotypic diversity of pea (*Pisum sativum* L.) varieties and the polyphenols, flavonoids, and antioxidant activity of their seeds. Ciência Rural 50:e20190196

Zorin EA, Sulima AS, Zhernakov AI, Kuzmina DO, Rakova VA, Kliukova MS, Romanyuk DA et al (2023) Genomic and Transcriptomic Analysis of Pea (Pisum sativum L.) Breeding Line 'Triumph'with High Symbiotic Responsivity. Plants, 13(1):78. https://doi.org/10.3390/plants13010078

Marker-Assisted Breeding in Faba Bean for Drought Tolerance

10

Zakir Ullah, Javed Iqbal, Banzeer Ahsan Abbasi, Shumaila Ijaz, Akhtar Munir, Tabassum Yaseen, Muhammad Majeed, Noshin Ilyas, Amin Fathi, Sobia Kanwal, and Tariq Mahmood

Abstract

Faba bean (*Vicia faba* L.) is a vital legume crop with high nutritional value and significant economic importance worldwide. However, its production is severely constrained by environmental stresses, particularly drought, which is becoming increasingly prevalent due to climate change. Developing faba bean varieties with enhanced drought tolerance is imperative to ensure food security and sustainable agriculture. Marker-assisted breeding (MAB) has emerged as a powerful tool to accelerate the selection process for complex traits like drought tolerance in faba beans. By identifying and utilizing genetic markers linked to drought-responsive genes, MAB enables the efficient selection of desirable characteristics at the molecular level, facilitating the development of drought-tolerant varieties with improved yield stability and resilience under water-limited conditions. This chapter provides a comprehensive overview of the principles and methodologies involved in marker-assisted breeding for drought tolerance in

Zakir Ullah and Javed Iqbal contributed equally with all other contributors.

Z. Ullah · T. Mahmood
Department of Plant Sciences, Faculty of Biological Sciences, Quaid-i-Azam University Islamabad, Islamabad, Pakistan

J. Iqbal (✉) · T. Yaseen
Department of Botany, Bacha Khan University, Charsadda, Khyber Pakhtunkhwa, Pakistan

B. A. Abbasi
Department of Botany, Rawalpindi Women University, Rawalpindi, Pakistan

S. Ijaz
School of Biomedical Engineering, Shenzhen University Medical School, Shenzhen University, Shenzhen, China

© The Author(s), under exclusive license to Springer Nature Singapore Pte Ltd. 2025

M. A. Nadeem et al. (eds.), *Marker-Assisted Breeding in Legumes for Drought Tolerance*, https://doi.org/10.1007/978-981-96-4112-3_10

faba beans. It discusses the identification of molecular markers associated with drought-related traits through various genomic approaches, such as quantitative trait loci (QTL) mapping and genome-wide association studies (GWAS). Furthermore, it explores integrating genomic information with traditional breeding techniques to enhance the efficiency and precision of selection for drought tolerance. Case studies highlighting successful applications of MAB in faba bean breeding programs across different regions are presented, demonstrating this approach's practical utility and impact in improving drought resilience and enhancing crop productivity. Overall, this chapter underscores the potential of marker-assisted breeding as a valuable strategy for accelerating the development of drought-tolerant faba bean varieties, thereby contributing to the resilience and sustainability of faba bean production systems in the face of climate change.

Keywords

Marker-assisted breeding · Drought tolerance · Faba bean · Genetic markers · Crop improvement · Molecular breeding · Abiotic stress adaptation

10.1 Introduction

The broad bean, often referred as the faba bean, fava bean, or vetch, is *Vicia faba*, a species of vetch that is a member of the Fabaceae family of peas and beans (Fig. 10.1) (İlker et al. 2023; Torres et al. 2010). It is extensively cultivated both for human consumption and as a cover crop. Field beans, tic beans, or tick beans are varieties with smaller, tougher seeds that are frequently used as animal feed (Schneider et al. 1997a, b). The officially recognized name for the horse bean variety is *Vicia faba* var. equina Pers. The cuisines of North Africa, East Asia, South America, and Northern and Southern Europe all make extensive use of this legume. On the other hand, some people have favism, a hemolytic reaction to eating broad beans that is linked to a metabolic condition known as G6PDD. If not, the beans can be cooked or eaten raw once the outer seed coat has been removed. The exterior seed coat of immature plants is edible, and in very immature plants, the seed pod itself is edible (Torres et al. 2010; Gutiérrez et al. 2023; Ali et al. 2016).

A. Munir
Department of Chemistry, Quaid-I-Azam University Islamabad, Islamabad, Pakistan

Department of Chemistry, University of Sialkot, Sialkot, Pakistan

M. Majeed
Department of Botany, University of Gujrat, Punjab, Pakistan

N. Ilyas
Department of Botany, PMAS Arid Agriculture University, Rawalpindi, Pakistan

A. Fathi
Department of Agronomy, Ayatollah Amoli Branch, Islamic Azad University, Amol, Iran

S. Kanwal
Department of Biology and Environmental Sciences, Allama Iqbal Open University, Islamabad, Pakistan

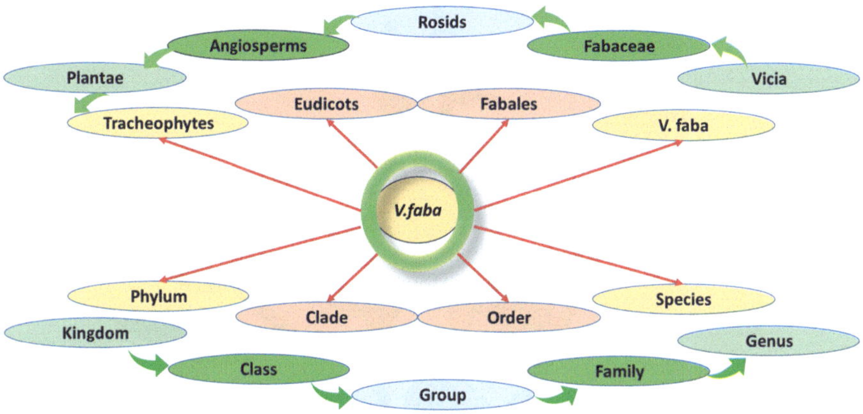

Fig. 10.1 The scientific description of *Vicia faba*

10.1.1 Description

10.1.1.1 Worldwide Broad Bean Production

Vicia faba is an upright annual plant with two to four square-stemmed branches that may grow to a height of 0.5–1.8 meters (1 ft. 8 to 5 ft. 11 in). The leaves are a gray-green (glaucous) color and are 10–25 centimeters (4–10 in) in length. They are pinnate, having two to seven leaflets. The leaves don't have climbing tendrils like those of most other vetches (Ali et al. 2016). The blooms have five petals and are 1–2.5 millimeters (0.5–1 in) in length. White makes up the keel petals, regular petals, and wing petals with a black patch (actual black, not the deep purple or blue found in many "black" colorings). There are also wide beans with crimson flowers that were just rescued from extinction (Lavania et al. 2015). Bees and other pollinators are drawn to the blossoms by their potent sweet aroma (Sabry et al. 2022). The fruit is a large, leathery pod with a densely downy surface that is green at first and turns dark blackish-brown as it ages. Pods of the natural species are 5–10 cm (2–4 in) long and 1 cm in diameter; many modern cultivars grown for food have pods that are 15–25 cm (6–10 in) long and 2–3 cm thick. Bean pods contain three to eight seeds apiece. In the wild plant, these seeds are round to oval and have a diameter of 5–10 mm; however, in food cultivars, they are often flattened and have dimensions of up to 20–25 mm, 15 mm in width, and 5–10 mm in thickness. With 6 homologous pairs making up its diploid chromosomal count of 12, 1 pair is metacentric, while the other 5 pairs are acrocentric (Fig. 10.2) (Singh et al. 2012; Maga 2021).

10.1.1.2 History and Cultivation

Broad Beans in the Pod

Since they are among the first cultivated plants and among the simplest to grow, broad beans have a long history in Old World agriculture. But their origin is unclear, and their wild ancestor is still unknown (Singh et al. 2012; Maga 2021). Three

Fig. 10.2 Diagrammatic depiction of the fava bean's different organs; the diagram is not scaled. (Source: Saikat Gantait's unpublished diagram)

Neolithic sites in Lower Galilee, Israel, have yielded charred remnants of faba beans (Caracuta, V(2017). These bones' radiocarbon date indicates that the crop may have been domesticated as early as 8250 BCE. Since broad beans are a legume that fixes nitrogen in the soil and can overwinter, they are still commonly used as a cover crop to reduce erosion. The broad bean is incredibly resilient and can withstand severe weather conditions (Carrillo-Perdomo et al. 2020). Though they prefer rich loam soils, wide beans may be cultivated on clay and high salt soils, unlike most other legumes. Large-seeded cultivars farmed for human use are referred to as "broad beans" in much of the English-speaking world, while smaller, harder-seeded cultivars used for animal feed are referred to as "horse beans" and "field beans" (Carrillo-Perdomo et al. 2020). Though their stronger flavor is desired in some human meals, such as falafel, these smaller seeds are more like the wild variety. While "fava bean" (from the Italian: fava for the bean) is used in other English-speaking countries like Canada and the USA, "broad bean" is the most used term in Commonwealth nations, including the UK, Australia, and New Zealand (Alghamdi et al. 2012).

10.1.2 Impact of Drought Stress

In both natural and agricultural environments, plants grow and develop in response to a variety of environmental stressors (Ahmad et al. 2022). Drought is one of these environmental disturbances that greatly affects plant productivity (Alghamdi et al. 2012; Ahmad et al. 2022). Water makes up between 80 and 95 percent of a plant's fresh biomass, and it is essential for many physiological functions, such as several facets of plant growth, development, and metabolism (Ahmad et al. 2022). As a result, some believe that drought, particularly in areas where it is prone to occur, is the main environmental stressor for different plants (Sallam and Ul-Allah 2019). It is considered the biggest danger to future global food security and has already caused severe famines (Ahmad et al. 2022). The loss of water supplies and the rise in food demand brought on by the worrisome global population growth compound the consequences of drought on agriculture. Evapotranspiration, the water-holding capacity around the rhizosphere, and the unequal and unexpected distribution of rainfall are some of the variables that contribute to the unpredictable character of drought (Alghamdi et al. 2012). Furthermore, there are instances where plants cannot absorb water from the soil while having enough moisture in the root zone; this is referred to as physiological drought or pseudo-drought (Ahmad et al. 2022).

Many molecular, biochemical, physiological, morphological, and ecological features and activities in plants are hampered by drought stress conditions (Ahmad et al. 2022). Water-deficient conditions hurt plant output and quality. Key variables impacting plant responses to drought include age, species, growth stages, and the intensity and length of the drought (Sallam and Ul-Allah 2019). Different plant species have different drought resistance mechanisms. As a result, to adapt to unfavorable environmental circumstances like drought, plants might decrease the number of resources they use and modify their development (Redden et al. 2014). Several molecular networks, including those about signal transduction, augment the responses of plants to drought stress. The molecular processes of plant response to drought stress also involve stomatal control in plants through increased ion transport, transcription factor activities, and abscisic acid (ABA) signaling. Water stress affects a plant's morphology, physiology, and biochemical dynamics (Redden et al. 2014; Ahmad et al. 2024a). It is necessary to increase plant resistance to drought in specific changing environmental circumstances. When physical adaptations of roots and leaves are insufficient to withstand drought, molecular signals, such as gene-coding regulatory proteins, express a multitude of other genes and signals through crosstalk under various regulatory mechanisms, therefore optimizing water-use efficiency (Ullah et al. 2024a, b, c). It will be crucial to support more research on drought-tolerant plants and the application of cost-effective and advantageous agricultural techniques to fulfill future food demand (Zong et al. 2019).

10.2 Causes of Drought Stress in Plants

Future climate change is expected to pick up speed because of the ongoing rise in atmospheric CO_2 levels and air temperature, which eventually modifies rainfall patterns and distribution (Maqbool et al. 2017). Drastic weather events, strong light, and dry winds can cause water to evaporate from soils, which might worsen an already-existing drought stress event. However, inadequate water intake from rainfall is typically the main cause of drought stress. Global climate change usually results in drought stress conditions that are pervasive across large portions of the planet. In addition to drought, salt stress is thought to be the main factor causing plants' water shortage. A few causes of drought stress are briefly discussed here (Maalouf et al. 2022).

10.2.1 Global Warming

Future projections indicate that the ongoing increase in air temperature and atmospheric CO_2 levels will accelerate global climate change, which will ultimately modify rainfall patterns and their distribution. Drastic weather events, strong light, and dry winds can cause water to evaporate from soils, which might worsen an already-existing drought stress event, even though low rainfall is often the main cause of drought stress (Maalouf et al. 2022). Global climate change usually results in drought stress conditions that are pervasive across large portions of the planet. In addition to drought, salt stress is thought to be a major contributor to plant water deficit (Arbaoui et al. 2008).

10.2.2 Rainfall Anomalies

Compared to places that are irrigated by canals, rivers, and water channels, areas whose agricultural production depends only on rainfall are predicted to experience more stress (Maalouf et al. 2022). As a result, in regions that get rain, the frequency of drought events and the annual rainfall distribution are highly associated, increasing the risk of water stress during the years. Three major human activities—industrialization, deforestation, and urbanization—influence rainfall patterns and, in turn, plant water availability by contributing to climate change (Arbaoui et al. 2008). In Pakistan, early spring and winter have unpredictable and frequent rainfall, but early fall and summer see little to no rainfall, making drier and hotter seasons more prevalent (Fouad et al. 2013). Drought stress negatively impacts plant growth and development during the summer months, especially because of the combination of greater transpiration and evaporation rates, decreased rainfall availability, and increased atmospheric water requirement for plants. Nonetheless, the distribution and intensity of rainfall both within and between years are critical in controlling plant water resources and, in most circumstances, the incidence of drought stress (Ullah et al. 2024a, b, c).

10.2.3 Shifts in Monsoon Patterns

Growing air temperatures and growing atmospheric CO_2 levels are expected to hasten global climate change, which will ultimately modify rainfall patterns and their distribution. Drought stress is usually caused by an inadequate inflow of water from rainfall (Maalouf et al. 2022). Drought conditions can, however, be made worse by evaporative water loss from soils caused by high temperatures, bright sunshine, and dry winds. In rain-fed agricultural settings, where crop production depends only on precipitation from nature, the distribution of rainfall is very important (Arbaoui et al. 2008). Comparatively speaking to places that are watered by rivers, canals, and water channels, these locations undergo a great deal of stress. Drought events are highly associated with the annual rainfall distribution in rain-fed locations, increasing the risk of water stress during specific times of the year (Fouad et al. 2013). Because they affect climate change, anthropogenic activities including industrialization, deforestation, and urbanization have a major influence on rainfall patterns and plant water availability (Ullah et al. 2024a, b, c). For example, in Pakistan, early spring and winter get unpredictable and heavier rainfall, but early fall and summer see little or no rainfall, making for drier and hotter seasons. The negative impacts of drought stress on plant development are made worse by a combination of increased atmospheric water demand, increased rates of transpiration and evaporation, and decreased availability of rainfall throughout the summer exacerbates the detrimental effects of drought stress on plant growth and development (Fouad et al. 2013; Ullah et al. 2024a, b, c).

10.2.4 Impact of Monsoon System on Rainfall and Agriculture

In many places of the world, the monsoon system plays a crucial role in providing rainfall throughout the summer months due to temperature (Ullah et al. 2024a, b, c). By the start of the twenty-second century, rain-fed areas will experience a 70% drop in summer precipitation if current trends continue. Higher rainfall is predicted to result from rising CO_2 levels, which might hurt crop productivity and result in severe flooding and financial losses in nations with dense populations (Fouad et al. 2013). Variability in monsoon rainfall alters the rhizosphere's moisture content and impacts plant production by varying the intensity, frequency, and length of rainfall. Owing to substantial changes in seasonal rainfall from dry to wet owing to monsoon shifts, two-thirds of the world's population currently confronts food insecurity. In some temperate zones, issues with water surplus and deficit may arise due to the unpredictability of rainfall patterns and recent changes in the climate (Liu et al. 2007). Crop planning, appropriate management, and sustainable crop production are the main goals of agricultural techniques that must be modified to the monsoon's behavior to meet these objectives (Ullah et al. 2024a, b, c).

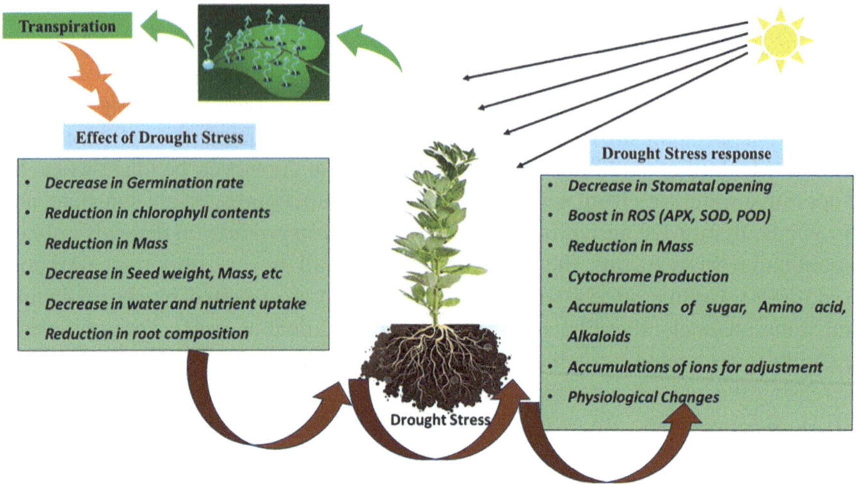

Fig. 10.3 Impact of drought stress and response of the plants

10.2.5 Drought Stress and Plant Adaptation

Plant development, growth, and yield components are all greatly impacted by drought stress, which results in severe productivity losses (Ahmad et al. 2024b). In plants such as the faba bean, it causes various physiological, biochemical, and molecular disturbances. Numerous enzymatic pathways are upset, and photosynthesis is hampered by drought stress (Kelly et al. 2003). The goal of modern tactics, molecular tools, and conventional breeding techniques is to increase plant resistance to environmental challenges. Stimulating natural plant defenses through the induction of resistance is regarded as the most effective approach among them. Techniques include foliar spraying or seed priming with antioxidants, vitamins, and osmoprotectant chemicals (Ahmad et al. 2024b). Exogenous administration of osmoprotectants and plant growth regulators has been demonstrated in studies to improve drought tolerance in crops, such as soybean, lucerne, maize, and sunflower (Kelly et al. 2003). Under drought stress, these substances support the build-up of antioxidants and suitable solutes, preserving osmotic balance and scavenging reactive oxygen species (ROS) to protect membrane structures, enzymes, and other macromolecules. Plants that receive these priming treatments are better equipped to react to biotic or abiotic stressors (Fig. 10.3) (Ahmad et al. 2024b).

10.2.6 Role of Polyamines and Non-protein Amino Acids in Drought Response

Plants respond to environmental variables through the accumulation of polyamines, such as putrescine, spermidine, and spermine, as well as non-protein amino acids like γ-aminobutyric acid (GABA), which are released in response to abiotic stressors (Kelly et al. 2003). By applying natural chemicals like GABA or its synthetic equivalent, β-aminobutyric acid (BABA) exogenously, one can induce priming that reduces oxidative damage by promoting osmotic equilibrium and increasing the activity of antioxidant enzymes. Plants treated with BABA exhibit increased abscisic acid (ABA) accumulation, a phytohormone essential for drought stress response and tolerance. Both ABA-dependent and independent processes are involved in BABA priming. Numerous plant species have demonstrated these defense mechanisms, which strengthen transcript accumulation of ABA-dependent genes and upregulate antioxidant enzymes to improve responses to drought stress. As a result, there is a decrease in water consumption during droughts, partial stomatal closure, and increased drought tolerance (Kelly et al. 2003).

10.2.7 Faba Bean and Its Global Significance

The faba bean, one of the cultivated grain legumes, is produced in large quantities and is ranked sixth in the world (Alharbi and Adhikari 2020). One of the earliest crops to be domesticated in the Middle East, it eventually made its way to Europe, Australia, the Americas, North Africa, China, India, and Central Asia. Carbonized remnants from 14,000 BC show that the faba bean has been around for a very long time, even if its wild origins are unknown (Alharbi and Adhikari 2020). Grain legumes, such as the faba bean, block the life cycles of some nematodes, weed species, and illnesses, which is a significant benefit in cropping systems based on cereals. Furthermore, the faba bean and its symbiotic rhizobacteria fix atmospheric nitrogen, contributing significantly to the leftover nitrogen in crop rotations and meeting around 80% of the plant's nitrogen requirements (Belachew et al. 2019). Rich in vitamins and protein, faba bean seeds are a healthy source of dietary fiber and energy. With an average protein level of 29%, this pulse is among the highest in terms of protein content. While young pods and seeds are used as cooked or salad vegetables, older beans are eaten as a paste or stew. Under unfavorable circumstances, whole-crop silage made from the complete plant provides a means of retaining crop value (Tiwari, J.(2022).

Challenges and Strategies for Improving Drought Tolerance
Faba beans are sensitive to water deficit, with significantly higher yields in irrigated agriculture compared to rainfed systems. Improving drought tolerance is crucial for enhancing yield performance and stability. Plant breeding programs face hurdles in the form of mixed breeding systems and efficient selection processes. Variations in drought timing, duration, and severity can make field phenotyping for drought responses costly and time-consuming (Sallam et al. 2016).

Controlled condition screening, which provides reliable information for selection and is based on secondary physiological features, reduces population size. Understanding how changes in the environment impact metabolite concentrations is possible through metabolic profiling. By using this information, faba bean breeders can enhance their plants' ability to withstand drought (Kelly and Bornowski 2018). Addressing the challenges posed by global climate change and drought stress on agriculture requires a multifaceted approach. Adaptation of agricultural practices, effective plant breeding programs, and the use of innovative strategies to enhance drought tolerance are crucial (Ali et al. 2022). By integrating these approaches, we can improve plant productivity, ensure food security, and mitigate the adverse effects of climate change on agriculture (Tiwari, J.(2022).

10.2.8 Strategies to Cope with Drought Stress in Faba Beans

V. faba faces significant challenges due to drought stress, particularly in regions with Mediterranean-type climates. There are three primary strategies plants use to cope with drought stress: drought escaping, drought dryness tolerance, and drought diversion.

10.2.8.1 Drought Escaping

Drought avoidance involves mechanisms that maintain a high water status within the plant under water-limited conditions. This is achieved by minimizing water loss and maximizing water uptake through enhanced root characteristics (Ali et al. 2022). Effective stomatal regulation and the presence of epicuticular wax also play crucial roles in reducing water loss.

10.2.8.2 Dehydration Tolerance

Dehydration tolerance refers to the plant's ability to sustain metabolic activities despite low-tissue water potential, thus surviving internal water deficits. This strategy includes osmotic adjustments and the accumulation of compatible solutes to maintain cellular functions (Varshney et al. 2011).

10.2.8.3 Drought Evade

The capacity of a crop to finish growing before the beginning of a drought or other unfavorable conditions is known as drought escape. When temperatures rise and rainfall decreases in Mediterranean climes, where faba beans are seeded in the fall and harvested in the spring, the crop frequently experiences "terminal" droughts (Rubiales and Khazaei 2022). By reaching maturity early and finishing their life cycles before the drought hits, adapted faba bean types can avoid terminal droughts (Varshney et al. 2011). But there's always a yield penalty associated with this approach (Rubiales and Khazaei 2022).

10.2.9 Challenges with Transient Droughts

In subtropical and temperate climates, faba beans frequently experience transitory droughts in addition to terminal droughts (Puspitasari et al. 2022). Plant height and the number of reproductive nodes are decreased by these droughts. Unlike terminal droughts, transient droughts are unpredictable and intermittent, making them harder to manage. Effective strategies for dealing with transient droughts include maintaining higher relative water content, which indicates efficient stomatal function, and enhancing root morphology and plasticity to improve water uptake (Rispail et al. 2010).

10.2.10 Drought Adaptation and Breeding Potential

Despite originating from semi-arid regions, faba beans generally exhibit poor drought adaptation due to their relatively shallow root systems, which cannot access deep soil moisture. When the topsoil dries during the reproductive period, this frequently results in moisture stress. Nonetheless, there are genetic resources that can adapt to changes in soil moisture content, suggesting that drought-tolerant cultivars might be bred (Maalouf 2017). Drought-tolerant cultivars and lines have been found via breeding, especially in areas with moderate moisture levels and uneven water distribution. Classical breeding approaches can leverage the wide genotypic variations to develop faba bean genotypes better suited for drought conditions (Singh et al. 2013).

10.2.11 Impact of Drought on Yield and Seed Quality

The degree of the water shortage and the stage of growth of the plant both have a major impact on faba bean yield loss. Drought largely affects the early stages of reproduction, which include blooming, early podding, and pod setting. Although it usually has no effect on seed size or quantity per pod, moderate drought stress has a detrimental effect on the number of pods per plant (Maalouf 2017). Seed quality is also impacted by drought. Heat and drought are two environmental variables that control the amount of protein in seeds. Faba beans in the winter tend to contain more protein than their spring counterparts, indicating that a lack of water might make the protein level higher. Additionally, water-deficit conditions can improve protein and sulfur content in faba beans, though they may impair nitrogen fixation, leading to reduced nodule numbers (Singh et al. 2013).

10.2.12 Physiological Responses and Management

Understanding the physiological responses of faba beans to drought is crucial for managing crop performance and improving yield quality. Identifying tools to quantify the physiological status of faba beans can assist in developing management practices and breeding programs aimed at mitigating the adverse effects of drought (Gupta et al. 2008). By focusing on these strategies and leveraging genetic diversity, it is possible to enhance the resilience of faba beans to drought stress.

10.2.12.1 Physiological Attributes Related to Stress Adaptation

Phenotypic selection with an emphasis on secondary adaptive features has been shown to increase productivity in water-limited situations (Hassan et al. 2023). These secondary qualities should be highly heritable, simple to quantify, and genetically associated with yield in the intended habitat. Large-scale population selection yields more substantial results when it comes to drought tolerance. Water usage efficiency (WUE), stomatal features, leaf temperature, and carbon isotope discrimination are among the physiological variables that are appropriate criteria for enhancing drought adaptation in faba bean breeding programs (Ijaz et al. 2024). Aluminum (Al^{+3}) toxicity, low temperatures, and salinity of the soil frequently combine with drought stress. Drought and salt both upset the osmotic balance similarly, whereas Al^{+3} toxicity prevents roots from growing, and cold temperatures alter water flow in roots, making it more difficult for plants to establish symbiotic associations. Among legumes, interactions between aluminum toxicity and drought have been studied in common beans and soybeans (Mir et al. 2012). Al^{+3} toxicity primarily worsens drought effects by inhibiting root growth. Conversely, drought can sometimes mitigate Al^{+3} injury by reducing the uptake of the toxic ion. However, the combination of aluminum toxicity and drought stress has a synergistic effect, making recovery difficult for crops (Bohra et al. 2014). The simultaneous presence of salinity and drought reduces symbiotic nitrogen fixation, possibly due to impaired photosynthate or oxygen supply to nodules and bacteroides (Bohra et al. 2014). Crosstalk in stress signaling and gene expression has been explored in responses to drought and cold stress, as well as cold and salinity stress. In faba beans, there is a minimal genetic correlation between drought response and freezing tolerance (Ouji et al. 2017).

10.2.12.2 Root Traits

Compared to aboveground traits, belowground traits have received relatively less attention in research. Analyzing root traits is challenging due to the complexity of root trait analysis systems, making it difficult to maintain precision in field-level screening (Nabateregga et al. 2019). Morphology and plasticity are the two primary categories of diversity seen in root architectural characteristics. Plants may find and absorb moisture and nutrients from the soil more easily when their roots are shaped differently due to genetic variances in root morphology. For instance, soybean genotypes with deep roots demonstrate a higher capacity for water absorption from deeper soil layers, resulting in greater yield potential (Carrillo-Perdomo et al. 2022). Furthermore, in response to stress and nutrition availability, roots can change in architecture a phenomenon referred to as plasticity. For instance, several lupin species showed an adaptive response when faced with a water deficit, as evidenced by an increase in the root-to-shoot biomass ratio (Sofi et al. 2021). Various studies have observed changes in root density and morphology in legume species under different environmental conditions. Drought tolerance is usually seen in faba bean genotypes with wide and prolific root features. Some legumes, including lentils, cowpeas, chickpeas, and common beans, have shown a similar degree of heterogeneity in their root properties and their root-based phenotyping for drought tolerance (Li et al. 2023).

Certain crops have shown more root development per unit area during the blooming stage when there is a water deficit. Nonetheless, dry root weight and root length density are frequently decreased because of drought stress. Researchers have used sophisticated phenotyping methods like GROWSCREEN-Rhizo boxes to identify root phenotypic indicators related to drought resistance features in early faba bean plants (Ouji et al. 2017). Some faba bean accessions that have greater root area coverage and deeper root systems have demonstrated superior drought avoidance characteristics by retaining their primary and tertiary root lengths. In faba beans, taproot development has a major impact on the overall rooting depth. Even in situations when surface moisture is present, taproots may find it difficult to obtain water for development (Mir et al. 2012). Shallow-dwelling lateral root development is essential for obtaining surface moisture and nutrients, which impacts the general health of the plant. To maintain dry root weight and root length density, faba bean breeding programs might potentially improve drought adaptability by selecting for fast taproot development and lateral root system expansion (Bohra et al. 2014). There hasn't been much research done on the genetic variability of faba bean root development and branching. Root mass, root mass fraction, rooting volume, root length, and root-to-shoot ratio are significant metrics of root systems. Among them, root mass makes up a sizable amount of the total dry matter and is quite simple to measure. In chickpeas, it is substantially linked with shoot mass (Ijaz et al. 2024; Bohra et al. 2014).

10.2.12.3 Shoot-Related Traits

Prioritizing a larger yield may not always be beneficial in stressful conditions since yield is complicated and impacted by several variables that are regulated by quantitative trait loci (QTL) (Li et al. 2023). Breeding efforts should thus focus on the underlying traits that contribute to drought tolerance rather than just yield when choosing genotypes for drought tolerance. These contributing traits must be causally linked to drought tolerance, with shoot-related traits being particularly crucial for maintaining water status under water-deficit conditions (Ouji et al. 2017).

Leaf/Canopy Temperature

The vapor pressure deficit directly affects the temperature of the outer canopy, which in turn affects the rate of transpiration and the resulting evaporative cooling. Non-disruptive measurement of canopy temperature (CT) can be achieved using infrared thermometers (IRTs), which can be mounted on drones to efficiently cover large areas and minimize environmental fluctuations (Hauggaard-Nielsen et al. 2011). CT was first presented as a criterion for dehydration tolerance in (Bohra et al. 2014). When there is a water shortage, closed stomata limit transpiration, which raises leaf temperatures (Ouji et al. 2017). Plant water absorption capacity under stress is often enhanced when a lower CT is maintained under field settings (Nabateregga et al. 2019). CT may be used as a stand-in selection criterion to filter genotypes of faba beans under various moisture conditions (Carrillo-Perdomo et al. 2022). It has proven to be an informative trait for distinguishing germplasm origins based on moisture levels for various legumes, including soybeans (Ghatak et al.

2017), chickpeas (Khatun et al. 2021), cowpeas (Wang and Zong 2011), and lentils (Haider et al. 2021a). The non-invasive nature of CT measurement allows for early evaluations across numerous genotypes, although careful adherence to IRT protocols is necessary to minimize variance errors and ensure repeatability (Li et al. 2023). Thermal infrared imaging of crop canopies, combined with robust image analysis tools, offers a valuable approach for studying stomatal responses and phenotyping large germplasm collections under field conditions (Haider et al. 2021a).

Leaf Water Relations

Plants can retain their water potential during dry spells by generating and storing cellular metabolites, or osmotica, which stomata perceive and regulate turgor pressure (Khatun et al. 2021). The types and functions of secondary metabolites are influenced by plant species and developmental phases. Only a select few substances can function as cellular osmotica because they must be compatible with cellular activities at high concentrations. These compounds consist of carbohydrates, polyols, polyamines, inorganic ions, and organic acids (Wang and Zong 2011). Carbohydrates and polyols make up more than half of the osmotic pool in chickpeas, although their share varies from 10% to 46% in field peas. Drought resistance may be effectively screened using leaf water potential (ψ), which is made up of osmotic potential (ψs) and turgor potential (ψp) (Bibi et al. 2023). Interpretations of plant water status are influenced by leaf ψ, a significant factor in leaf-level water deficit (Kondwakwenda et al. 2021). Mechanisms about both the root and the shoot help maintain leaf ψ during stress, which is frequently connected to avoiding drought. As a means of adaptation, several plant species show increased leaf turgor in response to water deprivation. In lentils, field peas, and chickpeas, osmotic adjustment has been shown to cause drought adaptation (Rauf et al. 2016). Although it has occasionally been noted that faba bean drought resistance through osmotic adjustment has not been thoroughly shown across large germplasm. Changes in metabolism in response to drought might highlight disrupted pathways and reveal important metabolite pools impacting physiology and metabolism (Cooper et al. 2017).

Relative Water Content

For evaluating the water condition of plants across various plant sections and species, the relative water content (RWC) is a reliable and straightforward selection criterion (Wang and Zong 2011). It contrasts the water content of tissues under typical circumstances with tissues under dehydration. RWC becomes essential in water-scarce situations for increasing water absorption through root plasticity or saving water through stomatal control, leaf area reduction, and leaf shedding (Wang and Zong 2011; Haider et al. 2021a; Kondwakwenda et al. 2021). RWC indicates solute concentrations and cell wall stiffness by reflecting variations in cell turgor and osmotic potential. Its effectiveness in detecting genotypes resistant to drought has been proven in a variety of legumes, such as common beans, chickpeas, and faba beans (Rauf et al. 2016). To produce drought-tolerant faba beans genotypes that retain a higher RWC in water deficiency are advantageous (Cooper et al. 2017).

Stomatal Conductance

To maintain gas exchange during drought stress, drought avoidance mechanisms either control stomatal closure to prevent water loss or maintain a greater stomatal conductance (gs) (Mandour et al. 2023). Photosynthesis and carbon absorption are directly impacted by this trade-off between gas exchange and stomatal closure. Comparable to canopy temperature (CT) readings when evaluated across a variety of plants, stomatal conductance has been suggested as an efficient selection criterion (Rauf et al. 2016). Evaluating gs relies on various stomatal features, often assessed through microscopy, although accurately measuring stomatal aperture size can be challenging due to its rapid stress response. While portable photosynthetic machines and leaf pyrometers measure gs, they might not be able to capture longer-term variations (Wang and Zong 2011). Environmental factors, time of day, and plant growth stage all affect gs readings. Early drought phases are associated with lower gs; however, this effect lessens as pods begin to fill (Bibi et al. 2023). Higher stomatal density genotypes of faba beans often showed poorer yields and reduced resistance to drought, whereas genotypes with lower stomatal density performed better under stress. The potential of gs as a selection characteristic is shown by the significant variance seen in it across faba bean accessions in non-stressful circumstances (Kondwakwenda et al. 2021). Drought treatments during early growth stages reduced biomass in Faba beans, and genetic mapping revealed co-located quantitative trait loci (QTLs) governing stomatal morphology, gs, and CT on chromosome 2 (Cooper et al. 2017). Studies suggest that selecting for CT and gs in early generations can maximize genetic gains in faba bean productivity. Enhancing stomatal control by restricting non-stomatal transpiration pathways, such as through the leaf cuticle's epicuticular wax (ECW), represents an important drought avoidance strategy (Wang and Zong 2011). Substantial variation in ECW among faba bean accessions offers opportunities for genetic enhancement of this trait in breeding programs aimed at improving drought adaptation (Rauf et al. 2016).

Carbon Isotope Discrimination

A popular proxy for plant physiological state, especially water use efficiency (WUE), is carbon isotope discrimination (Δ), which has been used to measure transpiration efficiency (TE) in cereals and legumes such as common beans, soybeans, chickpeas, cowpeas, and lentils (Mandour et al. 2023). Drought-tolerant genotypes in faba beans showed lower Δ than drought-susceptible genotypes, suggesting possible correlations with WUE (Berny Mier y Teran et al. 2019). However, controversy exists regarding the correlation between Δ and TE, with some studies suggesting no significant relationship across various grain legumes. Because Δ is measured from leaf tissue, which may not precisely represent short-term environmental changes, its applicability as an actual WUE predictor may be restricted (Ghatak et al. 2017). Furthermore, screening techniques based on carbon isotope discrimination are expensive and need specialist equipment like isotope ratio mass spectrometry (IRMS). To find genetic differences and prove that faba bean germplasm is a reliable screening tool for faba bean breeding programs, a more thorough examination of faba bean germplasm under various environmental circumstances is required (Fig. 10.4) (Bibi et al. 2023).

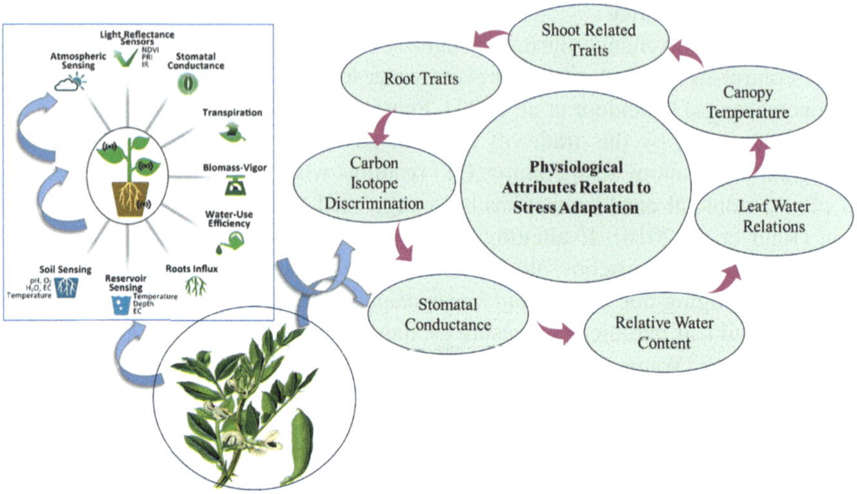

Fig. 10.4 Physiological attributes related to stress adaptation in bean

10.3 Metabolomics for Legume Breeding

Because metabolic products have a significant correlation with environmental circumstances, metabolomics, a rapidly developing discipline, combines gene expression, protein interaction, and other regulatory processes to use them as biomarkers for crop improvement (Berny Mier y Teran et al. 2019). Plants have been shown to have over 200,000 metabolites, each with a unique function and measurement technique. Understanding the processes behind inducible chemical and physiological responses may be gained by grouping these traits into subsets of metabolic activities (Haider et al. 2021a). Low molecular weight metabolites may be identified and quantified with exceptional precision and scale thanks to analytical methods such as nuclear magnetic resonance (NMR), liquid chromatography (LC), gas chromatography (GC), and high-performance liquid chromatography (HPLC) (Kondwakwenda et al. 2021). While research on the metabolic profile in grain legumes is still lacking, particularly in faba beans, recent studies have looked at the short-term impact of water scarcity on overall development and the metabolic changes across symbionts in *Medicago truncatula* (Berny Mier y Teran et al. 2019). Under stress, plants generate both primary and secondary metabolites; two important mechanisms are osmotic adjustment and solute accumulation (Bot 2016). Osmotica are metabolically active molecules, and they are generated from readily available precursors. Enhancing faba bean improvements using current metabolomic resources in legume research may reveal novel solutes and biochemical indicators for stress tolerance in breeding initiatives (Adhikari et al. 2022).

10.3.1 Accumulation of Metabolite in Legumes

10.3.1.1 Amino Acids

When plants, including many species of legumes like *M. truncatula* and soybeans, are under stress, the build-up of amino acids and nitrogen-based compounds is an essential biochemical signal (Blair et al. 2007). These vital metabolites tend to build up in response to stressors such as salt, cold, and drought, helping to improve stress-induced protein breakdown and allowing plants to endure harsh environments. Among these amino acids, proline stands out as one of the most prevalent in plant tissues under stress (Sharma and Crouch 2004). Its accumulation has been correlated with drought tolerance in barley cultivars, while other reports have shown mixed associations in legumes. For example, proline concentrations were higher in *Lotus* spp. and soybeans during the reproductive stage, but proline buildup was seen in peanuts during the pre-flowering stage when there was a water deficit (Bibi et al. 2023). The development stage of the plant and the water state both affect the accumulation of proline. Proline quantification can be used to identify short-term stress episodes since, in some experiments, a single day of water deprivation during the seedling stage caused its buildup. While faba beans exhibit increased proline content in response to intense stress, in ideal circumstances, there is no fluctuation in proline at the genotypic level. Interestingly, exogenous proline application has shown potential in decreasing stomatal opening under drought, positively impacting drought tolerance mechanisms (Kondwakwenda et al. 2021). Furthermore, it has been shown that the nonprotein amino acid β-aminobutyric acid (BABA) improves proline and soluble sugar accumulation, which helps faba beans adapt better to drought. Drought stress in lentils has been shown to increase not just proline but also methionine, isoleucine, valine, arginine, and histidine. Moreover, in *Arabidopsis thaliana*, soybeans, and common beans, drought stress has been connected to the manufacture of gamma-aminobutyric acid (GABA). Abiotic stressors raise cytosolic Ca^{2+} levels, which in turn stimulates the synthesis of GABA. Nevertheless, there isn't any proof as of now that GABA uses its ion transporter function to promote stomatal opening or closure (Rauf et al. 2016; Cooper et al. 2017).

10.3.1.2 Polyamines

Another way that plants respond to abiotic stimuli is by synthesizing polyamines (PAs), albeit there is still debate on the relationship between PA accumulation and stress tolerance (Blair et al. 2007). Putrescine, spermidine, and spermine are among the frequently detected PAs; stress-tolerant accessions often show greater PA levels than sensitive ones (Adhikari et al. 2022). However, not all three PAs accumulate simultaneously, with one type often showing a stronger response. For example, higher putrescine and spermidine levels were seen in lentil cotyledons and roots under PEG-osmotic stress conditions, although the presence of cadaverine was negligible (Ijaz et al. 2023a, b). Elevated PA concentrations were also seen in drought-adapted yellow lupin accessions, especially in spermidine and spermine. Interestingly, while leaves typically accumulate more PAs, seeds commonly exhibit the opposite trend, with reduced PA content (Berny Mier y Teran et al. 2019).

Numerous factors, such as plant species, the kind and intensity of stress, and the physiological state of the tissues under study, all have an impact on the accumulation of PA. Contradictory results might arise from the dynamics of PA accumulation under abiotic stress; nonetheless, in general, bigger PA pool sizes appear to correlate with stress tolerance, suggesting the role of PAs in stress protection (Cooper et al. 2017). There are currently few studies on the accumulation of PA in faba beans in response to drought stress; nevertheless, studies on other plants indicate that total polyamines, cadaverine, and spermine rise with salt levels, whereas putrescine and spermidine levels fall (Cooper et al. 2017; Mandour et al. 2023).

10.3.1.3 Organic Acids

Organic acids are vital to plant metabolism because they are an early source of photosynthesis and precursors to the synthesis of numerous compounds (Sharma and Crouch 2004). They play a role in the formation of ammonium, ion balance, stomatal control, and nutrient absorption. Organic acids make for around 20% of the identified metabolites in soybean leaf tissue (Ullah et al. 2023a, b, c, d). The concentration of organic acids can vary depending on the organ type; for instance, lentil root tissue generally exhibits lower concentrations compared to cotyledon tissue. Studies have shown alterations in organic acid levels under stress conditions, such as increased succinate levels in drought-treated soybean leaves and elevated concentrations of isocitric, nicotinic, and oxaloacetic acids in osmotically stressed lentils (Cooper et al. 2017). However, the specific functions of organic acids in faba beans under stress conditions have yet to be elucidated, representing a potential area for future research (Mandour et al. 2023; Berny Mier y Teran et al. 2019).

10.3.1.4 Carbohydrates

In many plant species, carbohydrates are essential for osmoregulation and frequently act as the right osmotica (Maalouf et al. 2017). The drought-induced metabolic profile of the model forage legume *Lotus japonicus* L. showed an increase of polyols such arabitol, ununitol, and galactitol as well as sugars like fructose, galactose, glucose, and maltose (Singh et al. 2021). Soluble sugars and sugar alcohols, constituting about 30% of identified metabolites in soybean leaves, are essential for plant adaptation to stresses, particularly water deficits (Ghatak, A (2018). Because of the intimate connection between primary metabolism and carbohydrates, it may be possible to improve stress tolerance by molecular alteration of the production and breakdown of these substances. The desiccation-tolerant plant *Myrothamnus flabellifolia* Wolw. has a large accumulation of trehalose, a nonreducing disaccharide that helps stabilize proteins and membranes. Through the promotion of trehalose production, rhizobium inoculation has been demonstrated to enhance dehydration tolerance in several legume species (Maalouf et al. 2017). Furthermore, a higher content of trehalose plays a major role in drought resistance. Another important class of metabolites, polyols, are also susceptible to environmental factors. For example, D-pinitol is impacted by the developmental stage and activated by stress from water and salinity. Although the broad knowledge of pinitol's role in plant systems is strong, there are still unanswered questions about the circumstances that lead to its accumulation and the molecular processes that regulate it (Ijaz et al. 2023a, b).

10.4 Understanding Drought Tolerance

A variety of physiological strategies have been developed by plants to deal with drought stress, which lowers water-use efficiency (WUE), alters water relations, and hinders normal development. Plants utilize many morphological, physiological, and biochemical mechanisms to counteract these impacts. These efforts include modifications to photosynthesis, respiration, ion exchange, transpiration, osmotic control, and the activation of antioxidant enzymes and drought-induced proteins (Haider et al. 2021b). At the morphological level, root and leaf responses are critical for reducing water loss and promoting WUE. The root system, including its length, weight, volume, and density, plays a vital role in drought resistance. Deep rooting allows plants to access deeper soil moisture, while an increased root-to-shoot ratio prioritizes water uptake over leaf growth (Kumawat et al. 2020). Leaf rolling, induced by turgor pressure, reduces water loss by minimizing the leaf surface area exposed to sunlight. Stomata, the pores responsible for gas exchange, close under drought conditions to reduce water loss, though this also limits the CO_2 uptake necessary for photosynthesis (Elshaer and Hellal 2020).

Abscisic acid (ABA) is produced by plants when their root cells sense a deficiency in soil water. By inducing morphological changes such as root-to-shoot ratio adjustments, stomatal closure, and leaf senescence, ABA increases water absorption and decreases water potential through osmotic adjustments (Kumar et al. 2021). The ABA signaling pathway is a complex series of receptors, phosphatases, kinases, G-proteins, proteins involved in the ubiquitin process, and transcription factors. According to research by Tryphone et al. (2013), rice drought tolerance is enhanced by the transcription factor IPA1, which activates the ABA pathway and results in a smaller leaf stomatal aperture and a more developed root system (Ullah et al. 2023a, b, c, d). A prompt response to water scarcity is mediated by ABA-mediated stomatal closure, whereas long-term adaptations like osmolyte accumulation, upregulation of adaptive genes, hypersensitive response (HR), and programmed cell death (PCD) are facilitated by secondary messengers such as ROS, NO, and Ca^{+2} (Singh et al. 2021).

Physiological and biochemical responses are also crucial for plant performance under drought. Photosynthesis is particularly affected, as decreased soil water content reduces photosynthetic and transpiration rates (Elshaer and Hellal 2020). Reduced carboxylation and internal CO_2 levels are the main causes of reduced photosynthetic rates under mild to severe drought stress since stomatal closure restricts CO2 inflow (Haider et al. 2021b). Additionally, when the electron transport chain (ETC) is disrupted by drought, more reactive oxygen species (ROS) such as hydrogen peroxide, superoxide, and hydroxyl radicals are produced. These ROS can harm thylakoids and photosystems (Haider et al. 2021b; Kumawat et al. 2020). By changing the way light energy is dissipated thermally, breaking down the PSII D1 protein, and separating light-harvesting complexes from photosynthetic reaction centers, plants can offset these impacts. Maintaining photosynthesis and respiration balance is crucial, as drought-induced reductions in photosynthesis can lower substrate supply to mitochondria, potentially decreasing respiration rates (Elshaer and Hellal 2020). Conversely, increased demand for respiratory ATP to support cellular metabolism can elevate respiration (Singh et al. 2021).

In drought conditions, ROS is produced in mitochondria, risking cellular damage. Plants utilize the mitochondrial ETC's alternative oxidase (AOX) pathway and photorespiration to prevent ROS accumulation and maintain cellular function (Faridi, 2022). The TCA cycle is modified to prevent excess reductant generation through γ-aminobutyric acid (GABA) synthesis, which accumulates under stress and helps modulate stomatal aperture and enhance WUE. Elshaer and Hellal (2020) discovered that while GABA does not directly alter stomatal pore size, it does so indirectly adversely affect the function of ALMT9 (aluminum-activated malate transporter 9) (Haider et al. 2021b). Stress tolerance is largely dependent on hormones, secondary metabolites, inorganic acids, carbohydrates, amino acids, and polyamines (Kumar et al. 2021). These substances accumulate to lower osmotic potential, which enhances cell water retention, stabilizes membranes, promotes osmoregulation, and scavenges reactive oxygen species (Blair et al. 2007). One important osmoprotectant that lowers ROS levels and supports membrane integrity is proline (Maalouf et al. 2017). Other amino acids like methionine, valine, glutarate, glutamate, and tryptophan also contribute to drought tolerance (Blair et al. 2007). According to recent research, drought tolerance can be increased by the exogenous application of plant growth regulators (PGRs), which include hormones, polyamines, organic acids, antioxidants, and osmoprotectants (Tryphone et al. 2013). The photosynthetic capacity, relative water content (RWC), gas exchange properties, and antioxidant and osmolyte buildup are all enhanced by these treatments. For instance Sharma and Crouch (2004) showed that, in tobacco plants grown during a drought, exogenous citric acid treatment enhanced the quantity of chlorophyll, net photosynthesis, RWC, and ABA while decreasing transpiration, stomatal conductance, and water loss. These priming techniques hold promise for creating crops resistant to drought (Ullah et al. 2023a, b, c, d).

10.4.1 Consequences in Developing the Drought-Tolerant Cultivars

A complicated quantitative polygenic characteristic, drought tolerance is regulated by several genes, each of which has little impact (Maalouf et al. 2017). This intricacy makes it difficult to understand the molecular processes of drought response and to improve crop yields in drought-prone areas. Other factors that contribute to this difficulty include poor heritability, strong genotype-by-environment (G × E) interactions, and the difficulty of mass screening plant characteristics and genes. Certain crops have genetically enhanced drought-tolerant cultivars thanks to conventional plant breeding; however, this process is expensive, time-consuming, and labor-intensive (Singh et al. 2021; Faridi 2022). Recent developments in omics technology, marker-assisted breeding, and crop physiology have offered fresh perspectives and methods for enhancing drought tolerance. Our understanding of drought tolerance mechanisms has improved because of marker-assisted selection (MAS), genome-wide association studies (GWAS), gene editing techniques, and other omics methods (genomics, transcriptomics, proteomics, and metabolomics) (Haider et al. 2021b). Expanding the field of plant biology, phenomics is an approach that uses robots, computers, image analysis, and machine vision to connect plant physiology and genomes. Phenomics has been effective in identifying features associated with drought (Kumawat et al. 2020). GWAS analysis and quantitative trait loci (QTL) are useful tools for locating genomic areas linked to genetic variations and alleles relevant to drought. However, it's still unclear which candidate genes these genetic variations use to affect characteristics (Elshaer and Hellal 2020). The production of drought-tolerant cultivars can be sped up by combining contemporary breeding methods with high-throughput phenotyping and multi-omics platforms to improve our understanding of how crops respond to drought stress (Ullah et al. 2023a, b, c, d).

10.5 Marker-Assisted Breeding (MAB)

Utilizing DNA alteration to promote desired characteristics in plants and animals, marker-assisted breeding (MAB) is a sophisticated breeding technique (Haider et al. 2021b). This covers methods such as genomic selection, molecular marker-assisted selection, genetic engineering, and gene modification (Singh et al. 2021). To enhance features based on genotypic tests, MAB typically uses molecular biotechnologies, especially molecular markers, in combination with linkage maps and genomics (Haider et al. 2021b). This method incorporates several contemporary breeding techniques, including genome-wide selection (GWS) or genomic selection (GS), marker-assisted backcrossing (MABC), marker-assisted recurrent selection (MARS), and marker-assisted selection (MAS) (Kumawat et al. 2020). The general ideas and techniques of marker-assisted plant breeding will be covered in this talk, along with its benefits over conventional breeding techniques and an examination of related processes and uses such as backcrossing, pyramiding of multiple genes, and marker-assisted selection (Elshaer and Hellal 2020).

10.5.1 Types of Molecular Markers

Molecular markers are crucial tools in plant breeding, allowing for the precise identification of genetic variations (Kumar et al. 2021). Different types of molecular markers include:

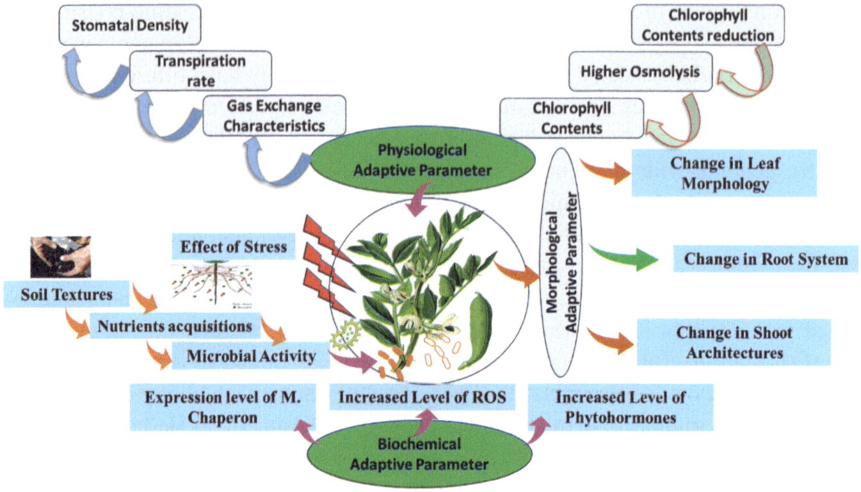

Fig. 10.5 Sophisticated marker-assisted breeding in *V. faba*

(a) *Simple Sequence Repeats (SSRs):* Also known as microsatellites, SSRs are repeating sequences of 1–6 base pairs of DNA. They are highly polymorphic and widely used in genetic mapping and diversity studies due to their abundance and variability (Elshaer and Hellal 2020).
(b) *Single Nucleotide Polymorphisms (SNPs):* SNPs are single base-pair variations in the DNA sequence. They are the most common type of genetic variation and provide high-resolution mapping, making them valuable for genome-wide association studies (GWAS) and genomic selection (Kumar et al. 2021).
(c) *Amplified Fragment Length Polymorphisms (AFLPs):* AFLPs are based on the selective PCR amplification of restriction fragments. They are useful for detecting polymorphisms across the genome and are employed in genetic mapping, diversity analysis, and marker-assisted selection (Fig. 10.5) (Tryphone et al. 2013).

10.5.1.1 QTL Mapping

Quantitative trait loci (QTL) mapping is a method used to identify regions of the genome associated with specific quantitative traits, such as drought tolerance (Kumawat et al. 2020). QTL mapping involves crossing individuals with contrasting traits, genotyping their progeny, and statistically analyzing the association between genetic markers and phenotypic variation (Elshaer and Hellal 2020). This approach helps in pinpointing genomic regions that influence complex traits, facilitating the selection and breeding of individuals with desired characteristics (Kumar et al. 2021).

10.5.1.2 Molecular Breeding (MB)

The term "molecular breeding" (MB) refers to the process of improving plant and animal features through genetic modification at the DNA level. It includes, among other things, genomic selection, molecular marker-assisted selection, genetic engineering, and gene modification (Tryphone et al. 2013). More precisely, the term "MB" frequently refers to "molecular marker-assisted breeding" (MAB), which is the process of applying genomics, linkage maps, and genetic markers to improve and alter characteristics based on genotypic tests (Ohm et al. 2024). This phrase encompasses several sophisticated breeding techniques, including genome-wide selection (GWS) or genomic selection (GS), marker-assisted backcrossing (MABC), marker-assisted recurrent selection (MARS), and marker-assisted selection (MAS) (Tryphone et al. 2013).

10.5.2 Advantages of MAB Over Traditional Breeding

(a) *Precision:* MAB allows for the precise identification and selection of desirable traits based on genetic markers, reducing the reliance on phenotypic selection.
(b) *Speed:* By using genetic markers, MAB accelerates the breeding process, enabling faster development of improved cultivars.
(c) *Efficiency:* MAB increases the efficiency of breeding programs by reducing the number of generations required to achieve desired traits.
(d) *Genetic Diversity:* MAB helps maintain genetic diversity by allowing the selection of traits from a broad genetic base without extensive backcrossing.

In summary, marker-assisted breeding is a powerful tool in modern plant breeding, leveraging molecular markers and advanced genetic techniques to enhance the selection and development of crops with improved traits (Tryphone et al. 2013). This approach integrates traditional breeding practices with cutting-edge biotechnology, offering a more efficient, precise, and rapid path to agricultural innovation (Ohm et al. 2024).

10.6 Genetic Markers in Plant Breeding

Allelic versions of genes or genetic loci, which are carried through generations, determine biological traits known as genetic markers (Reddy et al.,). These markers are used as tags or experimental probes to keep an eye on certain people, tissues, cells, chromosomes, nuclei, or genes. Genetic markers are divided into two primary groups in genetics and plant breeding: classical markers and DNA markers (Ullah et al. 2023a, b, c, d). Morphological, cytological, and biochemical indicators are all considered classical markers. Based on distinct methods for identifying polymorphisms, DNA markers have developed into a variety of systems, including Southern blotting (nucleic acid hybridization), PCR (polymerase chain reaction), and DNA sequencing (Rai and Pandey-Rai 2021). Systems for identifying DNA include RAPD, SSR, AFLP, RFLP, and SNP.

10.6.1 Classical Markers

Plant breeding has a long history, and markers have been employed to identify desired features in plants. Leaf form, blossom color, pubescence color, pod color, seed color, seed shape, hilum color, awn type and length, fruit shape, rind color and stripe, flesh color, and stem length were among the obvious qualities that early breeders primarily looked for (Ohm et al. 2024; Ullah et al. 2023a, b, c, d). These morphological markers are genetic variations that are easily recognized. They have been used in the construction of linkage maps using traditional two- and/or three-point tests. Some morphological markers are associated with other agronomic traits, which makes them useful as indirect selection criteria in practical breeding (Tryphone et al. 2013). For example, the Green Revolution's selection of semi-dwarfism in wheat and rice greatly aided the development of high-yielding cultivars. Rht10-controlled dwarfism was backcrossed into Taigu nuclear male-sterile wheat in wheat breeding, establishing a close relationship between Rht10 and the male-sterility gene Ta1 (Ullah et al. 2023a, b, c, d). Recurrent selection in wheat was later aided using dwarfism as a marker to identify and choose male-sterile plants in breeding populations. Morphological indicators are few, though, and many of them are unrelated to significant economic characteristics like yield and quality. Some may even harm the growth and development of plants (Rai and Pandey-Rai 2021).

10.6.2 Cytological Markers

Chromosome karyotypes and banding patterns are two ways to visualize the structural characteristics of chromosomes, which are involved in cytological markers (Nadeem et al. 2018). These patterns show variations in the distribution of heterochromatin and euchromatin based on color, breadth, order, and location (Bohra et al. 2020). Giemsa stain produces G bands, quinacrine hydrochloride produces Q bands, and R bands represent the reverse pattern of G bands. (Baum 2003). These markers help characterize healthy chromosomes and find chromosomal alterations. They are often used in physical mapping and linkage group identification. Although physical maps based on morphological and cytological markers form the foundation for genetic linkage mapping made feasible by molecular techniques, there has been minimal direct application of cytological markers in genetic mapping and plant breeding (Bohra et al. 2020).

10.6.3 Biochemical/Protein Markers

Isomers are structural variations or alternate versions of an enzyme with the same catalytic activity but differing molecular weights and electrophoretic mobility. These are known as protein markers, which are also referred to as molecular markers (Baum 2003). Isozymes may be genetically mapped onto chromosomes and utilized as genetic markers to map other genes since they reflect the results of distinct alleles rather than separate genes because of point mutations (Haider et al. 2022a). They are occasionally employed in plant breeding as well as in testing seed purity. However most crop species have a limited number of isozymes, and some are unique to certain strains. High molecular weight glutenin subunits (HMW-GS) in wheat are another example of a biochemical marker; they relate to both gluten strength and bread-making quality (Bohra et al. 2020). It is possible to select certain HMW-GS alleles in wheat to improve its bread-making quality, although differences in bread-making quality between types cannot be entirely attributed to differences in HMW-GS composition (Baum 2003).

10.6.4 DNA Markers

DNA markers are sections of DNA that show mutations or changes and can identify polymorphisms in a population between various genotypes or alleles of a gene (Nadeem et al. 2018; Tryphone et al. 2013). Molecular technologies are employed to identify them and link them to certain genomic regions (Ohm et al. 2024). Molecular breeding relies heavily on DNA markers, also known as molecular markers. High polymorphism, an even genome distribution, co-dominance in expression, distinct allelic features, a single copy without pleiotropic effects, low cost, simple assay/detection and automation, high availability, and no negative phenotypic effects are the ideal characteristics of DNA markers for marker-assisted breeding (Ullah et al. 2023a, b, c, d).

10.6.5 RFLP Markers

The first generation of DNA markers, known as restriction fragment length polymorphism (RFLP) markers, serves as a fundamental tool for plant genome mapping (Baum 2003). Their foundation is on Southern blotting methods (Haider et al. 2022a). DNA probing and electrophoresis are used to identify differences in restriction fragment sizes caused by mutations at restriction sites or between neighboring sites. RFLP markers exhibit locus-specificity, co-dominance, and good reproducibility. High-throughput marker creation is made possible by enhanced RFLP methods, such as PCR-RFLP and CAPS (cleaved amplified polymorphism sequence) (Haider et al. 2022a). Nevertheless, RFLP analysis has poor genotyping throughput, requires a lot of high-quality DNA, and is challenging to automate. Although less often used, it is nonetheless essential to many scientific investigations (Maiti and Singh 2016).

10.6.6 RAPD Markers

PCR is the basis for random amplified polymorphic DNA (RAPD) markers. An individual's whole genomic DNA may be amplified with a single, brief, random primer (Maurya et al. 2023). The length and size of the primer and the target genome determine the amplified fragments produced by PCR, and polymorphisms may be seen using electrophoresis. High polymorphism and dominance are found in RAPD markers. They don't require certain primers or DNA probes, are straightforward, and only need modest quantities of DNA. Nevertheless, RAPD markers are not very reproducible and are unable to identify allelic variations in heterozygotes (Ijaz et al. 2023a, b).

10.6.7 AFLP Markers

PCR and RFLP are combined in amplified fragment length polymorphism (AFLP) markers. In these, restriction fragments from a double-digest of genomic DNA are amplified selectively by PCR (Maurya et al. 2023). Adaptor sequences, restriction endonuclease recognition sequences, and arbitrary selection sequences make up AFLP primers, which are used to amplify fragments cut by restriction enzymes (Ijaz et al. 2023a, b). Technologies for high-resolution electrophoresis are used to separate the amplified fragments. AFLP markers are highly reproducible, require small DNA samples, and do not need sequence information or probe collection. However, AFLP assays are labor-intensive, not cost-efficient, and require high-quality DNA (Ijaz et al. 2023a, b; Assefa et al. 2019).

10.6.8 SSR Markers

Simple sequence repeats (SSR), also known as microsatellites, are PCR-based markers consisting of short nucleotide repeats (2–6 bp) randomly distributed throughout the genome. SSRs are highly polymorphic, co-dominant, and locus-specific (Rai and Pandey-Rai 2021). They require specific primers for PCR amplification and are detected by high-resolution electrophoresis (Baum 2003). Genetic linkage maps, QTL mapping, marker-assisted selection, and germplasm analysis can all benefit from the use of SSR markers. They have a labor-intensive development method and require nucleotide information for primer creation (Haider et al. 2022a).

10.6.9 SNP Markers

DNA sequences differing by a single nucleotide base are known as single nucleotide polymorphisms, or SNPs (Maurya et al. 2023). SNPs can be transversions (C/G, A/T, C/A, or T/G) or transitions (C/T or G/A), and they are often found in both plants and mammals (Assefa et al. 2019). With their co-dominant markers connected to genes, they offer the most markers possible for genetic research and breeding. Sequencing, PCR-RFLP, and other genotyping techniques like primer extension and allele-specific hybridization can all be used to find SNPs (Nkomo et al. 2021). SNPs are being utilized more and more as entire DNA sequences for

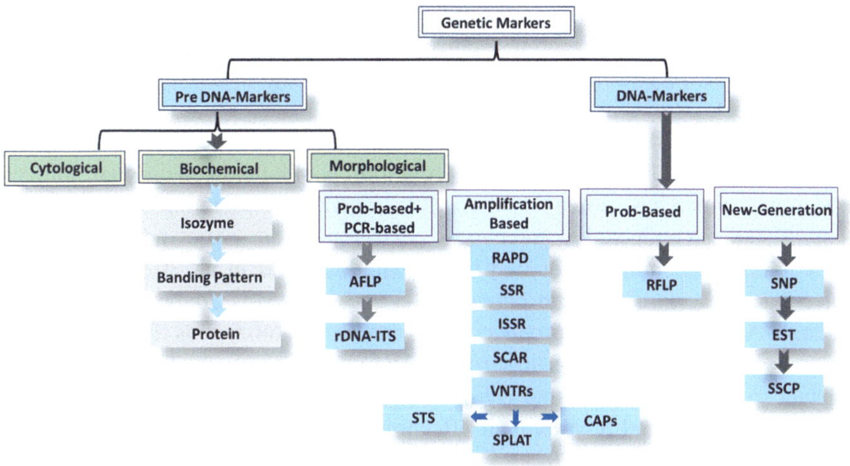

Fig. 10.6 Various molecular markers and marker-assisted breeding in bean

additional species become accessible since they are very effective at identifying polymorphism and are easily automated. However, their use in some labs and breeding programs is restricted by expensive start-up costs, high-quality DNA needs, and technological constraints (Ijaz et al. 2023a; b). Several variables, including financing, manpower, equipment, genetic resources, and research objectives, influence the selection and application of DNA markers in breeding and research. Breeders must select markers for their breeding operations based on their needs and the resources at their disposal. Each marker system has benefits and drawbacks (Fig. 10.6) (Assefa et al. 2019).

10.7 MAS for Major Genes or Improvement of Qualitative Traits

Major genes, or QTLs, govern a wide range of commercially significant qualities in agricultural plants, including self-incompatibility, disease and pest resistance, male sterility, and many characteristics of plant architecture, color, and form (Ijaz et al. 2023a, b). These characteristics frequently have a mono- or oligogenic inheritance pattern. A small number of important QTLs or genes can explain a sizable percentage of the phenotypic variance for some quality variables (Assefa et al. 2019; Nkomo et al. 2021). Introducing such genes into specific lines can significantly enhance the trait in the resulting cultivar (Maurya et al. 2023). When marker expression is greater or the target gene's allelic variation is hard to find, markers that are closely related to important genes can be more effective for selection than direct gene selection (Maiti and Singh 2016). *Heterodera glycines* Inchinoe, the soybean cyst nematode (SCN), offers a glaring illustration of MAS for key genes (Baum 2003). SCN is the most economically significant pest for soybeans, and resistant cultivars are the primary strategy for mitigating damage (Baum 2003; Haider et al. 2022a). Phenotypic assays for resistance are labor-intensive, taking five weeks, requiring extensive greenhouse space, and demanding significant labor (Ullah et al. 2023a, b, c, d).

However, the SSR marker Satt309, which is 1–2 cM from the resistance gene rhg1, offers a helpful stand-in for selection. A third population with different sources of SCN resistance has shown 80% accuracy in genotypic selection using Satt309, whereas studies conducted in greenhouse environments have shown 99% accuracy in predicting susceptibility for two test populations (Ijaz et al. 2023a, b). Using molecular markers in a hybrid between J05 and V94–5152 soybeans (Ijaz et al. 2023a, b; Assefa et al. 2019) produced five F4:5 lines homozygous for eight marker alleles linked to soybean mosaic virus (SMV) resistance genes. These lines were resistant to SMV strains G1 and G7, most likely because of having the resistance genes Rsv1, Rsv3, and Rsv4, which offer long-lasting and widespread protection against SMV (Nkomo et al. 2021). Overall, MAS leveraging tightly linked markers facilitates efficient and accurate selection for desirable traits, significantly advancing breeding programs.

10.7.1 MAS for Enhancement of Quantitative Traits

Multiple QTLs govern significant agronomic variables in crops, making marker-assisted selection (MAS) for these traits complicated (Ijaz et al. 2023a, b). The process is complicated by the participation of several genes or QTLs, QTL x environment (QTL x E) interactions, and epistasis. The phenotypic manifestation of a characteristic is often regulated by a combination of environmental factors and each gene to a lesser extent (Assefa et al. 2019). Because of this, phenotyping these characteristics becomes difficult, and to precisely define QTL effects and assess stability across settings, repeated field experiments are needed to determine marker-phenotype relationships. While epistasis can distort QTL effects on the trait, QTL x E interactions lower MAS efficiency (Nkomo et al. 2021). Even though QTL mapping has been done extensively in the last 10 years, plant breeding has not used it much in practice (Mwadzingeni et al. 2017). Appropriate field experimental designs and reliable data processing techniques are necessary to increase MAS efficiency for quantitative characteristics (Ijaz et al. 2023a, b). These include consistent experimental techniques, adequate replications over time and space, and thorough statistical analysis. For instance, composite interval mapping (CIM) estimates QTL-environment interactions by combining data from several sites, which aids in the identification of stable QTLs in a variety of contexts. To precisely locate connected and targeted QTLs throughout the coupling and repulsion phases, a saturated linkage map is essential (Raza et al. 2023).

Typically, in actual breeding, MAS targets a small number of important QTLs rather than many minor ones. For traits involving numerous minor-effect QTLs, breeders might employ gene pyramiding (Maurya et al. 2023). Combating the devastating disease Fusarium head blight (FHB) in wheat and barley, which is brought on by Fusarium species, is an example of effective MAS. FHB resistance is quantitatively inherited, and several QTLs have been found from different germplasm sources (Ullah et al. 2023a, b, c, d). Fhb1, a significant QTL that accounts for 20–40% of phenotypic variance, has been repeatedly identified in a variety of settings and populations (Maiti and Singh 2016; Ullah et al. 2023a, b, c, d). Wheat breeders prefer using Fhb1 to develop FHB-resistant cultivars. Ohm et al. 2024 found a 23% reduction in disease severity and a 27% reduction in kernel infection when comparing NIL pairs for Fhb1. Further studies demonstrated the successful implementation of MAS for this QTL (Ullah et al. 2023a, b, c, d). Researchers have also used MAS to include numerous QTLs (Rai and Pandey-Rai (2021) and Ullah et al. (2023a, b, c, d) demonstrated how choosing three FHB resistance QTLs at the same time significantly increased FHB resistance in German spring wheat, particularly when QTLs 3B and 5A were combined Ohm et al. (2024), Ullah et al. (2023a, b, c, d), and Rai and Pandey-Rai (2021) compared multiple-locus combinations in a RIL population from a cross between "Veery" and "CJ 9306." They found that genotypes with more favorable alleles exhibited higher resistance levels. The coefficients of determination between the number or percentage of diseased spikelets and the accumulated effects of alleles for four or five loci with favorable alleles were 0.33–0.41 ($P < 0.01$), indicating that selecting favorable marker alleles for multiple loci may enhance resistance (Baum 2003). Coordinated agricultural projects (CAPs) in the United States support cooperative efforts in molecular and applied plant genomics research on a variety of crops, including rice, wheat, barley, beans, potatoes, and tomatoes (Maiti and Singh 2016). A crucial CAP tactic is using MAS in plant breeding to create agricultural cultivars with increased yield, resistance, and quality. This involves utilizing vast genetic resources and infrastructure. To support breeding operations, these initiatives use plant introductions and hundreds of DNA markers (Maiti and Singh 2016; Ullah et al. 2023a, b, c, d; Assefa et al. 2019; Nkomo et al. 2021).

10.8 Marker-Assisted Backcrossing

10.8.1 MABC Procedure and Theoretical and Practical Considerations

In molecular breeding, marker-assisted or marker-based backcrossing (MABC) is a simple and popular technique (Raza et al. 2023). It aims to transfer specific genes or quantitative trait loci (QTLs) from an agronomically inferior donor parent (DP) into a superior recurrent parent (RP) to enhance targeted traits (Raza et al. 2023). Unlike traditional backcrossing, MABC relies on DNA markers linked to the target genes/QTLs rather than phenotypic performance. The procedure starts with selecting parents and making the cross (Ijaz et al. 2023a, b). While the DP has the required characteristics and related DNA marker alleles, the RP performs better overall. Assuming a suitably large population and no selection during backcrossing, the fraction of the RP genome after n backcross generations is theoretically provided by

1 – (1/2) n + 1 for a single locus and [1 – (1/2) n + 1] k for k loci. Full recovery of the RP genome typically requires 6–-8 generations, but this can be slowed by linkage drag, where the target gene is linked to undesirable traits (Haider et al. 2022a).

Foreground and backdrop selection are the two methods used by MABC. Up until the last backcross, when the chosen plants self and produce offspring with homozygous DP alleles, foreground selection keeps the target locus in a heterozygous condition in each backcross generation (Ullah et al. 2023a, b, c, d; Maurya et al. 2023). The number of genes involved, the distance between marker-gene linkage, and undesired links to the target gene all affect how successful foreground selection is (Assefa et al. 2019). To rapidly restore the RP genome and get rid of unwanted DP genes, background selection concentrates on looking for RP marker alleles in areas with desired features (Nkomo et al. 2021). The recovery rate increases with the number of markers utilized; nevertheless, more genotyping and bigger populations are required. By employing DNA markers to overcome linkage drag, background selection surpasses classical backcrossing in efficiency (Mwadzingeni et al. 2017).

In an MABC application, foreground and backdrop selection are typically utilized simultaneously. First, individuals with desired target trait marker alleles are selected (foreground) and then screened for other RP marker alleles (background) (Raza et al. 2023). This combined approach ensures efficiency since non-target allele carriers are discarded early (Disasa and Gebre 2018). The population size, marker-gene relationship, number of markers for the target trait and RP background, and linkage drag are some of the parameters that affect MABC efficiency. Simulations show that combined foreground and background selection accelerates RP genome recovery compared to traditional backcrossing. For example, selecting heterozygotes at the target locus with RP alleles for flanking and non-target chromosome markers can achieve faster RP genome recovery (Göl et al. 2017).

At least one genotype containing every beneficial allele for a QTL should be included in MABC programs (De la Rosa et al. 2020). The percentage of undesired genes because of linkage drag rises when more QTLs are targeted. In early backcross generations, many unwanted genes are found on non-target chromosomes and are quickly removed (Ullah et al. 2023a, b, c, d). Conversely, DP genes on the target chromosome gradually vanish; some remain after six backcrosses (Diaz et al. 2020). This emphasizes the significance of background selection, which selects individuals homozygous for RP alleles and heterozygous at the target gene using flanking markers (Mwadzingeni et al. 2017). Effective background selection requires determining optimal marker numbers and positions (Nkomo et al. 2021). Simulations suggest two to four markers per chromosome are sufficient for efficient selection (De la Rosa et al. 2020). Three to four backcross generations can achieve over 99% RP genome recovery, reducing time and costs. Background selection is more efficient in later backcross generations, so genotyping in BC3 rather than BC1 is recommended (De la Rosa et al. 2020). MABC is widely used in plant breeding for various traits in species like rice, wheat, maize, barley, soybean, and tomato. For instance, Bt transgenes for insect resistance have been integrated into maize using MABC. In rice, aroma controlled by a recessive gene has been selected using

MABC. Other examples include high lysine content in maize and disease resistance in various crops (Ullah et al. 2023a, b, c, d).

Even with its achievements, MABC still has problems with certain QTLs (Diaz et al. 2020). For instance, in soybeans, the introduction of two QTLs for seed protein led to the confirmation of just one QTL in subsequent generations (Sallam 2014). A method for introducing resistance genes from wild relatives into top germplasm is called advanced backcross-QTL (AB-QTL), and it works well for characteristics like disease resistance and fruit quality in tomatoes and other crops (Dhungana et al. 2021). In barley, MABC was used to select for resistance to barley yellow dwarf virus, resulting in higher yields and lighter leaf symptoms in infected plants. In maize, MABC improved complex traits like grain yield by transferring chromosomal segments into elite lines, resulting in hybrids with significantly higher yields. A cooperative project in the USA is using MABC to transfer high-oleic acid traits into locally superior soybean cultivars. This project illustrates the ongoing potential of MABC in improving crop traits (Sallam 2014). Overall, MABC is a powerful tool for plant breeders, offering faster and more efficient trait introgression compared to traditional methods. It combines the strengths of molecular markers and backcrossing to enhance desirable traits while minimizing undesirable ones, making it a cornerstone of modern plant breeding.

10.8.2 Gene Pyramiding and Recurrent Selection Supported by Markers and Convergent Crossing

One important use of DNA markers in plant breeding is marker-assisted gene pyramiding (MAGP), which selects numerous genes at once to increase resistance to pests and diseases (Dhungana et al. 2021). Schneider et al. (Schneider et al. 1997a, b) claimed results in pyramiding QTLs and qualitative genes for barley resistance to stripe rust. It is possible to identify QTL-allele-linked markers with the same phenotypic effects by using markers. Pyramiding numerous genes or QTLs is advised to improve quantitatively inherited features; the cumulative impacts of this strategy have been shown in crops such as soybean, wheat, and barley (Maalouf et al. 2011). (Dhungana et al. 2021). The hybrid of convergent crossing is self-pollinated and goes through multiple generations of marker-based selection for target characteristics until stable lines with desired qualities are produced (Sallam 2014). Different markers may be prioritized in subsequent generations to reduce population size and halt the loss of important genes or QTLs (Diaz et al. 2020).

10.8.3 MARS, or Marker-Assisted Recurrent Selection

For polygenic trait improvement, MARS combines genotypic selection and intercrossing within the same crop season, enhancing recurrent selection efficiency and accelerating progress (Sallam 2014). This scheme is particularly effective for integrating multiple favorable genes/QTLs from different sources (Ullah et al. 2023a, b, c, d). MAGP, though complex, is a potent strategy for developing crops with enhanced resistance and improved traits. Its practical application continues to expand, promising significant advances in crop breeding (Diaz et al. 2020).

10.8.4 The General Procedure MARS

MARS for "forward breeding" of native genes and pyramiding multiple QTLs have been proposed for complex characteristics like grain yield and resilience to biotic and abiotic stressors (Maalouf et al. 2011). According to definitions (Dubey et al. 2019), MARS is a recurrent selection technique that selects and identifies several genomic areas linked to complex features using molecular markers to construct the best-performing genotype within a single population or among related populations. Buitrago-Bitar et al. (2021) used a maize example to show how effective MARS is for quantitative attributes. Markers are widely utilized in bi-parental populations in maize MARS programs for initial QTL discovery and then yield-based MARS (Assefa et al. 2019). This raises the frequency of advantageous alleles and improves long-term selection efficiency through quick cycles of recombination and selection based on markers linked to yield (Mwadzingeni et al. 2017). Buitrago-Bitar et al. (2021) and Cortés and Blair (2018) reported that in some populations, the genetic gain via MARS in maize was about double that of phenotypic selection (PS). Improved populations had noticeably better resistance levels following recurrent selection in comparison to previous populations (Cortés and Blair 2018), according to a publication (Maalouf et al. 2011) on the considerable efficacy of MARS for resistance to *Helicoverpa armigera* in upland cotton.

10.9 Genomic Selection

Described by Tock (2017) as the simultaneous selection of several markers, spanning the entire genome densely such that all genes are in linkage disequilibrium with at least some markers, genomic selection (GS), also known as genome-wide selection (GWS), is a marker-based selection technique. Genotypic information from the entire genome in GS may accurately predict complex characteristics, allowing for selection based only on genomic estimated breeding value (GEBV) (Abd EL-Mageed et al. 2022). Genome-wide dense DNA markers are used to compute the expected breeding value, or GEBV (Cortés and Blair 2018). GS does not need considerable testing or the identification of a subset of markers linked to the trait, in contrast to standard MAS. Rather, it analyzes phenotypes and genome-wide genotypes in a training population to create GS models that predict GEBVs (Abd EL-Mageed et al. 2022).

GS depends on novel statistical techniques, high-performance computers, and high-throughput marker technologies (Dubey et al. 2019). This is now possible because of developments in genome sequencing and effective SNP genotyping techniques. Assuming a previous distribution of QTL effects, Buitrago-Bitar et al.

(2021) propose that the best way to estimate breeding values from genomic data is to compute the conditional mean of the breeding value given the genotype at each QTL. QTL genotypes are almost replaced by marker genotypes; nonetheless, the optimal approach will become closer as more sequence and SNP data become available. Studies in theory, simulation, and empirical methods have been carried out since Abd EL-Mageed et al. (2022) introduced GS, mostly in animals (Dubey et al. 2019). Although GS in plants has gained popularity, there isn't much extensive empirical research in the public plant breeding domain (Dhungana et al. 2021). Research shows that GS offers more accuracy than only pedigree data (Ijaz et al. 2023a, b). For instance, in terms of gain per unit cost and time in oil palm, GS performed better than MARS and phenotypic selection (Maurya et al. 2023). These results imply that, with proper population selection, GS may prove to be a useful tool in plant breeding (Ullah et al. 2023a, b, c, d).

Although GS is thought to be a potent tool for plant breeding, its use is still limited. The primary obstacle is a lack of practical understanding of GS (Assefa et al. 2019). Plant breeders frequently find that the intricate simulations and statistics involved in GS are too specialized to use in real-world initiatives. While GS may work for a small number of breeding populations with narrow objectives, it is not viable for large breeding programs that involve several crosses/populations (Nkomo et al. 2021). To produce compelling empirical data, GS must thus go from theory to practice, where its precision and economic viability are assessed in real-world breeding operations (Mwadzingeni et al. 2017). It would be easier to apply GS in plant breeding if software and GEBV formulas were developed with ease of use in mind. A thorough assessment of GS for plant breeding has been conducted (Raza et al. 2023). In conclusion, GS provides a potential method for MAS by effectively predicting complex features utilizing genome-wide dense markers. However, because of its intricacy and the requirement for additional empirical proof of its efficacy, the method's practical implementation in plant breeding is still restricted (Disasa and Gebre 2018). GS is anticipated to become a more useful and popular strategy in plant breeding projects as techniques and technology advance and more accessible tools become available (Göl et al. 2017).

10.10 Marker-Based Breeding and Conventional Breeding: Challenges and Perspectives

With the creation of several molecular markers in agricultural plants throughout the 1980s and 1990s, marker-assisted breeding, or MAB, came into being (Jacob et al. 2016). Through the facilitation of genetic mapping, map-based gene identification, trait characterization, and crop improvement, it has completely changed the field of plant breeding (Abd EL-Mageed et al. 2022). MAB has significant advantages over conventional breeding methods: Early Selection: MAB allows selection at the seedling stage, reducing the time required to identify desirable phenotypes (Tock 2017). This is particularly useful for traits expressed later in development, as undesirable genotypes can be quickly eliminated. Environmental Independence: Selection can occur under any environmental conditions, which is crucial for traits like disease resistance that only manifest under specific conditions. For low-heritability traits, MAB can be more effective than phenotypic selection (Asif et al. 2011).

(a) *Recessive Allele Selection:* Co-dominant markers (e.g., SSR and SNP) enable the effective selection of recessive alleles without the need for selfing or test crossing, thus speeding up the breeding process (Asif et al. 2011).
(b) *Gene Pyramiding:* MAB is suitable for traits controlled by multiple genes or QTLs, as it can identify and select these simultaneously. This is difficult with traditional phenotypic selection where one gene may mask others (Varshney 2016).
(c) *Efficiency and Accuracy:* Genotypic assays using molecular markers can be faster, cheaper, and more accurate than phenotypic assays, leading to higher effectiveness and efficiency in breeding. Since the 2000s, MAB research and use have increased, especially for simply inherited traits like monogenic or oligogenic resistance to diseases and pests (Bourgault 2009). However, the impact of MAB has been less than expected, particularly for complex agronomic traits such as yield and quality. Several challenges remain:
(d) *Breeder-Friendly Markers:* Not all markers are suitable for breeders. Converting non-breeder-friendly markers to more user-friendly types (e.g., RFLP to STS) can help.
(e) *Marker Polymorphism and Association:* Lack of polymorphism or solid marker-trait relationships may make markers inapplicable across populations. Understanding marker diversity and genetic background effects can be aided by using several mapping populations (Duc et al. 2010).
(f) *Recombination Issues:* Recombination between markers and target genes/QTLs can lead to false selection. This can be mitigated by using several markers or flanking genes for target genes/QTLs (Bourgault 2009).
(g) *QTL Location and Effects:* Accurate QTL estimations hinder development. More precise QTL estimations may be obtained using robust phenotyping across settings and high marker density fine mapping with large populations (Varshney 2016).
(h) *Infrastructure and Costs:* Many breeding programs lack the facilities for large-scale MAB adoption. High startup expenses and labor costs remain significant obstacles, especially in developing countries (Asif et al. 2011).

Despite these challenges, MAB offers great opportunities and prospects for plant breeding. It complements rather than replaces conventional breeding, which relies heavily on empirical selection and subjective evaluation (Asif et al. 2011). The integration of MAB into conventional breeding programs can enhance crop improvement (Jacob et al. 2016). As the theory, technology, and application of MAB advance, its drawbacks are likely to be overcome, leading to wider adoption in practical breeding programs worldwide (Fig. 10.7) (Abd EL-Mageed et al. 2022). In summary, MAB is a valuable addition to the plant breeding toolbox, offering precision and efficiency. While challenges remain, particularly in terms of costs and

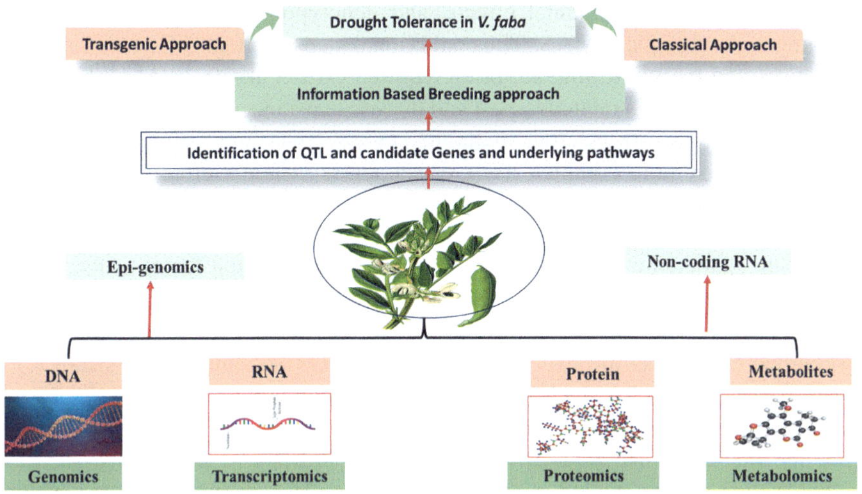

Fig. 10.7 Marker-based breeding and conventional breeding in *V. faba*

infrastructure, ongoing advancements in molecular technology are expected to facilitate the broader application of MAB in crop improvement (Tock 2017). Integrating MAB with conventional breeding methodologies presents a promising strategy for future agricultural advancements.

10.10.1 Marker Development in Faba Bean

Marker-assisted selection (MAS) has revolutionized plant breeding by enabling the efficient selection of desirable traits based on molecular markers associated with those traits (Duc et al. 2010). This approach has been particularly effective for traits controlled by major genes and has shown promise for some polygenic traits as well. MAS offers several advantages over conventional phenotypic selection, including simplicity, reliability, and the ability to perform screening at early seedling stages, thus saving time and resources (Duc et al. 2010). The success of MAS depends on various factors, including the genetic basis of the trait, the distance between markers and the target gene, the size and type of population analyzed, genetic background considerations, and technical options available at the marker level (Braich et al. 2017). While MAS has been predominantly employed for simply inherited traits, efforts are ongoing to extend its application to more complex traits controlled by multiple genes (Ullah 2017).

In the case of major gene traits, such as many disease resistances, MAS has been relatively straightforward due to minimal genetic background effects and the ease of phenotyping. However, challenges arise when markers are not polymorphic across breeding populations, necessitating the identification of alternative markers (Savita et al. 2020). For polygenic traits, factors such as population size, replication of phenotypic data, environmental effects, and genotyping errors complicate QTL mapping and subsequent MAS. Thus, additional steps such as QTL confirmation, validation, and marker testing are often required (Wu et al. 2021). MAS has been most effective for traits related to pest and disease resistance, with successful applications reported in various crops including cereals, roots, and tubers. However, its effectiveness for more complex traits varies across different crops. For legumes, MAS has shown promise for certain crops like soybean, common bean, and pea, but its application for more complex traits has been limited (Asfaw et al. 2012).

Despite challenges, advances have been made in applying MAS to faba bean breeding. Efforts have focused on developing molecular markers for disease resistance, particularly against fungal diseases like rust, ascochyta blight, and crenate broomrape (Ahmad et al. 2017). Several putative QTLs controlling resistance to these diseases have been identified, laying the groundwork for further research and marker development (Arya et al. 2021). In addition to disease resistance, there is growing interest in using MAS for other desirable traits in faba beans, such as growth habits and nutritional quality. Furthermore, the potential exists to develop molecular markers for traits like frost or drought resistance, which are difficult to breed conventionally (Ray and Georges 2010). Overall, the progress toward molecular breeding in faba beans has been significant, with the development of comprehensive genetic maps and the identification of markers associated with various traits (Arya et al. 2021). Further research and collaboration are needed to fully exploit the potential of MAS for genetic improvement in faba beans and other crops (Ray and Georges 2010). Breeding drought-tolerant cultivars of faba beans is crucial for adapting to water-limited environments. To achieve this, identifying genetic traits that enhance adaptation is essential. To maintain yields in regions with variable rainfall, Ortiz (2013) suggested selecting for early blooming and maturity, effective nutrition partitioning toward reproductive structures, and phenotypic plasticity. Ortiz (2013) showed that water loss is decreased, and soil moisture extraction efficiency is increased with a bigger root system about leaf area. To select faba bean genotypes suited to water shortage circumstances, physiological traits like transpiration efficiency, stomatal conductance, leaf temperature, and carbon isotope discrimination have been studied (Thudi et al. 2023).

Initial studies by ICARDA and subsequent research by Leonforte et al. (2013) aimed to genetically enhance drought tolerance in faba beans. Abid et al. (2021) suggested that genome-based approaches could facilitate accessing desirable alleles for drought tolerance (Li et al. 2018). Expressed sequence tagged sites (ESTs) related to drought avoidance were identified in chickpeas, peas, and *Medicago truncatula* (Li et al. 2018). Integrating these genes into faba bean maps via comparative genomics may enable targeted molecular breeding efforts (Prasanna et al. 2014). Physiological changes in *M. truncatula* during environmental stress were investigated by the European Grain Legumes Integrated Project (GLIP), which provided insights that might be used to faba beans (http://www.grainlegumes.com/index.php/aep/special_reports). Through marker-assisted selection (MAS), QTL maps enriched with candidate genes that combine agronomic, physiological, and molecular data can help generate cultivars that are resistant to drought. Our knowledge of the regulatory networks underpinning drought tolerance may be improved by the identification of eQTLs altering gene expression levels by transcriptome analysis (Abou-Khater et al. 2022; Liaqat et al. 2024).

Frost tolerance is crucial for autumn-sown faba beans in cool temperate climates. Winter tires, with superior grain yields, highlight the importance of winter hardiness (Leonforte et al. 2013). Recent studies identified QTLs associated with frost tolerance and leaf fatty acid content, offering prospects for MAS in breeding frost-tolerant faba beans (Abid et al. 2021). Plant architecture affects faba bean cultivation and harvesting. Determinate growth habit, characterized by a terminal inflorescence, reduces plant height and lodging, facilitating management and harvesting (Ray and Georges 2010). Leonforte et al. (2013) developed molecular markers for determining growth habits, aiding in the selection of this trait (Abid et al. 2021). Seed quality traits like tannin content and vicine and convincing levels impact faba bean utility for human and animal nutrition. The absence of tannins and low vicine and convincing content are desirable breeding objectives (Li et al. 2018). Marker-assisted selection (MAS) facilitates the identification and introgression of genes conferring these traits (Abou-Khater et al. 2022). In conclusion, breeding drought-tolerant, frost-tolerant, and high-quality faba bean cultivars involves identifying and integrating desirable genetic traits. Marker-assisted selection (MAS) streamlines the breeding process by enabling targeted selection for these traits, contributing to the development of resilient and nutritionally superior faba bean varieties (Liaqat et al. 2024). In the realm of molecular breeding for faba bean improvement, the development of SCAR markers linked to genes responsible for desirable traits like zero tannins, low vicine and convicine content, and determinate growth habit holds significant promise. These markers facilitate gene pyramiding, expediting the creation of new faba bean cultivars with enhanced nutritional value (Varshney et al. 2021a, b).

Moreover, the utilization of resistance gene analogs (RGAs) offers a strategic approach to isolating multiple resistance gene candidates (Roy et al. 2016). These genes, often encoding nucleotide-binding site (NBS) and leucine-rich repeat (LRR) regions, play pivotal roles in plant defense against various phytopathogens (Gaur et al. 2012). While large-genome crops like faba beans may not be sequenced soon, leveraging the knowledge from fully sequenced organisms can be transformative. By identifying and cloning homologs of candidate genes from model organisms, researchers can advance the understanding and manipulation of non-model species like faba beans (Varshney et al. 2021a, b). Liaqat et al. (2024) isolated and characterized RGAs belonging to the NBS gene families in faba beans and chickpeas, laying the groundwork for further research (Abou-Khater et al. 2022). These sequences, exhibiting strong similarities to known plant resistance genes, were grouped into ten classes based on their deduced structures. Although the initial amplification of RGAs using degenerate primers has been valuable, the next step involves developing polymorphic markers based on these sequences (Varshney et al. 2021a, b). Specific primers have been designed to genotype single nucleotide polymorphisms (SNPs) within the RGA sequences, enabling their mapping and assessment of potential linkage relationships with disease-resistance genes (Roy et al. 2016). To date, the polymorphism for RGAs in faba beans has been relatively low, potentially due to the narrow genetic basis of breeding populations (Gaur et al. 2012). However, ongoing studies aim to explore the presence of SNPs in ten RGA classes, mapping them in available faba bean, pea, and chickpea populations. The detection of genes and quantitative trait loci (QTLs) for resistance to various pathogens underscores the potential of RGAs in faba bean breeding (Polania et al. 2020). Comparative genomics offers another avenue for advancing our understanding of the faba bean genome (Roy et al. 2016). While anonymous markers like RAPDs and SSRs placed on faba bean linkage maps may not be orthologous among species, genome sequencing projects in model legume systems provide valuable resources (Varshney et al. 2021a, b). For instance, *Medicago truncatula* SSRs have been successfully transferred to cool-season legumes such as chickpeas, peas, and faba beans, facilitating comparative genomics (Roy et al. 2016). The extensive conservation of gene order among different legume crops implies that genetic and genomic tools developed in model species like Medicago can be readily applied to other legumes, including faba beans (Thudi et al. 2023). Within the European Grain Legumes Integrated Project (GLIP), efforts have been made to create a large set of markers from single or low-copy coding regions in Medicago, identifying potential orthologous genes in related crops like faba beans (Prasanna et al. 2014).

These efforts are complemented using expressed sequence tags (ESTs) as a source for identifying candidate genes or QTLs involved in specific traits (Varshney et al. 2021a, b). Cross-amplification and polymorphism assays of intron-targeted primer pairs have been conducted, furthering our understanding of faba bean genetics and breeding (Prasanna et al. 2014). Collaboration with research groups worldwide has enabled the development of gene-based genetic linkage maps, paving the way for comparative genomic studies and the identification of macrosyntenic relationships between faba beans and related legume species (Leonforte et al. 2013). In summary, the integration of molecular breeding strategies such as SCAR markers, RGAs, and comparative genomics holds tremendous potential for accelerating faba bean improvement efforts (Li et al. 2018). By leveraging knowledge and resources from model legume systems and employing advanced molecular techniques, researchers can expedite the development of faba bean cultivars with enhanced traits, ultimately benefiting both human and animal nutrition (Fig. 10.8) (Gaur et al. 2012).

10.11 Identification of Drought Tolerance Traits

10.11.1 State-of-the-Art Phenotyping

A wide range of facilities and methods have been created to address the dynamic character of field environments, making it possible to analyze how a plant responds phenotypically to its surroundings (Polania et al. 2020). These facilities vary in their level of environmental control and offer increasing options for collecting phenotypic data at a multifaceted resolution. Ranging from basic setups like rainout

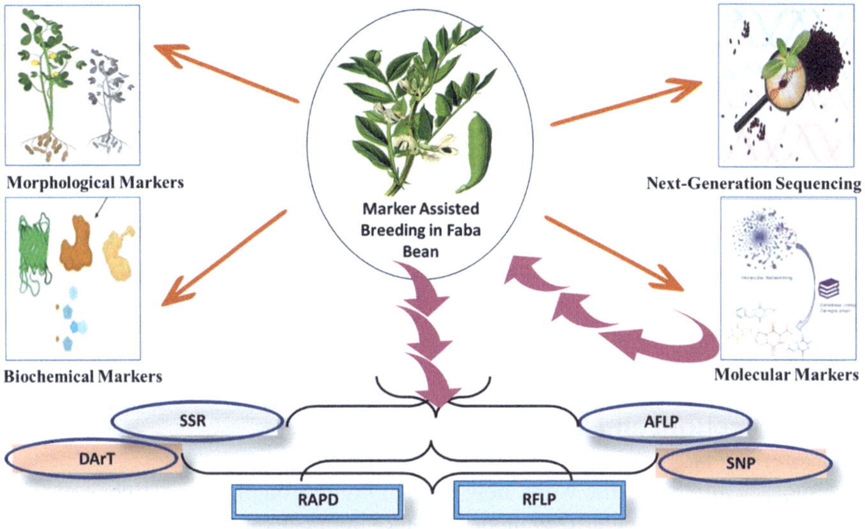

Fig. 10.8 Marker-assisted breeding in faba bean

shelters to sophisticated greenhouses and fully controlled environments, these facilities simulate various field-like scenarios (Roy et al. 2016). In dynamic field situations, cameras and sensors installed on mobile carrier platforms are the main tools used for prototyping. However, this method is susceptible to interference from environmental variables such as sunlight intensity and wind during the measurement process (Roy et al. 2016). This susceptibility can introduce noise into the recorded phenotypic data, making it challenging to differentiate between genotypic variation and environmental effects (Varshney et al. 2021a, b). Notwithstanding these difficulties, field phenotyping methods produce useful data that further our knowledge of the interactions between crops and their surroundings. Unmanned aerial vehicles (UAVs) or mobile tractors equipped with video systems provide non-invasive, high-throughput phenotyping of field stands (Liaqat et al. 2024). In a short amount of time, these technologies can quickly and precisely assess plant height and extract characteristics relating to color from crop canopies covering wide regions. For characteristics such as the normalized difference vegetation index, satellite photography provides the maximum throughput; it can capture whole fields in a single image and yields important information on the correlations between biomass and yield (Haider et al. 2022b). Stationary equipment centered on a field stand can be used to accomplish higher-resolution field phenotyping (Varshney et al. 2021a, b). For instance, a stationary crane outfitted with cameras at the UK's Rothamsted research site was able to accurately identify the phases of wheat heading and flowering by high-throughput phenotyping a field stand (Polania et al. 2020). Furthermore, field stands may be given a minimal amount of control using automated rainout shelters, which can also be used to track how a genotype reacts to water scarcity (Roy et al. 2016). However, field experiments are constrained by the present environmental conditions and cannot account for future changes predicted due to climate change, such as higher average temperatures. Increased night-time temperatures can negatively impact yield development (Trapp 2015). While attempts to artificially increase night-time temperatures have been made using meshes, these methods are only suitable for small research field sites. Although indoor phenotyping facilities provide more control over environmental variance, the artificiality of the growing circumstances may rise. Increased control and reproducibility are made possible by integrating high-throughput phenotyping devices into growth chambers and indoor glasshouses (Trapp 2015). Advances in sensor technology have further enhanced the applicability of indoor phenotyping platforms for both controlled and field conditions (Varshney et al. 2021a, b).

10.11.2 Drought and Salt Stress

Glasshouses all around the world now contain high-throughput shoot phenotyping facilities that use a standardized method of moving plants to image stations (Plant-to-sensor) (Trapp 2015). These stations, which have cameras ranging from visible light to near infrared, take pictures of important crop species like canola, rice, sorghum, maize, and barley (Polania et al. 2020). This setup enables the phenotyping of entire crop populations within glasshouses, allowing for the association of natural genomic variations with image-derived traits through genome-wide association studies, as demonstrated in rice, maize, and canola (Polania et al. 2020; Haider et al. 2022b). Glasshouses offer a controlled environment that protects plants from precipitation and external temperature variations. They also make it easier to apply abiotic stressors such as nitrogen or water shortages and raise soil salinity (Varshney et al. 2021a, b). Non-invasive phenotyping makes it possible to track plant responses to applied drought stress daily, including wilting, the turning point at rewatering, and the recovery stage (Haider et al. 2022b). Traits vary in heritability during drought stress, with some decreasing and recovering after rewatering, highlighting the importance of assessing multiple traits to characterize phenotypes accurately (Trapp 2015). The increase in global mean temperature is widely acknowledged, but predicting its impact on agriculture and drought severity remains complex. Robust drought indices are needed to characterize and predict drought severity, considering meteorological, agricultural, and socioeconomic factors (Mercati and Sunseri 2020). Crop yield responses to drought vary by species and region, further complicated by factors like soil salinity increase due to sea-level rise in coastal regions and prolonged droughts in arid and semi-arid areas (Gaur et al. 2012). Controlled-environment phenotyping helps elucidate genotype-environment interactions under defined drought conditions and soil salinity levels (Mercati and Sunseri 2020). Large-volume containers address constraints associated with typical pot sizes in controlled environments. These developments benefit the creation of robust crop types by improving our understanding of how crops react to shifting environmental circumstances (Mercati and Sunseri 2020).

10.11.3 Temperature

Crop output is significantly impacted by rising global temperatures; it is estimated that for each degree Celsius of increase, yield will decrease by three to seven percent on average for various species (Haider et al. 2022b). Studies conducted under controlled environments have shown that elevated temperatures during the reproductive stage can drastically reduce crop yield while increasing vegetative biomass (Dubey et al. 2019). However, the specific response of crop species to temperature varies, and sensitivity patterns against environmental factors have been deduced through experiments involving different crop species grown at various locations (Polania et al. 2020). The genetic architecture of temperature response during different developmental stages has been investigated, highlighting the importance of understanding growth dynamics and genetic factors underlying temperature response (Varshney et al. 2021a, b). Repetitive non-invasive high-throughput approaches coupled with controlled-environment (CE) phenotyping have made it possible to identify time-resolving quantitative trait loci (QTL) activity in a variety of crops, including *Arabidopsis*, canola, and maize (Mercati and Sunseri 2020). To understand development patterns and the genetic foundation of phenotypic features, this method highlights the need for recurrent phenotyping (De Ron et al. 2019). With the help of reasonably priced Raspberry Pi computers, any greenhouse may be transformed into a high-throughput phenotyping platform for CE phenotyping, thanks to recent technical breakthroughs (Boopathi 2013). This democratization of phenotyping technology enables stakeholders with lower budgets to conduct sophisticated experiments and generate high-quality data sufficient for deep neural networks (Haider et al. 2022b).

10.11.4 Carbon Fertilization

Current and projected climate change scenarios indicate a rise in global temperatures coupled with increased levels of atmospheric CO_2 due to human activities (Trapp 2015). While elevated CO_2 levels may offer some mitigation against the negative impacts of temperature increases on crop yields, the extent of this benefit varies between C_3 and C_4 plants (Attamah et al. 2024). However, the effects of elevated CO_2 on plant growth are nuanced and depend on factors such as regional environments, frequency of droughts, and potential decreases in nutritional value for both crops and vegetables (Varshney et al. 2010). Studies comparing yield increases under elevated CO_2 concentrations found substantial differences between C_3 and C_4 plants, with C_3 plants exhibiting lower yield increases in field conditions compared to indoor settings, and no increase observed for C_4 plants (Maalouf et al. 2016). Therefore, high-throughput phenotyping platforms must simulate dynamic environments comprehensively to capture the complexity of genotype-environment interactions required for modeling plant performance under future climate change scenarios (Maalouf et al. 2016; Abdelraouf et al. 2016). Although controlled-environment (CE) phenotyping facilities have significantly contributed to advancing knowledge in various domains, each approach has its strengths and limitations (Varshney et al. 2010). Some facilities excel in multifaceted data acquisition at an unprecedented level of detail but may lack comparability in plant growth conditions (Attamah et al. 2024). On the other hand, other systems are better at growing plants in environments like those seen in fields, but they might not have high-resolution sensors for accurate trait identification (Roy et al. 2016). As Polania et al. (2020) point out that the installations and technologies under discussion are just a small portion of the global pool of phenotypic resources, since 82 automated phenotyping platforms are recognized. Researchers from several disciplines and nations have collaborated to make significant advancements in phenotyping technology, both indoors and outdoors. Organizations like IPPN, EPPN, and DPPN are excellent tools for increasing the visibility of phenotyping resources throughout the world (Varshney et al. 2010).

10.11.5 Automated Digital Trait Assessment

Automated digital trait assessment in controlled-environment phenotyping facilities offers significant advancements by enabling high-throughput, accurate data collection with reduced manual labor and the ability to impose "stress on demand" conditions. Identifying relevant traits for assessing plant performance in adverse environments is crucial (Maalouf et al. 2016). Phenotyping methods are categorized into invasive and non-invasive approaches. Invasive methods are destructive, preventing subsequent measurements, while non-invasive methods allow repeated data collection without damaging the plant. Non-invasive image-derived trait acquisition is common, utilizing interactions between plants and photons across various spectra (Attamah et al. 2024). RGB cameras capture visible light reflections to measure traits like leaf area and digital biomass. Near-infrared spectrum reflections assess drought severity and water content (Boopathi 2013). Chlorophyll fluorescence detection provides insights into plant stress levels. Hyperspectral cameras capture reflections across a wide range of wavelengths, detecting plant stresses and diseases, often analyzed using machine learning. High-resolution root phenotyping uses X-ray computed tomography or magnetic resonance imaging to analyze root architecture and stress responses (De Ron et al. 2019). Image-derived traits integrate past genotype-environment interactions, making them multigenic traits that effectively analyze plant stress or health status (Mercati and Sunseri 2020). Light reflection spectroscopy predicts forage quality and bean texture quality (Attamah et al. 2024). Transpiration-related traits, like water use efficiency, support breeding for drought tolerance. Root phenotyping platforms using transparent gels or soil-filled rhizotrons enable non-invasive root trait measurements (Varshney et al. 2010). Translating controlled environment results to field conditions remains challenging, but technologies like ground-penetrating radar present promising avenues for in situ root structure assessment. Integrating various phenotyping approaches aids in developing stress-tolerant crop varieties, essential for food security in a changing climate (Roy et al. 2016).

10.11.6 Disease Resistance

The effects of climate change encourage the growth and reproduction of infections, increasing the strain of illness on crops (Varshney et al. 2010). Therefore, it may be necessary to identify novel genetic loci that contribute to resistance against both biotic and abiotic stresses, as well as their combinations, in the future as crops may encounter new and more severe combinations of biotic stressors (Boopathi 2013). To better understand the complexities of resistance and assess the effectiveness of individual resistances against specific viruses or fungal isolates under realistic stress scenarios and combinations, controlled inoculations within controlled environments are preferred over random natural infections (Maalouf et al. 2016). When it comes to repeatable and reliable phenotypic data, controlled-environment phenotyping of disease resistance and tolerance circumvents the seasonal restrictions of field trials, allowing for more frequent and consistent testing. In barley (Varshney et al. 2010) and wheat (Abdelraouf et al. 2016), studies using controlled settings have phenotyped mature plant resistance to leaf rust; the data exhibit a strong association with field trial results. Automated high-throughput techniques are used to quantify illness severity precisely and objectively (Salgotra and Stewart Jr 2020). Plant susceptibility to pathogens may be quickly and objectively assessed using imaging methods (Pandey et al. 2022). For example, a hyperspectral phenotyping device was used in an automated high-throughput method to quantify powdery mildew symptoms in barley plants grown in big containers (Mitache et al. 2023). Leveraging "omics" data is poised to enhance our understanding of pathogen infection strategies and interactions between different stress types, facilitating the identification of genetic components underlying resistances and tolerances. This approach aids in the discovery of genetic determinants of specific resistances (Mitache et al. 2023).

10.12 Application of MAB in Faba Bean

For many years, it has been known that using markers to indirectly select desirable features may have some benefits (Bohra et al. 2014). These days, we have access to a wide range of molecular markers that use various detection techniques including DNA sequencing, PCR, and hybridization. A new generation of DNA markers has emerged since the early days of restriction fragment length polymorphism (RFLP). These markers include but are not limited to single nucleotide polymorphisms (SNPs), sequence tagged sites (STS), microsatellites, sequence characterized amplified regions (SCARs), random amplified polymorphic DNA (RAPD) markers, amplified fragment length polymorphisms (AFLPs), and simple sequence repeats (SSRs or microsatellites). These markers have significantly enhanced our understanding of genetic diversity within various plant species and have greatly facilitated the mapping of genomic regions contributing to trait variation (Bohra et al. 2014).

Marker-assisted selection (MAS) is a breeding technique where DNA markers are used to select and combine favorable alleles in a genome. Compared to traditional phenotypic selection, MAS has benefits including simplicity, consistency, and the capacity to screen seedlings early on, which saves time and money (Kamaluddin et al. 2017). It improves the effectiveness of selection, especially for features that are hard to score or need certain environmental circumstances. The genetic basis of the trait, the distance between markers and target genes, and the number of individuals tested are some of the parameters that affect the effectiveness of MAS (Mitache et al. 2023). Because of its simple phenotyping and quick gene confirmation, it is mostly employed for simple inherited features like disease resistance (Bohra et al. 2014). However, MAS can be challenging for polygenic traits, requiring accurate QTL mapping, validation across environments, and reliable marker conversion (Afiah et al. 2016). Despite its greater success in major gene traits, MAS has shown promise in complex traits in crops like maize, rice, and barley. In legumes, while MAS has been less effective, some successful applications exist, including gene pyramiding in faba beans, highlighting the potential of MAS in enhancing legume breeding strategies despite limited resources (Kashiwagi et al. 2008).

RFLPs, RAPDs, and morphological features served as the foundation for the first faba bean linkage maps (Mitache et al. 2023; Bohra et al. 2014). Trisomic plants were employed in later research at IFAPA in Córdoba to map loci and linkage groups onto chromosomes (Afiah et al. 2016). Isozyme loci assignments (Kashiwagi et al. 2008) and additional F2 families identified QTLs for seed weight (Ray and Georges 2010). Linkage maps for crenate broomrape and ascochyta blight were developed using an F2 population (Bohra et al. 2014). A composite V. faba genome map, enhanced with SSRs (Roy et al. 2016), was published by (Leonforte et al. 2013). A review of linkage maps was provided by Varshney et al. (2010). Research has focused on developing markers for disease resistance, growth habit, and nutritional traits. Future efforts aim to develop markers for traits like frost and drought resistance, enhancing breeding efficiency for faba beans (Pandey et al. 2022).

10.12.1 Disease Resistance

Faba beans face various fungal diseases, with severity varying by region. Significant diseases include ascochyta blight, rust, chocolate spots, downy mildew, and foot rots (Salgotra & Stewart Jr., 2020). Crenate broomrape is particularly severe in the Mediterranean, especially in Northern Africa (Haider et al. 2022b). Traditional breeding for disease resistance, especially for broomrape, is challenging. Molecular breeding has targeted rust, crenate broomrape, and ascochyta blight using populations like F2 and advanced inbred lines. For rust resistance, Mercati and Sunseri (2020) used bulk segregant analysis to find markers linked to the Uvf-1 gene. To ensure durable resistance, gene pyramiding, which combines multiple resistance genes, is preferred. Bohra et al. (2014) identified QTLs for broomrape resistance, with ongoing efforts to validate and enhance these regions with additional markers. For ascochyta blight, multiple QTLs were found by Afiah et al. (2016), and new molecular maps are being developed. These markers streamline breeding by eliminating labor-intensive inoculation processes. Abiotic stress resistance, like drought tolerance, is less explored (Kashiwagi et al. 2008). Faba beans are highly sensitive to drought, affecting yield during flowering and pod filling. Breeding for drought tolerance involves selecting traits like early flowering, efficient nutrient partitioning, and larger root systems. Traits like stomatal conductance and transpiration efficiency are studied for drought adaptation (Alghamdi et al. 2012; Arbaoui et al. 2008; Bohra et al. 2014). Express sequence tags (ESTs) from related legumes may help identify drought-resistance genes (Wang and Zong 2011; Blair et al. 2007). Frost tolerance is crucial for autumn-sown faba beans in cool climates (Ijaz et al. 2023a, b; Ray and Georges 2010; Varshney et al. 2021a, b). Polyunsaturated fatty acid accumulation is linked to frost adaptation, with QTLs identified for frost tolerance (Mitache et al. 2023). These markers help introduce favorable alleles from exotic lines into European beans to improve winter hardiness (Afiah et al. 2016). Plant architecture, specifically determining growth habit, is also important. Determinate plants, which have terminal inflorescences, facilitate mechanical harvesting (Pandey et al. 2022). Markers linked to the ti gene, responsible for determinate growth, assist in breeding programs (Kashiwagi et al. 2008). Seed quality traits like low tannin and vicine-convicine content improve nutritional value (Leonforte et al. 2013). Tannins reduce digestibility, while vicine and convicine can cause health issues. Gene tagging has developed markers for these traits, enhancing selection efficiency in breeding. In summary, molecular breeding in faba beans has focused on disease resistance, abiotic stress tolerance, plant architecture, and seed quality, leading to the development of molecular markers for efficient breeding (Bohra et al. 2014). Further validation across diverse backgrounds is needed to integrate these markers into breeding programs effectively (Bohra et al. 2014).

10.13 Challenges and Solutions

Genomic analysis of crops with large genomes, like faba beans, is challenging. To overcome this, researchers use data from sequenced organisms to study non-model species (Thudi et al. 2023; Haider et al. 2022b). One approach is identifying and characterizing homologous genes from model organisms in non-model crops (Bohra et al. 2014; Wang and Zong 2011; Blair et al. 2007). Disease resistance genes (R genes), which encode proteins with nucleotide-binding site (NBS) and leucine-rich repeat (LRR) regions, have been cloned from model plants and show high sequence conservation (Bohra et al. 2014; Sharma and Crouch 2004; Ijaz et al. 2023a, b). Researchers have generated resistance-gene analogs (RGAs) from legumes such as soybean, common bean, lentil, cowpea, pea, chickpea, and alfalfa by using PCR using degenerate primers from conserved patterns in these areas (Ali et al. 2022; Bohra et al. 2014; Haider et al. 2021a). These RGAs, often near known resistance loci or QTLs, are turned into genetic markers using methods like RGA polymorphism (RGAP), NBS profiling, and derived CAPS (dCAPS) (Haider et al. 2021a; Ijaz et al. 2023a, b; Nkomo et al. 2021).

In faba beans, RGAs have been isolated and characterized, showing similarity to known plant R genes from tomato, *Arabidopsis*, and potato (Alghamdi et al. 2012; Varshney et al. 2011; Blair et al. 2007; Ijaz et al. 2023a, b). These sequences were grouped based on amino acid identity, with classes resembling non-TIR-type and TIR-type NBS genes (Bohra et al. 2014). Although degenerate primer PCR is valuable for R gene detection, it is labor-intensive and requires cloning and sequencing (Ohm et al. 2024). The next step involves developing polymorphic markers from cloned sequences for genetic mapping (Sallam 2014; Schneider et al. 1997a, b). Specific primers for RGA classes are being used to create co-dominant CAPS and dCAPS markers for genotyping SNPs within these sequences (Ray and Georges 2010). These markers are being tested in faba bean, pea, and chickpea populations to map disease-resistance genes and QTLs (Ray and Georges 2010; Haider et al. 2022b). Polymorphism in faba bean RGAs is low due to the narrow genetic basis of breeding populations, with only four RGA classes showing polymorphism (Kashiwagi et al. 2008). In peas, marker development and genotyping are in progress, while chickpea alleles show higher polymorphism, leading to the successful genotyping of seven markers in two chickpea progenies (Ray and Georges 2010). Combined RGA analysis in these legumes aims to map advanced populations, co-localize with functional R genes, and identify macrosyntenic relationships between resistance regions. Comparative genomics can help rapidly identify structurally similar genes in related species, leveraging conserved R gene cluster locations (Leonforte et al. 2013).

10.14 Comparative Genomics

A comprehensive understanding of the faba bean genome may be gained by comparative genomics and synteny investigations using model species. Comparative mapping is hindered by the lack of species orthology in many anonymous markers, such as SSRs and RAPDs (Varshney et al. 2011; Wang and Zong 2011; Blair et al. 2007). Legumes have sequenced genomes, including *Medicago truncatula*, *Lotus japonicus*, and *Glycine max*, which offer valuable bioinformatics resources and necessary tools (Pandey et al. 2022). *Medicago truncatula* data is particularly beneficial for faba bean genetic characterization and comparative mapping (Kamaluddin et al. 2017). Medicago SSRs and ESTs have shown significant transferability and high macrosynteny with legumes like chickpeas, peas, and faba beans. To facilitate comparative genomic investigations in European agricultural legumes, the GLIP project produced markers from Medicago single or low-copy coding regions (Kashiwagi et al. 2008). The first gene-based genetic linkage map of the faba bean was produced through collaboration with Afiah et al. (2016), who expanded these findings to a new faba bean population. By improving our knowledge of the macrosyntenic interactions between *Lens culinaris*, *M. truncatula*, and *V. faba*, our investigations lay the groundwork for upcoming genomic and breeding applications (Kashiwagi et al. 2008).

10.15 Prospects

In various regions worldwide, the frequency and severity of drought are projected to rise, posing significant challenges to global agricultural productivity (Ray and Georges 2010). Among various abiotic stresses, drought is expected to have the most pronounced impact on soil microorganisms and plants, as well as broader ecological systems. Plants, being immobile, are particularly susceptible to drought-induced osmotic stress, which hampers nutrient uptake due to soil heterogeneity, thereby limiting nutrient availability to the root system (Mitache et al. 2023). Drought tolerance, a multifaceted quantitative trait influenced by multiple genes, presents a formidable challenge in terms of study and classification. Fortunately, the advent of clustered regularly interspaced short palindromic repeat (CRISPR) technology represents a new frontier in crop enhancement, revolutionizing plant breeding practices (Kashiwagi et al. 2008). CRISPR systems have demonstrated transformative potential across various biological disciplines, notably in biomedicine and agriculture. This review underscores the fundamental principles and optimization of CRISPR systems, emphasizing their application in crop enhancement, particularly drought tolerance, yield improvement, and domestication (Pandey et al. 2022). Additionally, we explore how innovative genome editing tools can aid in identifying and modifying novel genes associated with drought tolerance (Haider et al. 2022b). We foresee the development of effective strategies for enhancing crop productivity in water-limited regions through collaborative efforts in the foreseeable future. In the realm of sustainable agriculture, the contribution of drought-tolerant varieties of faba bean to food security and sustainable farming practices is noteworthy (Polania et al. 2020; Haider et al. 2022b; Varshney et al. 2010).

10.16 Conclusion

Utilizing marker-assisted breeding (MAB) in faba beans to enhance drought tolerance marks a pivotal advancement in crop improvement strategies. One of the main factors limiting faba bean yield is drought stress, especially in dry and semi-arid areas. Due to the complexity of drought tolerance, which involves several physiological and biochemical processes, conventional breeding techniques have had difficulty producing drought-tolerant cultivars. MAB offers a promising approach by enabling the precise identification and selection of desirable traits at the genetic level. This method employs molecular markers linked to drought tolerance traits, facilitating the selection of superior genotypes even at early growth stages. Integrating MAB with traditional breeding methodologies accelerates the development of drought-tolerant faba bean varieties. The success of MAB depends on the availability of genetic diversity and reliable molecular markers associated with drought tolerance. Advances in genomic technologies, such as genome-wide association studies (GWAS) and quantitative trait locus (QTL) mapping, have been crucial in identifying these markers in faba beans. Incorporating MAB into faba bean breeding programs enhances the efficiency and precision of selecting drought-tolerant genotypes, reducing the breeding cycle duration and increasing the likelihood of developing resilient varieties. Integrating phenotypic data with genotypic information is essential for the effective deployment of MAB. High-throughput phenotyping techniques, coupled with advanced genomic tools, allow breeders to correlate specific markers with drought tolerance traits, ensuring a more targeted breeding process. Despite its potential, MAB in faba beans faces challenges such as the need for high-quality reference genomes, the complexity of drought tolerance traits, and the requirement for multidisciplinary collaboration. Future efforts should focus on overcoming these obstacles by developing more precise markers, enhancing genomic resources, and incorporating novel biotechnological tools. In conclusion, MAB holds significant potential for improving drought tolerance in faba beans. By leveraging genetic markers, breeders can more efficiently develop varieties capable of withstanding drought stress, thereby promoting food security and sustainable agriculture. Continued research and technological advancements will further enhance the promise of MAB, ensuring the cultivation of robust, drought-tolerant faba bean cultivars.

Author Contributions Z.U J.I, B.A.A, and S.I. designed and conceived the chapter idea. Z.U., J.I, B.A.A, and Z.K., A.M., T.Y., M.M., N.I. wrote the original draft. N.I., A.F., S.K., T.M. reviewed and edited the manuscript and provided valuable insights. All authors have made valuable revisions and edited the manuscript and approved the last version.

References

Abd EL-Mageed AM, Mahmoud SA, Emam MA, Abu-Elsaoud AM, Sabry SA (2022) Genetic Variability and ISSR Markers of some Faba bean (Vicia faba L.) cultivars under drought condition. Curr Sci Int 11(04):365–377

Abdelraouf EAA, Adss IAA, Dakroury MZ (2016) Effect of salinity on growth and genetic diversity of broad bean (Vicia faba L.) cultivars. Alexandria Sci Exchange J 37(July–September):467–479

Abid G, Saidi MN, Ouertani RN, Muhovski Y, Jebara SH, Ghouili E et al (2021) Differential gene expression reveals candidate genes for osmotic stress response in faba bean (Vicia faba L.) involved in different molecular pathways. Acta Physiol Plant 43:1–20

Abou-Khater L, Maalouf F, Rubiales D (2022) Status of faba bean (Vicia faba L.) in the Mediterranean and East African countries. In: Developing climate resilient grain and forage legumes. Springer Nature Singapore, Singapore, pp 297–327

Adhikari KN, Khater LA, Maalouf F (2022) Application of genetic, genomic strategies to address the biotic stresses in Faba bean. In: Genomic designing for biotic stress resistant pulse crops. Springer International Publishing, Cham, pp 353–380

Afiah SA, Zinab Z, El-Gawad AA, Mohamed TR, Al-Agwany HH (2016) Salt and water deficit tolerance in some Vicia faba L. genotypes in relation to pigments, ISSR-PCR markers and stress tolerance indices. Egypt J Bot 56(1):121–144

Ahmad F, Akram A, Farman K, Abbas T, Bibi A, Khalid S, Waseem M (2017) Molecular markers and marker assisted plant breeding: current status and their applications in agricultural development. J Environ Agric Sci 11:35–50

Ahmad Y, Haider S, Iqbal J, Abbasi BA, Yaseen T, Mahmood T (2022) The mechanisms of genome editing technologies in crop plants. In: Principles and practices of OMICS and genome editing for crop improvement. Springer International Publishing, Cham, pp 295–313

Ahmad Y, Haakim Z, Iqbal J, Abbasi BA, Mahmood T, Kazi M (2024a) Technological innovations for abiotic stress resistance in horticultural crops. In: OMICs-based techniques for global food security, pp 233–244. https://doi.org/10.1002/9781394209156.ch12

Ahmad Y, Haider S, Iqbal J, Naseer S, Attia KA, Mohammed AA, Mahmood T (2024b) In-silico analysis and transformation of OsMYB48 transcription factor driven by CaMV35S promoter in model plant–Nicotiana tabacum L. conferring abiotic stress tolerance. GM Crops Food 15(1):130–149

Alghamdi SS, Migdadi HM, Ammar MH, Paull JG, Siddique KHM (2012) Faba bean genomics: current status and future prospects. Euphytica 186:609–624

Alharbi NH, Adhikari KN (2020) Factors of yield determination in faba bean (Vicia faba). Crop Pasture Sci 71(4):305–321

Ali MB, Welna GC, Sallam A, Martsch R, Balko C, Gebser B et al (2016) Association analyses to genetically improve drought and freezing tolerance of faba bean (Vicia faba L.). Crop Sci 56(3):1036–1048

Ali I, Salah KBH, Sher H, Ali H, Ullah Z, Ali A, Mahmood T (2022) Drought stress enhances the efficiency of floral dip method of Agrobacterium-mediated transformation in Arabidopsis thaliana. Braz J Biol 84:e259326

Arbaoui M, Link W, Satovic Z, Torres AM (2008) Quantitative trait loci of frost tolerance and physiologically related trait in faba bean (Vicia faba L.). Euphytica 164:93–104

Arya H, Singh MB, Bhalla PL (2021) Towards developing drought-smart soybeans. Front Plant Sci 12:750664

Asfaw A, Blair MW, Struik PC (2012) Multienvironment quantitative trait loci analysis for photosynthate acquisition, accumulation, and remobilization traits in common bean under drought stress. G3 Genes|Genomes|Genetics 2(5):579–595

Asif M, Shaheen T, Tabbasam N, Zafar Y, Paterson AH (2011) Marker-assisted breeding in higher plants. In: Alternative farming systems, biotechnology, drought stress and ecological fertilisation. Springer, Berlin, pp 39–76

Assefa T, Mahama AA, Brown AV, Cannon EK, Rubyogo JC, Rao IM et al (2019) A review of breeding objectives, genomic resources, and marker-assisted methods in common bean (*Phaseolus vulgaris* L.). Mol Breeding 39:20

Attamah P, Kusi F, Kena AW, Awuku FJ, Lamini S, Mensah G et al (2024) Pyramiding aphid resistance genes into the elite cowpea variety, Zaayura, using marker-assisted backcrossing. Heliyon 10:e31976

Baum M (2003) Biotechnological approaches to crop improvement in the dry areas at the International Centre for Agricultural Research in the dry areas. Afr J Food Agric Nutr Dev 3(1):39–44

Belachew KY, Nagel KA, Poorter H, Stoddard FL (2019) Association of shoot and root responses to water deficit in young faba bean (Vicia faba L.) plants. Front Plant Sci 10:1063

Berny Mier y Teran JC, Konzen ER, Palkovic A, Tsai SM, Rao IM, Beebe S, Gepts P (2019) Effect of drought stress on the genetic architecture of photosynthate allocation and remobilization in pods of common bean (Phaseolus vulgaris L.), a key species for food security. BMC Plant Biol 19:1–15

Bibi S, Ullah R, Burni T, Ullah Z, Kazi M (2023) Impact of resorcinol and biochar application on the growth attributes, metabolite contents, and antioxidant systems of tomato (Lycopersicon esculentum Mill.). ACS Omega 8(48):45750–45762

Blair MW, Fregene MA, Beebe SE, Ceballos H (2007) Marker-assisted selection in common beans and cassava. Marker-assisted selection. In: Current status and future perspectives in crops, livestock, forestry and fish. Food and Agriculture Organization of the United Nations, Rome, pp 81–115

Bohra A, Pandey MK, Jha UC, Singh B, Singh IP, Datta D, Varshney RK (2014) Genomics-assisted breeding in four major pulse crops of developing countries: present status and prospects. Theor Appl Genet 127:1263–1291

Bohra A, Saxena KB, Varshney RK, Saxena RK (2020) Genomics-assisted breeding for pigeonpea improvement. Theor Appl Genet 133(5):1721–1737

Boopathi NM (2013) Genetic mapping and marker assisted selection. Springer, India
Bot EJ (2016) Salt and water deficit tolerance in some Vicia faba L. genotypes in relation to pigments. ISSR–PCR Markers and Stress Tolerance Indices
Bourgault M (2009) Legume production in semi-arid areas: comparative study of the physiology of drought tolerance in common bean («Phaseolus vulgaris L.») and mungbean «(Vigna radiata (L.) Wilczek)»
Braich S, Sudheesh S, Forster JW, Kaur S (2017) Characterisation of faba bean (Vicia faba L.) transcriptome using RNA-seq: sequencing, de novo assembly, annotation, and expression analysis. Agronomy 7(3):53
Buitrago-Bitar MA, Cortés AJ, López-Hernández F, Londoño-Caicedo JM, Muñoz-Florez JE, Muñoz LC, Blair MW (2021) Allelic diversity at abiotic stress responsive genes in relationship to ecological drought indices for cultivated tepary bean, Phaseolus acutifolius A. Gray, and its wild relatives. Genes 12(4):556
Caracuta V, Vardi J, Paz Y, Boaretto E (2017) Farming legumes in the pre-pottery Neolithic: New discoveries from the site of Ahihud (Israel). PloS one 12(5), e0177859
Carrillo-Perdomo E, Vidal A, Kreplak J, Duborjal H, Leveugle M, Duarte J et al (2020) Development of new genetic resources for faba bean (Vicia faba L.) breeding through the discovery of gene-based SNP markers and the construction of a high-density consensus map. Sci Rep 10(1):6790
Carrillo-Perdomo E, Magnin-Robert JB, Raffiot B, Deulvot C, Floriot M, Lejeune-Hénaut I et al (2022) A QTL approach in faba bean highlights the conservation of genetic control of frost tolerance among legume species. Front Plant Sci 13:970865
Cooper JW, Wilson MH, Derks MF, Smit S, Kunert KJ, Cullis C, Foyer CH (2017) Enhancing faba bean (Vicia faba L.) genome resources. J Exp Bot 68(8):1941–1953
Cortés AJ, Blair MW (2018) Genotyping by sequencing and genome–environment associations in wild common bean predict widespread divergent adaptation to drought. Front Plant Sci 9:128
De la Rosa L, Zambrana E, Ramirez-Parra E (2020) Molecular bases for drought tolerance in common vetch: designing new molecular breeding tools. BMC Plant Biol 20:1–18
De Ron AM, Kalavacharla V, Álvarez-García S, Casquero PA, Carro-Huelga G, Gutiérrez S et al (2019) Common bean genetics, breeding, and genomics for adaptation to changing to new agri-environmental conditions. In: Genomic designing of climate-smart pulse crops. Springer, pp 1–106
Dhungana SK, Park JH, Oh JH, Kang BK, Seo JH, Sung JS et al (2021) Quantitative trait locus mapping for drought tolerance in soybean recombinant inbred line population. Plan Theory 10(9):1816
Diaz S, Ariza-Suarez D, Izquierdo P, Lobaton JD, de La Hoz JF, Acevedo F et al (2020) Genetic mapping for agronomic traits in a MAGIC population of common bean (Phaseolus vulgaris L.) under drought conditions. BMC Genomics 21:1–20
Disasa T, Gebre E (2018). Harnessing modern breeding tools, techniques, and approaches to improve yield and quality in legume crops. Ethiopian J Crop SciEthiop. J. Crop Sci. Special Issue Vol. 6 No. 3 (2018)
Dubey A, Kumar A, Abd_Allah EF, Hashem A, Khan ML (2019) Growing more with less: breeding and developing drought resilient soybean to improve food security. Ecol Indic 105:425–437
Duc G, Bao S, Baum M, Redden B, Sadiki M, Suso MJ et al (2010) Diversity maintenance and use of Vicia faba L. genetic resources. Field Crop Res 115(3):270–278
Elshaer H, Hellal AG (2020) Genetic diversity based on SSR markers and morphological analysis of faba bean (Vicia faba L.) in Egypt. Taeckholmia 40(1):130–142
Faridi Myvan F, Jami Al-Ahmadi M, Eslami SV, Shojaei Noferest K (2022) Role of Potassium in modifying the potato physiological responses to irrigation regimes under different planting patterns. Potato Research 65(3):581–600
Fouad M, Mohammed N, Aladdin H, Ahmed A, Xuxiao Z, Shiying B, Tao Y (2013) Faba bean. In: Genetic and genomic resources of grain legume improvement. Elsevier, pp 113–136
Gaur PM, Jukanti AK, Varshney RK (2012) Impact of genomic technologies on chickpea breeding strategies. Agronomy 2(3):199–221

Ghatak A, Chaturvedi P, Weckwerth W (2017) Cereal crop proteomics: systemic analysis of crop drought stress responses towards marker-assisted selection breeding. Front Plant Sci 8:757

Ghatak A, Chaturvedi P, Weckwerth W (2018) Metabolomics in plant stress physiology. Plant genetics and molecular biology 187–236

Göl Ş, Doğanlar S, Frary A (2017) Relationship between geographical origin, seed size and genetic diversity in faba bean (Vicia faba L.) as revealed by SSR markers. Mol Gen Genomics 292:991–999

Gupta VS, Gowda M, Kadoo NY, Radhika P, Sainani MN (2008) Quantitative trait loci (QTL) mapping and marker assisted selection (MAS) in food legumes. In Proceedings of the fourth international food legumes research conference, vol 1, pp 277–288

Gutiérrez N, Pégard M, Balko C, Torres AM (2023) Genome-wide association analysis for drought tolerance and associated traits in faba bean (Vicia faba L.). Front Plant Sci 14:1091875

Haider S, Iqbal J, Naseer S, Yaseen T, Shaukat M, Bibi H, Mahmood T (2021a) Molecular mechanisms of plant tolerance to heat stress: current landscape and future perspectives. Plant Cell Rep 40:1–25

Haider S, Iqbal J, Shaukat M, Naseer S, Mahmood T (2021b) The epigenetic chromatin-based regulation of somatic heat stress memory in plants. Plant Gene 27:100318

Haider S, Raza A, Iqbal J, Shaukat M, Mahmood T (2022a) Analyzing the regulatory role of heat shock transcription factors in plant heat stress tolerance: a brief appraisal. Mol Biol Rep 49(6):5771–5785

Haider S, Iqbal J, Naseer S, Shaukat M, Abbasi BA, Yaseen T, Mahmood T (2022b) Unfolding molecular switches in plant heat stress resistance: a comprehensive review. Plant Cell Rep 41(3):775–798

Hassan MA, Dahu N, Hongning T, Qian Z, Yueming Y, Yiru L, Shimei W (2023) Drought stress in rice: morpho-physiological and molecular responses and marker-assisted breeding. Front Plant Sci 14:1215371

Hauggaard-Nielsen H, Peoples MB, Jensen ES (2011) Faba bean in cropping systems. Grain Legumes 56:32–33

Ijaz S, Iqbal J, Abbasi BA, Tufail A, Ullah Z, Sharifi-Rad J et al (2023a) Biofortification: lessons from the past and strategies for future food security. In: Legumes biofortification. Springer International Publishing, Cham, pp 521–545

Ijaz S, Iqbal J, Abbasi BA, Ullah Z, Yaseen T, Kanwal S et al (2023b) Rosmarinic acid and its derivatives: current insights on anticancer potential and other biomedical applications. Biomed Pharmacother 162:114687

Ijaz S, Iqbal J, Abbasi BA, Tufail A, Ullah Z, Yaseen T, Iqbal R (2024) Heavy metal–polluted arable land and its consequences: a global scenario. In: Biochar-assisted remediation of contaminated soils under changing climate. Elsevier, pp 71–99

İlker Y, Sarikaya MF, Tatar M, Kaddour RB, Karaköy T (2023) Characterization of some Turkish Faba bean (Vicia faba L.) genotypes for agro-morphological traits. ISPEC J Agric Sci 7(1):116–134

Jacob C, Carrasco B, Schwember AR (2016) Advances in breeding and biotechnology of legume crops. Plant Cell Tiss Organ Cult 127:561–584

Kamaluddin, Khan MA, Kiran U, Ali A, Abdin MZ, Zargar MY et al (2017) Molecular markers and marker-assisted selection in crop plants. In: Plant biotechnology: principles and applications. Springer, pp 295–328

Kashiwagi J, Krishnamurthy L, Gaur PM, Chandra S, Upadhyaya HD (2008) Estimation of gene effects of the drought avoidance root characteristics in chickpea (C. arietinum L.). Field Crop Res 105(1–2):64–69

Kelly JD, Bornowski N (2018) Marker-assisted breeding for economic traits in common bean. In: Biotechnologies of crop improvement, volume 3: genomic approaches. Springer, pp 211–238

Kelly JD, Gepts P, Miklas PN, Coyne DP (2003) Tagging and mapping of genes and QTL and molecular marker-assisted selection for traits of economic importance in bean and cowpea. Field Crop Res 82(2–3):135–154

Khatun M, Sarkar S, Era FM, Islam AM, Anwar MP, Fahad S et al (2021) Drought stress in grain legumes: effects, tolerance mechanisms and management. Agronomy 11(12):2374

Kondwakwenda A, Sibiya J, Zengeni R, Musvosvi C (2021) Aspects in breeding maize for drought tolerance: progress and modern breeding approaches. Aust J Crop Sci 15(4):510–517

Kumar J, Sen Gupta D, Baum M, Varshney RK, Kumar S (2021) Genomics-assisted lentil breeding: current status and future strategies. Legume Sci 3(3):e71

Kumawat G, Kumawat CK, Chandra K, Pandey S, Chand S, Mishra UN, Sharma R (2020) Insights into marker assisted selection and its applications in plant breeding. In: Plant breeding-current and future views. Intechopen

Lavania D, Siddiqui MH, Al-Whaibi MH, Singh AK, Kumar R, Grover A (2015) Genetic approaches for breeding heat stress tolerance in faba bean (Vicia faba L.). Acta Physiol Plant 37:1–9

Leonforte A, Sudheesh S, Cogan NO, Salisbury PA, Nicolas ME, Materne M et al (2013) SNP marker discovery, linkage map construction and identification of QTLs for enhanced salinity tolerance in field pea (Pisum sativum L.). BMC Plant Biol 13:1–14

Li Y, Ruperao P, Batley J, Edwards D, Khan T, Colmer TD et al (2018) Investigating drought tolerance in chickpea using genome-wide association mapping and genomic selection based on whole-genome resequencing data. Front Plant Sci 9:190

Li MW, He YH, Rong LIU, Guan LI, Dong W, Ji YS et al (2023) Construction of SNP genetic maps based on targeted next-generation sequencing and QTL mapping of vital agronomic traits in faba bean (Vicia faba L.). J Integrat Agric 22(9):2648–2659

Liaqat W, Altaf MT, Barutçular C, Mohamed HI, Ali Z, Khan MO (2024) Drought stress in sorghum: physiological tools, breeding technology, omics approaches and genomic-assisted breeding-A review. J Soil Sci Plant Nutr 24:1–27

Liu S, Banik M, Yu K, Park SJ, Poysa V, Guan Y (2007) Marker-assisted selection (MAS) in major cereal and legume crop breeding: current progress and future directions. Int J Plant Breed 1(2):75–78

Maalouf F (2017) Trends in science-Faba bean-improving faba bean for sustainable agriculture in dry areas

Maalouf F, Muhammad I, Kumar S, Malhotra R (2011). Breeding food legumes for enhanced drought and heat tolerance to cope with climate change. Food security and climate change in dry areas, p 244

Maalouf F, Ahmed S, Shaaban K, Bassam B, Nawar F, Singh M, Amri A (2016) New faba bean germplasm with multiple resistances to Ascochyta blight, chocolate spot and rust diseases. Euphytica 211:157–167

Maalouf F, Ahmed S, Patil S (2017) CH31-developing improved varieties of faba bean. In Review Meeting

Maalouf F, Khater LA, Babiker Z, Mohamed A (2022) Genetic and genomic research for abiotic stresses in Faba bean. In: Genomic designing for abiotic stress resistant pulse crops. Springer International Publishing, Cham, pp 249–270

Maga TJ (2021) Evaluation of faba bean (Vicia faba L.) genotypes for drought tolerance under the subtropical region of New South Wales (Doctoral dissertation)

Maiti RK, Singh VP (2016) Mechanisms of resistance to drought, temperature and salinity in bean crops-a review. Farming Manag 1(2):134–161

Mandour H, Khazaei H, Stoddard FL, Dodd IC (2023) Identifying physiological and genetic determinants of faba bean transpiration response to evaporative demand. Ann Bot 131(3):533–544

Maqbool MA, Aslam M, Ali H (2017) Breeding for improved drought tolerance in Chickpea (Cicer arietinum L.). Plant Breed 136(3):300–318

Maurya V, Singh N, Sharma A, Kumar R (2023) DNA markers-assisted crop improvement for biotic and abiotic stresses in legumes. In: Molecular marker techniques: a potential approach of crop improvement. Springer Nature Singapore, Singapore, pp 161–197

Mercati F, Sunseri F (2020) Genetic diversity assessment and marker-assisted selection in crops. Genes 11(12):1481

Mir RR, Zaman-Allah M, Sreenivasulu N, Trethowan R, Varshney RK (2012) Integrated genomics, physiology and breeding approaches for improving drought tolerance in crops. Theor Appl Genet 125:625–645

Mitache M, Baidani A, Zeroual A, Bencharki B, Idrissi O (2023) Session 6 Crop breeding: challenges, tools, strategies and achievements. Welcome Message 4 Key Speakers 5 Session Abstracts 15, 134

Mwadzingeni L, Figlan S, Shimelis H, Mondal S, Tsilo TJ (2017) Genetic resources and breeding methodologies for improving drought tolerance in wheat. J Crop Improv 31(5):648–672

Nabateregga M, Mukankusi C, Raatz B, Edema R, Nkalubo S, Alladassi BME (2019) Quantitative trait loci (QTL) mapping for intermittent drought tolerance in BRB 191× SEQ 1027 Andean Intragene cross recombinant inbred line population of common bean (Phaseolus vulgaris L.). Afr J Biotechnol 18(21)

Nadeem MA, Nawaz MA, Shahid MQ, Doğan Y, Comertpay G, Yıldız M et al (2018) DNA molecular markers in plant breeding: current status and recent advancements in genomic selection and genome editing. Biotechnol Biotechnol Equip 32(2):261–285

Nkomo GV, Sedibe MM, Mofokeng MA (2021) Production constraints and improvement strategies of cowpea (Vigna unguiculata L. Walp.) genotypes for drought tolerance. Int J Agron 2021(1):5536417

Ohm H, Åstrand J, Ceplitis A, Bengtsson D, Hammenhag C, Chawade A, Grimberg Å (2024) Novel SNP markers for flowering and seed quality traits in faba bean (Vicia faba L.): characterization and GWAS of a diversity panel. Front Plant Sci 15:1348014

Ortiz R (2013) Drought tolerance. In: Genomics and breeding for climate-resilient crops: Vol. 2 Target traits. Springer, pp 203–223

Ouji A, Naouari M, Mouelhi M, Ben Younes M (2017) Yield and yield components of faba bean (Vicia faba L.) as influenced by supplemental irrigation under semi-arid region of Tunisia. World J Agric Res 5:52–57

Pandey S, Singh A, Parida SK, Prasad M (2022) Combining speed breeding with traditional and genomics-assisted breeding for crop improvement. Plant Breed 141(3):301–313

Polania JA, Chater CC, Covarrubias AA, Rao IM (2020) Phaseolus species responses and tolerance to drought. In: The plant family Fabaceae: biology and physiological responses to environmental stresses. Springer Nature, Singapore, pp 319–336

Prasanna BM, Babu R, Nair S, Semagn K, Chaikam V, Cairns J et al (2014) Molecular marker-assisted breeding for tropical maize improvement. In: Genetics, genomics and breeding of maize. CRC Press, London, pp 89–119

Puspitasari W, Allemann B, Angra D, Appleyard H, Ecke W, Möllers C et al (2022) NIRS for vicine and convicine content of faba bean seed allowed GWAS to prepare for marker-assisted adjustment of seed quality of German winter faba beans. J Cultivated Plants 74(01–02)

Rai SK, Pandey-Rai S (2021) Marker-assisted breeding for abiotic stress tolerance in horticultural crops. In: Stress tolerance in horticultural crops. Woodhead Publishing, pp 63–74

Rauf S, Al-Khayri JM, Zaharieva M, Monneveux P, Khalil F (2016) Breeding strategies to enhance drought tolerance in crops. In: Advances in plant breeding strategies: agronomic, abiotic and biotic stress traits. Springer, pp 397–445

Ray H, Georges F (2010) A genomic approach to nutritional, pharmacological and genetic issues of faba bean (Vicia faba): prospects for genetic modifications. GM Crops 1(2):99–106

Raza A, Mubarik MS, Sharif R, Habib M, Jabeen W, Zhang C et al (2023) Developing drought-smart, ready-to-grow future crops. Plant Genome 16(1):e20279

Redden R, Paull J, Zong X, Sass O, Yang T, Ling L (2014) Faba bean. In: Broadening the genetic base of grain legumes. Springer, pp 75–93

Rispail N, Kaló P, Kiss GB, Ellis TN, Gallardo K, Thompson RD et al (2010) Model legumes contribute to faba bean breeding. Field Crop Res 115(3):253–269

Roy AK, Malaviya DR, Kaushal P (2016) Genetic improvement of fodder legumes especially dual purpose pulses. Indian J Genet Plant Breed 76(04):608–625

Rubiales D, Khazaei H (2022) Advances in disease and pest resistance in faba bean. Theor Appl Genet 135(11):3735–3756

Sabry SA, Abd EL-Mageed AM, Mahmoud SA, Emam MA (2022) Assessing drought tolerance of some Faba bean (Vicia faba L.) cultivars using genetic variability, tolerance indices and ISSR markers. J Plant Product Sci 11(1):111–121

Salgotra RK, Stewart CN Jr (2020) Functional markers for precision plant breeding. Int J Mol Sci 21(13):4792

Sallam A (2014) Detailed genetic approach to improve frost tolerance of German winter faba beans (Doctoral dissertation, Niedersächsische Staats-und Universitätsbibliothek Göttingen)

Sallam A, Ul-Allah S (2019) Genomics-aided breeding for climate-smart traits in faba bean. In: Genomic designing of climate-smart pulse crops. Springer, pp 359–395

Sallam A, Dhanapal AP, Liu S (2016) Association mapping of winter hardiness and yield traits in faba bean (Vicia faba L.). Crop Pasture Sci 67(1):55–68

Savita, Tomer A, Singh SK (2020) Drought stress tolerance in legume crops. In: Agronomic crops: volume 3: stress responses and tolerance. Springer, pp 149–155

Schneider KA, Brothers ME, Kelly JD (1997a) Marker-assisted selection to improve drought resistance in common bean. Crop Sci 37(1):51–60

Schneider KA, Rosales-Serna R, Ibarra-Perez F, Cazares-Enriquez B, Acosta-Gallegos JA, Ramirez-Vallejo P et al (1997b) Improving common bean performance under drought stress. Crop Sci 37(1):43–50

Sharma HC, Crouch JH (2004) Molecular marker-assisted selection: a novel approach for host plant resistance to insects in grain legumes

Singh AK, Bhatt BP, Upadhyaya A, Kumar S, Sundaram PK, Singh BK et al (2012) Improvement of faba bean (Vicia faba L.) yield and quality through biotechnological approach: A review. Afr J Biotechnol 11(87):15264–15271

Singh AK, Bharati RC, Manibhushan NC, Pedpati A (2013) An assessment of faba bean (Vicia faba L.) current status and future prospect. Afr J Agric Res 8(50):6634–6641

Singh CM, Singh P, Tiwari C, Purwar S, Kumar M, Pratap A et al (2021) Improving drought tolerance in mungbean (Vigna radiata L. Wilczek): morpho-physiological, biochemical and molecular perspectives. Agronomy 11(8):1534

Sofi PA, Rehman K, Gull M, Kumari J, Djanaguiraman M, Prasad PVV (2021) Integrating root architecture and physiological approaches for improving drought tolerance in common bean (Phaseolus vulgaris L.). Plant Physiol Rep 26:4–22

Thudi M, Samineni S, Li W, Boer MP, Roorkiwal M, Yang Z et al (2023) Whole genome resequencing and phenotyping of MAGIC population for high resolution mapping of drought tolerance in chickpea. Plant Genome 17(1):e20333

Tiwari JK, Yerasu SR, Rai N, Singh DP, Singh AK, Karkute SG, ... Behera TK (2022) Progress in marker-assisted selection to genomics-assisted breeding in tomato. Critical Reviews in Plant Sciences 41(5):321–350

Tock AJ (2017). Applying next-generation sequencing to enable marker-assisted breeding for adaptive traits in common bean (Phaseolus vulgaris L.) (Doctoral dissertation, University of Warwick)

Torres AM, Avila CM, Gutierrez N, Palomino C, Moreno MT, Cubero JI (2010) Marker-assisted selection in faba bean (Vicia faba L.). Field Crop Res 115(3):243–252

Trapp JJ (2015) Genetics of drought tolerance in common bean (Phaseolus vulgaris L.). Washington State University

Tryphone GM, Chilagane LA, Protas D, Kusolwa PM, Nchimbi-Msolla S (2013) Marker assisted selection for common bean diseases improvements in Tanzania: prospects and future needs. Plant breeding from laboratories to fields. INTECH 22:121–147

Ullah Z (2017) Estimation of genetic diversity in the germplasm of brassica Campestris (L.) using morphological and biochemical markers. J Bot 1(1):45–59

Ullah R, Ullah Z, Iqbal J, Chalgham W, Ahmad A (2023a) Aspartic acid-based nano-copper induces resilience in Zea mays to applied Lead stress via conserving photosynthetic pigments and triggering the antioxidant biosystem. Sustain For 15(16):12186

Ullah Z, Gul F, Iqbal J, Abbasi BA, Kanwal S, Chalgham W et al (2023b) Biogenic synthesis of multifunctional silver oxide nanoparticles (Ag2ONPs) using Parieteria alsinaefolia delile aqueous extract and assessment of their diverse biological applications. Microorganisms 11(4):1069

Ullah Z, Iqbal J, Abbasi BA, Akhtar W, Kanwal S, Ali I et al (2023c) Assessment of gus expression induced by anti-sense Os PPO gene promoter and antioxidant enzymatic assays in response to drought and heavy metal stress in transgenic Arabidopsis thaliana. Sustain For 15(17):12783

Ullah Z, Iqbal J, Abbasi BA, Ijaz S, Sharifi-Rad J, Yaseen T et al (2023d) How genome editing can be helpful in the biofortification of legumes. In: Legumes biofortification. Springer International Publishing, Cham, pp 207–232

Ullah S, Shakir L, Subhan G, Sohail M, Bilqees R, Khan Y, Ali S (2024a) Phytodiversity and conservation assessment of Ethnobotanically significant Flora in Khall Hagram Dara, lower Dir, Khyber Pakhtunkhwa. Pakistan Plant Protect 8(1):143–162

Ullah Z, Iqbal J, Abbasi BA, Gul F, Ijaz S, Kanwal S et al (2024b) Rhynchosia capitata driven bioproduction of Zinc oxide nanoparticles, characterization and multifaceted therapeutic applications

Ullah Z, Iqbal J, Gul F, Abbasi BA, Kanwal S, Elsadek MF et al (2024c) Biogenic synthesis, characterization, and in vitro biological investigation of silver oxide nanoparticles (AgONPs) using Rhynchosia capitata. Sci Rep 14(1):10484

Varshney RK (2016) Exciting journey of 10 years from genomes to fields and markets: some success stories of genomics-assisted breeding in chickpea, pigeonpea and groundnut. Plant Sci 242:98–107

Varshney RK, Thudi M, May GD, Jackson SA (2010) Legume genomics and breeding. Plant Breed Rev 33:257–304

Varshney RK, Pazhamala L, Kashiwagi J, Gaur PM, Krishnamurthy L, Hoisington D (2011) Genomics and physiological approaches for root trait breeding to improve drought tolerance in chickpea (Cicer arietinum L.). Root Genomics:233–250

Varshney RK, Barmukh R, Roorkiwal M, Qi Y, Kholova J, Tuberosa R et al (2021a) Breeding custom-designed crops for improved drought adaptation. Adv Genet 2(3):e202100017

Varshney RK, Bohra A, Yu J, Graner A, Zhang Q, Sorrells ME (2021b) Designing future crops: genomics-assisted breeding comes of age. Trends Plant Sci 26(6):631–649

Wang HF, Zong XX (2011) Advances in research of genetic resources, breeding and QTL for disease resistance and resistance to abiotic stresses on Vicia faba L. J Plant Genet Resources 12(2):259–270

Wu L, Chang Y, Wang L, Wu J, Wang S (2021) Genetic dissection of drought resistance based on root traits at the bud stage in common bean. Theor Appl Genet 134:1047–1061

Zong X, Yang T, Liu R (2019) Faba bean (Vicia faba L.) breeding. In: Advances in plant breeding strategies: legumes: Volume 7. Springer, pp 245–286

Marker-Assisted Breeding in Lentil for Drought Tolerance

11

Afifa Younas, Sajid Fiaz, Nadia Riaz, Madiha Rashid, Zahra Noreen, Aasma Tufail, Maha Noor, Sundas Fatima, and Hurmat Mehdi Mirrani

Abstract

Lentil is a winter sown crop for nutrition and food availability source. The legumes widely cultivated in the Mediterranean and semi-arid regions are exposed to various stresses. Drought stress is a major limiting factor which causes deleterious effects on lentil production. In this chapter, various factors and the morpho-physiological and biochemical effects of drought on lentil are discussed. Additionally, marker-assisted breeding intervention can improve lentil production under drought-stress conditions. The drought stress affects the early growth, photosynthesis, and root and shoot water relations, which leads to reduction in the yield of lentil. Conventional and modern breeding can improve the drought stress tolerance. Recent ecofriendly biotechnologies and agronomic interventions may also participate to improve the drought stress tolerance. Due

A. Younas (✉)
Department of Botany, Lahore College for Women University, Lahore, Pakistan

Department of Botany, Baba Guru Nanak University, Nankana Sahib, Nankana, Pakistan

S. Fiaz
Institute of Molecular Biology and Biotechnology, The University of Lahore, Lahore, Pakistan

N. Riaz
Department of Botany, Lahore College for Women University, Lahore, Pakistan

M. Rashid
Department of Botany, Division of Science and Technology, University of Education, Lahore, Pakistan

Department of Botany, Lahore College for Women University, Lahore, Pakistan

Z. Noreen · A. Tufail · M. Noor · S. Fatima · H. M. Mirrani
Department of Botany, Division of Science and Technology, University of Education, Lahore, Pakistan

© The Author(s), under exclusive license to Springer Nature Singapore Pte Ltd. 2025

M. A. Nadeem et al. (eds.), *Marker-Assisted Breeding in Legumes for Drought Tolerance*, https://doi.org/10.1007/978-981-96-4112-3_11

to all the latest biotechnological tools and agronomic practices can improve the drought tolerance improve crop yield.

Keywords

Lentil · Morphology · Physiology · Biochemical reaction · Marker-assisted breeding · Biotechnology · Omics approaches

11.1 Introduction

Lentils are a type of leguminous plant that is high in fiber and low in fat. They have more total soluble fiber than chickpeas and peas and contain higher dietary fiber content than beans and chickpeas. Lentils are also a great source of protein, with storage proteins found in the cotyledon. These proteins have a low percentage of sulfur-containing amino acids. The proteins in lentils are made up of albumins, globulins, glutelins, and prolamins. Globulins contain legumin- and vicilin-like proteins, with specific subunits and molecular weights. Based on centrifugation studies, these proteins are classified as 11S or 13S, and 7S proteins with specific characteristics have also been isolated from seed extracts. The isoelectric points of lentil proteins vary, and all protein fractions are glycosylated. Vicilins contain a significant amount of carbohydrates. Lentils' protein chemistry and structure are complex, with different protein groups and characteristics identified through various studies. Pulses like lentils have high antioxidant capacity due to their phenolic content. Lentils exhibit significant antioxidant activities and longer LDL oxidation lag times than other pulses. Peptides produced during the gastrointestinal digestion of lentils have dual antioxidant and ACE inhibitory activities. Hydrolysates from lentils show good ACE inhibitory activity. The protein quality of lentils can be improved through processes such as extrusion, baking, and cooking. Red lentils have higher protein digestibility scores compared to green lentils, and extruded lentil flour shows the highest scores. Lentils are linked to various health benefits, including reduced cholesterol levels, decreased risk of colon cancer, and effective management of type-2 diabetes. They contain high levels of phenolic compounds, flavonoids, and condensed tannins. Land plants evolved around 470 million years ago (during the Paleozoic era). The divergence and evolution of the Leguminosae plant family are speculative, based on fossils and phylogenetic records. Fossil records suggest that legumes originated in arid regions along the Tethys seaway (during the early Tertiary period). Despite documented legume diversity in tropical regions, fossil records oppose the theory of Mesozoic origin and diversification. The understanding of Leguminosae origin is complex due to the mass extinction event during the Cretaceous-Paleogene boundary, which led to increased plant origination and diversification (Kislev and Bar-Yosef 1988). The evolution and divergence of legumes into several clades and subclades are attributed to whole-genome duplication (WGD) events and allopolyploidy events among the earliest lineages. The Leguminosae family currently consists of over 19,000 extant species found in diverse ecological settings. Agriculture in the Near East saw the planting of food legumes alongside wheat and barley (Zohary and Hopf 1973). Lentils are believed to be the first grain legume domesticated around 11,000 BC, according to Bahl et al. (1993). Other pulses such as *Phaseolus vulgaris* and *Glycine max* have also been propagated as agriculture has expanded. Current domestications include pigeon pea and mung bean. Around 4000 years ago in South Asia, alfalfa and narrow-leaved lupine were domesticated. Alfalfa was domesticated in Roman times, while narrow-leaved lupine was propagated as a sweet lupine over the past century. The domestication of faba bean has been unclear due to the absence of a known compatible wild relative, but archaeological evidence is shedding light on the timeline of domestication.

Lentil (*Lens culinaris* Medik.) is one of the oldest cultivated crops globally. It is known by various names like split peas or red dhal and is a staple food rich in fiber, vitamins A, K, and B, and Fe. Lentils play a crucial role in crop rotation, improving soil fertility and promoting sustainable cereal-based production systems by fixing free nitrogen. Lentils are a significant pulse crop grown in 53 countries worldwide, with yields ranging from 227 kg/ha in Israel to 2567 kg/ha in China in 2018. The global cultivated area for lentils is approximately 6.10 M hectares, producing 6.33 million tons annually with an average yield of 1038 kg/ha. Lentil production contributes to environmental services and sustainable agriculture practices. The cultivation of lentils has a significant impact on global agriculture, spanning across numerous nations. Lentils are a valuable source of nutrients and play a vital role in maintaining soil health and fertility. The crop's versatility and nutritional benefits make it a crucial component in various diets worldwide. In recent years, there has been a notable increase in lentil production, mainly in Asian countries and North America. By 2030, global lentil consumption is estimated to reach 5.50 million metric tons, with an additional 2.0 million metric tons in production. Australia, Canada, the United States, and Turkey are the top exporting countries, accounting for 87.02% of the total world export trade in 2017. India is the biggest grower and consumer of lentils, with 5.21% of the total pulses area in the country dedicated to lentil cultivation. This accounts for 6.67% of the total pulses production in 2018–19. The majority of lentil cultivation in India takes place in Madhya Pradesh and Uttar Pradesh, which together contribute over 71% of the production. However, lentil cultivation in India faces challenges such as low yields, high variability, and is primarily carried out by smallholder farmers. Biotic stresses have been reported to cause a 20–25% reduction in lentil productivity in India, as highlighted by Maheshwari et al. in 2008.

Pulses crops are grown in about 1.3 hectors area in Pakistan. There are several pulses, e.g., major and minor pluses grown in our country. The main winter pulse chickpea is produced 18% and summer pulse mung bean is produced 16% of the total area of legume production. Other pluses are lentils, and black gram has grown on 5% of total area production.

11.2 Salt Stress on Lentils

The negative effects of salt stress on plants include reduced photosynthetic rates, plant height, leaf area, and relative water content. Sodium toxicity disrupts membrane stability, leading to decreased growth and yield. When plants experience salt stress, they respond by closing stomata, reducing photosynthesis, and producing reactive oxygen species. Salinity affects water and osmotic potential, causing plant cells to shrink and creating alternative pathways for salt movement within roots. Salinity causes changes in leaf structure, such as increased epidermal and mesophyll thickness, cell length, and diameter. It also affects cellular structures like mitochondrial cristae, endoplasmic reticulum, and tonoplast. The second phase of salt stress results in ion toxicity within plant cells, leading to the compartmentalization of salts in vacuoles and hindering enzymatic processes. Increased NaCl treatments can disrupt the balance of ions in plant leaves. Plants adapted to saline environments have developed biochemical and molecular strategies to cope with ion accumulation, including selective ion accumulation, control of ion uptake, and compartmentalization. These alterations affect enzyme activity and metabolism. Salinity accumulation in chloroplasts affects photosynthesis and growth rates. While some studies show that salinity inhibits PS-II, others indicate no effect on PS-II.

11.3 Morpho-physiological and Biochemical Impacts of Drought Stress on Lentils

Drought has a significant impact on the growth and development of plants at various stages, affecting processes such as germination, flowering, and seed formation. This ultimately leads to a reduction in grain yield in crops (Farooq et al. 2009). Early drought stress has a negative effect on lentil seed germination, root and shoot growth, as well as the activity of germination-related enzymes (Hojjat and Ganjali 2016). Furthermore, drought stress impedes shoot and root height, dry weight, and seedling growth rate in plants (Idrissi et al. 2020). During the reproductive stage, drought stress in lentil plants results in a decrease in dry matter production, plant height, and leaf area (Shrestha et al. 2006a, b). Different lentil varieties exhibit varying responses to drought stress, with wild accessions displaying unique root and shoot reactions regardless of their ecological origin (Kumar et al. 2013). The ability of wild lentil accessions to adapt to drought conditions is linked to the average rainfall levels in their native habitats (Gorim and Vandenberg 2017). Photosynthesis is a crucial process in plants, but it is significantly affected by drought stress in lentil plants, especially during the reproductive stage (Sehgal et al. 2017). Lentil cultivars that are sensitive to drought show reduced photosynthetic activity under stress conditions. They also exhibit changes in photochemical reactions after prolonged exposure to stress (Morgil et al. 2019). Plants respond to drought stress by closing their stomata to reduce water loss through transpiration, which in turn affects CO_2 absorption. The decrease in photosynthetic activity due to water stress is related to reduced water use efficiency in plants (Ullah and Farooq 2022). This reduction is caused by an excessive production of ROS that react with cell's biomolecules and components of photosynthesis (Farooq et al. 2009). Drought stress affects the relative water content of plants, which is a crucial indicator for assessing plant response and adaptation to water deficit (Sehgal et al. 2017; Lamaoui et al. 2018). Wild lentil genotypes have shown higher relative water content compared to cultivated lentils under progressive drought stress (Singh et al. 2019; Idrissi et al. 2013; Sehgal et al. 2017). Drought stress influences crucial traits like water use efficiency, leaf water potential, and transpiration rate. It adversely effects the leaf water potential and transpiration rate, resulting in higher canopy temperature in plants (Hasanuzzaman et al. 2013; Turner et al. 2001; Ghoulam et al. 2022; Biju et al. 2018). Decreasing leaf area and stomatal conductance is a crucial adaptation strategy for conserving water in environments under stress (Varshney et al. 2021). Abiotic stresses, such as drought, induce oxidative stress by increasing the production of reactive oxygen species in plant organelles (Hasanuzzaman et al. 2013; Fahad et al. 2017). Plants respond to ROS by upregulating antioxidant defense systems and osmoprotectants to maintain cellular functions (Fig. 11.1). Drought stress leads to lipid peroxidation and production of MDA in lentil plants due to oxidative stress. Plants synthesize compatible solutes like glycine betaine, proline, and soluble sugars to cope with drought-induced stress. Yield losses under stress conditions are influenced by stress intensity and duration, the crop's phenological stage, plant genotype tolerance, and the occurrence of other stresses (Hasanuzzaman et al. 2020; Ceritoglu and Erman 2021; Singh et al. 2017; Mishra et al. 2018; Muscolo et al. 2014). Drought stress at

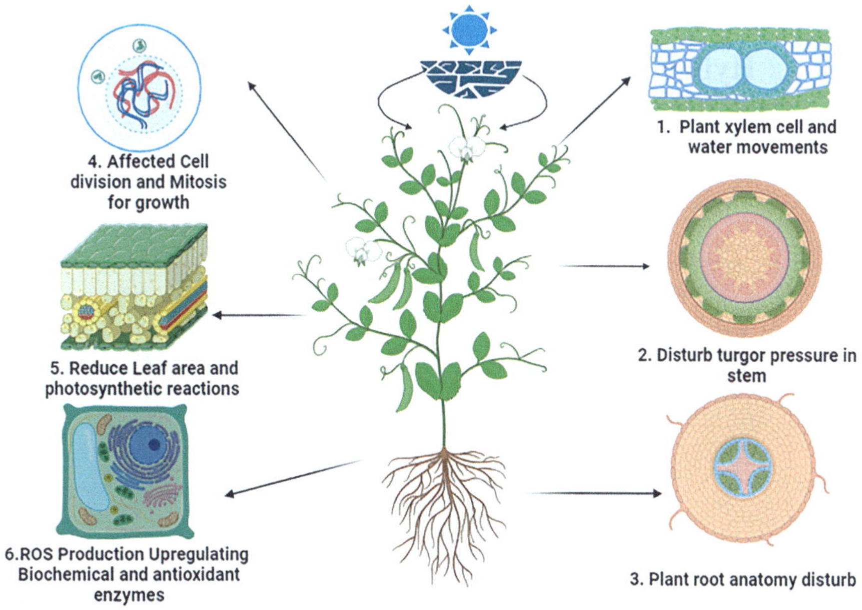

Fig. 11.1 Drought stress affects morphology, physiology, and biochemical processes in grain legumes

reproductive stage significantly reduces yield and harvest index in lentil plants (Shrestha et al. 2006a; b).

11.4 Conventional Breeding and Drought Tolerance in Lentils

Improving lentil production in areas prone to drought can be accomplished using both traditional and advanced breeding techniques (Fig. 11.2). Traditional methods have already been contributed to the development. The type of lentil plays a role in drought tolerance, as macrosperma-type lentils are less tolerant than microsperma-type lentils under drought conditions (Erskine 1996). However, the relationship between seed size and yield reduction is inconclusive (Shrestha et al. 2006a, b). Seed quality is crucial in determining the market value of lentils, and the effects of drought stress on lentil quality have not been extensively studied. Lentil plants are impacted by both drought stress and heat stress during the reproductive stage, which

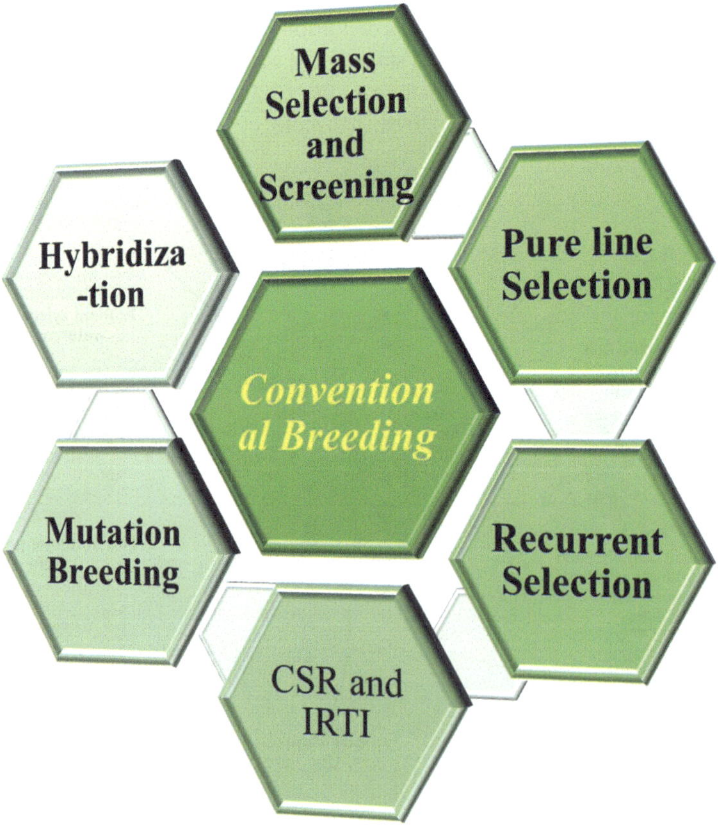

Fig. 11.2 Types of conventional breeding in plants to mitigate drought stress

affects traits associated with yield formation (Sehgal et al. 2017). The combination of heat and drought stress leads to a reduction in lentil yield and nutritional profile, with drought stress having more drastic consequences as compared to heat stress. These effects include reduced seed weight, protein content, minerals, and carbohydrates. Amino acid content is also altered under stress conditions, with genotype-specific variations. Drought-tolerant cultivars demonstrate improved adaptability to a range of stressful conditions (Sehgal et al. 2019), and there is a possibility of producing lentil cultivars that are adapted to combined stress. Terminal drought and heat stress significantly reduce Fe, B, and seed's protein contents (Venugopalan et al. 2022). Furthermore, the interplay of drought and heat stress causes a significant reduction in the essential contents of lentil grains (El Haddad et al. 2021; Choukri et al. 2020). Drought stress impacts several morpho-physiological and biochemical traits. Quality of lentil grain suffers due to declining biological processes like photosynthesis and nutrient use efficiency. Moreover, drought stress affects biological nitrogen fixation, potentially leading to negative impacts on yield and seed quality (Serraj 2003). Various studies have been conducted to develop high-yielding cultivars suitable for different environmental conditions (Idrissi et al. 2019; Ullah and Farooq 2022; Houasli et al. 2021). The National Institute of Agricultural Research in Morocco (INRA-Morocco) has successfully released improved lentil varieties that are stable in yield and adaptable to diverse Moroccan agro-environments through breeding efforts (Idrissi et al. 2020; Idrissi et al. 2019; Idrissi et al. 2015). Studying genetic variability for drought tolerance is vital for selecting genotypes that perform well in water-stressed environments. Research has been done on both cultivated lentil and its wild relatives to identify potential sources of tolerance (Sarker et al. 2005; Choukri et al. 2020; El Haddad et al. 2021; Akter et al. 2021; Idrissi et al. 2013; Gorim and Vandenberg 2017; Hamdi and Erskine 1996; Sánchez et al. 2019; Babayeva et al. 2014; Coyne et al. 2020; Dash et al. 2017; Kumar et al. 2012; Priya et al. 2021; Singh et al. 2018; Sarker et al. 2005). Traits like early flowering and maturity in lentil plants are beneficial for dealing with terminal drought stress, which can improve lentil propagation in areas prone to drought and heat stress (Gorim and Vandenberg 2017; Kumar et al. 2012; Silim et al. 1993). Seed yield under water-deficit conditions is influenced by factors such as seed size, number of pods and seeds per plant, seedling vigor, harvest index, and biological yield, all of which show positive correlations under drought stress (Kumar et al. 2012; Shrestha et al. 2006a, b). The length of the stem in lentil plants has a significant correlation with the seed yield per plant under rainfed conditions, indicating its importance in drought resilience.

There is variability among different lentil genotypes in agro-morphological traits like yield per plant, number of branches, pods, and seeds per plant, as well as seed weight under both irrigated and arid conditions. This highlights the potential for selecting drought-tolerant genotypes based on these traits (Akter et al. 2021). The root system of lentils, which consists of a slender taproot and fibrous lateral roots, shows genetic variability in traits such as taproot length, root dry weight, and root surface area (Saxena 2009). The genetic variability in root traits indicates the potential for selecting and improving underground traits for drought tolerance, which are often overlooked compared to traits of aerial parts (Saxena 2009). Physiological traits including stomatal conductance, water use efficiency, leaf area, and temperature of canopy are used to measure lentil's drought tolerance. By evaluating canopy temperature and water stress index, infrared thermal imaging (IRTI) is proposed as a quick and harmless method for evaluating drought tolerance in lentil. It has been discovered that traits including earliness, leaf area, high stomatal density, and antioxidant capacity help lentil plants avoid early drought stress (Idrissi et al. 2015a, b; Sarker et al. 2005; Akter et al. 2021; Kumar et al. 2012; Singh et al. 2013). Other indicators, such as cell membrane stability, osmotic adjustment, and epicuticular wax, can be used to evaluate drought tolerance. Photosynthesis traits, like chlorophyll content, photosynthetic rate, and enzyme activity, are important indicators of performance under drought stress. Specific indicators like chlorophyll stability index (CSI) are linked to drought tolerance in lentil genotypes, making them potential targets for screening processes (Idrissi et al. 2015a, b; Ghoulam et al. 2022; Biju et al. 2018; Sánchez et al. 2019; Idrissi et al. 2015a, b; Gorim and Vandenberg 2017). Despite the significance of physiological traits in assessing drought tolerance, they are not widely used in breeding programs due to limitations in screening methods for a large number of genotypes. Establishing effective phenotyping methods is crucial for developing next-generation drought-tolerant lentil cultivars. Drought stress leads to biochemical changes in plants, and compounds like glycine betaine, proline, and soluble sugars help the plant tolerate water deficit. Studying these metabolic changes and biochemical traits can assist in identifying drought-tolerant genotypes and improving screening processes for drought tolerance in lentils (Ghoulam et al. 2022). Lentil genotypes display varying levels of drought tolerance, with drought-tolerant genotypes exhibiting higher contents of proline, anthocyanin, total soluble sugars, and enzymatic antioxidants compared to drought-sensitive genotypes. Osmoprotectants help reduce the adverse effects of oxidative stress and support growth under stressful conditions (Sánchez et al. 2019). Various drought tolerance indices, including drought susceptibility index (DSI), stress tolerance index (STI), geometric mean productivity (GMP), harmonic mean (HM), and mean productivity (MP), are used to evaluate drought tolerance in lentil. MP, GMP, and STI are considered the best indices for genotype's selection that can withstand combined heat and drought stress (Ghoulam et al. 2022; Mishra et al. 2018). Traditional breeding methods for developing drought-tolerant lentil varieties are slow in addressing current climate change challenges. Therefore, the integration of biotechnological tools provides new opportunities to expedite varietal development and enhance lentil productivity globally, ensuring food security.

11.5 Manipulation of Genetic Variation for Electing Potential Sources of Water Tolerance

Traditional breeding techniques play a vital role in the development of high-yielding cultivars which are present in different conditions of the environment. In the drought-stress environment, manipulation of genetic variability for water stress plays a significant role in electing genotypes to perform better. Genetic changes for water tolerance had been studied in wild as well as cultivated lentil, and potential sources of drought stress were also identified (Akter et al. 2021; Choukri et al. 2020; Coyne et al. 2020).

11.5.1 Selection and Breeding Approaches

These procedures are necessary for the large amount of heritable diversity of lentils to improve drought tolerance. The heritability in arid regions is poor due to variations in amount and timing of precipitation with significant environment and genotype interactions. Due to which it is difficult to identify the essential characters that are responsible for yield potential and stability under water stress. Furthermore, to modify the utilization of features of interest, accurate characters of environment are required (Mir et al. 2012).

11.5.2 Mass Selection and Screening

Techniques such as mass selection and screening are important for gaining desired phenotypic features which are connected to yield of lentils. It is difficult to obtain desired phenotypic characters from plants because their physiological variables are connected to drought stress where new methodology is applied only on a small number of genotypes. There are two tiers of selection: the first tier is based on the traits which are quick, straightforward, and simple to quantify, and in the second tier, the different tests are performed on genotypes. However, it results in the mass selection based on heritable character, making it reasonably straightforward and cost-effective to quantify, whereas heritable trait gives positive results under favorable condition (Khatun et al. 2021).

The mass selection is one of the simplest selection procedures, which is beneficial for improving the whole population of crops by positive and negative mass selection. This selection is used within the populations for the crop traits with high and narrow sense heritability. The major inappropriate of mass selection is the high influence of the environment on phenotype, performance, and development of single plants. It is considered one of the advantages in varieties which are selected for local performance (Rauf et al. 2016).

11.5.3 Pure Line Selection

This method is essential for segregating the individuals in plants. For developing the segregation populations, drought resistant parents crossed with high yielding parents through pure line selection. We obtained a new combination of alleles due to the segregation and recombination of traits. Plants which are performing best in F_2 generation are selected, in both irrigated and drought-stress condition. New progenies of plants are raised from collected seeds of chosen plants. New types of plants are developed to screen good progenies, and the plants which are superior are among population to raise the pure lines (Ranjith and Rao 2021; Rauf et al. 2016).

11.5.4 Recurrent Selection

This method is mostly used to increase yield resistance and ability to drought in plants. The genetic variability increased after each precise selection. In this selection, the individual plants are chosen from the original population and moved to testing of progeny, and then these individuals are crossed in each individual containing all the possible combinations to grow seeds to get the new base population. An increase in gaining of high yield is proved in many studies (Giménez et al. 2001).

The heritability of lentils under water stress is low due to high environmental and genotype interactions, changes in the timing, and amount of precipitation occurred under drought stress. Apart from this, it is important to find the traits that are related to yield stability and potential under dry condition. For better effectiveness of target traits in lentils, there should be better environmental characters. To get the desired phenotypic characters which are based on correlated traits with yield, the mass selection and screening strategies are used. Traits including podding, maturity, and early flowering are used for mass screening and escape mechanism. The high conductance of stomata and cooler canopies often cause the high lentil yield under water stress, and due to these traits, selection criteria can be provided (Duc et al. 2015).

11.5.5 Canopy Spectral Reflectance (CSR) and Infrared Thermal Imaging (IRTI)

It is the high-throughput phenotyping and non-invasive approach, which plays beneficial role for straightforward and quick assessment of parameters of dynamic complex including plant biomass accumulation, canopy size and carbon assimilation (Zhu et al. 2021). This strategy is used for the mass screening of grain legumes for water stress. The other methods, such as thermal infrared imaging which is also known as infrared thermography, help in the estimation of canopy temperature also applicable for the screening of lentils for resistance to drought. At the vegetative stage of plant, the canopy temperature is variably measured which is closely linked with canopy conductance, which gives insight into the plant water status. Thermal infrared is thought to be used for the conductance of complete canopy and plant level over time (Hou et al. 2018).

The IRTI is considered a simple, harmless, and rapid tool for controlling drought stress in lentil which is based on canopy temperature depression (CTD), crop water stress index (CWSI), and canopy temperature which is the indicator of status of plant water. It was suggested that IRTI is considered a strong tool for screening a huge amount of germplasm of lentil (Biju et al. 2018).

To explain the response of lentil landraces, elite advanced lines and wild relatives to water stress (Ghoulam et al. 2022) used the infrared thermal imaging (IRTI) and morpho-physiological parameters (water losing rate, RWC, cell membrane stability, and harvest index).

11.5.6 Hybridization

Hybridization is the process in which two or more parents with desirable characters combine together in a single cultivar. To visualize the combining ability and diversity of genotypes including in hybridization, biometrical strategies are applied to ensure a scientific and rational basis for selecting the parents. Other than morpho-physiological markers providing assistance in characterization of germplasm, DNA-based markers help in the estimation of genetic diversity. In lentil, cross-breeding is a hectic job, and 20–50% success of artificial hybridization depends upon involved genotypes and climatic conditions such as humidity and temperature. Induced mutation method is a complementary cost effective approach and quick method to enhance the genetic base in lentils (Laskar et al. 2019).

Hybridization is a breeding approach to get the particular traits within or between the species of lentil. Lot of legumes had interspecific crosses along with distinct degree of success. Hybridization has great potential in breeding programs used for improving legumes from drought stress along with breeding success (Zhu et al. 2021).

When legumes are crossed with their wild relatives, which have more osmotic variations. Therefore, they demand interspecific hybridization transfer to the cultivated generation (Nadeem et al. 2019).

11.5.7 Mutation Breeding in Lentil

The "mutation" word was first introduced by De Vries H (1901), which means causing a sudden change in genome. During the last era of the eighteenth century, the Van Harten (1998) observed and documented about the spontaneous mutation which creates new alleles. It took about 30 years to create the mutations artificially by using the chemical and physical mutagens. After that, the mutations were used to improve different crop species and about 2200 cultivars of plants were released (Kharkwal et al. 2004; Rauf et al. 2016).

For a long period of time, traditional breeding methods have been used with narrow genetic variability. But crop species need enhanced novel genetic variability for improvement. According to this point, mutations may be beneficial for plant breeding programs. The induced mutation method is a complementary breeding technique, which is a cost-effective and quick method to enhance the genetic base in lentil. Mutation breeding helps in the recreation of genetic variability for the improvement of many crop species. Crop genetics improvements depend upon the mutation induction which can be controlled by nature and efficacy of screening and accuracy of induced mutation (Laskar et al. 2019).

There are limited attempts for lentil to induce mutation. Lentil is affected by both physical and chemical mutagens. In the sub-continent the mutation breeding was started in Indian agricultural institute Delhi, which releases mutant species of various crops of lentils (Solanki et al. 2011).

The traditional approaches of breeding drought tolerant lentil varieties accomplished success. These methods progress very slowly in addressing the challenge posed by current climate change. Therefore, the biotechnological approaches are used, which may provide better chances to evade the limitations of conventional tools and enhance the different developments aimed at boosting the productivity of lentil worldwide, thereby ensuring global food security (Zeroual et al. 2022a; b).

11.6 Biotechnological Methods for Enhancing Drought Tolerance in Lentils

11.6.1 Quantitative Trait Loci (QTLs)

Identification of quantitative trait loci (QTLs) is crucial for improving drought tolerance in lentils. Drought tolerance is a polygenic trait influenced by genotype-environment interactions, and it can be enhanced by identifying and transferring genes governing relevant traits using molecular approaches such as marker-assisted selection (Gupta et al. 2020). Advances in genomics and high-throughput genotyping technologies have facilitated the identification and integration of key genetic factors for drought tolerance in lentil breeding programs (Gupta et al. 2018). While various traits such as fusarium wilt resistance, anthracnose resistance, and tolerance to boron and aluminum toxicities have been successfully identified in lentil plants, our understanding of the genetic basis of drought tolerance is limited.

Studies by Singh et al. and Idrissi et al. focused on genetic control and the association of molecular markers with drought tolerance in lentils, identifying SSR and AFLP markers linked to seedling survival, relative water content, water loss rate, and wilting severity. Researchers have made progress in lentil genomics and breeding, but advanced technologies for studying drought tolerance are not as widely used compared to other food legumes like chickpea. In chickpea, marker-assisted backcrossing has resulted in improved lines with better drought tolerance and yield. In a study by Idrissi et al. QTL mapping in lentil identified multiple QTLs controlling various shoot and root traits associated with drought tolerance. The study emphasizes the need for further research to validate these QTLs across different genetic backgrounds and environmental conditions (Bett 2016; Ramsay et al. 2021). Genome-wide association studies (GWAS) are crucial for identifying genetic variations associated with agronomic traits in plants. These studies have been particularly useful for understanding drought-adaptive traits in important crops such as soybean, wheat, maize, and barley. They also have the potential to enhance lentil breeding programs by aiding in the identification of QTLs (quantitative trait loci) linked to complex traits like drought tolerance.

Integrating conventional QTL mapping with GWAS approaches has been demonstrated to improve the detection of QTLs associated with complex traits. For example, Ma et al. integrated both mapping strategies in a study on Aphanomyces root rot resistance in lentils and identified a total of 57 QTLs. Several studies (Saha et al. 2010, 2013; Dadu et al. 2021; Gupta et al. 2012; Polanco et al. 2019; Hamwieh et al. 2005; Gela et al. 2021) have shown the effectiveness of this integrated approach. The lentil reference genome availability and advances in genotyping platforms show promise for using GWAS in lentil breeding programs to better understand the genetic architecture of traits such as drought tolerance. Traditional methods for evaluating traits like drought tolerance in lentils are prone to errors and are time-consuming, limiting the identification of QTLs/genes for marker-assisted breeding (Chen et al. 2019). Automated high-throughput phenotyping platforms provide opportunities to efficiently screen large germplasm collections for drought adaptive traits, integrating phenomics and genomics data for precise breeding programs (Marsh et al. 2021).

Multidisciplinary approaches are necessary for phenomics studies, to efficiently gather comprehensive phenotypic data that measure plant growth, development, and response to environmental stresses (Yang et al. 2020). High-throughput phenotyping has been successfully employed in crops including maize, rice, wheat, and chickpea. However, there have been limited studies conducted in lentils, particularly in assessing Aphanomyces root rot resistance and salt tolerance using image-based tools (Wu et al. 2021; Jiang et al. 2021; Condorelli et al. 2018; Li et al. 2020; Rane et al. 2021a, b; Sivasakthi et al. 2018; Marzougui et al. 2019). The Canopeo app is a potent tool that can be used on smartphones to quickly assess green canopy cover in crops like lentils for traits such as drought tolerance and waterlogging tolerance. The Canopeo app effectively measures green canopy cover and correlates it with wilting score and relative water content. This demonstrates its usefulness in quantifying drought tolerance in lentils under both controlled and field conditions (Zeroual et al. 2022a, b).

Screening diverse lentil genetic materials, such as recombinant inbred line populations, landraces, and wild accessions, using high-throughput phenotyping platforms reveals high genetic variation and heritability for root biomass. The use of advanced phenotyping tools, like high-throughput platforms and image-based techniques, is crucial for accurately evaluating complex traits like drought tolerance in lentils. This approach leads to the identification of genes and markers beneficial for breeding programs. The use of various omics tools, e.g., metabolomics, transcriptomics, and proteomics, is significant in the study of drought stress-associated proteins, genes, and metabolites. These tools help in connecting genotype and phenotype. Transcriptomics plays an important role in decoding gene regulatory networks responsible for plant responses and tolerance to stress. This is achieved through technologies like RNA sequencing (RNA-seq), which allows for the investigation of drought tolerance mechanisms in lentil cultivars (Jha et al. 2020; Singh et al. 2017). Transcriptomic analysis of lentil cultivars (drought-tolerant & drought-sensitive) under drought-stress conditions reveals differential gene expressions related to glucose metabolism, TCA cycle, protein folding, and ion channel transport between the two genotypes (Singh et al. 2017). Investigation of the root and leaf transcriptomic responses in a drought-sensitive lentil cultivar under short- and long-term drought stress treatments identifies differentially expressed genes involved in various biological processes including response to abscisic acid, protein phosphorylation, and cell wall organization in roots and leaves under different drought conditions (Morgil et al. 2019). The response of lentils to several types of stress like heat, drought, salt stress, and alkalinity was studied using RNA sequencing. The study revealed genes that were uniquely affected by combined stress, showing notable enrichment in pathways associated with ribosome and RNA transport. The analysis also showed distinct differentially expressed genes for each type of stress, with common upregulated genes under salinity and alkalinity stress linked to ABA signaling and cation transport regulation (Hosseini et al. 2021; Singh et al. 2022). Specific genes such as citrate synthase, ribosomal protein, and malic enzyme were found to be upregulated under heat and drought stress, while chalcone synthase genes were downregulated (Singh et al. 2022). The overexpression of the citrate synthase gene improved stress tolerance in tobacco plants, highlighting the importance of functional characterization of these genes in lentils for drought tolerance (Yi et al. 2022).

Proteomic studies provide insights into the proteome of plants under stress conditions, helping to identify stress-responsive proteins crucial for plant adaptation. The application of proteomics in studying the response of lentils to drought stress can provide valuable information on mechanisms of tolerance and identify new factors involved in stress response. Unlike in other food legumes like chickpeas, soybeans, pigeon peas, peas, and common beans, where proteomics has offered insights into proteins altered by various stress conditions, the proteomic response in lentils under abiotic or biotic stresses has not been extensively studied (Raza et al. 2022).

Metabolomics is a valuable tool for studying the metabolites of biological samples under specific conditions, with a focus on how plants adapt to abiotic stress such as drought (Jan et al. 2022). It can aid in identifying important metabolites and pathways that contribute to drought tolerance in lentils. Integrating metabolomics with genomics studies can help in discovering genes for developing climate-resilient crops (Baloch et al. 2023; Malik et al. 2022) Metabolomics studies have identified key metabolites, including sugars, proline, and γ-aminobutyric acid, that contribute to drought tolerance in food legumes (Raza et al. 2022). Although there are limited studies on lentils, biomarkers such as ornithine and asparagine for drought tolerance, and alanine and homoserine for salt stress tolerance have been suggested (Muscolo et al. 2015). More research is essential to identify robust biomarkers for developing customized lentil cultivars. Genetic engineering presents a promising approach to developing drought-resistant lentil cultivars (Gupta et al. 2020). Genetic manipulation studies have shown improved salt stress tolerance in lentils by introducing genes like DREB1A (Khatib et al. 2011). We can enhance drought tolerance in lentils by targeting genes that encode antioxidants, osmoregulators, and transcription factors through genetic engineering. Further research is necessary to confirm the effectiveness of using transgenic approaches to improve lentil resilience to drought stress.

Genome editing technologies, such as CRISPR/Cas9, offer precise alteration of plant genomes to enhance drought tolerance in various crops, including food legumes like lentils (Rasheed et al. 2021). Despite challenges in genetic transformation, promising results have been achieved in terms of lentil regeneration and transformation using novel techniques like decapitated embryo explants and in vitro flowering (Sarker et al. 2003). It is crucial to optimize in vitro regeneration protocols through the examination of different culture medium ingredients for the future application of transgenic and genome editing tools in lentil genetic improvement (Bhowmik et al. 2021). The implementation of protocols for producing transgenic hairy roots in lentils can aid in evaluating root development under various stress conditions (Foti and Pavli 2020). The combination of biotechnological tools with speed breeding methods has the potential to expedite breeding cycles by allowing for multiple generations per year (Watson et al. 2018). In lentils, extended photoperiods, plant growth regulators, and hydroponic systems have facilitated the achievement of several generations annually. This has enabled the screening and selection of desirable traits such as disease resistance (Lulsdorf and Banniza 2018). The manipulation of environmental factors, like temperature, day length, and light quality, in speed breeding can speed up plant growth, prompt early flowering, and enhance photosynthetic activity. These advancements have made breeding activities, such as crossing, gene mapping, and trait exploitation, faster in various crops, including lentils (Mobini et al. 2016).

Integrating omics approaches with bioinformatics pipelines helps rapidly identify genes linked to drought tolerance in lentils. Genes recognized for drought tolerance can be used through marker-assisted selection and genetic engineering to develop improved lentil varieties. Speed breeding techniques, when combined with marker-assisted breeding approaches, lead to faster cultivar development by swiftly integrating traits of interest. Marker-assisted speed breeding is a novel concept that enables rapid integration of genes or QTLs responsible for drought tolerance in lentils. Robust markers closely linked to drought tolerance traits in lentils can be developed through QTL mapping and genome-wide association studies (GWAS) to expedite breeding. Molecular markers can help in identifying offspring with specific traits of interest during the breeding process. The establishment of "marker-assisted speed breeding" holds promise for practical lentil breeding programs focusing on complex and polygenic traits like drought tolerance.

11.6.2 High-Throughput Phenotyping (HTP)

High-throughput phenotyping (HTP) involves utilizing advanced technologies and methods to swiftly and accurately measure and analyze numerous physical and biochemical traits (phenotypes) in plants, animals, or microorganisms. This approach is especially beneficial in agriculture, plant breeding, and genetics, as it helps elucidate the connections between genetic variations and phenotypic traits. This understanding is crucial for developing enhanced species with desirable characteristics (Furbank and Tester 2011; Fiorani and Schurr 2013). Numerous high-throughput phenomics techniques and platforms have been invented over the intervening 10 years, and these are currently being used to phenotype a range of plant properties, such as growth, phenology, physiology, prevalence of diseases, infestations, and resilience to drought. These platforms also gather information on the characteristics of other plant parts, including the roots, seeds, and shoots (Cobb et al. 2013).

Highly accurate and precise demographic mapping, breeding material, germplasm gathering, and screening of large plant populations are made possible by high-throughput phenotyping technology (Kumar et al. 2014). By using automation, remote control, and effective data (picture) analysis pipelines made for high-throughput processing, these platforms lessen the need for manpower (Furbank and Tester 2011). Plant responses to a range of abiotic stressors, including drought, heat, cold tolerance, salt, nitrogen limitation, and UV light, have also been studied using high-throughput techniques. According to Mir et al. (2012), these are some of the most significant and frequently employed techniques for accurate high-throughput phenotyping of the intricate quantitative attribute of drought tolerance.

11.6.3 Field Phenotyping

High-throughput phenotyping platforms (HTPPs) facilitate the comprehensive and non-invasive acquisition of data throughout the entirety of a plant's life cycle within controlled environments. However, the quantitative trait loci and candidate genes identified under these conditions have generally failed to translate into enhanced grain yield when applied in field environments (Araus et al. 2008). Field conditions exhibit noteworthy variability, rendering the control of environmental parameters challenging and complicating the interpretation of experimental results. Phenotyping for drought stress is particularly arduous due to the reduction of soil moisture in the field, which results in heightened mechanical resistance—a phenomenon that is difficult to replicate in potted plants (Cairns et al. 2011). Furthermore, rather than flourishing alone, plants in the field are shaded by canopies. Multilocation screening within the target environment has been critical to varietal advancement (Cairns et al. 2013). Because features of the environment are frequently not observed, they are poorly understood and challenging to duplicate in controlled environments. For improving breeding efforts that field phenotyping showed substantial difficulty (Cabrera-Bosquet et al. 2012; Cairns et al. 2012). Depending on the goal of the phenotyping, trait heritability, and logistical data collection concerns, one may choose to phenotype in a field setting or under controlled conditions (Cobb et al. 2013).

11.6.4 Field-Based High-Throughput Phenotyping Platforms

Advancements in aeronautics, remote sensing, and high-performance computing are propelling the development of field-based high-throughput phenotyping platforms (HTPPs), resulting in both terrestrial and aerial systems (Araus and Cairns

2014). Ground-based HTPPs, frequently referred to as "phenomobiles," are specially adapted vehicles outfitted with GPS and an array of sensors. Recent innovations in this domain include systems capable of measuring canopy height, reflectance, and temperature in cotton fields across four rows simultaneously, with a data acquisition rate of 0.84 h per hour. These platforms facilitate direct data capture at the plot level and necessitate minimal post-processing (Comar et al. 2012; Andrade-Sanchez et al. 2013).

Aerial high-throughput phenotyping platforms (HTPPs) offer a rapid way to characterize numerous plots, overcoming limitations of ground-based systems. Early aerial HTPPs used small airplanes but were costly and impractical for high-definition imaging at low altitudes. Modern alternatives, such as "phenotowers" and blimps, have emerged, with blimps providing a stationary platform for heavy payloads and multiple sensors. Despite their advantages, blimps require significant space for storage and multiple operators for control (Rascher et al. 2011).

Crop yields must double by 2050 in response to the world's promptly expanding requirement for food supply (Tilman et al. 2011). Increases in output from agriculture, endurance to stress, and genetic improvement through breeding and agronomy have been limited by the expense and difficulty of HTP. In order to assess important characteristics including plant height, biomass, blooming period, and grain production, recent advancements in high-throughput, high-resolution phenotyping combine plant science, engineering, and computation (Tanger et al. 2017). Sensor technologies deployed in the field improve agricultural research by increasing germplasm testing and automating data collection. These innovations quicken the breeding process, increasing crops' genetic gain and resistance to disease (Mahlein 2016).

Drought stress significantly cuts lentil's seed yield by altering morpho-physiological and biochemical characteristics. This stress disrupts photosynthesis and nutrient efficiency, thereby degrading grain quality. Furthermore, it impairs biological nitrogen fixation, adversely affecting both yield and quality (Serraj and Sinclair 2003). Improving drought tolerance in lentils involves conventional breeding by selecting and cross-breeding plants with traits like root depth and water use efficiency. Biotechnological methods, such as identifying quantitative trait loci (QTLs) and using high-throughput phenotyping, further enhance the selection process, leading to more efficient and accurate breeding of drought-resistant lentil varieties (Zeroual et al. 2022a, b).

The development of lentil cultivars exhibiting improved drought tolerance via conventional breeding techniques has frequently come across limited success. This is attributable to the intricate nature of the trait and the difficulties in identifying dependable phenotyping methods. For instance, well-developed root systems correlate with drought tolerance, functioning as an avoidance strategy to maintain plant productivity under water-deficient conditions (Idrissi et al. 2016).

11.6.5 High-Throughput Phenotyping in Lentil

High-quality phenotypic data plays a crucial role in the precision and accuracy of QTL/gene recognition. Breeders consequently are interested in phenotyping elements such as drought resiliency, and advances in technology that provide swift reliable screening of huge amounts of germplasm and population mapping must be used. Traditional phenotyping approaches are frequently labor-intensive, error-prone, and tedious. Additionally, utilizing destructive and low-throughput traditional phenotyping strategies makes it harder to screen for specific drought-adaptive features, like root system design (Idrissi et al. 2016). Evaluating a hefty number of genotypes for root individualities presents significant challenges, which hinders progress in identifying QTLs/genes for marker-assisted breeding (Chen et al. 2019). Recent advancements in automated high-throughput phenotyping platforms present significant potential to transcend the constraints of conventional methodologies. These platforms, which are currently being developed across various international breeding programs, facilitate the precise and effective assessment of drought-adaptive traits within extensive germplasm collections. Furthermore, the data generated by these phenomic technologies can be merged with genomic information through sophisticated bioinformatics tools, thereby augmenting the precision and efficacy of breeding programs (Marsh et al. 2021).

Vast-scale screening of characteristics conferring drought tolerance in vast germplasm collections and mapping populations under controlled and field circumstances are made possible by high-throughput phenotyping (Wu et al. 2021). Numerous crops, including maize (Wang et al. 2023), rapeseed (Li et al. 2020), mung bean (Rane et al. 2021a, b), and chickpea (Bharadwaj et al. 2021), have already had their drought tolerance assessed using these cutting-edge technologies. Notably, the findings revealed a moderate to great association between image-derived features and traditional scoring, emphasizing the efficacy of image-based phenotyping as a strong method for assessing biotic stress tolerance in lentils (Marzougui et al. 2019).

HTP intended for drought stress tolerance in lentils leverages advanced technologies to efficiently and rapidly assess traits related to drought tolerance. This methodology involves several key components: imaging systems use cameras and sensors to capture detailed images of lentil plants, enabling the analysis of traits like leaf area, plant height, and biomass; spectroscopy and remote sensing employ spectral sensors to measure crucial indicators like chlorophyll content and water status for evaluating drought stress; automated data collection systems improve efficiency by handling large plant datasets and streamlining phenotypic data acquisition; data analytics applies computational tools and machine learning to analyze extensive datasets, identify patterns, and correlate traits with drought tolerance; and genotype-phenotype association connects phenotypic data with genetic information to identify drought tolerance markers, aiding in the development of more resilient lentil varieties (Idrissi et al. 2015a, b).

At the moment, the majority of genomic selection (GS) models use phenotypic data specific to the target trait, concentrating on single variables (Jia and Jannink

2012). Although these single-trait models have been successful (Heffner et al. 2009), Jia and Jannink (2012) argue that they do not take advantage of the connections between various qualities. While grain yield is often predicted by GS models, it is important to note that various physiological processes contribute to this yield (Pask et al. 2012), and there are documented examples of physiological phenotypes being connected with yield. Measurements of features like spectral reflectance and canopy temperature (CT), which have a significant association with grain production and are useful as selection tools, have shown particular efficacy using HTP techniques (Gutiérrez-Rodríguez et al. 2004).

HTP has emerged as an efficacious methodology that provides quick, precise, and automated data collection techniques to hasten the creation of drought-resistant cultivars (Kim et al. 2020). HTP facilitates the swift assessment of a large number of plants or genotypes, which drastically cuts down on the amount of time needed for data collection and processing (Tayade et al. 2022). Furthermore, it permits the screening of desirable traits during early growth stages, obviating the necessity to await full plant maturity (Jangra et al. 2021). Contemporary imaging technologies have transformed the ability to non-destructively monitor plant physiological responses to drought stress, functioning as early indicators for the identification of drought-resistant cultivars (Moustakas et al. 2022). HTPPs have been indispensable in evaluating water stress tolerance in tomato genotypes (Genangeli et al. 2023) and forecasting drought and heat resilience in *Brassica* species (Chen et al. 2019). Consequently, HTP provides a rapid, dependable, and efficient platform for assessing, detecting, and tracking an array of traits across different growth stages (Gill et al. 2022).

11.7 Marker-Assisted Breeding in Lentil for Drought Tolerance

11.7.1 Transcriptomics

Charles Auffray first used the word "transcriptomics" in 1999, describing the complete set of transcripts expressing in either a cell, a tissue, or an organ at a particular time, under some specific conditions. Plant response and stress tolerance may be understood through the use of transcriptomics, which allows the unraveling of genes and gene regulatory networks (Jha et al. 2020). Various methods, such as RNA sequencing (RNA-seq), microarray, serial analysis of gene expression (SAGE), and expressed sequence tags (ESTs), have been designed to explore the variation at transcriptomics level under certain conditions. The use of RNA-seq technology demonstrated prospects as a low-cost and innovative tool for understanding the fundamental processes of drought resilience (Jha et al. 2020; Singh et al. 2017). Therefore, it is crucial to investigate the processes behind lentil drought tolerance using this kind of technique. In order to do this, a sequencing technology, Illumina HiSeq 2500, was used to analyze the transcriptome response of lentil including both drought-sensitive (JL3) and drought-tolerant (PDL2) cultivars, which were exposed to a drought stress during its seedling stage (Singh et al. 2017). An investigation assessing the differentially expressed genes (DEGs) under the conditions of drought stress revealed that 11,435 transcripts were upregulated and 6934 were downregulated. Furthermore, under drought stress, transcriptome analysis of both tolerant and sensitive genotypes showed elevated DEGs associated with the complete metabolism of glucose, ion channel transport, and ABC family protein. In contrast to the sensitive genotype, the tolerant genotype showed downregulation of genes related to GABA, synthesis of cell wall protein, and negative regulation of abscisic acid. While the upregulation was shown by the genes related to TCA cycle, organ senescence, electron transport chain, redox process, proper folding of proteins, and decline in stomatal conductance (Singh et al. 2017).

Researchers have more recently examined how lentil responds to various abiotic stresses, such as heat (35/33 °C), alkalinity (40 mM $NaHCO_3$), drought (exposing lentil seedlings to air for 4 h for 3 days), and salinity stress (120 mM NaCl), by using various physicochemical indicators (Singh et al. 2022). The cDNA of seedlings given salinity, alkalinity, and drought stress was sequenced using the Illumina HiSeq 2500 sequencing platform, whereas cDNA of seedlings given heat stress was sequenced through Illumina HiSeq 2000 platform. Compared to sensitive genotypes, transcripts from tolerant genotypes showed that there were 51 DEGs with drought stress, 1060 DEGs with salt treatment, 720 DEGs with heat stress, and 1358 DEGs with alkalinity stress. The study revealed that genes associated with ABA signaling, cation transport regulation at the root symplast-xylem interface, sucrose transport, epigenetic regulation, and vesicular trafficking were frequently upregulated under salinity and alkalinity stress. On the contrary, upregulation was found in the genes related to organic acid synthesis and translational regulation under both heat and drought stress. The genes for chalcone synthase were downregulated in drought and heat, whereas the genes for citrate synthase, the NADP-dependent malic enzyme, and 50S ribosomal protein were increased (Singh et al. 2022). An essential component of the tricarboxylic acid cycle (TCA cycle) is citrate synthase, which catalyzes the first condensation reaction of the TCA cycle, in between acetyl coenzyme A and oxaloacetate to produce citric acid. The transgenic tobacco's resistance to several abiotic stress factors, such as drought, was found to be enhanced, in a recent study, by overexpressing the citrate synthase gene of *Rhododendron micranthum* Turcz. Moreover, the transgenic tobacco plants had greater levels of proline and organic acids (tartaric, malic, and citric acids) in the roots; but, when compared to wild-type counterparts, they were shorter, had fewer leaves, and had lower MDA contents (Yi et al. 2022). Therefore, functional analysis of the lentil citrate synthase genes at various developmental stages and stress conditions is required. Moreover, the identification of molecular pathways underlying lentil drought resistance may benefit from the functional characterization of the chalcone synthase gene, 50S ribosomal protein, and NADP-dependent malic enzyme. These drought-responsive genes might be used to create gene-based markers that could have been helpful in the molecular breeding of lentils.

11.7.2 Proteomics

Proteomics is the collection of proteins within a cell, tissue, or organ expressed under particular circumstances. It is also an omics technique that helps in identifying the function, structure, interactions, and role of proteins under stressful circumstances (Raza et al. 2022). It is a potential approach for examining plant response to drought stress, as it can reveal novel factors involved in plant adaptation to adverse circumstances and provide insight into response and tolerance mechanisms at the proteome level. The description of abiotic stress-responsive proteins that may be modified to control lentil behavior under stressful conditions and maintain the level of grain production requires the use of proteomics. The proteomics has been successfully applied to different food legumes like pea, chickpea, pigeon pea, soybean, and common bean representing the documentation of proteins that had altered due to a variety of abiotic stresses (Jan et al. 2022). To date, no study has been reported in relation to the lentil proteomic response under abiotic or biotic stresses (Tiwari et al. 2022). Therefore, a comprehensive study was strongly recommended for its proteome response toward drought stress.

11.7.3 Metabolomics

Studying the variation of a biological sample's metabolome—a collection of metabolites obtained under particular circumstances—has been made possible by the development of metabolomics. Examining the metabolites linked to a plant's ability to adapt to abiotic stress is made easier by metabolomics (Malik et al. 2022). Plants typically react to abiotic stress by initiating a variety of biochemical processes. Quantifying these chemical compounds through the application of metabolomics is a practical means of conducting an accurate and efficient study of lentil genetic materials in order to identify the most significant metabolites and their pathways that confer drought tolerance, as well as to identify a superior genotype. Additionally, breeding techniques supported by high-throughput metabolomics technology can assist in identifying characteristics associated with high yield and stress tolerance (Razzaq et al. 2019). A promising method to speed up gene discovery and create crops resistant to climate change may be metabolic-assisted breeding, which may combine the metabolomics and other genomics research (Raza et al. 2022). Metabolomics has enabled the thorough identification of metabolites that contribute to drought resistance in dietary legumes, including sugars, proline, and γ-aminobutyric acid (GABA) (Jha et al. 2020). There are, however, few research on lentils that address the metabolomics response toward drought stress. For example, morphophysiological characteristics and gas chromatography-mass spectrometry (GC-MS) were used to evaluate the genotype response of lentils to drought stress (18% PEG 6000) and salt stress (150 mM NaCl) (Muscolo et al. 2014). Alanine and homoserine were identified by the authors as important indicators for salt stress tolerance, whereas asparagine and ornithine were proposed as the biomarkers for drought tolerance (Muscolo et al. 2015). In a recent study, the metabolomics response of drought-sensitive (Flip03–24 L) and drought-tolerant (Elpida) lentil cultivars was investigated at the seedling stage under the drought stress (2.5% and 5% PEG-6000). Potential biomarkers used in evaluating the lentil response to drought stress included D-fructose, L-tryptophan α,α-trehalose, and myoinositol (Foti et al. 2021). However, a number of factors, including the analytical technique utilized, the plant tissue examined, the growth stage, the genotype, the stress applied, and the statistical tools employed for data processing and interpretation, have a significant impact on plant metabolite fluctuation. Thus, more investigation is required to identify strong biomarkers that can be used to create lentil cultivars with optimized metabolic pathways. Moreover, by combining metabolomics and transcriptomics, it is possible to have a better understanding about the metabolic pathways and molecular mechanisms that control how lentils respond to and tolerate drought stress.

11.8 Biotechnological Tools Improve Breeding Programs

Speed breeding technology has demonstrated in recent years that it can shorten the cycle of breeding by enabling the production of many generations annually. Crops including wheat, pea, chickpea, and barley can produce up to six generations per year under speed breeding circumstances, compared to one or two generations in traditional breeding programs (Watson et al. 2018). With the use of this technology, plants can grow more quickly, increase photosynthetic activity, and trigger early flowering by controlling temperature, light intensity, light quality, and day length. A variety of breeding processes, including crossing, population mapping, and the exploration of genes linked to desirable traits, can be improved using speed breeding (Hickey et al. 2019). By using a prolonged photoperiod, a plant growth regulator, and the nurturing of immature seeds, it was possible to produce up to eight generations of lentils annually (Mobini et al. 2015). A similar technique allowed for the screening and the selection of lentils resistant to Aphanomyces root rot disease by utilizing an extended photoperiod (20 h of light with 4 h of darkness) and immature seed germination, which was provided six times a year (Lulsdorf and Banniza 2018). Furthermore, growing lentils in a hydroponic system with an extended photoperiod of 20 h of light and 4 h of darkness with 23 °C temperature and supplementing with 0.6 μM flurprimidol enabled the production of six generations per harvest (Maglia et al. 2020).

In general, high-throughput evaluation of genetic resources of lentil and quick identification of its genes associated with drought resistance can be made possible by integrating omics techniques with reliable bioinformatics tools. Genetic engineering techniques and marker-assisted selection could be used to improve drought-tolerant crops by utilizing identified genes. Furthermore, the whole breeding process could be sped up through the use of speed breeding; for instance, combining speed breeding with marker-assisted breeding techniques (also known as "marker-assisted speed breeding") might speed up the identification and introgression of desirable traits into new cultivars (Nadeem et al. 2018). This integrated approach is a creative and effective method for introducing QTLs or genes that influence drought tolerance in lentils rapidly. This can be accomplished by using both QTL mapping and GWAS to produce strong markers closely connected to characteristics that confer drought tolerance in lentils. Crossing can then be carried out in a fast-breeding environment to speed up generation turnover. Furthermore, molecular markers can be used to identify progenies carrying desirable traits. A procedure like this demonstrates the potential of implementing marker-assisted speed breeding in the empirical breeding programs for lentils, particularly for traits that are complicated and polygenic, like drought tolerance.

11.8.1 Agronomic Mediation Improves Drought Stress Tolerance

Implementing appropriate agronomic techniques is an additional viable approach to minimize decreased yields in lentil fields affected by drought stress. The detrimental effects of drought stress can actually be reduced by the use of agronomic interference like optimizing sowing time and sowing rate, enhancing seeds, applying plant growth regulators and the osmoprotectants exogenously, supplementing adequate nutrients, applying arbuscular mycorrhizal fungi (AMF) and plant growth promoting rhizobacteria (PGPR), and implementing conservation agricultural practices like crop rotation and crop residue management (Ullah and Farooq 2022; Bamouh et al. 2020). Planting at the appropriate time of year is an effective way to avoid drought stress and maintain ideal growing conditions for crops. Optimal planting timing facilitates optimal utilization of precipitation during the crop cycle. Moreover, at the flowering and pod-felling stages, a timely seeding can mitigate the consequences of a drought (Farooq et al. 2017).

Plants can be made more resilient to the negative effects of drought stress by exogenously applying phytohormones, like ethylene, melatonin, salicylic acid, brassinosteroids, jasmonic acid, abscisic acid, cytokinins, and gibberellins, and osmoprotectants, like glycine, proline, betaine, trehalose, and polyamines (Jogawat et al. 2021; Iqbal et al. 2022). However, there have been few studies done on lentil in this field; therefore, more investigation is required to assess the possible contributions of growth regulators and osmoprotectants to the induction of drought tolerance in lentil. Improved lentil output in both optimal and sub-optimal circumstances depends on adequate nutrient management. Drought stress is known to reduce the efficiency with which nutrients are absorbed and used; yet, sufficient zinc treatment has enhanced chickpea resistance to heat stress and drought (Ullah et al. 2019). Additionally, micronutrients like iron, zinc, and selenium applied superficially and in the soil increase plant resistance to abiotic stress (Tavanti et al. 2021). When 2 mM silicon was added to lentils during the germination stage, it improved drought tolerance of the genotype of lentils that were affected by drought via regulating the metabolism of osmolytes and antioxidant defense system (Biju et al. 2017).

Another productive technique for enhancing seeds is seed priming, which also helps with crop stand establishment and stress tolerance (Ghassemi-Golezani et al. 2008). Osmopriming with 1% Calcium dichloride ($CaCl_2$) showed better results in germination, chlorophyll content, amylase activity and soluble sugars (Farooq et al. 2020). Furthermore, hydropriming improved field-sown lentil stand establishment in water-stressed environments (Sağlam et al. 2010). Due to the lower costs involved with tillage operations, conservation agriculture techniques including crop residue conservation and reduced or zero-till farming have made it possible to produce crops efficiently. According to a study in Morocco, switching to no-till farming enhanced the physicochemical quality and the water-holding capacity of soil, decreased soil erosion, and produced better and more consistent yields (Mrabet et al. 2012). However, the crop output under no-till is dependent on a number of additional variables, including crop rotation, soil type, climate, and residue assimilation. When compared to traditional tillage, minimum and no-till lentil farming with residue retention preserved soil moisture and improved the crop's RWC, biomass, growth rate, and chlorophyll content while producing a comparatively greater yield (Bandyopadhyay et al. 2018). When mulch was applied instead of residue removal, the yield increased by 36.5–66.6% (Das et al. 2019). Furthermore, no-till improved the yield of lentils by increasing their leaf area index, carbohydrate content, and RWC and favored the soil microbial activity, which was linked to soil moisture conservation. If combined with other agronomic practices like sowing timing, no-till could therefore act as a buffer system to prevent drought-related negative effects on rain-fed lentils (Saha et al. 2020).

11.9 Conclusion

Due to climatic changes the drastic impacts on agricultural crops. All plant morphology, physiology and biochemical reactions, antioxidant enzymes, and pigments are badly affected by drought stress. So, breeding programs and biotechnological tools can help to improve all these factors and produce drought-tolerant crops. Latest practices in QTL mapping, GWAS, and next-generation sequencing can advance drought tolerance. High throughput phenotyping and biotechnology can deliver data that dissect the traits, Omics, transcriptomics and metabolomics can improve the crop genetics. The agronomic mediators which are growth regulators e.g. arbuscular mycorrhizal, plant growth promoting rhizobacteria, phytohormones, seed priming, biotechnology tools can speed up cultivar development and improve lentil harvest.

Author Contribution A.Y. and S.F. conceptualized and collected the data. N.R. and M.R. designed and analyzed the book chapter, A.Y. was responsible for Supervision and editing the book chapter. M.N., S.F., and H.M.M. assisted in writing, review, and editing, and A.T. and Z.N. revised the book chapter. All these authors have substantial contributions to the final book chapter and approved it for submission.

References

Akter S, Jahan I, Hossain MA, Hossain MA (2021) Laboratory-and field-phenotyping for drought stress tolerance and diversity study in lentil (Lens culinaris Medik.). Phyton 90:949–970

Andrade-Sanchez P, Gore MA, Heun JT, Thorp KR, Carmo-Silva AE, French AN et al (2013) Development and evaluation of a field-based high-throughput phenotyping platform. Funct Plant Biol 41(1):68–79

Araus JL, Cairns JE (2014) Field high-throughput phenotyping: the new crop breeding frontier. Trends Plant Sci 19(1):52–61

Araus JL, Slafer GA, Royo C, Serret MD (2008) Breeding for yield potential and stress adaptation in cereals. Crit Rev Plant Sci 27(6):377–412

Babayeva S, Akparov Z, Damania A, Izzatullayeva V, Abbasov M (2014) Genetic diversity for drought tolerance in lentils from Central Asia and the Caucasus: CACLentil. Albanian J Agric Sci 13:1–8

Bahl A, Loitsch S, Kahl G (1993) Air pollution and plant gene expression. current topics in plant molecular biology: plant responses to the environment. CRC press, Boca Raton, FL, 71–6

Baloch FS, Altaf MT, Liaqat W, Bedir M, Nadeem MA, Cömertpay G et al (2023) Recent advancements in the breeding of sorghum crop: current status and future strategies for marker-assisted breeding. Front Genet 14:1150616

Bamouh A, Azizi I, Srour M (2020) Evaluation of productivity, profitability and farmer's adoption potential of direct seeding of lentils in Zaer region (Morocco). Mor J Agric Sci 1(4)

Bandyopadhyay PK, Halder S, Mondal K, Singh KC, Nandi R, Ghosh PK (2018) Response of lentil (lens culinaries) to post-rice residual soil moisture under contrasting tillage practices. Agric Res 7:463–479

Bett K (2016) The lentil genome–from the sequencer to the field

Bharadwaj C, Tripathi S, Soren KR, Thudi M, Singh RK, Sheoran S et al (2021) Introgression of "QTL-hotspot" region enhances drought tolerance and grain yield in three elite chickpea cultivars. Plant Genome 14(1):e20076

Bhowmik P, Konkin D, Polowick P, Hodgins CL, Subedi M, Xiang D, Yu B, Patterson N, Rajagopalan N, Babic V (2021) CRISPR/Cas9 gene editing in legume crops: opportunities and challenges. Legume Sci 3:e96

Biju S, Fuentes S, Gupta D (2017) Silicon improves seed germination and alleviates drought stress in lentil crops by regulating Osmolytes, hydrolytic enzymes and antioxidant defense system. Plant Physiol Biochem 119:250–264

Biju S, Fuentes S, Gupta D (2018) The use of infrared thermal imaging as a non-destructive screening tool for identifying drought-tolerant lentil genotypes. Plant Physiol Biochem 127:11–24

Cabrera-Bosquet L, Crossa J, von Zitzewitz J, Serret MD, Luis Araus J (2012) High-throughput phenotyping and genomic selection: The frontiers of crop breeding converge F. Journal of integrative plant biology, 54(5), 312–320

Cairns JE, Impa SM, O'Toole JC, Jagadish SVK, Price AH (2011) Influence of the soil physical environment on drought stress and its implications for drought research. Field Crop Res 121:303–310

Cairns JE, Sanchez C, Vargas M, Ordoñez RA, Araus JL (2012) Dissecting maize productivity: ideotypes associated with grain yield under drought stress and well-watered conditions. J Integr Plant Biol 54:1007–1020

Cairns JE, Crossa J, Zaidi PH, Grudloyma P, Sanchez C, Araus JL et al (2013) Identification of drought, heat, and combined drought and heat tolerant donors in maize. Crop Sci 53(4):1335–1346

Ceritoglu M, Erman M (2021) Effect of silicon priming treatments on germination and some agronomic traits in lentil. In 3rd International African Conference on Current Studies (pp. 27–28)

Chen Y, Palta JA, Wu P, Siddique KH (2019) Crop root systems and rhizosphere interactions. Plant Soil 439:1–5

Choukri H, Hejjaoui K, El-Baouchi A, El Haddad N, Smouni A, Maalouf F et al (2020) Heat and drought stress impact on phenology, grain yield, and nutritional quality of lentil (Lens culinaris Medikus). Front Nutr 7:596307

Cobb JN, DeClerck G, Greenberg A, Clark R, McCouch S (2013) Next-generation phenotyping: requirements and strategies for enhancing our understanding of genotype–phenotype relationships and its relevance to crop improvement. Theor Appl Genet 126:867–887

Comar A, Burger P, De Solan B, Baret F, Daumard F, Hanocq JF (2012) A semi-automatic system for high throughput phenotyping wheat cultivars in-field conditions: description and first results. Funct Plant Biol 39(11):914–924

Condorelli GE, Maccaferri M, Newcomb M, Andrade-Sanchez P, White JW, French AN, Sciara G, Ward R, Tuberosa R (2018) Comparative aerial and ground based high throughput phenotyping for the genetic dissection of NDVI as a proxy for drought adaptive traits in durum wheat. Front Plant Sci 9:893

Coyne CJ, Kumar S, Wettberg EJB, Marques E, Berger JD, Redden RJ, Ellis THN, Brus J, Zablatzká L, Smýkal P (2020) Potential and limits of exploitation of crop wild relatives for pea, lentil, and chickpea improvement. Legume Sci. 2:e36

Dadu RHR, Bar I, Ford R, Sambasivam P, Croser J, Ribalta F, Kaur S, Sudheesh S, Gupta D (2021) Lens Orientalis contributes quantitative trait loci and candidate genes associated with Ascochyta blight resistance in lentil. Front Plant Sci 12:703283

Das A, Layek J, Ramkrushna GI, Rangappa K, Lal R, Ghosh PK, Choudhury BU, Mandal S, Ngangom B, Dey U (2019) Effects of tillage and Rice residue management practices on lentil root architecture, productivity and soil properties in India's lower Himalayas. Soil Tillage Res 194:104313

Dash AP, De DK, Mohanty S, Lenka D (2017) Screening of lentil (Lens culinaris Medik.) genotypes and correlation analysis under PEGImposed water stress condition. Int J Bio-Resour Stress Manag 8:539–547

De Vries H (1901) Die mutations theorie. Leipzig: Veit (Vol. 2, pp. 166–166)

Duc G, Agrama H, Bao S, Berger J, Bourion V, De Ron AM et al (2015) Breeding annual grain legumes for sustainable agriculture: new methods to approach complex traits and target new cultivar ideotypes. Crit Rev Plant Sci 34(1–3):381–411

El Haddad N, Choukri H, Ghanem ME, Smouni A, Mentag R, Rajendran K et al (2021) High-temperature and drought stress effects on growth, yield and nutritional quality with transpiration response to vapor pressure deficit in lentil. *Plants 11*(1):95

Erskine W (1996) Seed-size effects on lentil (Lens culinaris) yield potential and adaptation to temperature and rainfall in West Asia. J Agric Sci 126:335–341

Fahad S, Bajwa AA, Nazir U, Anjum SA, Farooq A, Zohaib A, Sadia S, Nasim W, Adkins S, Saud S (2017) Crop production under drought and heat stress: plant responses and management options. Front Plant Sci 8:1147

Farooq M, Wahid A, Kobayashi N, Fujita D, Basra SMA (2009) Plant drought stress: effects, mechanisms and management. Agron Sustain Dev 29:185–212

Farooq M, Gogoi N, Barthakur S, Baroowa B, Bharadwaj N, Alghamdi SS, Siddique KH (2017) Drought stress in grain legumes during reproduction and grain filling. J Agron Crop Sci 203(2):81–102

Farooq M, Romdhane L, Al Sulti MK, Rehman A, Al-Busaidi WM, Lee DJ (2020) Morphological, physiological and biochemical aspects of osmopriming-induced drought tolerance in lentil. J Agron Crop Sci 206(2):176–186

Fiorani F, Schurr U (2013) Future scenarios for plant phenotyping. Annu Rev Plant Biol 64(1):267–291

Foti C, Pavli OI (2020) High-efficiency agrobacterium Rhizogenes-mediated transgenic hairy root induction of Lens culinaris. Agronomy 10:1170

Foti C, Kalampokis IF, Aliferis KA, Pavli OI (2021) Metabolic responses of two contrasting lentil genotypes to PEG-induced drought stress. Agronomy 11:1190

Furbank RT, Tester M (2011) Phenomics–technologies to relieve the phenotyping bottleneck. Trends Plant Sci 16(12):635–644

Gela T, Ramsay L, Haile TA, Vandenberg A, Bett K (2021) Identification of anthracnose race 1 resistance loci in lentil by integrating linkage mapping and genome-wide association study. Plant Genome 14:e20131

Genangeli A, Avola G, Bindi M, Cantini C, Cellini F, Grillo S et al (2023) Low-cost hyperspectral imaging to detect drought stress in high-throughput phenotyping. Plan Theory 12(8):1730

Ghassemi-Golezani K, Aliloo AA, Valizadeh M, Moghaddam M (2008) Effects of hydro and Osmo-priming on seed germination and field emergence of lentil (Lens culinaris Medik.). Not Bot Horti Agrobot Cluj-Napoca 36:29–33

Ghoulam SB, Zeroual A, Baidani A, Idrissi O (2022) Réponse au déficit hydrique progressif chez la lentille: vers une différentiation morpho-physiologique entre des accessions sauvages (Lens orientalis), populations locales et lignées avancées (Lens culinaris). Botany 100(1):33
Gill T, Gill SK, Saini DK, Chopra Y, de Koff JP, Sandhu KS (2022) A comprehensive review of high throughput phenotyping and machine learning for plant stress phenotyping. Phenomics 2(3):156–183
Giménez C, De Garcia E, De Enrech NX, Blanca I (2001) Somaclonal variation in banana: cytogenetic and molecular characterization of the somaclonal variant CIEN BTA-03. In Vitro Cell Dev Biol Plant 37:217–222
Gorim LY, Vandenberg A (2017) Evaluation of wild lentil species as genetic resources to improve drought tolerance in cultivated lentil. Front Plant Sci 8:1129
Gupta D, Taylor PWJ, Inder P, Phan HTT, Ellwood SR, Mathur PN, Sarker A, Ford R (2012) Integration of EST-SSR markers of Medicago Truncatula into intraspecific linkage map of lentil and identification of QTL conferring resistance to Ascochyta blight at seedling and pod stages. Mol Breed 30:429–439
Gupta DS, Kumar J, Gupta S, Dubey S, Gupta P, Singh NP, Sablok G (2018) Identification, development, and application of cross-species intron-spanning markers in lentil (Lens culinaris Medik.). The Crop Journal, 6(3), 299–305
Gupta D, Dadu RH, Sambasivam P, Bar I, Azad M, Beera N, Ford R, Biju S (2020) Conventional and biotechnological approaches for targeted trait improvement in lentil. In: Accelerated plant breeding, volume 3. Springer, Berlin/Heidelberg, pp 67–107
Gutiérrez-Rodríguez M, Reynolds MP, Escalante-Estrada JA, Rodríguez-González MT (2004) Association between canopy reflectance indices and yield and physiological traits in bread wheat under drought and well-irrigated conditions. Aust J Agric Res 55(11):1139–1147
Hamdi A, Erskine W (1996) Reaction of wild species of the genus lens to drought. Euphytica 91:173–179
Hamwieh A, Udupa SM, Choumane W, Sarker A, Dreyer F, Jung C, Baum M (2005) A genetic linkage map of lens Sp. based on microsatellite and AFLP markers and the localization of Fusarium vascular wilt resistance. Theor Appl Genet 110:669–677
Hasanuzzaman M, Nahar K, Gill SS, Fujita M (2013) Drought stress responses in plants, oxidative stress, and antioxidant defense. In: Climate change and plant abiotic stress tolerance, pp 209–250
Hasanuzzaman M, Bhuyan MB, Zulfiqar F, Raza A, Mohsin SM, Mahmud JA et al (2020) Reactive oxygen species and antioxidant defense in plants under abiotic stress: revisiting the crucial role of a universal defense regulator. Antioxidants 9(8):681
Heffner EL, Sorrells ME, Jannink JL (2009) Genomic selection for crop improvement. Crop Sci 49(1):1–12
Hickey LT, Hafeez AN, Robinson H, Jackson SA, Leal-Bertioli S, Tester M, Gao C, Godwin ID, Hayes BJ, Wulff BB (2019) Breeding crops to feed 10 billion. Nat Biotechnol 37:744–754
Hojjat SS, Ganjali A (2016) The effect of silver nanoparticle on lentil seed germination under drought stress. Int J Farming Allied Sci 5:208–212
Hosseini SZ, Ismaili A, Nazarian-Firouzabadi F, Fallahi H, Nejad AR, Sohrabi SS (2021) Dissecting the molecular responses of lentil to individual and combined drought and heat stresses by comparative transcriptomic analysis. Genomics 113:693–705. [CrossRef] [PubMed]
Hou M, Tian F, Zhang L, Li S, Du T, Huang M, Yuan Y (2018) Estimating crop transpiration of soybean under different irrigation treatments using thermal infrared remote sensing imagery. Agronomy 9(1):8
Houasli C, Sahri A, Nsarellah N, Idrissi O (2021) Chickpea (Cicer arietinum L.) breeding in Morocco: genetic gain and stability of grain yield and seed size under winter planting conditions. Euphytica 217:159
Idrissi O, Chafika H, Nsarellah N (2013) Comparaison de lignées avancées de lentille sous stress hydrique durant la phase de f loraison et formation des gousses. Nat Technol 10:53–61

Idrissi O, Houasli C, Udupa SM, De Keyser E, Van Damme P, De Riek J (2015) Genetic variability for root and shoot traits in a lentil (Lens culinaris Medik.) recombinant inbred line population and their association with drought tolerance. Euphytica, 204:693–709

Idrissi O, Houasli C, Udupa SM, De Keyser E, Van Damme P, De Riek J (2015a) Genetic variability for root and shoot traits in a lentil (Lens culinaris Medik.) recombinant inbred line population and their association with drought tolerance. Euphytica 204:693–709

Idrissi O, Udupa SM, Houasli C, De Keyser E, Van Damme P, De Riek J (2015b) Genetic diversity analysis of Moroccan lentil (Lens culinaris Medik.). In: Landraces using simple sequence repeat and amplified fragment length polymorphisms reveals functional adaptation towards agro-environmental origins, vol 134. Plant Breed, pp 322–332

Idrissi O, Udupa SM, De Keyser E, McGee RJ, Coyne CJ, Saha GC et al (2016) Identification of quantitative trait loci controlling root and shoot traits associated with drought tolerance in a lentil (Lens culinaris Medik.) recombinant inbred line population. Front Plant Sci 7:1174

Idrissi O, Sahri A, Houasli C, Nsarellah N (2019) Breeding progress, adaptation, and stability for grain yield in Moroccan lentil improved varieties. Crop Sci 59(3):925 936

Idrissi O, Houasli C, Amamou A, Nsarellah N (2020) Lentil genetic improvement in Morocco: state of art of the program, major achievements and perspectives. Mor J Agric Sci 1(1)

Iqbal S, Wang X, Mubeen I, Kamran M, Kanwal I, Díaz GA, Abbas A, Parveen A, Atiq MN, Alshaya H (2022) Phytohormones trigger drought tolerance in crop plants: outlook and future perspectives. Front Plant Sci 12:799318

Jan N, Rather AM-U-D, John R, Chaturvedi P, Ghatak A, Weckwerth W, Zargar SM, Mir RA, Khan MA, Mir RR (2022) Proteomics for abiotic stresses in legumes: present status and future directions. Crit Rev Biotechnol 2:1–20

Jangra S, Chaudhary V, Yadav RC, Yadav NR (2021) High-throughput phenotyping: a platform to accelerate crop improvement. Phenomics 1(2):31–53

Jha UC, Bohra A, Nayyar H (2020) Advances in "omics" approaches to tackle drought stress in grain legumes. Plant Breed 139:1–27

Jia Y, Jannink JL (2012) Multiple-trait genomic selection methods increase genetic value prediction accuracy. Genetics 192(4):1513–1522

Jiang Z, Tu H, Bai B, Yang C, Zhao B, Guo Z, Liu Q, Zhao H, Yang W, Xiong L (2021) Combining UAV-RGB high-throughput field phenotyping and genome-wide association study to reveal genetic variation of Rice germplasms in dynamic response to drought stress. New Phytol 232:440–455

Jogawat A, Yadav B, Chhaya, Lakra N, Singh AK, Narayan OP (2021) Crosstalk between phytohormones and secondary metabolites in the drought stress tolerance of crop plants: a review. Physiol Plant 172(2):1106–1132

Kharkwal MC, Pandey RN, Pawar SE (2004) Mutation breeding for crop improvement. Plant breeding: Mendelian to molecular approaches, 601–645

Khatib F, Makris A, Yamaguchi-Shinozaki K, Kumar S, Sarker A, Erskine W, Baum M (2011) Expression of the DREB1A gene in lentil (Lens culinaris Medik. Subsp Culinaris) Transformed with the Agrobacterium System. Crop Pasture Sci 62:488

Khatun M, Sarkar S, Era FM, Islam AM, Anwar MP, Fahad S et al (2021) Drought stress in grain legumes: effects, tolerance mechanisms and management. Agronomy 11(12):2374

Kim SL, Kim N, Lee H, Lee E, Cheon KS, Kim M et al (2020) High-throughput phenotyping platform for analyzing drought tolerance in rice. Planta 252(3):38

Kislev M. E, Bar-Yosef, O (1988) The legumes: the earliest domesticated plants in the Near East?. Current anthropology, 29(1), 175–179

Kumar J, Basu PS, Srivastava E, Chaturvedi SK, Nadarajan N, Kumar S (2012) Phenotyping of traits imparting drought tolerance in lentil. Crop Pasture Sci 63:547

Kumar S, Barpete S, Kumar J, Gupta P, Sarker A (2013) Global lentil production: constraints and strategies. SATSA Mukhapatra-Annual Technical, 17, 1–13

Kumar P, Huang C, Cai J, Miklavcic SJ (2014) Root phenotyping by root tip detection and classification through statistical learning. Plant Soil 380:193–209

Lamaoui M, Jemo M, Datla R, Bekkaoui F (2018) Heat and drought stresses in crops and approaches for their mitigation. Front Chem 6:26

Laskar RA, Khan S, Deb CR, Tomlekova N, Wani MR, Raina A, Amin R (2019) Lentil (Lens culinaris Medik.) diversity, cytogenetics and breeding. In: Advances in plant breeding strategies: legumes: volume 7, pp 319–369

Li H, Feng H, Guo C, Yang S, Huang W, Xiong X et al (2020) High-throughput phenotyping accelerates the dissection of the dynamic genetic architecture of plant growth and yield improvement in rapeseed. Plant Biotechnol J 18(11):2345–2353

Lulsdorf MM, Banniza S (2018) Rapid generation cycling of an F2 population derived from a cross between Lens culinaris Medik. And lens Ervoides (Brign.) Grande after Aphanomyces root rot selection. Plant Breed 137:486–491

Maglia F, Bermejo C, Palacios LT, Cointry E (2020) Speed breeding para La Multiplicación de Colecciones Activas En Lenteja (Lens culinaris Medik.). J Basic Appl Genet 31:123–145

Mahlein AK (2016) Plant disease detection by imaging sensors–parallels and specific demands for precision agriculture and plant phenotyping. Plant Dis 100(2):241–251

Malik JA, Mishra G, Hajam YA, Lone R, Quazi S (2022) Metabolome Analyses in Response to Diverse Abiotic Stress. In: Omics Approach to Manage Abiotic Stress in Cereals. Springer, Berlin/Heidelberg, pp 103–117

Marsh JI, Hu H, Gill M, Batley J, Edwards D (2021) Crop breeding for a changing climate: integrating Phenomics and genomics with bioinformatics. Theor Appl Genet 134:1677–1690

Marzougui A, Ma Y, Zhang C, McGee RJ, Coyne CJ, Main D, Sankaran S (2019) Advanced imaging for quantitative evaluation of Aphanomyces root rot resistance in lentil. Front Plant Sci 10:383

Mir RR, Zaman-Allah M, Sreenivasulu N, Trethowan R, Varshney RK (2012) Integrated genomics, physiology and breeding approaches for improving drought tolerance in crops. Theor Appl Genet 125:625–645

Mishra BK, Srivastava JP, Lal JP (2018) Drought resistance in lentil (Lens culinaris Medik.) in relation to morphological, physiological parameters and Phenological developments. Int J Curr Microbiol Appl Sci 7:2288–2304

Mobini SH, Lulsdorf M, Warkentin TD, Vandenberg A (2015) Plant growth regulators improve in vitro flowering and rapid generation advancement in lentil and Faba bean. Vitro Cell Dev Biol Plant 51:71–79

Mobini SH, Lulsdorf M, Warkentin TD, Vandenberg A (2016) Low red: far-red light ratio causes faster in vitro flowering in lentil. Can J Plant Sci 96:908–918

Morgil H, Tardu M, Cevahir G, Kavakli IH (2019) Comparative RNA-Seq analysis of the drought-sensitive lentil (Lens culinaris) root and leaf under short- and long-term water deficits. Funct Integr Genomics 19:715–727. [CrossRef] [PubMed]

Moustakas M, Sperdouli I, Moustaka J (2022) Early drought stress warning in plants: color pictures of photosystem II photochemistry. Climate 2022(10):179

Mrabet R, Moussadek R, Fadlaoui A, Van Ranst E (2012) Conservation agriculture in dry areas of Morocco. Field Crop Res 132:84–94

Muscolo A, Sidari M, Anastasi U, Santonoceto C, Maggio A (2014) Effect of PEG-induced drought stress on seed germination of four lentil genotypes. J Plant Interact 9(1):354–363

Muscolo A, Junker A, Klukas C, Weigelt-Fischer K, Riewe D, Altmann T (2015) Phenotypic and metabolic responses to drought and salinity of four contrasting lentil accessions. J Exp Bot 66:5467–5480. [CrossRef]

Nadeem MA, Nawaz MA, Shahid MQ, Doğan Y, Comertpay G, Yıldız M et al (2018) DNA molecular markers in plant breeding: current status and recent advancements in genomic selection and genome editing. Biotechnol Biotechnol Equip 32(2):261–285

Nadeem M, Li J, Yahya M, Sher A, Ma C, Wang X, Qiu L (2019) Research progress and perspective on drought stress in legumes: a review. Int J Mol Sci 20(10):2541

Pask AJD, Sylvester-Bradley R, Jamieson PD, Foulkes MJ (2012) Quantifying how winter wheat crops accumulate and use nitrogen reserves during growth. Field Crops Res *126*:104

Polanco C, Sáenz de Miera LE, González AI, García P, Fratini R, Vaquero F et al (2019) Construction of a high-density interspecific (Lens culinaris x L. odemensis) genetic map based on functional markers for mapping morphological and agronomical traits, and QTLs affecting resistance to Ascochyta in lentil. PloS One *14*(3):e0214409

Priya S, Bansal R, Kumar G, Dikshit HK, Kumari J, Pandey R et al (2021) Root trait variation in lentil (Lens culinaris Medikus) germplasm under drought stress. Plan Theory 2021(10):2410

Ramsay L, Koh CS, Kagale S, Gao D, Kaur S, Haile T et al (2021) Genomic rearrangements have consequences for introgression breeding as revealed by genome assemblies of wild and cultivated lentil species. *Biorxiv*:2021–2007

Rane J, Raina SK, Govindasamy V, Bindumadhava H, Hanjagi P, Giri R, Jangid KK, Kumar M, Nair RM (2021a) Use of phenomics for differentiation of Mungbean (Vigna Radiata L. Wilczek) genotypes varying in growth rates per unit of water. Front. Plant Sci. 12:692534

Rane J, Raina SK, Govindasamy V, Bindumadhava H, Hanjagi P, Giri R et al (2021b) Use of phenomics for differentiation of mungbean (Vigna radiata L. Wilczek) genotypes varying in growth rates per unit of water. Front Plant Sci 12:692564

Ranjith P, Rao MS (2021) Breeding for drought resistance. In: Plant breeding-current and future views. IntechOpen

Rascher U, Blossfeld S, Fiorani F, Jahnke S, Jansen M, Kuhn AJ et al (2011) Non-invasive approaches for phenotyping of enhanced performance traits in bean. Funct Plant Biol 38(12):968–983

Rasheed A, Gill RA, Hassan MU, Mahmood A, Qari S, Zaman QU, Ilyas M, Aamer M, Batool M, Li H (2021) A critical review: recent advancements in the use of CRISPR/Cas9 technology to enhance crops and alleviate global food crises. Curr Issues Mol Biol 43:1950–1976

Rauf S, Al-Khayri JM, Zaharieva M, Monneveux P, Khalil F (2016) Breeding strategies to enhance drought tolerance in crops. Advances in plant breeding strategies: agronomic, abiotic and biotic stress traits, 397–445

Raza A, Tabassum J, Fakhar AZ, Sharif R, Chen H, Zhang C, Ju L, Fotopoulos V, Siddique KH, Singh RK (2022) Smart reprograming of plants against salinity stress using modern biotechnological tools. Crit Rev Biotechnol 15:1–28

Razzaq A, Sadia B, Raza A, Khalid Hameed M, Saleem F (2019) Metabolomics: A way forward for crop improvement. Meta 9(12):303

Sağlam S, Sibel DAY, Gamze KAYA, Gürbüz A (2010) Hydropriming increases germination of lentil (Lens culinaris Medik.) under water stress. Not Sci Biol 2(2):103–106

Saha GC, Sarker A, Chen W, Vandemark GJ, Muehlbauer FJ (2010) Inheritance and linkage map positions of genes conferring resistance to Stemphylium blight in lentil. Crop Sci 50:1831–1839

Saha GC, Sarker A, Chen W, Vandemark GJ, Muehlbauer FJ (2013) Inheritance and linkage map positions of genes conferring agromorphological traits in Lens culinaris Medik. Int J Agron 2013(1):618926

Saha M, Bandyopadhyay PK, Sarkar A, Nandi R, Singh KC, Sanyal D (2020) Understanding the impacts of sowing time and tillage in optimizing the micro-environment for rainfed lentil (Lens culinaris Medik) production in the lower indo-Gangetic plain. J Soil Sci Plant Nutr 20:2536–2551

Sánchez-Gómez D, Cervera MT, Escolano-Tercero MA, Vélez MD, de María N, Diaz L, Sánchez-Vioque R, Aranda I, Guevara MÁ (2019) Drought escape can provide high grain yields under early drought in lentils. Theor Exp Plant Physiol 31:273–286

Sarker A, Erskine W, Singh M (2003) Regression models for lentil seed and straw yields in Near East. Agricultural and forest meteorology, 116(1-2), 61–72

Sarker A, Erskine W, Singh M (2005) Variation in shoot and root characteristics and their association with drought tolerance in lentil landraces. Genet Resour Crop Evol 52:89–97

Saxena MC (2009) Plant morphology, anatomy and growth habit. In: Erskine W, Muehlbauer FJ, Sarker A, Sharma B (eds) The lentil: botany, production and uses. CABI Press, Wallingford, pp 34–46

Sehgal A, Sita K, Kumar J, Kumar S, Singh S, Siddique KH, Nayyar H (2017) Effects of drought, heat and their interaction on the growth, yield and photosynthetic function of lentil (Lens culinaris Medikus) genotypes varying in heat and drought sensitivity. Front Plant Sci 8:1776

Sehgal A, Sita K, Bhandari K, Kumar S, Kumar J, Vara Prasad PV, Siddique KH, Nayyar H (2019) Influence of drought and heat stress, applied independently or in combination during seed development, on qualitative and quantitative aspects of seeds of lentil (Lens culinaris Medikus) genotypes, differing in drought sensitivity. Plant Cell Environ 42:198–211

Serraj R (2003) Effects of drought stress on legume symbiotic nitrogen fixation: physiological mechanisms. Indian J Exp Biol 41:1136–1141

Serraj R, Sinclair TR (2003) Evidence that carbon dioxide enrichment alleviates ureide-induced decline of nodule nitrogenase activity. Ann Bot 91(1):85–89

Shrestha R, Turner NC, Siddique KHM, Turner DW (2006a) Physiological and seed yield responses to water deficits among lentil genotypes from diverse origins. Aust J Agric Res 57:903

Shrestha R, Turner NC, Siddique KHM, Turner DW, Speijers J (2006b) A water deficit during pod development in lentils reduces flower and pod numbers but not seed size. Aust J Agric Res 57(4):427–438

Silim SN, Saxena MC, Erskine W (1993) Adaptation of lentil to the Mediterranean environment. I. Factors affecting yield under drought conditions. Exp Agric 29:9–19. [CrossRef]

Singh D, Dikshit HK, Singh R (2013) A new phenotyping technique for screening for drought tolerance in lentil (Lens culinaris Medik.). Plant Breed 132:185–190

Singh D, Singh CK, Taunk J, Tomar RSS, Chaturvedi AK, Gaikwad K, Pal M (2017) Transcriptome analysis of lentil (Lens culinaris Medikus) in response to seedling drought stress. BMC Genomics 18:206

Singh M, Sharma SK, Singh B, Malhotra N, Chandora R, Sarker A, Singh K, Gupta D (2018) Widening the Genetic Base of cultivated Gene Pool following introgression from wild lens taxa. Plant Breed 137:470–485. [CrossRef]

Singh A, Dikshit H. K, Mishra G. P, Aski M, Kumar S (2019) Association mapping for grain diameter and weight in lentil using SSR markers. Plant Gene, 20, 100204

Singh D, Taunk J, Singh CK, Chaudhary P, Gaikwad K, Yadav RK et al (2022) Comparative RNA sequencing for deciphering nodes of multiple abiotic stress tolerance in lentil (Lens culinaris Medikus). Plant Gene 31:100373

Sivasakthi K, Thudi M, Tharanya M, Kale SM, Kholová J, Halime MH et al (2018) Plant vigour QTLs co-map with an earlier reported QTL hotspot for drought tolerance while water saving QTLs map in other regions of the chickpea genome. BMC Plant Biol 18:1–18

Solanki R, Gill R, Verma P, Singh S (2011) Mutation breeding in pulses: an overview. Breeding of pulse crops. Kalyani Publishers, Ludhiana, pp 85–103

Tanger P, Klassen S, Mojica JP, Lovell JT, Moyers BT, Baraoidan M et al (2017) Field-based high throughput phenotyping rapidly identifies genomic regions controlling yield components in rice. Sci Rep 7(1):42839

Tavanti TR, de Melo AAR, Moreira LDK, Sanchez DEJ, dos Santos Silva R, da Silva RM, Dos Reis AR (2021) Micronutrient fertilization enhances ROS scavenging system for alleviation of abiotic stresses in plants. Plant Physiol Biochem 160:386–396

Tayade R, Yoon J, Lay L, Khan AL, Yoon Y, Kim Y (2022) Utilization of spectral indices for high-throughput phenotyping. Plan Theory 2022(11):1712

Tilman D, Balzer C, Hill J, Befort BL (2011) Global food demand and the sustainable intensification of agriculture. Proc Natl Acad Sci 108(50):20260–20264

Tiwari M, Singh B, Min D, Jagadish SK (2022) Omics path to increasing productivity in less-studied crops under changing climate—lentil a case study. Front Plant Sci 13:813985

Turner NC, Wright GC, Siddique KHM (2001) Adaptation of grain legumes (pulses) to water-limited environments. Adv Agron 71:193–231

Ullah A, Farooq M (2022) The challenge of drought stress for grain legumes and options for improvement. Arch Agron Soil Sci 68:1601–1618

Ullah A, Romdhane L, Rehman A, Farooq M (2019) Adequate zinc nutrition improves the tolerance against drought and heat stresses in chickpea. Plant Physiol Biochem 143:11–18

Van Harten AM (1998) Mutation breeding: theory and practical applications (Vol. 1). Cambridge University Press

Varshney R, Barmukh R, Roorkiwal M, Qi Y, Kholova J, Tuberosa R, Reynolds M, Tardieu F, Siddique K (2021) Breeding custom-designed crops for improved drought adaptation. Adv Genet 2:e202100017

Venugopalan V, Nath R, Sengupta K, Pal A, Banerjee S, Banerjee P, Chandran M, Roy S, Sharma L, Hossain A (2022) Foliar spray of micronutrients alleviates heat and moisture stress in lentil (Lens culinaris Medik) grown under Rainfed field conditions. Front Plant Sci 13:47743

Wang W, Guo W, Le L, Yu J, Wu Y, Li D et al (2023) Integration of high-throughput phenotyping, GWAS, and predictive models reveals the genetic architecture of plant height in maize. Mol Plant 16(2):354–373

Watson A, Ghosh S, Williams MJ, Cuddy WS, Simmonds J, Rey M-D, Hatta AM, Hinchliffe A, Steed A, Reynolds D (2018) Speed breeding is a powerful tool to accelerate crop research and breeding. Nat Plants 4:23–29. [CrossRef]

Wu X, Feng H, Wu D, Yan S, Zhang P, Wang W, Zhang J, Ye J, Dai G, Fan Y (2021) Using high-throughput multiple optical phenotyping to decipher the genetic architecture of maize drought tolerance. Genome Biol 22:185

Yang W, Feng H, Zhang X, Zhang J, Doonan JH, Batchelor WD et al (2020) Crop phenomics and high-throughput phenotyping: past decades, current challenges, and future perspectives. Mol Plant 13(2):187–214

Yi S, Zhou X, Sun Z, Li X-P, Li W (2022) Cloning and functional characterization of a citrate synthase gene of rhododendron Micranthum Turcz. S Afr J Bot 147:915–925. [CrossRef]

Zeroual A, Baidani A, Idrissi O (2022a, March 19–20) Use of canopeo, a smart phone based application, as a nondestructive and simple tool to estimate drought tolerance in lentil. In Proceedings of the ISPEC 9th international conference on agriculture, animal sciences and rural development, Burdur

Zeroual A, Baidani A, Idrissi O (2022b) Drought stress in lentil (Lens culinaris, medik) and approaches for its management. Horticulturae 9(1):1

Zhu G, Gu L, Shi Y, Chen H, Liu Y, Lu F et al (2021) Plant hydraulic conductivity determines photosynthesis in rice under PEG-induced drought stress. Pak J Bot 53(2):409–417

Zohary D, Hopf M (1973) Domestication of pulses in the old world: legumes were companions of wheat and barley when agriculture began in the Near East. Science, 182(4115), 887–894

Marker-Assisted Breeding of Soybean for Drought Tolerance

12

Jaber Bin Azim, Md. Mobinur Rahman, Md. Yousuf Ali, Touhidul Islam, Md. Parvez Kabir, Shahin Imran, and Akbar Hossain

Abstract

Soybean is an oilseed legume that plays a vital role in sustainable agriculture through the biological fixation of atmospheric nitrogen. However, global warming is currently worsening the drought stress on which productivity depends. In this scenario, marker-assisted breeding (MAB) is an appealing tool for improving tolerance against water-deficit condition in soybean by detecting and incorporating favourable alleles into elite cultivars. This chapter discusses the impacts of drought stress on the sustainability of soybean under the changing climate. This chapter also explores advanced techniques for enhancing the soybean plants' ability to withstand drought. This chapter primarily focuses on the recognition of genetic markers that are linked to drought tolerance features using QTL and association mappings and genomic selection, utilising advanced molecular methods. This chapter also searches the significance of genomics to accelerate the development of drought-tolerant soybean cultivars that can be integrated with conventional breeding approaches. Moreover, this chapter discusses the limitations and potential of using MAB to enhance water stress tolerance in breeding programmes. Due to the multigenic nature of drought resistance, a mixed selec-

J. B. Azim · M. M. Rahman · Md. Y. Ali
Maize Breeding Division, Bangladesh Wheat and Maize Research Institute, Dinajpur 5200, Bangladesh

T. Islam
On-Farm Research Division, Bangladesh Wheat and Maize Research Institute, Dinajpur 5200, Bangladesh

Md. P. Kabir · A. Hossain (✉)
Soil Science Division, Bangladesh Wheat and Maize Research Institute, Dinajpur 5200, Bangladesh

S. Imran
Department of Agronomy, Khulna Agricultural University, Khulna 9100, Bangladesh

© The Author(s), under exclusive license to Springer Nature Singapore Pte Ltd. 2025

M. A. Nadeem et al. (eds.), *Marker-Assisted Breeding in Legumes for Drought Tolerance*, https://doi.org/10.1007/978-981-96-4112-3_12

tive strategy is important for designing new soybean varieties with high yield stability under water-limited conditions. Molecular genetics and genomics provide significant opportunities for MAS, which can increase the rate of drought-tolerant soybean development for better food security and economic sustainability under current unpredictable climatic conditions.

Keywords

Soybean · Drought stress · Yield-related traits · Marker-assisted breeding (MAB) · Quantitative trait locus (QTL)

12.1 Introduction

The soybean (*Glycine max* Merrill) originated in China and it was first cultivated from the wild soybean (*Glycine soja* Sieb. & Zucc.) about 5000 years ago. It was later introduced to the Western Hemisphere (United States) and Europe (Wilson 2008). Soybean is an important cash crop, whose global output in 2022–23 was over 378 million metric tons. Brazil holds the title of being the greatest global producer of soybean, accounting for 44% of the total production. The United States follows closely after with a 31% share, while Argentina contributes 7% to the global production (SoyStats 2024; www.soystats.com). This crop has significant economic value due to its high content of seed protein (approximately 40%) and oil (approximately 20%). It is a rich source of starch, protein lipids, dietary fibre, and essential minerals. Additionally, it supplies high-level biomass that provides additional feed components to farm animals (Chaudhary et al. 2015). Soybean is also an important industrial crop because it has great commercial value and can be extracted in the form of edible oils, wax, soya food, paints, dyes, and fibres (Raghuvanshi and Bisht 2010).

Drought is an extreme environmental condition that significantly decreases worldwide crop productivity and the availability of food (Seleiman et al. 2021). According to Zia et al. (2021), the current trends in climate change anticipated that drought stress will be more intense and damaging in the future due to the extreme event of climate changing. Increasing global temperatures and the uneven nature of precipitation are also major constraints to soybean cultivation, especially in low-irrigation areas, which mainly rely on rainfall (Cotrim et al. 2021). On average, in the USA, the yield of soybeans decreases by 2 to 4% for every single degree raise in temperature throughout the crop's growth season. This results in an annual economic loss of $11 billion (Mourtzinis et al. 2015). The soybean yield can decrease up to 40% under dry conditions (Marinho et al. 2022), but this loss in production may reach as high as 80%, depending on the plant growth stage and stress intensity (Guimarães-Dias et al. 2012). Therefore, drought is an important climatic risk, and efficient mitigation strategies are needed to sustain global soybean availability.

To reduce the impact of drought stress, plants can scape drought stress by using many strategies, for example, tolerance, escape, and avoidance (Turner et al. 2001). When enough water is available, plants can finish their life cycle through drought escape. To avoid drought, plants use mechanisms based on efficient water uptake and a decrease in the evapotranspiration rate to retain their water status at a high level under stress. Drought tolerance refers to the appearance of strategies by which crops withstand drought conditions (Basu et al. 2016). The conventional approaches employed to enhance drought tolerance in crops include back-cross, single-pod-descent, pedigree methods, and bulk population methods. These methods primarily focus on evaluating genetic resources to identify potential sources for enhancing crop yield under water stress conditions (Poehlman et al. 1995). Some of these methods may be useful for enhancing the quality and productivity of soybeans, but there is a requirement for contemporary methods, such as MAB (Manavalan et al. 2009), as well as genomics-based approaches.

Recent advancements in genome mapping, genetics, and molecular biology techniques, along with other genomic tools, have greatly enhanced our ability to study biological processes. These tools are particularly valuable for identifying QTLs/genes that have the potential to enhance drought tolerance in soybean (Ku et al. 2013). Marker-assisted breeding (MAB) has been regarded as a means to enhance drought tolerance. The MAB employs DNA markers, linkage maps, and genomics to enhance plant attributes through tests conducted at the thread count level (Qiu et al. 2006). MAB strategies encompass various techniques such as marker-assisted selection (MAS), marker-assisted introgression (MAI), marker-assisted backcrossing (MABC), marker-assisted recurrent selection (MARS), gene stacking (previously known as pyramiding of genes), genome-wide selection (GWS), and genomic selection. These approaches have been considerably examined and documented by several earlier studies as reported by , Varshney et al. (2014), and Liu et al. (2005). Due to its effectiveness in delivering traits in soybean breeding, particularly under drought stress, MAB has been verified as a high-throughput approach (Wang et al. 2004; Liu et al. 2005; Abdel-Haleem et al. 2011; Arya et al. 2021). This chapter examines the present advancements and forthcoming prospects of MAB in enhancing drought stress adaptability characteristics in soybean plants.

12.2 Consequences of Drought Stress on Sustainable Soybean Production Under a Changing Climate

Water-deficit stress (drought) is one of the abiotic environmental factors that has a widespread effect on the productivity of soybean crops worldwide (Dubey et al. 2019). The occurrence of drought episodes caused by climate change exacerbates this issue and is a significant worry for soybean plants as well (Poudel et al. 2023; Basal 2020; Hasanuzzaman et al. 2013). Water-deficit stress (drought) induced oxidative stress generally leads to decreased stomatal conductance, total biomass, and yield, as do other physiological traits (Table 12.1). As climate wars and droughts become more frequent and intense due to human-induced global warming, this

Table 12.1 Effects of drought stress on various attributes of soybean

Consequences	References
Decrease in the process of sprouting of seeds, and the length of the shoot and root, together with their respective amounts of fresh and dry weight	Kosturkova et al. (2008)
Diminished plant stature, reduced leaf expanse, and decreased aboveground biomass in response to intense drought conditions	Wei et al. (2018)
The leaf's relative expansion rate, stomatal conductance, and leaf turgor decreased, but the ABA concentration in both the leaf and xylem increased	Liu et al. (2003)
Reduced development of lateral roots and an increased biomass of root-to-shoot ratio	Basu et al. (2016)
Decreased chlorophyll contents	Sadeghi et al. (2021)
Induced floral abortion and reduced pollen germination, fertile pods, and seed dry weight	Kobraee et al. (2011)
The seed-filling period was shortened, with R7 occurring up to 7 days' sooner. This led to a decrease in seed size by up to 32% and a fall in yields by up to 44%	de Souza et al. (1997)
The grain yield of soybean was reduced due to drought stress	Chandra et al. (2021)
Seed quality is impacted by drought	Dornbos and Mullen (1991)
Limited metabolism of sugar, N concentration, and protein concentrations	Du et al. (2020)
Lowered nutrient storage reserves	Wijewardana et al. (2019)
Decreased number of nodules and dry mass under drought conditions	Serraj et al. (1999)

danger is likely becoming increasingly imminent (Liaqat et al. 2022). Plants including soybean are generally highly susceptible to water-deficit stress, particularly at seedling and reproductive stages, when water demand of all plants is double as compared to vegetative stage (Poudel et al. 2023; Dubey et al. 2019). The decrease in photosynthetic activity and interruption of the assimilate supply chain at the time of the grain-formation stage are responsible for this phenomenon (Rasheed et al. 2020). Nevertheless, a reduced chlorophyll pigment in the leaf of any plants is directly associated with reduced yield and quality due to the drought-induced changes of physiological and metabolic change (Basal 2020; Hasanuzzaman et al. 2013*)*. They induce oxidative damage in plants by disrupting various intrinsic processes, gene expression, protein digestion, and enzyme inhibition (Mahajan and Tuteja 2005). Excessive levels of ROS can be harmful to the cell, leading to an embarrassment of soybean development (Dowling and Simmons 2009).

Water-scarcity stress (drought) severely reduced the dry mass reservation and photosynthesis efficiency (Fig. 12.1) (Cui et al. 2021), lead to alter the composition of fats and proteins in seeds, which affects grain yield (Xiong et al. 2021), and also decreases photo-chemical efficiency at the time of maturation, accompanied by a decrease in the rate of oil christening soybean grain development, which decreases the content at harvest (Basal 2020). P accumulation and translocation to seeds are reduced by drought (Jin et al. 2006) so that lipid synthesis is reduced, and

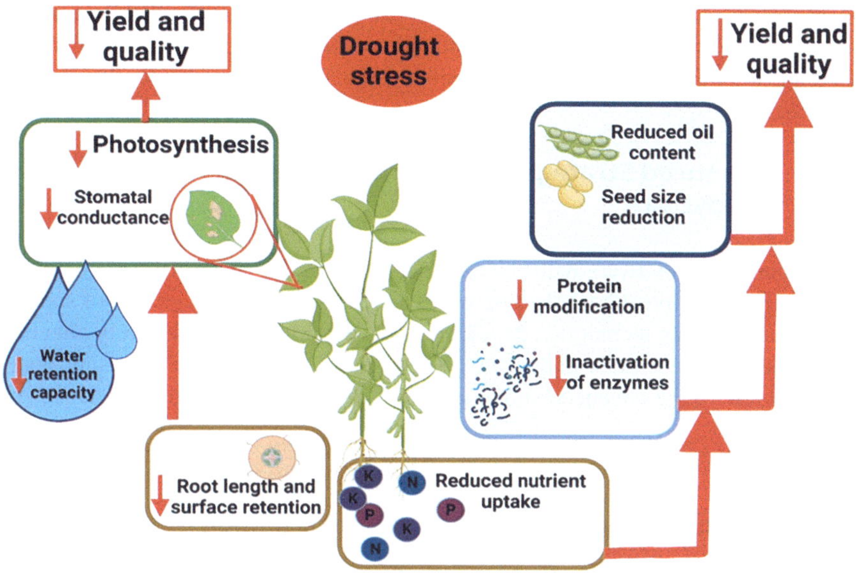

Fig. 12.1 The negative effect of drought stress on soybean production involves a nonlinear decrease of 11.1% in the photosynthesis rate, nutrient uptake, and protein and oil concentrations. It also reduces root length and lowers the enzymatic activities. (Rasheed et al. 2020)

membranes can be digested back into fatty acids and terpenoids by hydrolysation of lipases, leading to lower seed dry weight under stress conditions when only free Pi is present at low concentrations; as a result, all membrane phospholipids may decline considerably (Wei et al. 2015). Moreover, water imbalance in drought-sensitive crop cultivars decreases the relative water content, and leaf water potential reduces the productivity and quality of end products (Chowdhury et al. 2017).

Drought reduces soybean yields, but the magnitude of drought effects is often time- and duration dependent. Kpoghomou et al. (1990) observed that water-deficit stress during the pod-filling (R4) stage results in maximum yield loss as compared to vegetative stage. Desclaux et al. (2000) also stated that the highest susceptibility to this kind of stress, which reached 30–50%, was found at the R4 stage. It has a significant effect on the R4 stage, although it is damaging at all stages if not detrimental. The resiliency of soybeans to short-term dry periods during vegetative growth can limit reductions in yield. However, severe or prolonged water-deficit stress can result in dehydration in the cell of vegetative parts, leading to plant mortality (Licht et al. 2013). The largest and most recent droughts had heavy economic impacts. One example is Argentina, where a drought in 2018 reduced soybean yield by 31%, causing an estimated $3.4 billion US$ (Voiland 2018). The year 2010–2011 had the most severe drought in China in several decades, impacting seven million hectares land and causing a loss of 2.3 billion US$ (Buckley 2011). Simultaneously, East Africa had the most severe drought in six decades, resulting in a decrease in crop production by as much as 20% and causing the deaths of tens of thousands of people (Riebeek 2011). With continued climate change, these conditions are predicted to worsen, placing soybean production even more at risk. Climate-resilient soybean varieties are needed for an hour to ensure food security across the globe.

12.3 Advanced Strategies to Alleviate Drought Stress in Soybean

Developing novel cultivars with enhanced and long-lasting tolerance to drought is a highly effective approach for dealing with water scarcity. The integration of sophisticated phenotyping techniques, MAB methods, and MAS can effectively alleviate the shock of water scarcity–induced oxidative stress in soybean. In this framework, we detail few of the advanced strategies employed for drought stress management in soybean.

12.3.1 Breeding and High-Throughput Phenotyping

Soybean is a highly diverse wild species that can be crossbred with desirable features to simplify the selection of tolerant cultivars that are highly suitable to grow in drought condition (Kajiya-Kanegae et al. 2021). According to several reports, the traits associated with water deficit include slow canopy wilting, a more conservative transpiration response (Basu et al. 2016), and high root surface area density to avoid drought and low stomatal conductance. Nevertheless, traditional breeding methods for developing drought-tolerant soybean varieties pose challenges and require significant time and effort due to their quantitative characteristics (Charlson et al. 2009). High-throughput phenotyping (HTP) has been employed to expedite trait identification and reduce the time required for manual assessment of people (Crusiol et al. 2021; Zhou et al. 2021). Sensors involving remote sensing, e.g. thermographic cameras and spectral or hyperspectral imaging, can also provide digestible genotype analysis without destroying the plant (Li et al. 2014). Crusiol et al. (2021) distinguished between different types of soybean genotypes under various water conditions and stages of growth by analysing the spectral reflectance of their leaves. In contrast, Zhou et al. (2021) employed multispectral and RGB imaging techniques to forecast soybean production in the presence of drought-induced stress.

12.3.2 Marker-Assisted Breeding

Based on QTLs, discovery of marker-assisted breeding has emerged as a potential strategy for creating soybean plants that are drought tolerant (Arya et al. 2021). The constraints of this strategy stem from the challenges in connecting large populations and accurately identifying quantitative trait loci (QTL) due to the labour-intensive study required for selecting agronomic features linked with drought tolerance. However, molecular markers may permit the transfer of the above QTLs into genetically recalcitrant commercial cultivars through marker-assisted breeding so long that they are very closely linked to target subsequent trait genes (Arya et al. 2021). Identifying drought-tolerant quantitative trait loci (QTLs) is of utmost importance for breeding programmes. This strategy is advantageous as it enables the identification and choice of genotypes that possess desirable alleles (Zhang et al. 2015), and this method has been particularly successful in soybean, where there is relatively high linkage disequilibrium between fewer intervals due to low recombination frequency (Lam et al. 2010). In a previous study conducted by Abdel-Haleem et al. (2011), four QTLs related to root morphology were discovered using 629 SSR markers. This finding suggests that QTLs connected with fibrous roots may have an impact on the ability of soybean plants to withstand drought and influence seed yield attributes. Furthermore, Carter et al. (1999) stated that delayed canopy wilting of crop cultivars is linked to drought spell tolerance.

12.3.3 Genetic Engineering

The promising potential of some genome engineering methods that form an arsenal for upgrading plants against water-deficit stress conditions. Arya et al. (2021) employed high-throughput genome and transcriptome sequencing techniques to identify the primary transcriptional regulators involved in the soybean to drought tolerance. Transcription factors (TFs) are a significant category of genes that primarily influence other genes, particularly those involved in responding to stress. The following transcription factors were utilised for drought stress: DREB (Zhou et al. 2020), AREB (Fuganti-Pagliarini et al. 2017), NAC (Hussain et al. 2017), MYB (Chen et al. 2021), and WRKY (Ning et al. 2017).

The process of introducing foreign genetic material into soybean plants, known as genetic transformation, was initially documented in 1988 by Hinchee et al. (1988). The first genetically modified (GM) crop that exhibited higher tolerance to drought was created by introducing a P5CR gene from *Arabidopsis* into soybean, resulting in increased drought resistance. *Agrobacterium*-mediated transformation and particle bombardment are commonly employed techniques for creating transgenic soybean plants. These procedures are effective for introducing several genes and transcription factors (TFs) to grow soybeans that are resistant to drought. (See Table 12.2 for further details.)

12.3.4 Genome Editing

The development of genome editing tool has provided a powerful, affordable, and efficient solution for DNA manipulation to speed up the rate at which desired traits can be introduced into plants compared to traditional breeding (Altaf et al. 2024; Ali et al. 2024). The CRISPR/Cas9 genome editing method has been extensively utilised in various plant species due to its exceptional efficiency (Bao et al. 2020). This technology has provided researchers with the capability to precisely customise any

Table 12.2 Genes/transcription factors (TFs) linked to drought tolerance in soybean

Genes/TFs	Role	References
P5CR	Elevated levels of water content and proline concentrations	de Ronde et al. (2004)
LOS5/ ABA3	Water loss was minimised by reducing the size of stomatal apertures and the rate of transpiration	Li et al. (2013)
BADH	Biosynthesis of proline and reduced malondialdehyde contents	Qin et al. (2017)
AtDREB1D	Reduced total leaf area and greater leaf cell membrane stability	Guttikonda et al. (2014)
GmbZIP1	Improved tolerance to prolonged periods of dryness	Gao et al. (2011)
GmXTH1	Enhanced root system, higher water content, and intensified endeavour of antioxidants	Zhang et al. (2022)
GmMYB14	Reduced leaf dimensions, decreased length of leaf stalk, decreased angle of leaf stalk, and increased crop productivity	Chen et al. (2021)
WRKY20	Increased crop productivity, taller plant growth, and more seed quantity	Ning et al. (2017)

genetic sequence to specifically target genes of interest. In 2015, Li et al. demonstrated the successful use of CRISPR/Cas9 for knockout and DNA homology-directed recombination (HDR) in soybeans. The CRISPR/Cas9 system is being extensively utilised for functional investigations in soybean, as demonstrated by the studies conducted by Du et al. 2016 and Cai et al. 2018. Another instance involves the use of CRISPR/Cas9 to induce mutations in soybean circadian rhythm genes (GmL CLs), resulting in mutants that exhibit reduced water loss under conditions of dehydration stress (Yuan et al. 2021). When drought signalling genes and transcription factors are overexpressed, it leads to the buildup of both signalling molecules and metabolites that aid plants in surviving unfavourable situations like drought. Furthermore, drought stress tolerance genes are reduced in soybean plants by increasing antioxidant levels and hormonal balance (Rasheed et al. 2020).

12.3.5 Omics Strategy

In recent times, the application of omics-based technologies have been utilised to recognise and characterise genes linked with favourable qualities, as well as their precise activities. Omics refers to a collection of scientific disciplines, including transcriptomics, genomics, metabolomics, and proteomics. These disciplines study many aspects of biological molecules and their interactions. Additionally, they are closely related to phenomic techniques, which examine the observable characteristics of organisms (Khatun et al. 2021). For example, a transcriptome atlas was prepared for soybean by the sequencing of RNAs in samples obtained from 14 different drought stress conditions using the next-generation sequencing (NGS) method (Joshi et al. 2010). According to Wang et al. (2018), RNA sequencing offers valuable information about the molecular pathways involved in soybean's response to drought. In their study, Strozycki et al. (2007) performed a proteomic analysis on soybean roots that were exposed to drought stress. They successfully discovered a total of 35 proteins. Metabolomics studies have shown that sugar and N and phytochemical metabolisms play a critical role in soybean plants when they are subjected to water-deficit stress circumstances (Das et al. 2017).

12.3.6 Agronomic Techniques

While breeding programmes take time to implement, several agronomic options can ameliorate drought stress in the short term. Seed hydration, known as seed priming, is done before sowing and enhances the germination rate, percentage, and uniformity (Farooq et al. 2009). Studies have shown that using external hormones such as abscisic acid (ABA), salicylic acid, jasmonic acid, uniconazole, and brassinolide can enhance soybean yields in both normal and drought-stressed circumstances (Zhang et al. 2004; Alimento et al. 2021). In addition, the usage of bioactive hormones like melatonin (Imran et al. 2021) can enhance the ability of soybean plants to tolerate drought. Therefore, it is vital to apply these hormones to the leaves to reduce the impacts of drought stress. Hansel et al. (2017) stated that P fertiliser can enhance root growth in soybean plants while they are experiencing drought stress. This is because the plants are able to absorb and utilise more phosphorus, leading to improved root development. These findings highlight the significance of using proper fertilisation techniques. The presence of arbuscular mycorrhizal fungi (*Glomus mosseae* and *G. intraradices*) in the roots of soybean plants help improve the condition of drought-induced nodule senescence, as observed in a study by Porcel and Ruiz-Lozano 2004. The application of biochar can also benefit soil by enhancing water availability, reducing ionic and osmotic factors of toxicity (sodicity and salinity stress), and enhancing drought-stressed soil. Therefore, it is feasible to enhance water use efficiency and nutrient uptake, which are less troublesome because of the increased resilience to these factors (Thomas et al. 2013; Atkinson et al. 2010).

12.4 Concept and Prospects of Marker-Assisted Breeding for Enhancing Drought Tolerance in Soybean Cultivars

The application of MAB is an effective way to strategically exploit this genetic information to increase the rate at which drought-tolerant soybean cultivars are developed (Thư et al. 2014). The identification of drought-associated DNA markers can provide an efficient tool for breeders to select and introgress desired alleles into elite soybean lines without confronting the complex genetics of yield improvement under drought stress.

12.4.1 Marker-Assisted Selection

MAS employs molecular markers that are tightly linked to certain features or QTLs, allowing for the selection of individuals possessing desired alleles (Varshney et al. 2014). Orf et al. (2004) suggested several possible uses for MAS in soybean breeding, including the identification of superior parents with desirable gene combinations, tracking genetic regions during backcrossing, restoring recurrent parents, selecting segregating populations, and discovering new advantageous alleles from exotic or wild germplasm. When conducting QTL (quantitative trait locus) research on drought, it is important to take into account factors such as the size of the population being studied, doing many trials in different settings, and selecting traits based on both morphological and agronomical qualities. The progress in DNA marker technology and QTL identification has greatly facilitated the analysis of the genetic factors underlying drought tolerance. This approach relies on the idea that DNA markers are in close proximity to genes that influence specific qualities. Consequently, these genes will co-occur with the desired genes throughout the process of crossing. Recombinant inbred lines (RILs), near isogenic lines (NILs), back crosses, F2/F4 populations, and doubled haploid lines (DHs) are frequently employed mapping populations to efficiently narrow down QTL regions through chromosome segment introgression. While soybean displays 1174 quantitative trait loci (QTLs) for various characteristics, only the data on drought volume are provided in Table 12.3 (source: www.soybase.ncgr.org; accessed on 14 July 2024).

QTL mapping is important for determining the chromosomal region linked with phenotypic variations in drought resistance traits, as well as identifying favourable alleles to be exploited through marker-assisted breeding. This property has limited genetic gains for yield and has attracted increased amounts of attention to secondary traits linked to yield stability. Altered expression of several secondary traits, such as early seedling vigour, canopy wilting, root system architecture (RSA), heat tolerance at the level of canopy cooling or leaf temperature depression and partitioning, carbon isotope discrimination changes in photosynthesis under stress conditions, and alterations in nitrogen fixation characteristics, has also been reported for crops grown under water-deficit conditions (Zhao et al. 2018). Examples of promising secondary traits under drought in soybean include RSAs, water use efficiency, slow canopy wilting, and biological nitrogen fixation (Valliyodan et al. 2016). The desired root characteristics increase drought tolerance due to the deep and dense rooting depth (Fenta et al. 2011), which provides plants with the ability to extract soil water at greater depths during periods of stress, regardless of whether the rate or duration changes.

A fibrous root structure improves resilience to drought by increasing the surface area and root tips (Abdel-Haleem et al. 2011). A cross between the Benning variety, which has a small fibrous root, and the PI 416937 variety, which has a large fibrous root, resulted in a population of recombinant inbred lines (RILs). This population had five QTLs that were associated with fibrous rooting systems. These QTLs were located on chromosomes 1, 3, 4, 8, and 20. Benning supplied a single QTL, whereas PI 416937 contributed beneficial alleles for a total of four QTLs, as reported by

Table 12.3 Identification of QTLs are linked with drought tolerance in soybean

Traits	QTL	Mapping populations	Marker linked to QTL	References
Germination and seedling stages	4	Hongfeng11 × Clark, 44 BILs	Satt449, Satt499, Satt440, and Sat_180	Qiu PengCheng et al. (2011)
Water use efficiency (WUE)	5	Young × PI416937, 120 F_4 population	B031-1, A089-1, cr497-1, K375-1, A063-1	Mian et al. (1996)
	2	S-100 × Tokyo, 116 F_2 population	A489H	Mian et al. (1998)
Canopy-wilting	1	Jackson × KS4895, 81 RILs	Sat_044	Bhatnagar et al. (2005)
	7	PI 416937 × Benning, 150 RILs	Sat_254-Satt296, Satt646, Satt302, Satt066, Satt276, Satt135, Satt462	Abdel-haleem et al. (2012)
	4	KS4895 × Jackson, 92 RILs	Satt072, Satt319, Satt362	Charlson et al. (2009)
Leaf hydraulic conductance	4	PI 416937 × Benning, 147 RILs	Satt339, Satt300, Satt478, Sat_175	Carpentieri-Pipolo et al. (2012)
Water uptake	5	AC Colibri×OT91–3, 200 RILs	Sat_220, Satt651, Satt333, Satt456, Sat_244	Molnar et al. (2012)
Yield	1	Minsoy × Noir 1, 236 RILs	Satt205-satt489	Specht et al. (2001)
	2	PI 68658 × LD00–3309, 117 RILs	Satt300, Satt474	Fox et al. (2015)
	3	Hutcheson × PI471938, 140 F_4 population	Satt226, Sat_375, Sat_074	Monteros et al. (2006)
Root architectural traits	1	Dunbar × PI 326582A, 251 BILs	Satt315	Manavalan et al. (2015)
	5	PI 416937 × Benning, 240 RILs	Satt383, Satt339, Satt713, Satt228, Sat_420	Abdel-haleem et al. (2011)
	4	V71–370 × PI407162, 160 RILs	4222.1.S1_10, 8398.1.S1_11, 59884.1.S1_8	Prince et al. (2015)
Nitrogen fixation	4	KS4895 × Jackson, 97 RILs	BARC-060299-16,598, Satt561, BARC-014657-01608, BARC-057467-14,765	Hwang et al. (2013)

Abdel-Haleem et al. 2011. Manavalan et al. (2015) employed linkage mapping to detect a group of QTLs associated with root length and lateral root number. The researchers examined 251 BC_2F_5 backcrossed inbred lines and discovered that these lines possessed advantageous genes derived from Dunbar. Table 12.3 contains the specific information regarding their discoveries.

Canopy wilting is a drought tolerance trait that can be used to quickly screen soybean germplasm under water-limiting conditions. Thirteen QTLs linked to slow wilting under water-limited conditions were detected in five biparental populations. Eleven of these QTLs exhibited beneficial alleles derived from PI 416937, as reported by Charlson et al. 2009. Only a restricted number of research have been carried out to identify QTLs in nitrogen-fixing plants, such as soybean plants. Three QTLs have been identified for nodule number, accounting for 13% of the phenotypic variation, as reported by Santos et al. 2013. Additionally, the study identified two large QTLs for shoot ureide on chromosomes 9 and 19. Furthermore, two significant loci for shoot nitrogen concentration under water stress were found on chromosomes 13 and 17. These loci explained the phenotypic variance between 0.11 and 0.31, as reported by Hwang et al. 2013. Pyramiding QTLs for drought-tolerant yield in drought-susceptible cultivars can improve productivity. The trait most severely affected by drought stress is yield; thus, the discovery of QTLs that confer increased yield (i.e. enhanced stability under such conditions) will further assist in the improvement and deployment of stress-tolerant lines for cultivation in stressed environments. In their study, Lee et al. (2021) identified two QTLs on chromosomes 13 and 17 that are associated with yield under rain-fed circumstances. They found that lines containing the PI-471938 allele had higher yields compared to lines holding Hutcheson alleles. The QTLs, along with their associated markers, can be utilised in MAS to generate soybean genotypes that exhibit resistance to drought stress (Dhungana et al. 2021).

12.4.2 MABC and Recurrent Selection (MAS)

Major QTLs are defined when a large portion of phenotypic variation is contributed by the identified QTLs associated with drought tolerance traits. These well-characterised QTLs can be useful in MABC to introduce donor values for drought tolerance into elite cultivars or breeding lines with low susceptibility to drought, thereby decreasing linkage drag (Mir et al. 2012). While the use of MABC has been widespread in developing resistance to various biotic stresses, there is minimal information regarding its application for enhancing drought tolerance. For example, MABC was employed to transfer root trait QTLs from the elite rice cultivar IR64 to enhance drought tolerance (Steele et al. 2006), and favourable alleles affecting yield components and flowering traits improving grain production under water-limited conditions were introgressed through MAS into maize lines, enhancing grain yield under a stress environment in both upland area hybrids and hybrids grown in subtropical zones (Ribaut and Ragot 2007).

The complexity of drought may have hindered the success rate of improving these traits and is one reason why MABC has been relatively unsuccessful in enhancing drought tolerance (review by Zheng, 2010). The MARS strategy, as recommended by Ribaut and Ragot (2007) and Bernardo (2008), involves intermitting specific individuals in each selection cycle. This approach is suggested to address the limits of partial genome coverage that is associated with QTL mapping, particularly when a complex trait is influenced by numerous QTLs. This method commonly utilises an F_2 base population and is suitable for self-pollinated crops such as wheat, barley, and chickpea. Its purpose is to generate pure lines that exhibit superior performance as a cohesive entity (Bernardo 2008). The practical use of MARS in soybean has been achieved (Eathington et al. 2007).

12.4.3 Genome-Wide Selection (GWS)

Genome-wide selection (GWS) or genomic selection (GS) is a crucial technique employed to develop enhanced germplasm lines that demonstrate extraordinary performance under particular circumstances (Mir et al. 2012). Singh et al. (2015) state that GS has received significant interest from researchers due to its ability to address complex qualities, such as yield, that involve a high degree of quantification. Genomic selection (GS) does not depend on prior knowledge of genes or large-effect QTLs, unlike marker-assisted selection (MAS). However, GS utilises marker data that has a higher density, achieved using affordable genotyping technologies that cover the complete genome (Varshney et al. 2014). The reduced cost of sequencing has facilitated the acquisition of millions of single nucleotide polymorphisms (SNPs) across the entire genome, significantly enhancing the potential for genomic selection (GS) (Goddard and Hayes 2007). This phenomenon has been ascribed to the fact that GS reduces the duration of breeding cycles, so saving time and consequently enhancing the rate of genetic improvement per unit of time (Crossa et al. 2017). GS models have recently improved the accuracy of forecasting complex traits in grain legumes under drought stress conditions (Jha et al. 2020).

As an illustration, Dhanapal et al. (2015) discovered 21 specific locations in soybean where the carbon isotope ratio is linked to water use efficiency using genome-wide association studies (GWS). A further study conducted by GWS examined 345 different varieties of soybeans and revealed that 52 single nucleotide polymorphisms (SNPs) were strongly associated with canopy temperature. Canopy temperature is a significant physiological indicator utilised to evaluate the plant's response to drought stress (Kaler et al. 2018). Genome-wide selection (GWS) has been utilised in soybean to improve certain traits such as plant height and maturity (Sonah et al. 2015), yield and its components (Hao et al. 2012), as well as protein content or oil content. Consequently, systematic investigation of this strategy could combine various aspects associated with multigenic drought tolerance in soybean.

12.5 Challenges in Developing Drought-Tolerant Soybean Cultivars via MAB

Although marker-assisted breeding holds potential for developing drought-tolerant soybean cultivars, there are several challenges to this approach. There are several reasons why it is technically difficult to bring relief across the board from phenomena such as drought. First, drought seldom occurs alone and is often coupled with other biotic or abiotic stresses, such as temperature extremes, diseases, and nutrient deficiencies (Ceccarelli et al. 2004). Encouraging tolerance to many stresses at once is a major challenge because plants often rely on different pathways. Second, performance in one dry year does not continue well to another, and tolerance to one type of drought (e.g. terminal drought) may result in poorly suited genotypes for others (e.g. intermediate or early-season drought) (Ceccarelli et al. 2004). This unpredictability has made drought a "moving target," which complicates the design of breeding approaches. Third, high environmental variability results in differential genotype responses at the QTL level (Jarso and Keneni 2003).

Consequently, increased drought tolerance does not necessarily result in an environment-independent level of yield protection (e.g. Sinclair 2011), leading to data aggregation challenges. Fourth, the question of selecting genotypes for drought tolerance is complicated by the requirement that field environments in which the moisture-holding capacity of soil is poor but uniform and the probability or frequency of each year facing wheat crops at serious risk due to water deficit at a consistent level each year are needed. Environments of this type are unique at universities and field experiment stations (Pathan et al. 2007). These challenges demonstrate the need for elaborate methodologies and custom-built areas to optimise both phenotyping and genotyping of drought-tolerant soybean lines.

Achieving drought tolerance in soybean plants is difficult because it depends on complex interactions among many genes (quantitative trait) that may change depending upon the environment, between seasons under similar conditions and even with crop age or period of day (Sinclair 2011). This is a challenge for identifying molecular markers linked to water-deficit responses, as there are still incomplete data on the genomic regions involved in drought tolerance (Marinho et al. 2022). Consequently, there is a pressing requirement to develop genetic markers that might be utilised to ascertain soybean reactions to drought stress. The soybean genome sequence, as completed by Dubey et al. in 2018, has the potential to enhance marker-assisted selection by expediting the identification of pathways and QTL genes through linkage mapping endeavours. Marker-assisted selection (MAS) is most effective in traditional soybean breeding programmes when it is used for selection in the F_2 or F_3 generation, when the population is still undergoing segregation (Gupta et al. 2010).

In addition, the potential for successful breeding of drought-tolerant plants depends mostly on the amount of genetic variety present in the germplasm (Dubey et al. 2019). An additional challenge is combining MAB technology with more advanced methods (genomic selection, genome editing). However, the integration and coordination of these diverse tools and technologies within breeding programmes can be complex (Mir et al. 2012). While there have been several reviews on water-deficient stress in legume's grain, a complete evaluation specifically addressing the challenges of breeding for enhancing soybean tolerance to varying levels of water deficit and improving productivity has not yet been published (Farooq et al. 2017). Therefore, overcoming the problems in drought breeding of soybean plants will be possible through detailed knowledge of the use of state-of-the-art breeding tools to isolate genetic mechanisms that control tolerance against water-limiting stress.

12.6 Conclusion

The Green Revolution has helped increase plant production dramatically, but the complexity of feeding an ever-growing population and climate change have encouraged scientists to free these substances. The occurrence of climate change and its consequent increase in the frequency of droughts remains a substantial menace to world food supply (Liaqat et al. 2022). Soybean (*Glycine max*), a prominent legume and oilseed crop, is very susceptible to drought and has a significant impact on grain yield. This chapter reviews several drought tolerance mechanisms in soybean plants considering the current state of research and discusses future prospects for achieving them. Drought stress, a common issue in cultivation, has been extensively researched due to its negative impact on soybean crop output caused by agricultural drought. Significant advancements have been achieved in addressing the effects of drought on soybean plants, particularly in understanding and enhancing their ability to withstand drought through several mechanisms. These processes encompass the plant's reaction to drought at different levels, including its morphological, physiological, and molecular adaptations to water scarcity. A deeper understanding of the complex networks that regulate root growth, water utilisation efficiency, and nitrogen fixation during periods of drought stress would offer essential resources for developing new cultivars with enhanced tolerance.

Drought tolerance is a complicated trait and causes considerable difficulties in breeding projects for creating drought-tolerant soybean cultivars. With respect to yield, quality, and stress tolerance, traditional soybean breeding has made many remarkable accomplishments over the last 100 years. However, these are too few to cope with the increased global population and climate change. The introduction of molecular breeding has opened new avenues for lifting the bottlenecks faced during conventional breeding. Drought decreases the economic productivity of legumes such as soybean, which is recognised as a crop with worldwide importance but frequently experiences significant water stress through all stages from germination to seed setting. Multiple levels of abiotic stress acclimatisation mechanisms are mediated by hydroxyl-free radicals/reactive oxygen species, and genetic top-down re-engineering towards better climate adaptation would imply drought-responsive SNP target site mining and SSR high-throughput genotyping data crunching.

The discovery of novel QTLs/genes under drought stress along with genetic information will be very helpful for enhancing the level of tolerance. Combining genes at multiple loci can improve drought tolerance in soybean, and pyramiding MABs will be essential for improving drought tolerance. Genetic engineering, OMICS-based technologies, and CRISPR/Cas9 will further contribute to improving soybean drought tolerance. The systematic utilisation of genetic resources, a clear interpretation of the mechanisms underlying tolerance and/or resistance processes with marker-assisted selection (MAS), the construction of dense genome map kits, effective marker technology-aided selection techniques, and properly designed breeding strategies are necessary to further improve drought tolerance in soybean varieties. Therefore, a joint policy combining a conventional breeding programme along with MAS and genetic engineering approaches is essential for the development of drought-tolerant soybean genetics.

References

Abdel-Haleem H, Lee GJ, Boerma RH (2011) Identification of QTL for increased fibrous roots in soybean. Theor Appl Genet 122:935–946. https://doi.org/10.1007/s00122-010-1500-9

Abdel-Haleem H, Carter TE, Purcell LC, King CA, Ries LL, Chen P, Schapaugh W, Sinclair TR, Boerma HR (2012) Mapping of quantitative trait loci for canopy-wilting trait in soybean (*Glycine max* L. Merr). Theor Appl Genet 125:837–846

Ali A, Altaf MT, Bedir M, Liaqat W, Jamil A, Nadeem MA, Baloch FS (2024) Biotechnological advancements toward sugarcane crop improvement. In: Industrial crop plants. Springer, Singapore, pp 333–352

Alimento JJA, Abello NFH, Pascual VU, Pascual PRLP (2021) Morphological and yield response of soybean (*Glycine max*) to foliar application of antioxidants under drought condition. Plant Cell Biotechnol Mol Biol 22(19–20):103–110

Altaf MT, Liaqat W, Ali A, Jamil A, Bedir M, Nadeem MA et al (2024) Conventional and biotechnological approaches for the improvement of industrial crops. In: Industrial crop plants. Springer, Singapore, pp 1–48

Arya H, Singh MB, Bhalla PL (2021) Towards developing drought-smart soybeans. Front Plant Sci 12:750664

Atkinson CJ, Fitzgerald JD, Hipps NA (2010) Potential mechanisms for achieving agricultural benefits from biochar application to temperate soils: a review. Plant Soil 337:1–18

Bao A, Zhang C, Huang Y, Chen H, Zhou X, Cao D (2020) Genome editing technology and application in soybean improvement. Oil Crop Sci 5(1):31–40

Basal O (2020, July 30) Physiology, yield and quality of soybean as affected by drought stress. Life Sci Soc Pakistan 8(3):247–252. https://doi.org/10.35495/ajab.2019.11.505

Basu S, Ramegowda V, Kumar A, Pereira A (2016) Plant adaptation to drought stress. F1000 Res 5:1554. https://doi.org/10.12688/f1000research.7678.1

Bernardo R (2008) Molecular markers and selection for complex traits in plants: learning from the last 20 years. Crop Sci 48:1649–1664

Bhatnagar S, King CA, Purcell L, Ray JD (2005) Identification and mapping of quantitative trait loci associated with crop responses to water-deficit stress in soybean [*Glycine max* (L.) Merr.]. The ASACSSA-SSSA international annual meeting poster abstract. Salt Lake City

Buckley C (2011) Q&A: What are the effects of China's drought? Reuters. https://www.reuters.com/article/us-China-water-facts/qa-what-are-the-effects-of-chinas-drought-idUSTRE7520KS20110603. Accessed July 2019

Cai Y, Chen L, Sun S, Wu C, Yao W, Jiang B et al (2018) CRISPR/Cas9-mediated deletion of large genomic fragments in soybean. Int J Mol Sci 19:3835

Carpentieri-Pipolo V, Pipolo A, Abdel-Haleem H, Boerma H, Sinclair T (2012) Identification of QTLs associated with limited leaf hydraulic conductance in soybean. Euphytica 186:679–686

Carter TE, DeSouza PI, Purcell LC (1999) Recent advances in breeding for drought and aluminum resistance in soybean. Superior Printing

Ceccarelli S, Grando S, Baum M, Udupa SM (2004) Breeding for drought resistance in a changing climate. In: Challenges and strategies of dryland agriculture, vol 32. CSSA Special Publications, pp 167–190

Chandra P, Wunnava A, Verma P, Chandra A, Sharma RK (2021) Strategies to mitigate the adverse effect of drought stress on crop plants—influences of soil bacteria: a review. Pedosphere 31(3):496–509
Charlson DV, Bhatnagar S, King CA, Ray JD, Sneller CH, Carter TE, Purcell LC (2009) Polygenic inheritance of canopy wilting in soybean [*Glycine max* (L.) Merr.]. Theor Appl Genet 119:587–594
Chaudhary J, Patil GB, Sonah H, Deshmukh RK, Vuong TD, Valliyodan B, Nguyen HT (2015) Expanding omics resources for improvement of soybean seed composition traits. Front Plant Sci 6:1021
Chen L, Yang H, Fang Y, Guo W, Chen H, Zhang X et al (2021) Overexpression of GmMYB14 improves high-density yield and drought tolerance of soybean through regulating plant architecture mediated by the brassinosteroid pathway. Plant Biotechnol J 19(33098207):702–716. https://doi.org/10.1111/pbi.13496
Chowdhury JA, Karim MA, Khaliq QA, Ahmed AU, Mondol ATMAI (2017) Effect of drought stress on water relation traits of four soybean genotypes. SAARC J Agric 15:163
Cotrim MF, Gava R, Campos CNS, De David CHO, Reis IDA, Teodoro LPR et al (2021) Physiological performance of soybean genotypes grown under irrigated and rainfed conditions. J Agron Crop Sci 207:34–43. https://doi.org/10.1111/jac.12448
Crossa J, Pérez-Rodríguez P, Cuevas J, Montesinos-López O, Jarquín D, De Los Campos G et al (2017) Genomic selection in plant breeding: methods, models, and perspectives. Trends Plant Sci 22(11):961–975
Crusiol LGT, Nanni MR, Furlanetto RH, Sibaldelli RNR, Cezar E, Sun L et al (2021) Classification of soybean genotypes assessed under different water availability conditions and at different phenological stages using leaf-based hyperspectral reflectance. Remote Sens 13:172. https://doi.org/10.3390/rs13020172
Cui Y, Ning S, Jin J, Jiang S, Zhou Y, Wu C (2021). Quantitative lasting effects of drought stress at a growth stage on soybean evapotranspiration and aboveground BIOMASS. Water, 13(1),18. https://doi.org/10.3390/w13010018
Das A, Rushton PJ, Rohila JS (2017) Metabolomic profiling of soybeans (*Glycine max* L.) reveals the importance of sugar and nitrogen metabolism under drought and heat stress. Plan Theory 6(2):21
De Ronde J, Laurie R, Caetano T, Greyling M, Kerepesi I (2004) Comparative study between transgenic and nontransgenic soybean lines proved transgenic lines to be more drought tolerant. Euphytica 138:123–132
de Souza PI, Egli DB, Bruening WP (1997) Water stress during seed fi lling and leaf senescence in soybean. Agron J 89:807–812
Desclaux D, Huynh TT, Roumet P (2000) Identification of soybean plant characteristics that indicate the timing of drought stress. Crop Sci 40:716–722. https://doi.org/10.2135/cropsci2000.403716x
Dhanapal AP, Ray JD, Singh SK, Hoyos-Villegas V, Smith JR, Purcell LC et al (2015) Genome-wide association study (GWAS) of carbon isotope ratio (δ 13 C) in diverse soybean [*Glycine max* (L.) Merr.] genotypes. Theor Appl Genet 128:73–91
Dhungana SK, Park JH, Oh JH, Kang BK, Seo JH, Sung JS et al (2021) Quantitative trait locus mapping for drought tolerance in soybean recombinant inbred line population. Plan Theory 10(9):1816
Dornbos DL, Mullen RE (1991) Influence of stress during soybean seed fill on seed weight, germination, and seedling growth rate. Can J Plant Sci 71(2):373–383
Dowling DK, Simmons LW (2009) Reactive oxygen species as universal constraints in life-history evolution. Proceedings of the Royal Society B: Biological Sciences 276(1663):1737–1745. https://doi.org/10.1098/rspb.2008.1791
Du H, Zeng X, Zhao M, Cui X, Wang Q, Yang H et al (2016) Efficient targeted mutagenesis in soybean by TALENs and CRISPR/Cas9. J Biotechnol 217:90–97
Du Y, Zhao Q, Chen L, Yao X, Xie F (2020) Effect of drought stress at reproductive stages on growth and nitrogen metabolism in soybean. Agronomy 10(2):302

Dubey A, Kumar A, Abd Allah EF, Hashem A, Khan ML (2019, October 1) Growing more with less: breeding and developing drought resilient soybean to improve food security. Ecol Indic 105:425–437. https://doi.org/10.1016/j.ecolind.2018.03.003
Eathington SR, Crosbie TM, Edwards MD, Reiter RS, Bull JK (2007) Molecular markers in a commercial breeding program
Farooq M, Wahid A, Kobayashi N, Fujita D, Basra S (2009) Plant drought stress: Effects, mechanisms and management. Agron Sustain Dev 29:185–212. https://doi.org/10.1051/agro:2008021
Farooq M, Gogoi N, Barthakur S, Baroowa B, Bharadwaj N, Alghamdi SS, Siddique KH (2017) Drought stress in grain legumes during reproduction and grain filling. J Agron Crop Sci 203(2):81–102
Fenta BA, Schluter U, Marquez-Garcia B, DuPlessis M, Foyer CH, Kunert KJ (2011) Identification and application of phenotypic and molecular markers for abiotic stress tolerance in soybean. In: Krezhova D (ed) Soybean- genetics and novel techniques for yield enhancement, pp 181–195
Fox CM, Cary TR, Nelson RL, Diers BW (2015) Confirmation of a seed yield QTL in soybean. Crop Sci 55(3):992–998
Fuganti-Pagliarini R, Ferreira LC, Rodrigues FA, Molinari HBC, Marin SRR, Molinari MDC et al (2017) Characterization of soybean plants genetically modified for drought tolerance under field conditions. Front Plant Sci 8:448. https://doi.org/10.3389/fpls.2017.00448
Gao SQ, Chen M, Xu ZS, Zhao CP, Li L, Xu HJ et al (2011) The soybean GmbZIP1 transcription factor enhances multiple abiotic stress tolerances in transgenic plants. Plant Mol Biol 75:537–553
Goddard ME, Hayes BJ (2007) Genomic selection. J Anim Breed Genet 124(6):323–330
Guimarães-Dias F, Neves-Borges AC, Viana AAB, Mesquita RO, Romano E, Grossi-de-Sá MDF et al (2012) Expression analysis in response to drought stress in soybean: shedding light on the regulation of metabolic pathway genes. Genet Mol Biol 35:222–232
Gupta PK, Kumar J, Mir RR, Kumar A (2010) Marker-assisted selection as a component of conventional plant breeding. Plant Breed Rev 33:145–217. https://doi.org/10.1002/9780470535486.ch4
Guttikonda SK, Valliyodan B, Neelakandan AK, Tran LSP, Kumar R, Quach TN et al (2014) Overexpression of AtDREB1D transcription factor improves drought tolerance in soybean. Mol Biol Rep 41:7995–8008
Hansel FD, Amado TJ, Ruiz Diaz DA, Rosso LH, Nicoloso FT, Schorr M (2017) Phosphorus fertilizer placement and tillage affect soybean root growth and drought tolerance. Agron J 109(6):2936–2944
Hao D, Cheng H, Yin Z, Cui S, Zhang D, Wang H, Yu D (2012) Identification of single nucleotide polymorphisms and haplotypes associated with yield and yield components in soybean (*Glycine max*) landraces across multiple environments. Theor Appl Genet 124:447–458
Hasanuzzaman M, Nahar K, Gill SS, Fujita M (2013) Drought stress responses in plants, oxidative stress, and antioxidant defense. In: Climate change and plant abiotic stress tolerance, pp 209–250
Hinchee MA, Connor-Ward DV, Newell CA, McDonnell RE, Sato SJ, Gasser CS et al (1988) Production of transgenic soybean plants using agrobacterium-mediated DNA transfer. Biotechnology 6:915–922. https://doi.org/10.1038/nbt0888-915
Hussain RM, Ali M, Feng X, Li X (2017) The essence of NAC gene family to the cultivation of drought-resistant soybean (*Glycine max* L. Merr.) cultivars. BMC Plant Biol 17:1–11. https://doi.org/10.1186/s12870-017-1001-y
Hwang S, King CA, Davies MK, Ray JD, Cregan PB, Purcell LC (2013) QTL analysis of shoot ureide and nitrogen concentrations in soybean [*Glycine max* (L.) Merr.]. Crop Sci 53(6):2421–2433
Imran M, Latif Khan A, Shahzad R, Aaqil Khan M, Bilal S, Khan A et al (2021) Exogenous melatonin induces drought stress tolerance by promoting plant growth and antioxidant defence system of soybean plants. AoB Plants 13(4):plab026
Jarso M, Keneni G (2003) Expression of quantitative trait loci (QTL) in diverse spring barley (*Hordeum vulgare* L.) recombinant inbred lines under different nitrogen environments
Jha UC, Bohra A, Nayyar H (2020) Advances in "omics" approaches to tackle drought stress in grain legumes. Plant Breed 139(1):1–27

Jin J, Wang G, Liu X, Pan X, Herbert SJ, Tang C (2006) Interaction between phosphorus nutrition and drought on grain yield, and assimilation of phosphorus and nitrogen in two soybean cultivars differing in protein concentration in grains. J Plant Nutr 29(8):1433–1449

Joshi T, Yan Z, Libault M, Jeong DH, Park S, Green PJ et al (2010) Prediction of novel miRNAs and associated target genes in *Glycine max*. BMC Bioinfor 11:1–9

Kajiya-Kanegae H, Nagasaki H, Kaga A, Hirano K, Ogiso-Tanaka E, Matsuoka M et al (2021) Whole-genome sequence diversity and association analysis of 198 soybean accessions in mini-core collections. DNA Res 28:dsaa032. https://doi.org/10.1093/dnares/dsaa032

Kaler AS, Ray JD, Schapaugh WT, Asebedo AR, King CA, Gbur EE, Purcell LC (2018) Association mapping identifies loci for canopy temperature under drought in diverse soybean genotypes. Euphytica 214:1–18

Khatun M, Sarkar S, Era FM, Islam AM, Anwar MP, Fahad S et al (2021) Drought stress in grain legumes: Effects, tolerance mechanisms and management. Agronomy 11(12):2374

Kobraee S, Shamsi K, Rasekhi B (2011) Soybean production under water deficit conditions. Ann Biol Res 2:423–434

Kosturkova G, Todorova R, Sakthivelu G, Akitha Devi MK, Giridhar P, Rajasekaran T, Ravishankar GA (2008) Response of Bulgarian and Indian soybean genotypes to drought and water deficiency in field and laboratory conditions. Gen Appl Plant Physiol 34(3–4):239–250

Kpoghomou BK, Sapra VT, Beyl CA (1990) Sensitivity to drought stress of three soybean cultivars during different growth stages. J Agron Crop Sci 164:104–109

Ku YS, Au-Yeung WK, Yung YL, Li MW, Wen CQ, Liu X, Lam HM (2013) Drought stress and tolerance in soybean. In: A comprehensive survey of international soybean research—Genetics, physiology, agronomy and nitrogen relationships, pp 209–237

Lam H-M, Xu X, Liu X, Chen W, Yang G, Wong F-L et al (2010) Resequencing of 31 wild and cultivated soybean genomes identifies patterns of genetic diversity and selection. Nat Genet 42:1053–1059. https://doi.org/10.1038/ng.715

Lee GJ, Lee S, Carter TE Jr, Shannon G, Boerma HR (2021) Identification of soybean yield QTL in irrigated and rain-fed environments. Agronomy 11(11):2207

Li Y, Zhang J, Zhang J, Hao L, Hua J, Duan L et al (2013) Expression of an Arabidopsis molybdenum cofactor sulphurase gene in soybean enhances drought tolerance and increases yield under field conditions. Plant Biotech J 11(6):747–758

Li L, Zhang Q, Huang D (2014) A review of imaging techniques for plant phenotyping. Sensors 14:20078–20111. https://doi.org/10.3390/s141120078

Liaqat W, Barutcular C, Farooq M, Ahmad H, Jan M, Ahmad Z et al (2022) Climate change in relation to agriculture: A review. Span J Agric Res 20(2):e03R01

Licht M, Wright D, Lenssen AW (2013) Soybean response to drought. Agriculture and Environment Extension Publications, p 190. https://lib.dr.iastate.edu/extension_ag_pubs/190

Liu F, Andersen MN, Jensen CR (2003) Loss of pod set caused by drought stress is associated with water status and ABA content of reproductive structures in soybean. Funct Plant Biol 30:271–280

Liu Y, Gai JY, Lu HN, Wang YJ, Chen SY (2005) Identification of drought tolerant germplasm and inheritance and QTL mapping of related root traits in soybean (*Glycine max* (L.) Merr.). Acta Genet Sin 32(8):855–863

Mahajan S, Tuteja N (2005) Cold, salinity and drought stresses: an overview. Archives of biochemistry and biophysics, 444(2):139–158. https://doi.org/10.1016/j.abb.2005.10.018

Manavalan LP, Guttikonda SK, Tran LSP, Nguyen HT (2009) Physiological and molecular approaches to improve drought resistance in soybean. Plant Cell Physiol 50(7):1260–1276

Manavalan LP, Prince SJ, Musket TA, Chaky J, Deshmukh R, Vuong TD, Song L, Cregan PB, Nelson JC, Shannon JG (2015) Identification of novel QTL governing root architectural traits in an interspecific soybean population. PLoS One 10:e0120490

Marinho JP, Pagliarini RF, Molinari MDC, Marcolino-Gomes J, Caranhoto ALH, Marin SRR et al (2022) Overexpression of full-length and partial DREB2A enhances soybean drought tolerance. Agron Sci Biotechnol 8:1–21

Mian M, Bailey M, Ashley D, Wells R, Carter T, Parrott W, Boerma H (1996) Molecular markers associated with water use efficiency and leaf ash in soybean. Crop Sci 36:1252–1257
Mian M, Ashley D, Boerma H (1998) an additional QTL for water use efficiency in soybean. Crop Sci 38:390–393
Mir RR, Zaman-Allah M, Sreenivasulu N, Trethowan R, Varshney RK (2012) Integrated genomics, physiology and breeding approaches for improving drought tolerance in crops. Theor Appl Genet 125:625–645
Molnar SJ, Charette M, Cober ER (2012) Mapping quantitative trait loci for water uptake in a recombinant inbred line population of natto soybean. Can J Plant Sci 92:257–266
Monteros MJ, Lee G, Missaoui AM, Carter TE, Boerma HR (2006) Identification and confirmation of QTL conditioning drought tolerance in Nepalese soybean. The 11th Biennial conference on the molecular and cellular biology of the soybean, Abstract PI471938. August 5–8, Lincoln
Mourtzinis S, Specht JE, Lindsey LE, Wiebold WJ, Ross J, Nafziger ED, Kandel HJ, Mueller N, Devillez PL, Arriaga FJ et al (2015) Climate-induced reduction in US-wide soybean yields underpinned by region-and in-season-specific responses. Nat Plants 1:1–4
Ning W, Zhai H, Yu J, Liang S, Yang X, Xing X et al (2017) Overexpression of Glycine soja WRKY20 enhances drought tolerance and improves plant yields under drought stress in transgenic soybean. Mol Breed 37:19. https://doi.org/10.1007/s11032-016-0614-4
Orf JH, Diers BW, Boerma HR (2004) Genetic improvement: conventional and molecular-based strategies. In: Boerma HR, Specht JE (eds) Soybeans: improvement, production, and uses, agronomy monographs, vol No. 16, 3rd edn. ASA-CSSA-SSSA, Madison, pp 417–450
Pathan MS, Lee JD, Shannon JG, Nguyen HT (2007) Recent advances in breeding for drought and salt stress tolerance in soybean. In: Jenks MA, Hasegawa PM, Jain SM (eds) Advances in molecular breeding toward drought and salt tolerant crops. Springer, Dordrecht. https://doi.org/10.1007/978-1-4020-5578-2_30
Poehlman JM, Sleper DA, Rudd J (1995) Breeding field crops, 4th edn. Iowa State University Press, Ames, Iowa
Porcel R, Ruiz-Lozano JM (2004) Arbuscular mycorrhizal influence on leaf water potential, solute accumulation, and oxidative stress in soybean plants subjected to drought stress. J Exp Bot 55:1743–1750. https://doi.org/10.1093/jxb/erh188
Poudel S, Vennam RR, Shrestha A, Reddy KR, Wijewardane NK, Reddy KN, Bheemanahalli R (2023, January 23) Resilience of soybean cultivars to drought stress during flowering and early-seed setting stages. Nat Portfolio 13(1):1277. https://doi.org/10.1038/s41598-023-28354-0
Prince SJ, Song L, Qiu D, dos Santos JVM, Chai C, Joshi T, Patil G, Valliyodan B, Vuong TD, Murphy M (2015) Genetic variants in root architecture-related genes in a Glycine soja accession, a potential resource to improve cultivated soybean. BMC Genomics 16:132
Qin D, Zhao CL, Liu XY, Wang PW (2017) Transgenic soybeans expressing betaine aldehyde dehydrogenase from *Atriplex canescens* show increased drought tolerance. Plant Breed 136(5):699–709
Qiu PengCheng QP, Zhang WenBo ZW, Li CanDong LC, Jiang HongWei JH, Liu ChunYan LC, Fan DongMei FD et al (2011) Genetic overlap of drought-tolerance loci between germination stage and seedling stage analysed using introgression lines in soybean. Acta Agron Sci 37:477
Qiu LJ, Chang RZ, Liu ZX, Guan RX, Li YH (2006) Descriptors and data standard for soybean (Glycine spp.)
Raghuvanshi RS, Bisht K (2010) Chapter 18- Uses of Soybean: Products and Preparation. In: Singh G (ed) The soybean: botany, production and uses. CAB International, pp 404–426
Rasheed A, Hassan MU, Aamer M, Batool M, Sheng FANG, Ziming WU, Huijie LI (2020) A critical review on the improvement of drought stress tolerance in rice (*Oryza sativa* L.). Not Bot Horti Agrobo 48(4):1756–1788
Ribaut JM, Ragot M (2007) Marker-assisted selection to improve drought adaptation in maize: the backcross approach, perspectives, limitations, and alternatives. J Exp Bot 58:351–360
Riebeek H (2011) Severe drought causes famine in East Africa. NASA. https://earthobservatory.nasa.gov/images/51411/severe-drought-causes-famine-in-east- Africa. Accessed July 2019

Sadeghi L, Rafiee M, Daneshian J (2021) Effect of drought stress and aerosols on yield and some physiological traits of soybean (*Glycine max* L.). J Plant Process Funct 10(41):263–278
Santos MA, Geraldi IO, Garcia AAF, Bortolatto N, Schiavon A, Hungria M (2013) Mapping of QTLs associated with biological nitrogen fixation traits in soybean. Hereditas 150:17–25. https://doi.org/10.1111/j.1601-5223.2013.02275.x
Seleiman MF, Al-Suhaibani N, Ali N, Akmal M, Alotaibi M, Refay Y et al (2021) Drought stress impacts on plants and different approaches to alleviate its adverse effects. Plan Theory 10(2):259
Serraj R, Sinclair TR, Purcell LC (1999) Symbiotic N2 fixation response to drought. J Exp Bot 50(331):143–155
Sinclair TR (2011) Challenges in breeding for yield increase for drought. Trends Plant Sci 16:289–293. https://doi.org/10.1016/j.tplants.2011.02.008
Singh B, Bohra A, Mishra S, Joshi R, Pandey S (2015) Embracing new-generation 'omics' tools to improve drought tolerance in cereal and food-legume crops. Biol Plant 59:413–428
Sonah H, O'Donoughue L, Cober E, Rajcan I, Belzile F (2015) Identification of loci governing eight agronomic traits using a GBS-GWAS approach and validation by QTL mapping in soya bean. Plant Biotechnol J 13(2):211–221
SoyStats (2024) International: world soybean production. http://soystats.com/international-worldsoybean-production. Accessed July 2024
Specht JE, Chase K, Macrander M, Graef GL, Chung J, Markwell JP et al (2001) Soybean response to water: a QTL analysis of drought tolerance. Crop Sci 41:493–509
Steele KA, Price AH, Shashidhar HE, Witcombe JR (2006) Marker assisted selection to introgress rice QTLs controlling root traits into an Indian upland rice variety. Theor Appl Genet 112:208–221
Strozycki PM, Szczurek A, Lotocka B, Figlerowicz M, Legocki AB (2007) Ferritins and nodulation in Lupinus luteus: iron management in indeterminate type nodules. J Exp Bot 58(12):3145–3153
Thomas SC, Frye S, Gale N, Garmon M, Launchbury R, Machado N et al (2013) Biochar mitigates negative effects of salt additions on two herbaceous plant species. J Environ Manag 129:62–68
Thư NBA, Nguyen QH, Hoang XLT, Thảo NP, Tran LP (2014, January 1) Evaluation of drought tolerance of the vietnamese soybean cultivars provides potential resources for soybean production and genetic engineering. Hindawi Publishing Corporation, pp 1–9. https://doi.org/10.1155/2014/809736
Turner NC, Wright GC, Siddique KHM (2001) Adaptation of grain legumes (pulses) to water limited environments. Adv Agron 71:193–123
Valliyodan B, Ye H, Song L, Murphy M, Shannon JG, Nguyen HT (2016) Genetic diversity and genomic strategies for improving drought and waterlogging tolerance in soybeans. J Exp Bot 68:1835–1849. https://doi.org/10.1093/jxb/erw433
Varshney RK, Terauchi R, McCouch SR (2014) Harvesting the promising fruits of genomics: applying genome sequencing technologies to crop breeding. PLoS Biol 12(6):e1001883
Voiland A (2018) A costly drought in Argentina. NASA. https://earthobservatory.nasa.gov/images/91999/a-costlydrought-in-argentina. Accessed July 2019
Wang M, Cheung CY, Ma TF, Yao WC (2004) Studies on the drought resistance of seedling in soybean. Chinese J Oil Crop Sci 26(3):29–32
Wang L, Dong S, Liu L, Ma Y, Li S, Zu W (2018) Transcriptome profiling reveals PEG-simulated drought, heat and combined stress response mechanisms in soybean. Comput Biol Chem 77:413–419
Wei Z, Paredes P, Liu Y, Chi WW, Pereira LS (2015) Modelling transpiration, soil evaporation and yield prediction of soybean in North China Plain. Agric Water Manag 147:43–53
Wei Y, Jin J, Jiang S, Ning S, Liu L (2018) Quantitative response of soybean development and yield to drought stress during different growth stages in the Huaibei Plain. China Agron 8:97
Wijewardana C, Reddy KR, Krutz LJ, Gao W, Bellaloui N (2019) Drought stress has transgenerational effects on soybean seed germination and seedling vigor. PLoS One 14(9):e0214977

Wilson RF (2008) Soybean: market driven research needs. In: Genetics and genomics of soybean. Springer, pp 3–15
Xiong R, Liu S, Considine MJ, Siddique KH, Lam HM, Chen Y (2021) Root system architecture, physiological and transcriptional traits of soybean (Glycine max L.) in response to water deficit: A review. Physiologia Plantarum 172(2):405-418. https://doi.org/10.1111/ppl.13201
Yuan L, Xie GZ, Zhang S, Li B, Wang X, Li Y et al (2021) GmLCLs negatively regulate ABA perception and signalling genes in the soybean leaf dehydration response. Plant Cell Environ 44:412–424. https://doi.org/10.1111/pce.13931
Zhang M, Duan L, Zhai Z, Li J, Tian X, Wang B et al (eds) (2004) Effects of plant growth regulators on water deficit-induced yield loss in soybean. In: Proceedings of the 4th international crop science congress, Brisbane, pp 252–256
Zhang M, Jin ZQ, Zhao J, Zhang G, Wu F (2015) Physiological and biochemical responses to drought stress in cultivated and Tibetan wild barley. Plant Growth Regulation 75:567–574. https://doi.org/10.1007/s10725-014-0022
Zhang Y, Zhang HZ, Fu JY, Du YY, Qu J, Song Y, Wang PW (2022) The GmXTH1 gene improves the drought stress resistance of soybean plants. Mol Breed 42(1):3
Zhao T, Aleem M, Sharmin RA (2018) Adaptation to water stress in soybean: morphology to genetics. Plant, abiotic stress and responses to climate change. Intech Open, London, pp 33–68
Zheng J, Fu J, Gou M, Huai J, Liu Y, Jian M, Huang Q et al (2010) Genome-wide transcriptome analysis of two maize inbred lines under drought stress. Plant Molecular Biology 72:407–421. https://doi.org/10.1007/s11103-009-9579-6
Zhou Y, Zhou W, Liu H, Liu P, Li Z (2020) Genome-wide analysis of the soybean DREB gene family: identification, genomic organization and expression profiles in response to drought stress. Plant Breed 139:1158–1167. https://doi.org/10.1111/pbr.12867
Zhou J, Zhou J, Ye H, Ali ML, Chen P, Nguyen HT (2021) Yield estimation of soybean breeding lines under drought stress using unmanned aerial vehicle-based imagery and convolutional neural network. Biosyst Eng 204:90–103. https://doi.org/10.1016/j.biosystemseng.2021.01.017
Zia R, Nawaz MS, Siddique MJ, Hakim S, Imran A (2021) Plant survival under drought stress: Implications, adaptive responses, and integrated rhizosphere management strategy for stress mitigation. Microbiol Res 242:126626

Marker-Assisted Breeding in Mung Bean (*Vigna radiata* L.) for Drought Stress

13

Meliha F. Sarıkaya, Muhammed Tatar, Amjad Ali, Muhammad T. Altaf, Parnaz Mortazavi, Sarmad A. Qureshi, Mehmet Bedir, and Kağan Kökten

Abstract

Mung bean (*Vigna radiata* L.) is a significant legume considered for its exceptional nutritional composition and ecological benefits. This crop is fundamental to sustainable agriculture due to its drought resistance, brief growth cycle, and nitrogen-fixing capability. However, its productivity is frequently impeded by constraints from pests, illnesses, and water deficiency. Marker-assisted breeding (MAB) emerges as a transformative approach, leveraging molecular markers to enhance mung bean's genetic resilience. Drought tolerance is achieved through advanced methods such as quantitative trait loci (QTL) mapping, molecular markers like SSR and SNP, and conventional phenotypic selection. By making it possible to precisely identify features, these technologies hasten the development of cultivars that are resistant to drought. Research indicates that drought stress negatively impacts physiological processes, such as photosynthesis and nitrogen fixation, leading to diminished yields. The use of molecular markers enhances breeding strategies, alleviating negative effects and augmenting crop performance in water-scarce environments. Future research must prioritize integrating genomics and biotechnological advancements, such as CRISPR/Cas9, to unlock mung bean's potential. Furthermore, participatory breeding approaches

M. F. Sarıkaya (✉) · P. Mortazavi · S. A. Qureshi · M. Bedir · K. Kökten
Department of Plant Production and Technologies, Faculty of Agricultural Sciences and Technologies, Sivas University of Science and Technology, Sivas, Turkey
e-mail: fsarikaya@sivas.edu.tr

M. Tatar (✉) · A. Ali
Department of Plant Protection, Faculty of Agricultural Sciences and Technologies, Sivas University of Science and Technology, Sivas, Turkey

M. T. Altaf
Department of Field Crops, Faculty of Agriculture, Recep Tayyip Erdoğan University, Rize/Pazar, Turkey

© The Author(s), under exclusive license to Springer Nature Singapore Pte Ltd. 2025

M. A. Nadeem et al. (eds.), *Marker-Assisted Breeding in Legumes for Drought Tolerance*, https://doi.org/10.1007/978-981-96-4112-3_13

involving stakeholders will enhance adoption and ensure global food security. This chapter underscores the importance of MAB in combating drought stress in mung beans, fostering sustainable agricultural practices, and addressing climate change challenges.

Keywords

Mung bean · Drought · QTLs · Sustainable agriculture · Molecular markers

13.1 Introduction

Plant and animal protein has an important place in a balanced and regular human diet. Animal protein is the first choice in the human diet due to its nutritional value and flavor. However, animal protein production may have some negative impacts on the environment. The most important of these effects is the impact of animal husbandry on global climate change. This is because it emits two powerful greenhouse gases: methane from enteric fermentation and anaerobic fermentation of manure by ruminants and nitrous oxide from manure left over from grazing on pastures (Faverdin et al. 2022). The increase in the concentrations of greenhouse gases prevents the reflection of long-wavelength rays back into the atmosphere (Sarıkaya and Tatar 2023). This causes significant effects on the climate, especially on temperature increase, and thus on agricultural production. Population growth, urbanization, and industrialization may increase meat consumption demand, while unbalanced economic development in developing countries creates an imbalance in animal protein supply (Kumar et al. 2023). Certain individuals, particularly in industrialized nations, require a more healthful lifestyle. This encompasses the reduction of cholesterol and lipids. Limitations on meat production and its constituents may not simultaneously meet these dietary protein needs. For all these reasons, there is increasing demand for a further expansion of protein sources and alternative food proteins that can potentially be utilized instead of animal protein. In this sense, plant protein is regarded as the superior option to animal protein because of its reduced carbon footprint and greater environmental sustainability. There is a need to cultivate additional plant protein sources to achieve a positive outcome (Hou et al. 2023; Huang et al. 2024).

Legumes constitute the second most significant food source following grains. Legumes are crucial in addressing global food security issues and are regarded as a viable alternative to animal protein (Ali et al. 2022). Mung bean has garnered human interest due to its substantial nutritional content and significant economic and environmental advantages. Furthermore, MB is a leguminous seed with medicinal and nutritional similarities, extensively cultivated and consumed globally (Dai et al. 2024). Mung bean is a high-quality protein source, suitable for whole grain, lentil, or sprouted consumption, providing balanced nutrition and preserving soil fertility (Fig. 13.1).

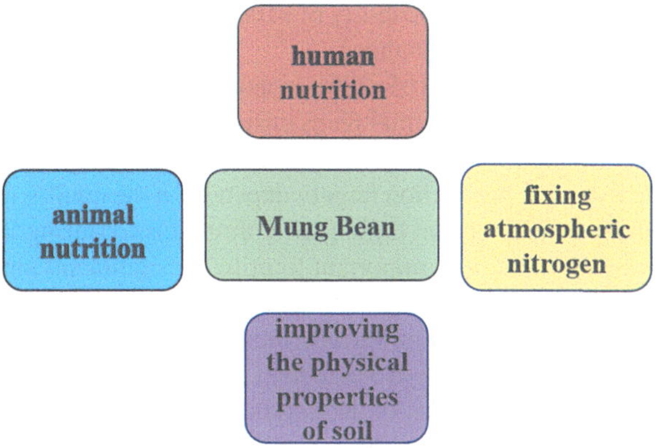

Fig. 13.1 Biological and environmental effects of MB

Mung bean is a significant annual, edible, warm-season, legume crop, representing around 8.4% of the overall legume acreage, exceeding six million hectares (Tarahi et al. 2024). It is extensively cultivated in Asian nations (about 90%) and temperate portions of Western countries owing to its distinctive attributes, including a brief growth cycle (70–90 days), little input necessities, elevated 400 kg/ha, drought resilience, and nitrogen fixation capabilities (Hou et al. 2019). Furthermore, mung bean seeds contain balanced macro- and micronutrients, vitamins, and minerals, including proteins, dietary fibers, as well as a high amount of bioactive components. Traditionally, they were ingested alongside cereals by the Indians and Chinese to mitigate the deficiency of lysine in cereal grains; nevertheless, they are now frequently utilized as fresh, cooked, processed, and canned food items. Mung beans are suitable for vegetarian and vegan diets as they contain a significant amount of easily digestible protein. They are also an exceptional source of protein for children thanks to their low allergenic and gaseous components (Sehrawat et al. 2020).

In the last decade, mung bean protein has gained considerable interest and importance as an alternative to traditional protein sources (animal protein and soy protein) due to its superior nutritional bioavailability. In addition, mung beans have a balanced nutrient profile that includes fiber and fat (Feng et al. 2024). Mung bean is a legume plant with high protein content (14.6–32.6%). Mung bean is known to have a digestible amino acid ratio of 93, which is higher than most other legumes. In general, mung bean proteins can be classified as albumin (16.2%), prolamin (0.9%), glutelin (13.4%), and globulin (61%) depending on their solubility. Its high quality, enhanced digestibility, and absorption suggest that it may serve as a suitable alternative to animal protein (Brishti et al. 2021). Additionally, mung bean proteins exhibit enhanced functional features, such as gelling, water and oil absorption capacity, emulsification, and foaming capabilities (Feng et al. 2024; Tarahi et al. 2024). Mung bean protein significantly supports human health by providing anti-hyperglycemic, antihypertensive, antioxidant, and anti-aging benefits (Xie et al. 2019). These are beneficial for broadening the applicability of MB protein. Another area where mung bean protein is used is animal husbandry.

The main factor and main cost (65–70%) in animal production is feed (Boonkoed et al. 2018). Quality and cheap animal feeds ensure the profitable operation of livestock enterprises. Malnutrition leads to low quantity and quality of animal productivity. Profitable livestock production largely depends on the quality of animal feed. Therefore, animals should receive high-quality animal feed during the whole production period. In this case, it is important to utilize a significant amount of available crop residues efficiently. The utilization of crop residues, agricultural by-products, and co-products from various industries for ruminant production is important in ensuring feed supply (Boonkoed et al. 2018).

Mung bean flour is a by-product of the noodle industry. After starch is separated from mung bean seeds to make noodles (vermicelli), the remainder is called mung bean flour or noodle residue. Mung bean flour has 88–91% dry matter content and about 15–21% crude protein. It also contains all essential amino acids, especially lysine. Because of its high protein content and low cost, it is an alternative feed source that can be used instead of protein source for animal feed. Many researchers have reported that the addition of mung bean meal can provide cost reduction in animal feed production. The utilization of mung bean by-products as a source of protein in animal feed is an example of sustainable production. Sustainable practices of MB productions have a positive impact on the environment.

Nitrogen (N) makes up about 78% of the Earth's atmosphere. It exists as a dimolecule (N_2). Nonetheless, it is an inapplicable form for flora, fauna, and microbes. All living creatures utilize nitrogen in the form of ammonia to synthesize essential macromolecules, including amino acids, proteins, and nucleic acids (Adams et al. 2018). N is the limiting reaction agent from an agricultural point of view. Farmers utilize excessive fertilizers to mitigate deficiencies, resulting in heightened environmental degradation and increased input costs (Adams et al. 2018; Salim and Raza 2020). Only a limited number of bacteria, known as diazotrophs, possess the capability to cleave the bonds between two nitrogen atoms and transform them into a form that can be utilized by living organisms. Nitrogen is an essential macronutrient, the second most important element for plants after water and the second most important element after carbon. Since N is present in the soil, it has played a fundamental role in the development, growth, and production of plants. Most soils are facing low levels of soil plant nutrients, which are essential for food production. Therefore, restoration and improvement of soil fertility has become an important issue nowadays. Legumes enhance soil fertility by facilitating the activity of microorganisms, which are essential for modifying soil characteristics, encompassing biological, physical, and chemical aspects.

Studies in many countries have shown that mung bean fixes atmospheric nitrogen into plant-favorable ammonium form by forming nodules with bacteria of the genera *Bradyrhizobium* (Zhang et al. 2008), *Rhizobium* (Yang et al. 2008; Zhang et al. 2008), *Ensifer* (Hakim et al. 2020a), and *Mesorhizobium* (Yang et al. 2008). However, due to the diversity of microsymbionts, Mung bean is recognized as a legume showing weak symbiotic specificity. Many pot and field studies have utilized *Bradyrhizobium* strains (Christopher et al. 2018). Studies show that *Bradyrhizobium* and *Ensifer* genera can coexist in mung bean nodules (Hakim et al. 2018). In a field inoculation study, *Ensifer* strains were reported to perform similar to *Bradyrhizobium* (Hakim et al. 2020b). However, in Brazilian tropical soils, almost 99% of bacterial genomes derived from MB nodules were identified as belonging to the species *Bradyrhizobium* (Favero et al. 2021).

13.2 Production Constraints

The ultimate goal of most plant breeders is the development of quality and productive cultivars that are well adapted to different growing environments. An ideal variety always shows high yield and stability in variable environments (Eberhart and Russell 1966). However, significant fluctuations are observed in the yield performance of genotypes grown in different environments. G×E interactions are crucial in breeding programs for identifying stable genotypes that are broadly or narrowly adapted to distinct environments, ensuring low fluctuation in yield (Verma et al. 2008). Breeding programs require assessing breeding line stability and adaptation to biotic and abiotic stresses. Stability is desirable, as yield varies based on cultivar adaptability to diverse growth conditions (Kamannavar and Vijaykumar 2011). Understanding G×E interaction and yield stability is crucial for breeding new varieties with better adaptation to environmental constraints, but limitations exist due to costly and labor-intensive yield and quality trials (Basford and Cooper 1998). This is because it is necessary to establish trials over many years and locations to identify a stable and high-yielding mung bean variety. Despite its enormous importance, green bean area and production are adversely affected by various abiotic and biotic constraints (Chauhan et al. 2010). Under stress conditions, a plant response occurs in a certain sequence (Fig. 13.2).

Biological pests are the primary biotic stressors that hinder the yield of mung bean crops. Among these constraints, diseases impose the most severe constraints,

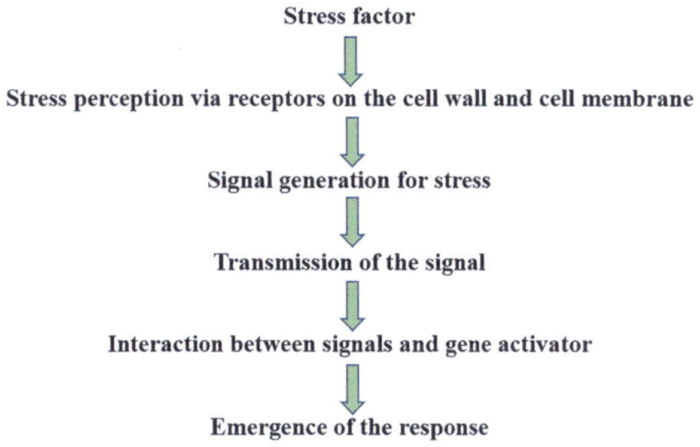

Fig. 13.2 Arrangement of stress responses in plants

affecting the physical quality of the seeds and limiting their productivity. Mung bean is susceptible to several viral, fungal, and bacterial illnesses (Fig. 13.3) (Noble et al. 2019). Mung bean yellow mosaic disease (MYMD) is the most significant MB disease, caused by Begomovirus and transmitted by whitefly (Singh and De 2006; Kitsanachandee et al. 2013). Although 35 fungal diseases are reported to affect beans globally, some of them are widespread and economically important (Pandey et al. 2018). Bacterial diseases are *P. savastanoi* pv. *phaseolicola* (halo blight), *C. flaccumfaciens* pv. *Flaccumfaciens* (tan spot), and bacterial leaf spot caused by *X. campestris* pv. *vigna radiata* (Ryley and Tatnell 2011).

Insect pests represent a significant limitation in mung bean cultivation, adversely impacting productivity. They assault several sections of the mung bean plant, encompassing roots, shoots, blooms, and pods at distinct growth phases. The damage caused by insect pests can be direct or indirect. Insects such as pod boring beetles, tobacco aphids, and aphids cause direct damage. Insects acting as vectors of some serious diseases such as mung bean yellow mosaic virus (MYMV), peanut bud necrosis virus (PBNV), and bean common mosaic virus (BCMV) cause indirect damage (Pratap et al. 2020). Abiotic stresses affecting green bean production include drought, temperature, waterlogging, and salinity. The plant experiences physiological changes when exposed to these stress conditions (Fig. 13.4) (Kumari et al. 2022). Among the abiotic stress factors mentioned, drought stress is regarded as a significant danger to agricultural output among the abiotic stress factors identified. It negatively impacts the biochemical, morpho-physiological, and molecular processes of plants, particularly in least-duration crops like mung bean (Singh et al. 2021a, b).

Drought stress significantly affects the biochemical processes and molecular functions of plants (Farooq et al. 2009; Jaleel et al. 2009). Drought resistance is a

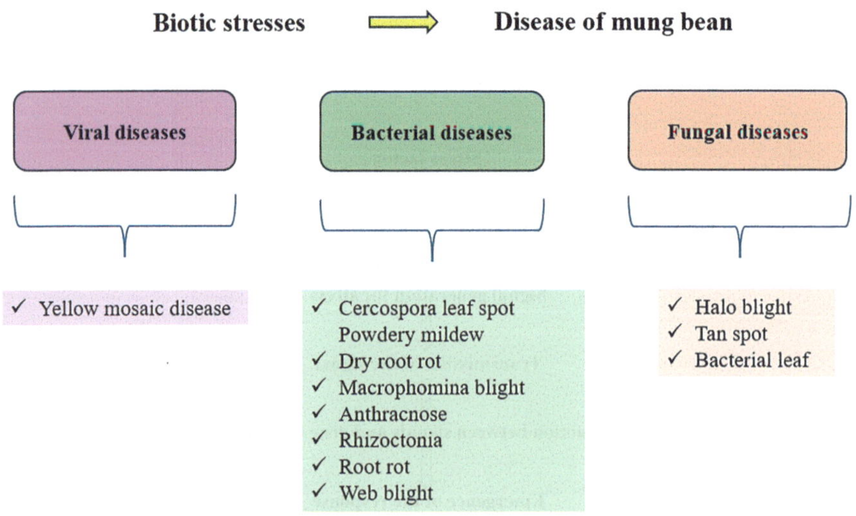

Fig. 13.3 Diseases seen in mung bean

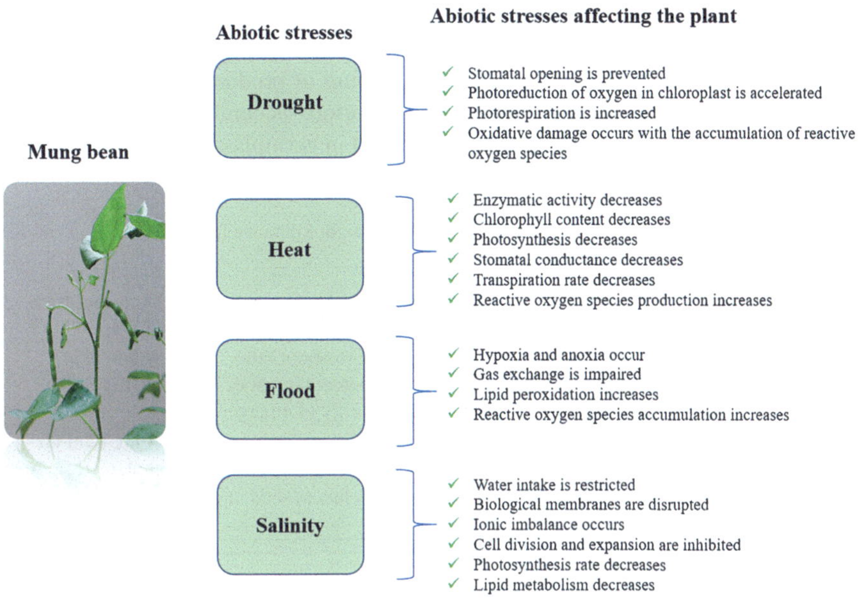

Fig. 13.4 Physiological effects of abiotic stress conditions on plants (Kumari et al. 2022)

multifaceted quantitative feature that entails interactions among numerous metabolic pathways associated with stress-resistant genes (Ahmad et al. 2015). Drought stress alters various physiological and biochemical activities by causing excessive proline and ROS (reactive oxygen species) production. In drought stress, photosynthesis decreases due to decreased leaf area and increased leaf senescence (Farooq et al. 2009). Another point where photosynthesis is affected is the negative effect on stomatal gas exchange parameters (Moradi et al. 2008). It also impairs the performance of stress-related enzymes, leading to a decrease in dry matter accumulation and seed yield (Fahad et al. 2017). As a result, plant water relations, nutrient uptake capacity, photosynthetic activity, DNA and protein damage, and current yield and quality are significantly reduced (Umer Chattha et al. 2021).

Drought stress affects morpho-physiological traits as well as biochemical processes of plants. Yield is a complex character influenced by the interaction of genotype and environmental factors (Singh et al. 2014, 2016). It is also contingent upon the manifestation of specific morpho-physiological functions. Drought stress significantly impacts these morpho-physiological systems, hence influencing yield. Under drought stress circumstances in MB, characteristics such as seed weight, plant height, pod filling index, leaf size, root structure, and crop production are markedly diminished (Khan et al. 2015). Drought stress significantly impacts plant root system nutrient uptake and water use, affecting nitrogen fixation and growth. Symbiotic relationships in legumes, including drought escape, avoidance, and tolerance, enable plants to survive, collect biomass, and generate seeds (Bista et al. 2018;

Levitt 1980; Singh et al. 2021a, b). Researchers have carried out many studies on this topic (Table 13.1).

Traditional breeding has delayed the production of productive and quality new mung bean varieties (Shahwar et al. 2019). Therefore, accurate information on the genetic diversity of mung bean collections stored in germplasm needs to be used to

Table 13.1 Essential biological traits for improving drought tolerance in MB

Stage of tolerance	Key traits	References
Seedling stage	Seedling biomass, stress tolerance index	Dutta and Bera (2008)
Seedling stage	Length of shoot, length of root, root-to-shoot ratio, diameter of stem, weight of shoot, percentage of dead leaves, percentage of emergence, energy of emergence	Aslam et al. (2013)
Seedling stage	Seedling survival rate, wilt index, relative water content (RWC), stress index	Wang LanFen et al. (2015a, b)
Seedling stage	Germination rate, germination stress index, seed vigor	Jincy et al. (2021)
Seedling/ vegetative stage	Early vigor, RWC	Biradar et al. (2010)
Vegetative/ reproductive stage	Stomatal conductance, photosynthesis rate, relative leaf water content, transpiration rate, (RLWC), leaf temperature	Moradi et al. (2008)
Vegetative/ reproductive stage	Early flowering, specific leaf area	Swathi (2013)
Vegetative/ reproductive stage	Biological trait enhancement	Prakash et al. (2017)
Vegetative/ reproductive stage	Length of shoot, length of root, quantity of roots, diameter of root	Kumar et al. (2015)
Vegetative/ reproductive stage	Relative injury (RI), specific leaf area (SLA), chlorophyll content	Paramesh et al. (2016)
Vegetative/ reproductive stage	SPAD chlorophyll meter measurements and SLA	Swathi et al. (2017)
Vegetative/ reproductive stage	Membrane stability index, RWC	Bangar et al. (2019)
Vegetative/ reproductive stage	Biological trait enhancement	Sadeghipour (2019)
Reproductive stage	Biomass, harvest index	Aslam et al. (2013)
Reproductive stage	Root length	Ranawake et al. (2012)
Reproductive stage	Yield components	Paramesh et al. (2016)

ensure the appropriate use of germplasm and hence the success of MB crop breeding programs (Mburu et al. 2016). This germplasm enhancement allows useful genes and their combinations to be identified and used in various improvement programs (Arias et al. 2015).

Phenotypic evaluations (Bisht et al. 1998; Wang et al. 2009; Abna et al. 2012) or genotype-related (Schafleitner et al. 2015; Lestari et al. 2014) diversity provides useful information for better utilization of germplasm collections, which are essential material for genetic improvement and breeding. Assessments of local germplasm indicate the presence of some elite genes that provide resilience to biotic and abiotic stressors; thus, the genetic diversity of modern cultivars has become progressively limited (Wang et al. 2018). Some studies have shown that alien or wild species can be of great value in local cultivation (Lee et al. 2000). For example, mung bean from the AVRDC has contributed to greatly increase the yield of local cultivars in China compared to the end of the last century (Wang et al. 2018). Furthermore, the bruchid resistance gene has been effectively transmitted from wild mung bean to cultivated varieties (Zhu et al. 2012; Chen et al. 2015).

13.3 Marker-Assisted Breeding (MAB) and Its Introduction to Mung Bean

Widely grown in Southern and Southeast Asia, mung bean is a vital legume crop, valued for its short life cycle of 60–75 days and relative drought resistance (Alam et al. 2010). Mung bean stands out in agricultural systems for its numerous advantages, such as rich nutritional value, a shorter growth cycle, adaptability to various climates, soil improvements through nitrogen fixation, high biomass yield, and wide use in many food products (Nair and Schreinemachers 2020; Somta and Srinives 2007). Although mung bean is a crop of significant economic value, its productivity remains plateaued as a result of unstable weather conditions combined with effects of biotic and abiotic challenges (Basu et al. 2019). Among the abiotic stresses, drought is the parameter for mung bean advancement that obstructs its progress and return. Various growth-related morpho-physiological and molecular procedures are influenced by drought stress, which ultimately causes lower grain yield (Baroowa et al. 2016; Maheswari et al. 2016). Following this, marker-assisted breeding (MAB) utilizing molecular markers can be used as a powerful tool for enhancing mung bean resilience, especially against abiotic stresses like drought. Ideal for a range of applications, molecular markers are utilized in genetic variation assessment, identification of QTLs, fingerprinting, naming beneficial genes, constructing linkage maps, map-based discovery of advantageous genes, transformational research, and marker-assisted selection (Koebner et al. 2001; Gupta and Rustgi 2004; Maqbool et al. 2016). Through the integration of MAB, breeders can create resilient cultivars that contribute to food supply and cope with climate change.

Marker-assisted breeding (MAB) is a contemporary breeding process that utilizes molecular markers, genetic maps, and genomic information to select targeted traits and enhance plant or animal traits based on genotypic assay, thereby improving their genetic composition in a more targeted and efficient manner (Jiang 2015). MAB encompasses a range of innovative breeding strategies such as MAS, marker-assisted backcrossing, marker-assisted recurrent selection, and genome-wide selection, which leverage molecular markers to identify and track desirable genetic traits in the breeding process (Jiang 2015). MAB offers several benefits, such as: the ability to select desirable gene alleles without being affected by the environment or other genetic factors, the ability to distinguish between homozygous and heterozygous plants, the ability to track the integration of a desired allele during backcrossing, and the identification of recombinants with minimal linkage drag (donor DNA) (Rajcan et al. 2011). Compared to traditional breeding methods, which often rely on time-consuming and labor-intensive processes, MAB offers several advantages. For example, MAB can drastically reduce the number of back-cross generations required to introgress a desired trait into an elite cultivar, thereby accelerating the development of stress-tolerant plant varieties. (Kumari and Prakash 2020).

13.4 Potential for Molecular Markers in Drought Resistance Breeding

Molecular markers are critical tools in genetics research, enabling quick and precise identification of genetic differences between genotypes, thus facilitating the breeding of improved crop varieties. These tools are crucial for various applications such as genome mapping, gene tagging, genetic diversity analysis, phylogenetic analysis, and forensic investigations due to their stability, affordability, and ease of use (Baloch et al. 2024; Yalinkiliç et al. 2024; Altaf et al. 2024a, b). By tracking specific markers through multiple generations, breeders can accurately identify and select for the desired alleles, even if they are recessive in nature. This approach has numerous applications, including estimation of genetic variances, prediction of hybrid performance, detection of genetic differences between parents, and association of QTLs with key traits (Sharma and Crouch 2004). Over the past few decades, numerous DNA-based marker technologies have been developed, but only a few, dominant and codominant (such as RFLP, RAPD, AFLP, ISSR, SSR, and SNP), have achieved widespread adoption (Grover and Sharma 2016). In crop improvement, molecular markers are classified into two categories: dominant and co-dominant. Multi-locus markers such as ISSR, RAPD, and AFLP are considered dominant, while single-locus markers including SSR and RFLP belong to the co-dominant category. This classification is primarily based on the ability of the markers to discriminate between homozygous and heterozygous genotypes. While dominant markers are unable to differentiate between the two genotypes, co-dominant markers possess the capacity to distinguish them, making them a valuable tool in crop improvement (Asati et al. 2022). Among these marker systems, RAPD, RFLP, and AFLP have traditionally been used in pulses for association and diversity analyses, but they are not preferred for marker-assisted selection (MAS) due to their limitations, including poor reproducibility, complex handling procedures, high skill requirements, and the use of radioactive elements (Gupta et al. 2010). PCR-based SNP and SSR markers are the preferred choice for plant breeders due to their easy application, low cost, and simplicity, which allows for efficient genotyping of large populations with minimal infrastructure. Additionally, their high reproducibility and user-friendliness make them ideal for MAS in traditional plant breeding programs (Basak et al. 2023). Legumes, from the heartiness of chickpeas to the versatility of lentils, boast a genetic diversity that is now being harnessed by the advancements in molecular marker technology. These tools are enabling breeders to efficiently select for desirable traits, such as disease resistance, drought tolerance, and high protein content, all while accelerating the development of sustainable and high-quality legume varieties. With the surge of genomic resources, such as genome assemblies, genome-wide markers, and genes/QTLs controlling agriculturally important traits, a window of opportunity has been thrown open for legume breeding. This not only facilitates advancements within the realm of legumes but also lays the groundwork for cross-pollinating the findings with other crops, ultimately resulting in hybrids that blend the most coveted traits of legumes and other plants. With the surge of genomic resources, such as genome assemblies, genome-wide markers, and genes/QTLs controlling agriculturally important traits, a window of opportunity has been thrown open for legume breeding (Bohra et al. 2020). This not only facilitates advancements within the realm of legumes but also lays the groundwork for cross-pollinating the findings with other crops, ultimately resulting in hybrids that blend the most coveted traits of legumes and other plants. Thousands of markers for each crop have been developed, making high-resolution genetic mapping, marker-assisted selection, and molecular breeding possible. With a focus on improving markers for orphan legumes, recent research has turned to developing SSR markers that can be transferred among legume species, taking advantage of their close phylogenetic similarity. Various types of markers, such as RFLP, RAPD, AFLP, SSR, and ISSR, have been utilized to assess genetic diversity and construct linkage maps in legumes (Gosal et al. 2020). Molecular markers, specifically AFLP and SCAR, have been successfully applied in crop improvement against abiotic stresses. Ouédraogo et al. (2002) identified an AFLP marker, E-AGA/M-CTA460, linked to the parasitic weed resistance gene Rsg3 in cowpea. Similarly, the molecular marker SCAR (SEACTMCAC83/85 was reported in cowpea (Boukar et al. 2004), and is associated with the parasitic weed resistance gene Rsg1.

Molecular markers play a critical role in improving drought tolerance in mung bean breeding. By identifying drought-tolerant genotypes, tracing drought-tolerance genes, and developing genetic linkage maps, breeders can efficiently utilize drought-tolerance traits to improve mung bean cultivars' agricultural productivity and environmental resilience (Nair et al. 2019). In the context of breeding drought-tolerant MB varieties, SSR markers have proven to be the most widely used marker type, providing an efficient means for identifying and manipulating drought tolerance-related traits at the molecular level. Yuliasti and Reflinur (2015) successfully identified three mutant lines (PSJ30, PSJ31, and PSJ32) with superior grain yields and drought tolerance. PSJ31 was officially released as Muri, exhibiting high yield and drought tolerance. SSR marker evaluations distinguished these mutant lines using nine polymorphic markers, providing valuable genetic variation information at the molecular level, essential for MB variety improvement. Yadav and Punia (2017) used SSR markers to identify nine clusters among 60 MB genotypes under limited moisture conditions, exhibiting high PIC values (0.84) and providing insight into genetic diversity for drought tolerance, with notable genotypes exhibiting superior yield and drought tolerance traits.

13.4.1 Importance of Drought Tolerance in MB

The phenotype of drought tolerance is exceedingly complex, governed by various interacting traits. It includes early maturity, deep root system for better water absorption, minimizing water loss through controlled stomatal conductance, and sustaining metabolic activity at low water potential (Manavalan et al. 2009). Drought is among the most crucial abiotic stresses impacting MB growth and productivity. Several studies have highlighted the negative effects of drought on crop performance, resulting in stunted growth, disrupted physiological functions, and lower yields in MB (Nadeem et al. 2019; Fathy et al. 2018; Raina et al. 2016; Kumar and Sharma 2009; Moradi et al. 2009; Thomas et al. 2004). These findings emphasize the crucial need for developing drought-tolerant varieties to mitigate such adverse effects.

13.4.2 Role of Molecular Markers in Identifying Drought-Tolerant Genotypes

The application of molecular marker techniques has enabled breeders to locate the loci correlated with favorable phenotypic traits. Utilizing molecular markers to trace the inheritance of a DNA segment with recognized benefits provides greater precision and efficiency than standard breeding techniques (Collard and Mackill 2008; Nadeem et al. 2018; Iqbal et al. 2023; Altaf et al. 2023). Molecular markers play an important role in the identification and breeding of drought-tolerant MB varieties. DNA-based molecular markers like simple sequence repeats (SSRs), which are short tandem repeats of DNA comprising one to five base pairs (Lung'aho et al. 2011) and single nucleotide polymorphisms (SNPs) (Ganal et al. 2009), have been established as robust instruments for investigating genetic diversity and population structure across species. Owing to their benefits including co-dominance and high level of polymorphisms, SSRs also known as microsatellite markers are especially appealing for exploring genetic structures and species relationships (Saxena et al. 2010), allowing breeders to effectively screen and select genotypes with enhanced drought tolerance.

13.5 Breeding Strategies for Integrating Drought Tolerance

13.5.1 Conventional Breeding

Conventional breeding, which involves in selection of favorable genotypes based on observable phenotypes and crossing the superior lines, is regarded as the most basic, accessible, and uncomplicated technique for enhancing crops (Noble et al. 2020). This approach, established and reliable since the advent of agriculture, has been instrumental in the creation of the majority of current crops and livestock (Borlaug 1983). Conventional breeding strategies, which rely on phenotypic selection, have successfully contributed to MB improvement; however, despite their simplicity in principle, they are labor-intensive and time-consuming to implement. In a phenomenon known as linkage drag, undesirable traits from parent lines are inherited alongside desired ones, which demands multiple generations of selection and backcrossing before a cultivar can be approved for release (Flint-Garcia et al. 2003). Traits of qualitative nature, such as particular forms of disease resistance, can be encoded by one gene or small set of genes (Moody et al. 2003). In contrast, quantitative traits like abiotic stress tolerance are regulated by numerous genes, leading to a variety of phenotypic variations (Hill 2010).In abiotic stress tolerance breeding, phenotype alone is insufficient, so integrating more discriminating markers is essential (Noble et al. 2020).

13.5.2 Marker-Assisted Selection (MAS)

Marker-assisted selection is one way to precise trait selection through markers associated with the desired trait like drought in MB. Markers are typically categorized as morphological, cytological, biochemical, and molecular or DNA markers, but the term marker-assisted selection (MAS) is solely linked to molecular/DNA markers (Jiang 2013). By using DNA for screening, MAS achieves enhanced efficiency and accuracy, making it independent of plant maturity and less susceptible to environmental factors. The ability to differentiate between alleles in genotypes enables breeders to make genetic selections independent of phenotypic assessment (Ota et al. 2007). Commonly used for MAS, five types of DNA markers include RFLP, AFLP, RAPD, SSR, and SNP. Each marker has its benefits and restrictions, entirely explored in a review by Jiang (2013).

13.5.3 QTL Mapping

The capacity to tolerate abiotic stress is a quantifiable attribute with the complicated mode of inheritance, making it stretching to determine, and often changes across diverse ecological conditions (Deshmukh et al. 2014). The quantitative trait loci technique aids in pinpointing specified genomic regions aligned with water-stress tolerance, enabling the selection of genotypes with boosted resistance to water

Table 13.2 Identified drought-tolerant varieties of MB

Abiotic stress	Varieties	Country	Reference
Drought	V-2013, V-1281, V-3372	Taiwan	AVRDC (1979)
	TCR 21	India	Tripathy et al. (2016)
	VC-6173-C, IC-325770, ML2082	India	Raina et al. (2016)
	K-851	India	Dutta and Bera (2008) and Dutta et al. (2016)
	ML 267	India	Swathi et al. (2017)
	SML-1136, SML-1436	India	Kaur et al. (2017)
	VC 2917	China	Wang et al. (2014) and Wang LanFen et al. (2015a, b)
	VC 2778, V 1381	China	He et al. (1988)

scarcity through marker-assisted breeding. Molecular biology and genetic methods are often used to identify quantitative trait loci (QTLs) and contestant genes related to drought-adaptation features (Mir et al. 2012). Several QTLs connected with drought tolerance have been revealed in several leguminous crops along with chickpeas (Kushwah et al. 2022), common beans (Diaz et al. 2020), soybean (Ren et al. 2020), lentils (Idrissi et al. 2016), and cowpeas (Muchero et al. 2009). However, a finite number of genetic investigations and research have lectured on the identification of QTLs under aridity resistance in MB (Liu et al. 2017; Sholihin 2002). The varieties developed against drought are listed in Table 13.2.

13.6 Conclusion, Future Research, and Technological Advancements

MB (*Vigna radiata*) research highlights its immense potential as a vital crop for sustainable agriculture, food security, and human nutrition. With its rich nutrient profile and adaptability to diverse agro-ecological zones, mung bean is particularly valued for its resilience to drought and role in nitrogen fixation, making it an ideal candidate for cultivation in challenging environments. However, despite these advantages, productivity remains constrained by various biotic and abiotic stresses, and the application of advanced breeding tools is limited in many regions. To unlock MB's full potential, future research must focus on developing stress-resilient cultivars, particularly those with enhanced tolerance to drought, heat, and salinity, which are becoming more critical due to climate change. Abiotic stresses have been shown to negatively impact mung bean's biochemical and physiological processes, reducing yield and quality. Traditional breeding methods have made progress in mitigating these issues, but the development of drought-resistant varieties remains a challenge. Marker-assisted breeding (MAB), a more targeted and efficient approach, offers significant potential by enabling the identification and selection of stress-resilient traits through molecular markers. As a future direction, integrating MAB into mung bean breeding programs should be prioritized, with a particular focus on traits that promote stress tolerance and improve overall crop productivity, ensuring mung bean's continued role in sustainable agriculture and global food security.

Advancements in genomics and bioinformatics will play a key role in the future of mung bean research. The sequencing of the mung bean genome has opened new possibilities for understanding the genetic basis of important traits such as stress tolerance, yield, and nutritional quality. The use of genome-wide association studies (GWAS) and genomic selection (GS) will enable the identification of beneficial alleles and accelerate the breeding process. In addition to molecular breeding, biotechnology can provide tools for the precise modification of MB's genetic makeup. Technologies such as CRISPR/Cas9 offer the potential for targeted genome editing, allowing researchers to introduce or remove specific genes associated with desirable traits (Ali et al. 2023, 2024). Future research should focus on optimizing these techniques for MB and ensuring that the resulting cultivars meet regulatory requirements for genetically modified crops. Lastly, participatory breeding approaches that involve farmers and other stakeholders in the selection process will be crucial to the successful adoption of new MB varieties. Engaging farmers in field trials and incorporating their preferences for traits such as yield, taste, and cooking quality will ensure that new varieties are both high-performing and acceptable to end-users.

References

Abna F, Golam F, Bhassu S (2012) Estimation of genetic diversity of mungbean (*Vigna radiata* L. Wilczek) in Malaysian tropical environment. Afr J Microbiol Res 6(8):1770–1775. https://doi.org/10.5897/AJMR11.1133

Adams MA, Buchmann N, Sprent J, Buckley TN, Turnbull TL (2018) Crops, nitrogen, water: are legumes friend, foe, or misunderstood ally? Trends Plant Sci 23(6):539–550. https://doi.org/10.1016/j.tplants.2018.02.009

Ahmad A, Selim MM, Alderfasi AA, Afzal M (2015) Effect of drought stress on mung bean (*Vigna radiata* L.) under arid climatic conditions of Saudi Arabia. In: Ecosystems and sustainable development, vol 192, pp 185–193. https://doi.org/10.2495/ECO150171

Alam AKMM, Podder R, Haque AHMM, Hossain AH (2010) Handbook of pulses. Publication no. 28. Bangladesh Agricultural Research Institute

Ali A, Altaf MT, Nadeem MA, Karaköy T, Shah AN, Azeem H et al (2022) Recent advancement in OMICS approaches to enhance abiotic stress tolerance in legumes. Front Plant Sci 13:952759

Ali A, Ölmez F, Altaf MT, Liaqat W, Umar UUD, Iqbal J (2023) Advanced and sustainable approaches in sugarcane crop improvements with reference to environmental stresses. In: Biotechnology and Omics approaches for bioenergy crops. Springer Nature Singapore, Singapore, pp 155–182

Ali A, Altaf MT, Bedir M, Liaqat W, Jamil A, Nadeem MA, Baloch FS (2024) Biotechnological advancements toward sugarcane crop improvement. In: Industrial crop plants. Springer Nature Singapore, Singapore, pp 333–352

Altaf MT, Liaqat W, Iqbal J, Baig MMA, Ali A, Nadeem MA, Baloch FS (2023) Exploring omics approaches to enhance stress tolerance in soybean for sustainable bioenergy production. In: Biotechnology and omics approaches for bioenergy crops. Springer Nature Singapore, Singapore, pp 123–154

Altaf MT, Nadeem MA, Ali A, Liaqat W, Bedir M, Baran N et al (2024a) Applicability of start codon targeted (SCoT) markers for the assessment of genetic diversity in bread wheat germplasm. Genet Resour Crop Evol 72:1–14

Altaf MT, Liaqat W, Ali A, Jamil A, Bedir M, Nadeem MA et al (2024b) Conventional and biotechnological approaches for the improvement of industrial crops. In: Kumar N (ed) Industrial crop plants. Interdisciplinary biotechnological advances. Springer, Singapore. https://doi.org/10.1007/978-981-97-1003-4_1

Arias D, González M, Prada F, Ayala-Diaz I, Montoya C, Daza E, Romero HM (2015) Genetic and phenotypic diversity of natural American oil palm (*Elaeis oleifera* (HBK) Cortés) accessions. Tree Genet Genomes 11:1–13. https://doi.org/10.1007/s11295-015-0946-y

Asati R, Tripathi MK, Tiwari S, Yadav RK, Tripathi N (2022) Molecular breeding and drought tolerance in chickpea. Life 12(11):1846

Asian Vegetable Research and Development Center (AVRDC) (1979) AVRDC progress report for (1978)

Aslam M, Maqbool MA, Zaman QU, Latif MZ, Ahmad RM (2013) Responses of mungbean genotypes to drought stress at early growth stages. Int J Basic Appl Sci 13(5):22–22

Baloch FS, Ali A, Tajibayev D, Nadeem MA, Ölmez F, Aktaş H et al (2024) Stripe rust resistance gene Yr15 in Turkish and Kazakhstan wheat germplasms and the potential of Turkish wild emmer for stripe rust breeding. Genet Resour Crop Evol 71(6):2699–2719

Basak D, Chakraborty A, Das A, Banerjee J (2023)Molecular marker-assisted crop improvement in pulses. In Molecular Marker Techniques: A Potential Approach of Crop Improvement (pp. 199-231). Singapore: Springer Nature Singapore.

Bangar P, Chaudhury A, Tiwari B, Kumar S, Kumari R, Bhat KV (2019) Morphophysiological and biochemical response of mungbean [*Vigna radiata* (L.) Wilczek] varieties at different developmental stages under drought stress. Turk J Biol 43(1):58–69. https://doi.org/10.3906/biy-1801-64

Baroowa B, Gogoi N, Farooq M (2016) Changes in physiological, biochemical and antioxidant enzyme activities of green gram (Vigna radiata L.) genotypes under drought. Acta Physiol Plant 38:1–10

Basford K, Cooper M (1998) Genotype× environmentinteractions and some considerations of their impli-cations for wheat breeding in Australia this review isone of a series commissioned by the AdvisoryCommittee of the journal. Aust J Agric Res 49:153–174. https://doi.org/10.1071/A97035

Basu PS, Pratap A, Gupta S, Sharma K, Tomar R, Singh NP (2019) Physiological traits for shortening crop duration and improving productivity of greengram (Vigna radiata L. Wilczek) under high temperature. Front Plant Sci 10:1508

Biradar KS, Salimath PM, Ravikumar RL (2010) Genetic studies in greengram and association analysis. Karnataka J Agric Sci 20:843–844

Bisht IS, Mahajan RK, Patel DP (1998) The use of characterisation data to establish the Indian mungbean core collection and assessment of genetic diversity. Genet Resour Crop Evol 45:127–133. https://doi.org/10.1023/A:1008670332570

Bista DR, Heckathorn SA, Jayawardena DM, Mishra S, Boldt JK (2018) Effects of drought on nutrient uptake and the levels of nutrient-uptake proteins in roots of drought-sensitive and-tolerant grasses. Plan Theory 7(2):28. https://doi.org/10.3390/plants7020028

Bohra A, Mir RR, Jha R, Maurya AK, Varshney RK (2020) Advances in genomics and molecular breeding for legume improvement. In: Advancement in crop improvement techniques. Woodhead Publishing, pp 129–139

Boonkoed S, Suphalucksana W, Sitthigripong R, Srikijkasemwat K, Mitchaothai J, Lukkananukool A (2018) The effect of adding mung bean meal supplementation on Napier Pakchong 1 silage on fermentation quality and nutrient composition

Borlaug NE (1983) Contributions of conventional plant breeding to food production. Science 219(4585):689–693

Boukar O, Kong L, Singh BB, Murdock L, Ohm HW (2004) AFLP and AFLP-derived SCAR markers associated with Striga gesnerioides resistance in cowpea. Crop Sci 44:1259–1264

Brishti FH, Chay SY, Muhammad K, Ismail-Fitry MR, Zarei M, Saari N (2021) Texturized mung bean protein as a sustainable food source: effects of extrusion on its physical, textural and protein quality. Innovative Food Sci Emerg Technol 67:102591. https://doi.org/10.1016/j.ifset.2020.102591

Chauhan Y, Douglas S, Rachaputi C, Agius RCN, Martin P, King WK (2010) Physiology of mungbean and development of the mungbean crop model. In: Proceedings of the 1st Australian summer grains conference Australia, Gold Coast, pp 21–24

Chen HT, Yuan XX, Zhang HM, Liu XQ, Cui XY, Chen X, Gu HP (2015) Breeding and cultivation techniques of a new mungbean variety Sulu 5 resistant to Bruchid. Crop Res 29:428–430

Christopher M, Macdonald B, Yeates S, Ziegler D, Seymour N (2018) Wild bradyrhizobia that occur in the burdekin region of Queensland are as effective as commercial inoculum for mungbean (*Vigna radiata* (L.)) and black gram (*Vigna mungo* (L.)) in fixing nitrogen and dry matter production. Appl Soil Ecol 124:88–94. https://doi.org/10.1016/j.apsoil.2017.11.005

Collard BC, Mackill DJ (2008) Marker-assisted selection: an approach for precision plant breeding in the twenty-first century. Philos Trans R Soc Lond Ser B Biol Sci 363:557–572

Dai Y, Li C, Liu J, Xing L, Zhu T, Liu S et al (2024) Enhancing the stability of mung bean-based milk: insights from protein characteristics and raw material selection. Int J Biol Macromol 265:131030. https://doi.org/10.1016/j.ijbiomac.2024.131030

Deshmukh R, Sonah H, Patil G, Chen W, Prince S, Mutava R et al (2014) Integrating omic approaches for abiotic stress tolerance in soybean. Front Plant Sci 5:244

Diaz S, Ariza-Suarez D, Izquierdo P, Lobaton JD, de La Hoz JF, Acevedo F et al (2020) Genetic mapping for agronomic traits in a MAGIC population of common bean (Phaseolus vulgaris L.) under drought conditions. BMC Genomics 21:1–20

Dutta P, Bera AK (2008) Screening of mungbean genotypes for drought tolerance. Legume Res 31(2):145–148

Dutta P, Bandopadhyay P, Bera AK (2016) Identification of leaf based physiological markers for drought susceptibility during early seedling development of mungbean. Am J Plant Sci 7(14):1921–1936

Eberhart ST, Russell W (1966) Stability parametersfor comparing varieties 1. Crop Sci 6:36–40. https://doi.org/10.2135/cropsci1966.0011183X000600010011x

Fahad S, Bajwa AA, Nazir U, Anjum SA, Farooq A, Zohaib A et al (2017) Crop production under drought and heat stress: plant responses and management options. Front Plant Sci 8:1147. https://doi.org/10.3389/fpls.2017.01147

Farooq M, Wahid A, Kobayashi NSMA, Fujita DBSMA, Basra SM (2009) Plant drought stress: effects, mechanisms and management. In: Sustainable agriculture. Springer Netherlands, Dordrecht, pp 153–188. https://doi.org/10.1007/978-90-481-2666-8_12

Fathy NE, Ismail SM, Basahi JM (2018) Optimizing mungbean productivity and irrigation water use efficiency through the use of low water-consumption during plant growth stages. Legum Res 41:108–113

Faverdin P, Guyomard H, Puillet L, Forslund A (2022) Animal board invited review: specialising and intensifying cattle production for better efficiency and less global warming: contrasting results for milk and meat co-production at different scales. Animal 16(1):100431

Favero VO, Carvalho RH, Motta VM, Leite ABC, Coelho MRR, Xavier GR et al (2021) Bradyrhizobium as the only rhizobial inhabitant of mung bean (*Vigna radiata*) nodules in tropical soils: a strategy based on microbiome for improving biological nitrogen fixation using bioproducts. Front Plant Sci 11:602645. https://doi.org/10.3389/fpls.2020.602645

Feng Q, Niu Z, Zhang S, Wang L, Qun S, Yan Z et al (2024) Mung bean protein as an emerging source of plant protein: a review on production methods, functional properties, modifications and its potential applications. J Sci Food Agric 104(5):2561–2573

Flint-Garcia SA, Thornsberry JM, Buckler ES IV (2003) Structure of linkage disequilibrium in plants. Annu Rev Plant Biol 54(1):357–374

Ganal MW, Altmann T, Röder MS (2009) SNP identification in crop plants. Curr Opin Plant Biol 12(2):211–217

Grover A, Sharma PC (2016) Development and use of molecular markers: past and present. Crit Rev Biotechnol 36(2):290–302
Gupta PK, Rustgi S (2004) Molecular markers from the transcribed/expressed region of the genome in higher plants. Funct Integr Genomics 4:139–162
Gupta PK, Kumar J, Mir RR, Kumar A (2010) 4 Marker-assisted selection as a component of conventional plant breeding. Plant Breed Rev 33(4):145–217
Hakim S, Mirza BS, Zaheer A, Mclean JE, Imran A, Yasmin S, Sajjad Mirza M (2018) Retrieved 16S rRNA and nif H sequences reveal co-dominance of bradyrhizobium and ensifer (sinorhizobium) strains in field-collected root nodules of the promiscuous host vigna radiata (L.) R. Wilczek. Appl Microbiol Biotechnol 102:485–497. https://doi.org/10.1007/s00253-017-8609-6
Hakim S, Imran A, Mirza MS (2020a) Phylogenetic diversity analysis reveals Bradyrhizobium yuanmingense and Ensifer aridi as major symbionts of mung bean (*Vigna radiata* L.) in Pakistan. Braz J Microbiol 52:311–324. https://doi.org/10.1007/s42770-020-00397-9
Hakim S, Mirza BS, Imran A, Zaheer A, Yasmin S, Mubeen F et al (2020b) Illumina sequencing of 16S rRNA tag shows disparity in rhizobial and non-rhizobial diversity associated with root nodules of mung bean (*Vigna radiata* L.) growing in different habitats in Pakistan. Microbiol Res 231:126356. https://doi.org/10.1016/j.micres.2019.126356
He X, He T, Xiong Y, Jiao C (1988) Research and use of mungbean germplasm resources in Hubei, China. In Mungbean: Proceedings of the Second International Symposium. AVRDC, Shanhua, pp 35–41
Hill WG (2010) Understanding and using quantitative genetic variation. Philos Trans R Soc Lond B Biol Sci 365(1537):73–85
Hou D, Yousaf L, Xue Y, Hu J, Wu J, Hu X et al (2019) Mung bean (*Vigna radiata* L.): bioactive polyphenols, polysaccharides, peptides, and health benefits. Nutrients 11(6):1238. https://doi.org/10.3390/nu11061238
Hou D, Feng Q, Niu Z, Wang L, Yan Z, Zhou S (2023) Promising mung bean proteins and peptides: a comprehensive review of preparation technologies, biological activities, and their potential applications. Food Biosci 55:102972. https://doi.org/10.1016/j.fbio.2023.102972
Huang Z, Li Y, Fan M, Qian H, Wang L (2024) Recent advances in mung bean protein: from structure, function to application. Int J Biol Macromol 273:133210. https://doi.org/10.1016/j.ijbiomac.2024.133210
Idrissi O, Udupa SM, De Keyser E, McGee RJ, Coyne CJ, Saha GC et al (2016) Identification of quantitative trait loci controlling root and shoot traits associated with drought tolerance in a lentil (Lens culinaris Medik.) recombinant inbred line population. Front Plant Sci 7:1174
Iqbal J, Altaf MT, Jan MF, Raza W, Liaqat W, Haq I et al (2023) Exploring genetic diversity in cotton genotypes using EST-SSR and ISSR markers: a comparative study. Sarhad J Agric 39(4):800–814
Jaleel CA, Manivannan P, Wahid A, Farooq M, Al-Juburi HJ, Somasundaram R, Panneerselvam R (2009) Drought stress in plants: a review on morphological characteristics and pigments composition. Int J Agric Biol 11(1):100–105
Jiang GL (2013) Molecular markers and marker-assisted breeding in plants. In: Plant breeding from laboratories to fields, vol 3, pp 45–83
Jiang GL (2015) Molecular marker-assisted breeding: a plant breeder's review. In: Advances in plant breeding strategies: breeding, biotechnology and molecular tools, pp 431–472
Jincy M, Prasad V, Jeyakumar P, Senthil A, Manivannan N (2021) Evaluation of green gram genotypes for drought tolerance by PEG (polyethylene glycol) induced drought stress at seedling stage. Legume Res 44(6):684–691. https://doi.org/10.18805/LR-4149
Kamannavar PY, Vijaykumar AG (2011) Genotype x environment interaction in Mungbean (*Vigna radiata* (L.) Wilczek) cultivars grown in different agro-climatic zones of Karnataka. Electron J Plant Breed 2(4):501–505
Kaur R, Kaur J, Bains TS (2017) Screening of mungbean genotypes for drought tolerance using different water potential levels. J Adv Agric Technol 4(2):159

Khan MB, Hussain M, Raza A, Farooq S, Jabran K (2015) Seed priming with CaCl2 and ridge planting for improved drought resistance in maize. Turk J Agric For 39(2):193–203. https://doi.org/10.3906/tar-1405-39

Kitsanachandee R, Somta P, Chatchawankanphanich O, Akhtar KP, Shah TM, Nair RM et al (2013) Detection of quantitative trait loci for mungbean yellow mosaic India virus (MYMIV) resistance in mungbean (*Vigna radiata* (L.) Wilczek) in India and Pakistan. Breed Sci 63(4):367–373. https://doi.org/10.1270/jsbbs.63.367

Koebner RM, Powell W, Donini P (2001) Contributions of DNA molecular marker technologies to the genetics and breeding of wheat and barley. Plant Breed Rev 21:181–220

Kumar A, Sharma KD (2009) Physiological responses and dry matter partitioning of summer mungbean (*Vigna radiata* L.) genotypes subjected to drought conditions. J Agron Crop Sci 95:270–277. https://doi.org/10.1111/j.1439-037X.2009.00373.x

Kumar BS, Sathiyanarayanan G, Prakash M (2015) Influence of drought stress on root distribution in mungbean (*Vigna radiata* (L.) Wilczek). Int J Trop Agric 33:837–843

Kumar S, Singh P, Devi U, Yathish KR, Saujanya PL, Kumar R, Mahanta SK (2023) An overview of the current fodder scenario and the potential for improving fodder productivity through genetic interventions in India. Anim Nutr Feed Technol 23(3):631–644. https://doi.org/10.5958/0974-181X.2023.00054.9

Kumari P, Prakash S (2020) Marker assisted backcrossing (MAB) an approach for selection by using molecular markers. In: Recent advances in chemical sciences and biotechnology, vol 161

Kumari VV, Banerjee P, Verma VC, Sukumaran S, Chandran MAS, Gopinath KA et al (2022) Plant nutrition: an effective way to alleviate abiotic stress in agricultural crops. Int J Mol Sci 23(15):8519. https://doi.org/10.3390/ijms23158519

Kushwah A, Bhatia D, Barmukh R, Singh I, Singh G, Bindra S et al (2022) Genetic mapping of QTLs for drought tolerance in chickpea (Cicer arietinum L.). Front Genet 13:953898

Lee YH, Moon JK, Park KY, Ku JH, Yun HT, Chung WK et al (2000) A new mungbean cultivar with bruchid resistance,'Jangannogdu'. Korean J Breed 32(3):296–297

Lestari P, Kim SK, Kang YJ, Dewi N, Lee SH (2014) Genetic diversity of mungbean (*Vigna radiata* L.) germplasm in Indonesia. Plant Genet Res 12(S1):S91–S94. https://doi.org/10.1017/S1479262114000343

Levitt J (1980) Chilling, freezing, and high temperature stresses. In: Responses of plants to environmental stress, vol 2. Academic Press, New York, pp 23–445

Liu C, Wu J, Wang L, Fan B, Cao Z, Su Q et al (2017) Quantitative trait locus mapping under irrigated and drought treatments based on a novel genetic linkage map in mungbean (Vigna radiata L.). Theor Appl Genet 130:2375–2393

Lung'aho C, Chemining'wa GN, Fu YB, Shibairo SI, Hutchinson MJ, Paniagua HG (2011) Genetic diversity of Kenyan potato germplasm revealed by simple sequence repeat markers. Am J Potato Res 88:424–434

Maheswari M, Tekula VL, Yellisetty V, Sarkar B, Yadav SK, Singh J et al (2016) Functional mechanisms of drought tolerance in maize through phenotyping and genotyping under well watered and water stressed conditions. Eur J Agron 79:43–57

Manavalan LP, Guttikonda SK, Phan Tran LS, Nguyen HT (2009) Physiological and molecular approaches to improve drought resistance in soybean. Plant Cell Physiol 50(7):1260–1276

Maqbool MA, Aslam M, Ali H, Shah TM (2016) Evaluation of advanced chickpea (Cicer arietinum L.) accessions based on drought tolerance indices and SSR markers against different water treatments. Pak J Bot 48(4):1421–1429

Mburu SW, Koskey G, Kimiti JM, Ombori O, Maingi JM, Njeru EM (2016) Agrobiodiversity conservation enhances food security in subsistence-based farming systems of Eastern Kenya. Agric Food Secur 5:1–10. https://doi.org/10.1186/s40066-016-0068-2

Mir RR, Zaman-Allah M, Sreenivasulu N, Trethowan R, Varshney RK (2012) Integrated genomics, physiology and breeding approaches for improving drought tolerance in crops. Theor Appl Genet 125:625–645

Moody DB, Emebiri LC, Spackman M, Mekuria GT (2003) Breeding for qualitative and quantitative traits using markers. In: Proceedings of a joint meeting of the 11th Australian barley technical symposium and the 53rd Australian cereal chem conference, Glenelg

Moradi A, Ahmadi AHMD, Hossain Zadeh A (2008) The effects of different timings and severity of drought stress on gas exchange parameters of mungbean. Desert 13(1):59–66. https://doi.org/10.22059/jdesert.2008.27176

Moradi A, Ahmadi A, Hoseinzadeh A (2009) Agronomic and physiological interaction of mung bean (Partov) to sever and light stress in different stages. J Agric Res 12:659–671

Muchero W, Ehlers JD, Close TJ, Roberts PA (2009) Mapping QTL for drought stress-induced premature senescence and maturity in cowpea [Vigna unguiculata (L.) Walp.]. Theor Appl Genet 118:849–863

Nadeem MA, Nawaz MA, Shahid MQ, Doğan Y, Comertpay G, Yıldız M et al (2018) DNA molecular markers in plant breeding: current status and recent advancements in genomic selection and genome editing. Biotechnol Biotechnol Equip 32(2):261–285

Nadeem M, Li J, Yahya M, Sher A, Ma C, Wang X et al (2019) Research progress and perspective on drought stress in legumes: a review. Int J Mol Sci 20:1–32. https://doi.org/10.3390/ijms20102541

Nair R, Schreinemachers P (2020) Global status and economic importance of mungbean. In: The mungbean genome. Springer, Cham, pp 1–8

Nair RM, Pandey AK, War AR, Hanumantharao B, Shwe T, Alam AKMM et al (2019) Biotic and abiotic constraints in mungbean production—progress in genetic improvement. Front Plant Sci 10:1340

Noble TJ, Young AJ, Douglas CA, Williams B, Mundree S (2019) Diagnosis and management of halo blight in Australian mungbeans: a review. Crop Pasture Sci 70(3):195–203. https://doi.org/10.1071/CP18541

Noble TJ, Williams B, Hoang TML, Das Bhowmik SS, Tan GZH, Mundree S (2020) Genomic approaches to abiotic stresses in mungbean. In: The mungbean genome. Springer, Cham, pp 169–179

Ota M, Fukushima H, Kulski JK, Inoko H (2007) Single nucleotide polymorphism detection by polymerase chain reaction-restriction fragment length polymorphism. Nat Protoc 2(11):2857–2864

Ouédraogo JT, Tignegre JB, Timko MP, Belzile FJ (2002) AFLP markers linked to resistance against Striga gesnerioides race 1 in cowpea (Vigna unguiculata). Genome 45:787–793

Pandey AK, Burlakoti RR, Kenyon L, Nair RM (2018) Perspectives and challenges for sustainable management of fungal diseases of mungbean [*Vigna radiata* (L.) R. Wilczek var. radiata]: a review. Front Environ Sci 6:53. https://doi.org/10.3389/fenvs.2018.00053

Paramesh M, Reddy DM, Priya MS, Sumathi P, Sudhakar P, Reddy KHP (2016) GT biplot analysis for yield and drought related traits in mung bean (*Vigna radiata* L. Wilczek). Electron J Plant Breed 7(3):538–543

Prakash M, Sunilkumar B, Sathiyanarayanan G, Gokulakrishnan J (2017) Screening for drought tolerance in mungbean. Legume Res 40(3):423–428

Pratap A, Douglas C, Prajapati U, Kumari G, War AR, Tomar R et al (2020) Breeding progress and future challenges: biotic stresses. In: The mungbean genome, pp 55–80. https://doi.org/10.1007/978-3-030-20008-4_5

Raina SK, Govindasamy V, Kumar M, Singh AK, Rane J, Minhas PS (2016) Genetic variation in physiological responses of mungbeans (*Vigna radiata* (L.) Wilczek) to drought. Acta Physiol Plant 38:263. https://doi.org/10.1007/s11738-016-2280-x

Rajcan I, Boersma JG, Shaw EJ (2011) 4.11—Plant genetic techniques: plant breeder's toolbox. In: Comprehensive biotechnology, 2nd edn. Academic Press, Burlington, pp 133–147

Ranawake AL, Dahanayaka N, Amarasingha UGS, Rodrigo WDRJ, Rodrigo UTD (2012) Effect of water stress on growth and yield of mung bean (*Vigna radiata* L). Trop Agric Res Ext 14(4). https://doi.org/10.4038/tare.v14i4.4851

Ren H, Han J, Wang X, Zhang B, Yu L, Gao H et al (2020) QTL mapping of drought tolerance traits in soybean with SLAF sequencing. Crop J 8(6):977–989

Ryley MJ, Tatnell JR (2011, January) Management of the major foliar diseases of mungbeans and peanuts in Australia. In: Proceedings of the 4th Asian conference on plant pathology and the 18th biennial Australasian Plant Pathology Society Conference (ACPP/APPS 2011). University of Southern, Queensland

Sadeghipour O (2019) Polyamines protect mung bean [*Vigna radiata* (L.) Wilczek] plants against drought stress. Biol Futura 70(1):71–78. https://doi.org/10.1556/019.70.2019.09

Salim N, Raza A (2020) Nutrient use efficiency (NUE) for sustainable wheat production: a review. J Plant Nutr 43(2):297–315. https://doi.org/10.1080/01904167.2019.1676907

Sarıkaya MF, Tatar M (2023) Global climate change and crop production. Int J Sivas Univ Sci Technol 2(2):65–78

Saxena RK, Saxena KB, Kumar RV, Hoisington DA, Varshney RK (2010) Simple sequence repeat-based diversity in elite pigeonpea genotypes for developing mapping populations to map resistance to Fusarium wilt and sterility mosaic disease. Plant Breed 129(2):135–141

Schafleitner R, Nair RM, Rathore A, Wang YW, Lin CY, Chu SH et al (2015) The AVRDC–the world vegetable center mungbean (*Vigna radiata*) core and mini core collections. BMC Genomics 16:1–11. https://doi.org/10.1186/s12864-015-1556-7

Sehrawat N, Yadav M, Kumar S, Upadhyay SK, Singh M, Sharma AK (2020) Review on health promoting biological activities of mungbean: a potent functional food of medicinal importance. Plant Arch 20(2):2969–2975

Shahwar D, Ansari MYK, Choudhary S (2019) Induction of phenotypic diversity in mutagenized population of lentil (*Lens culinaris* Medik) by using heavy metal. Heliyon 5(5):e01722. https://doi.org/10.1016/j.heliyon.2019.e01722

Sharma HC, Crouch JH (2004) Molecular marker-assisted selection: a novel approach for host plant resistance to insects in grain legumes

Sholihin HD (2002) Molecular mapping of drought resistance in mungbean (Vigna radiata): 2. QTL linked to drought resistance. J Bioteknologi Pertanian 7:55–61

Singh RA, De RK (2006) Major diseases and their management. In: Advances in mungbean and urdbean, pp 283–334

Singh CM, Mishra SB, Pandey A, Arya M (2014) Morphological characterization and discriminant function analysis in mungbean [*Vigna radiata* (L.) Wilczek] germplasm. Electron J Plant Breed 5(1):87–96

Singh CM, Singh AK, Mishra SB, Pandey A (2016) Generation mean analysis to estimate the genetic parameters for yield improvement and inheritance of seed colour and lusture in mungbean [*Vigna radiata* (L.) Wilczek]. Legume Res 39(4):494–501. https://doi.org/10.18805/lr.v0iOF.10762

Singh CM, Singh P, Tiwari C, Purwar S, Kumar M, Pratap A et al (2021a) Improving drought tolerance in mungbean (*Vigna radiata* L. Wilczek): morpho-physiological, biochemical and molecular perspectives. Agronomy 11(8):1534. https://doi.org/10.3390/agronomy11081534

Singh P, Mishra AK, Singh CM (2021b) Genome-wide identification and characterization of lectin receptor-like kinase (LecRLK) genes in mungbean (Vigna radiata L. Wilczek). J Appl Genet 62(2):223–234. https://doi.org/10.1007/s13353-021-00613-8

Somta P, Srinives P (2007) Genome research in mungbean [Vigna radiata (L.) Wilczek] and blackgram [V. Mungo (L.) Hepper]. ScienceAsia 33(Suppl 1):69–74

Swathi L (2013) Genetic diversity analysis using morpho-physiological and molecular markers for breeding yield and drought tolerance in mungbean (*Vigna radiata* (L.) Wilczek). M.Sc. (Ag.). Thesis, Acharya N.G. Ranga Agricultural University, Hyderabad

Swathi L, Reddy DM, Sudhakar P, Vineela V (2017) Screening of Mungbean (Vigna radiata L. Wilczek) genotypes against water stress mediated through polyethylene glycol. Int J Curr Microbiol App Sci 6(10):2524–2531

Tarahi M, Abdolalizadeh L, Hedayati S (2024) Mung bean protein isolate: extraction, structure, physicochemical properties, modifications, and food applications. Food Chem 444:138626. https://doi.org/10.1016/j.foodchem.2024.138626

Thomas M, Robertson J, Fukai S, Peoples MB (2004) The effect of timing and severity of water deficit on growth development, yield accumulation and nitrogen fixation of mung bean. Field Crop Res 86:67–68. https://doi.org/10.1016/S0378-4290(03)001205

Tripathy SK, Mohanty P, Jena M, Dash S, Lenka D, Mishra D et al (2016) Identification of seed storage protein markers for drought tolerance in mungbean. Res Biotechnol 7

Umer Chattha M, Arif W, Khan I, Soufan W, Bilal Chattha M, Hassan MU et al (2021) Mitigation of cadmium induced oxidative stress by using organic amendments to improve the growth and yield of mash beans [*Vigna mungo* (L.)]. Agronomy 11(11):2152. https://doi.org/10.3390/agronomy11112152

Verma SK, Tuteja OP, Monga D (2008) Evaluation for genotype× environment interactions in relation to stable genetic male sterility-based Asiatic cotton (*Gossypium arboreum*) hybrids of north zone. Indian J Agric Sci 78(4):375–378

Wang LanFen WL, Wu Jing WJ, Jing RuiLian JR, Cheng XuZhen CX, Wang ShuMin WS (2015a) Drought resistance identification of mungbean germplasm resources at seedlings stage. Acta Agron Sin 41:145–153

Wang LanFen WL, Wu Jing WJ, Jing RuiLian JR, Xu C, Zhen CX, Wang ShuMin WS (2015b) Drought resistance identification of mungbean germplasm resources at seedlings stage. Acta Agron Sin 41:145

Wang LX, Cheng XZ, Wang SH, Li JB, Li JX (2009) Genetic variability of agronomic traits in mungbean applied core collection of China. J Plant Genet Resour 10:589–593

Wang LF, Jing WU, Jing RL, Cheng XZ, Wang SM (2014) Drought resistance identification of mungbean germplasm resources at bud stage. J Plant Genet Resour 15:498–503

Wang L, Bai P, Yuan X, Chen H, Wang S, Chen X, Cheng X (2018) Genetic diversity assessment of a set of introduced mung bean accessions (*Vigna radiata* L.). Crop J 6(2):207–213. https://doi.org/10.1016/j.cj.2017.08.004

Xie J, Ye H, Du M, Yu Q, Chen Y, Shen M (2019) Mung bean protein hydrolysates protect mouse liver cell line Nctc-1469 cell from hydrogen peroxide-induced cell injury. Food Secur 9(1):14. https://doi.org/10.3390/foods9010014

Yadav R, Punia MS (2017) Genetic divergence studies in mungbean (Vigna radiata L. Wilczek) using morpho-physio and molecular markers to identify drought tolerant genotypes. Indian J Genet Plant Breed 77(04):574–578

Yalinkiliç NA, Başbağ S, Altaf MT, Ali A, Nadeem MA, Baloch FS (2024) Applicability of SCoT markers in unraveling genetic variation and population structure among sugar beet (Beta vulgaris L.) germplasm. Mol Biol Rep 51(1):584

Yang JK, Yuan TY, Zhang WT, Zhou JC, Li YG (2008) Polyphasic characterization of mung bean (*Vigna radiata* L.) rhizobia from different geographical regions of China. Soil Biol Biochem 40:1681–1688. https://doi.org/10.1016/J.SOILBIO.2008.02.002

Yuliasti Y, Reflinur R (2015) Evaluation of mungbean mutant lines to drought stress and their genetic relationships using SSR markers. At Indones 41(3):161–167

Zhang YF, Wang ET, Tian CF, Wang FQ, Han LL, Chen WF, Chen WX (2008) Bradyrhizobium elkanii, Bradyrhizobium yuanmingense and Bradyrhizobium japonicum are the main rhizobia associated with *Vigna unguiculata* and *Vigna radiata* in the subtropical region of China. FEMS Microbiol Lett 285(2):146–154. https://doi.org/10.1111/j.1574-6968.2008.01169.x

Zhu HJ, Zhao XY, Yan HB, Gao W, Zhang YW (2012) Breeding of bruchid resistant mung bean variety Jinlyudou 7. J Shanxi Agric Sci 40:606–607

Lablab (*Lablab Purpureus* L. Sweet): Current Developments in Breeding and Prospects for Drought Tolerance, Improved Yield, and Sustainable Production in Sub-Saharan Africa (SSA)

14

Julius S. Missanga and Pavithravani B. Venkataramana

Abstract

Drought is the largest threat to crop production, especially in sub-Saharan Africa (SSA). Since we have a few crop species performing well in dry areas , knowledge on their responsesto drought stress could help us to identify the drought-tolerant genotypes. Lablab (*Lablab purpureus* L. Sweet) is the most drought-tolerant crop with a wide range of benefits as compared to other legumes. It is a farmers' resource crop with a multiple range of physiological, agronomic and nutritional values. It excels well in the sustainability of soil fertility and conservation agriculture. Farmers in arid and semi-arid regions rely on it for human consumption, animal feeding and income. Despite its potential, Lablab is still an underutilized crop in many countries, especially in SSA where the effects of climate change are most visible. The main reason why farmers in SSA are not benefiting much from this crop is a lack of improved varieties, particularly the high-yield , drought-tolerant varieties. The present advancements in plant breeding and their potential in research for drought tolerance, improved yield, and sustainable production in Lablab have become the main topic of this review study. The morphological, biochemical, and genomic potential of this crop have all been thoroughly examined in this study in order to endorse them for further breeding in drought tolerancee and yield improvement. With a small genome size (367 Mb) and a longer coding sequence (1276 bp) to exhibit more gene expression, Lablab becomes a crop of choice for the future sustainability in present effect of climate change. . Some of the ongoing research initiatives in Lablab

J. S. Missanga (✉)
Department of Biology, University of Dodoma (UDOM), Dodoma, Tanzania

P. B. Venkataramana
Department of Sustainable Agriculture, Nelson Mandela African Institution of Science and Technology (NM-AIST), Arusha, Tanzania

© The Author(s), under exclusive license to Springer Nature Singapore Pte Ltd. 2025

M. A. Nadeem et al. (eds.), *Marker-Assisted Breeding in Legumes for Drought Tolerance*, https://doi.org/10.1007/978-981-96-4112-3_14

suggest more screening of germplasm for drought tolerance especially at the seedling and reproductive stages so that to develop moisture stress-tolerant varieties . Additional research initiatives should also consider advanced gene analysis to assess DNA and RNA expressions on drought conditions and more studies on sequence-based markers-polymorphism contrasting for responses to the drought stress. It is finally proposed that the transferred quantitative traits loci (QTL) from other legumes could help to achieve the suitable varieties so that to shift Lablab to the industrial and economic crop in SSA.

Keywords

Dolichos lablab · Drought · Yield · Climate change · Sustainable production

14.1 Introduction

A use of drought-tolerant crops to improve food productivity in the current phase of climate change stands as the most important production strategy that needs widespread attention (FAO 2019; UN 2019). Figuring out which crop species (sp.) thrive in arid climates and how they react to drought stress could help to speed up the effort to develop varieties resistant to drought stress (Missanga et al. 2021). Lablab (*Lablab purpureus* L. Sweet) is multipurpose and the most drought-tolerant crop unlike its relatives (Chapman 2015; Naeem et al. 2020). It is used as human food and animal feed (Mugerwa et al. 2024) nourishing them with essential nutrients. Dry and green beans, tender pods, and leafy vegetables are among the foodstuffs acquired from Lablab. Flour from Lablab bean can be used to make processed food staff such as biscuits (Rana et al. 2021). Feed obtained from Lablab includes fresh leaves, silage, or hay (Murphy and Colucci 1999). Lablab produces cutting-edge medications and nutraceuticals with useful application in human and animal health (Purwanti et al. 2021). Farmers prefer to incorporate Lablab into their fields in order to improve soil fertility (Mkonda and He 2017), soil structure, and conservation agriculture (Bakari and Pauline 2020). Given the long-term sustainability of various farming systems (Chakoma et al. 2016), Lablab is one of the most significant resource crops for maintaining biodiversity, minimizing carbon footprints, and generating very little greenhouse gases (Wang et al. 2021). It is one of the few crops that significantly increase farmers' income and access to food in situations when regularly produced crops get lost from prolonged drought stress (Nord et al. 2020). Based on its potential from the morphological, biochemical, and genomic features, Lablab has a highly diverse genetic makeup (Ramesh and Byregowda 2016) which contributes to its broad distribution and adaptation in various climatic and environmental conditions (Vidigal et al. 2018; Vaijayanthi et al. 2019). As a result, Lablab differs from its relatives as it can withstand a wide range of temperatures (18–35 °C) and rainfall (200–2500 mm). However, following the introduction of *Phaseolus* beans in the early 1900s during the colonial era, Lablab, a crop with African origin, lost its preference among the farmers. Due to its short duration in maturity and cooking compared to Lablab, the *Phaseolus* bean gained priority among farmers and consumers (Maass et al. 2010; Massawe et al. 2016). Relatively few farmers continued to plant Lablab in small and dispersed areas in Africa. According to Miller et al. (2018), the northern area of Tanzania and the central Kikuyu in Kenya were two of these unique locations. The genetic loss of Lablab and the loss of farmers' knowledge about the crop were brought oout by the decrease in Lablab production in several areas. Since then, Lablab has not seen much improvement in SSA and has continued to be an underutilized crop in many areas. Farmers have been depending muchon their own landraces, which lack improved characteristics, particularly high yield and drought tolerance (Karanja 2016; D'Alessandro and Molina 2021). In the efforts to improve breeding and prospects of Lablab for drought tolerance, and improved yield, along with its commercial production, the University of Dodoma (UDOM) (https://udom.ac.tz/) and the Nelson Mandela African Institution of Science and Technology (NM-AIST) (https://nm-aist.ac.tz/) in Tanzania have been working together to develop drought-tolerant and yield-improved varieties (Missanga et al. 2021, 2023a, b), which are still missing in many African nations. Genetic improvement of Lablab has been primarily supported by the conventional breeding as the principal research technique. This kind of breeding ignored drought stress tolerance and high yield in favor of morphological characterization, forage, influence of production on soil features, and production challenges. Due to the polygenic inheritance of factors associated with breeding traits, sustainable improvement of Lablab for drought tolerance and high-yield should consider genomic, biochemical, and morphological parameters into account (Rai et al. 2018a, b; Missanga et al. 2021). Current studies show that Lablab lacks a sufficient amount of information on morphological, biochemical, and genomic performance to assist enhanced breeding on drought resistance and high yield for sustainable and commercial production (Maass and Chapman 2022; Wang et al. 2018). Therefore, this review study focused on current developments in Lablab breeding and prospects for drought tolerance, improved yield, and sustainable production in SSA.

14.2 Increase in Drought Stress and Trend in Production and

A significant increase in drought stress is predicted to cause a strong food demand in many parts of the world, especially in SSA (UNCC 2020) where farmers are most vulnerable to climate change (Shiferaw et al. 2014). According to Kapoor et al. (2020), drought has affected about 40% of the world' s fertile land with roughly 50% of crop' resources being lost in the field each year as a result of drought . The drought stress has also a detrimental effect on the crops' nutritional value (FAO 2017). Some other consequences of the extended drought stress include hunger, famine and malnutrition, disputes over natural resources, land degradation, little income, poverty, and low economic development (UN 2019; Beltran-Pea et al. 2020). According to FAO (2019), many areas in SSA will experience drought stress because of the rising effects of climate change. As a result, a number of crops vis-à-vis legumes that farmers prefer for food and income such as maize (*Zea mays* L.), rice (*Oryza sativa* L.), common beans (*Phaseolus vulgaris* L.), cowpeas (*Vigna unguiculata* L.), and Soybeans (*Glycine max* L.) are no longer doing relatively well in many areas with drought stress conditions. Lablab is among the leguminous crop with high promise on drought tolerance. It is reported as the most drought-resilient crop compared to its relatives (Guretzki and Papenbrock 2014; Sennhenn et al. 2017). The prosperity of Lablab as the drought-tolerant crop lies on its great genetic diversity that enables its adaptation to many areas with low rainfall (about 200 mm) and high temperature (above 35 °C) (Maass et al. 2010; Missanga et al. 2021). Its morphological, biochemical, and genomic potential supports the growth of the crop during very little rainfall (Robotham and Chapman 2017; Miller et al. 2018).

14.3 Lablab

14.3.1 Domestication and Adaptation

Lablab, known by a variety of common names, including Dolichos lablab, Lablab bean, Hyacinth bean, and Egyptian kidney bean (in English) (Maass et al. 2010; Athmaselvi et al. 2020), Fiwi or Ngwara (in Tanzanian Swahili), and Njahe or Njahi (in Kikuyu, Kenya), is the leguminous crop in the family *Fabaceae,* similarly to other legumes (Missanga et al. 2021). Sequencing and assembly of Lablab genome has made it possible to compare this crop among different legume sp. (Table 14.1). Unlike other legumes, Lablab has a smaller genome (367 Mb), protein-coding genes (20946), and a scaffold assembly (395.47 Mb). It has also genes with longer coding sequences, exons, and introns. This exhibits greater gene expression and appropriate genomic exploitation in Lablab compared to other legumes (Missanga et al. 2021; Njaci et al. 2023). With chromosome numbers $2n = 2x = 20$, 22, and 24 (Gupta et al. 2017), Lablab is characterized into three subsp., i.e., *Lablab purpureus* subsp. *purpureus*; *Lablab purpureus* subsp. *bengalensis* (Jacq.) Verdc; and *Lablab purpureus* subsp. *uncinatus* Verdc (Ambrose et al. 2016; Mihailović et al. 2016). Both *purpureus* and *bengalensis* are two cultivated forms of Lablab while the *uncinatus* has both wild and cultivated subsp. The *purpureus* subsp. is pre-dominated in Asia where it is grown as pulses, fodder, or seed crop, whereas the *bengalensis* subsp. is grown in tropical areas of India, South East Asia, Africa, and America often as large podded subsp. The *uncinatus* subsp. prevails in East Africa mainly as wild subsp. (Shivashankar and Kulkarni 1989; Islam 2008). After several hypotheses and scientific experiments, the Eastern and Southern Africa was traced and discovered as the origin Centre of Lablab (Maass et al. 2017). The trace also showed Ethiopia was the potential Centre of Lablab domestication (Abdallah et al. 2015; Maass et al. 2017). Through double domestication events, Lablab was divided into two distinct groups i.e. haplotypes A (a group of two seeded pods) and haplotypes B (a group of four seeded pods) (Kongjaimun et al. 2022). Ethiopia was inhabited by haplotype A, while the rest of the world had haplotype B (Massawe et al. 2016).

Lablab is presently found in the tropical and sub-tropical regions mainly at 30° North and 30° South at 0–2300 meters above sea level (Athmaselvi et al. 2020).

Table 14.1 Genomic features (de novo sequenced genome) of Lablab and other related species

S/n	Common name/ scientific name	Chromosome number (2*n*)	Genome size (Mb)	Scaffold assembled lengths (Mb)	Number of proteins coding genes	Average gene length (bp)	Average length in coding sequence (bp)	Exon length (bp)	Intron length (bp)	Reference
1	Lablab (*L. purpureus*)	20, 22, 24	367	395.47	20,946	3696	1276	239	557	Iwata et al. (2013), Chang et al. (2018) and Njaci et al. (2023)
2	Common bean (*P. vulgaris*)	22	587	473	27,197	–	–	–	–	Schmutz et al. (2014)
3	Cowpea (*V. unguiculata*)	22	613	519	29,773	3881	–	313	–	Lonardi et al. (2019)
4	Soybean (*Glycine max*)	40	1115	950	55,137	3144	1169	232	488	Valliyodan et al. (2017) and Chang et al. (2018)
5	Adzuki beans (*V. angularis*)	22	542	466.7	34,183	–	–	–	–	Kang et al. (2015) and Yang et al. (2015)
6	Mung beans (*V. radiata*)	22	579	~431	22,427	–	–	–	–	Kang et al. (2014)
7	Bambara nut (*V. subterranean*)	22	864	535.05	31,707	3287	1163	222	501	Chang et al. (2018)
8	Pigeon pea (*C. cajan*)	22	833	605.78	48,680	2348	959.35	–	–	Singh et al. (2017)

(continued)

Table 14.1 (continued)

S/n	Common name/ scientific name	Chromosome number (2*n*)	Genome size (Mb)	Scaffold assembled lengths (Mb)	Number of proteins coding genes	Average gene length (bp)	Average length in coding sequence (bp)	Exon length (bp)	Intron length (bp)	Reference
9	Groundnut (*A. hypogaea*)	20, 40	2552	2530	83,709	4275	226	233.21	578	Pandey et al. (2020)
10	B. trefoil (*L. japonicus*)	12	480	554.08	29,598	–	–	417.54	527.12	Kamal et al. (2020)
11	Acacia (*F. albida*)	26	–	653.73	28,979	3396	1207	226	504	Chang et al. (2018)
12	Marula (*S. birrea*)	28	–	330.98	18,937	3561	1343	239	479	Bationo-kando et al. (2016) and Chang et al. (2018)
13	Drumstick tree (*M. oleifera*)	28	–	216.76	18,451	3308	1238	232	478	Tian et al. (2015) and Chang et al. (2018)
14	Barrel medic (*M. truncatula*)	16	390	388	50,358	2334	986	243	440	Young et al. (2005), Chang et al. (2018), Young and Zhou (2019)

Source: Missanga et al. (2021)

Despite having African origin, the genetic and breeding identity of this crop has not been widely proclaimed (Miller et al. 2018). African scientists have now an obligation to present the breeding advancements of this crop from its point of origin to provide more chances for producing sustainable outputs in the future. The results of such interventions could provide more benefits of the crop especially on the present climatic variability. More programmes on genetic resource collections, their conservation initiatives, and breeding improvements on Africa's climate challenges are recommended. Few countries in SSA, such as Tanzania (Missanga et al. 2023a) and Kenya (Kamotho et al. 2016), have been appreciated for completing a wide gathering of indigenous landraces to support the breeding studies.

14.3.2 A Minor Legume and an Indigenous Crop

Lablab is one of the leguminous crops that has not undergone significant genetic improvement in several areas, and therefore, termed as minor legume and an indigenous crop (Chivenge et al. 2015; Chibarabada et al. 2017). These indigenous sp. of crops are those that farmers are no longer producing like in the past (Padulosi and Hoeschle-Zeledon 2004). As they are mostly found in scattered forms of wild or semi-domesticated sp., only a few marginal farmers are involved in their production, especially closer or in their own center of origin. Many of the indigenous crops have not been improved much compared with other crops. The reason behind their poor research and production is a relatively narrow genetic base and a few breeding resources, especially in SSA. Since they have little economic importance, their systems of seed supply, production, extension services, post-harvest technologies, and governing policies are still very poor over the continent (CTA 1991). Apart from poor documentation, the attention of funding agencies on their research and development is exclusively very little, especially in SSA. Even the roles of various local research institutions and organizations in their improvement are insignificant. Therefore, the indigenous crops have been ignored among many users for some genetic, agronomic, nutritional, and socio-economic factors. Even many consumers are not familiar with them leading to their weak marketing channels in many areas (Baa-Poku 2018). As a result, these indigenous crops have been referred to as "underutilized sp." (Padulosi and Hoeschle-Zeledon 2004). Synonymous terms such as neglected, underdeveloped, orphan, lost, promising, forgotten, traditional, abandoned, and local crops have been used to refer to the indigenous crops (IPGRI 2002; Padulosi and Hoeschle-Zeledon 2004). Although it is an underutilized crop, Lablab has a great genetic diversity with germplasms found almost in every continent (Table 14.2). Such resources own a wide range of physiological parameters (Venkatesha et al. 2007, 2013) which include plant structures with different sizes, shapes, color, and textures observed in stems, leaves, flowers, fruits (pods), and seeds (Byre Gowda et al. 2015). These plant traits have been improved continuously through domestication and conventional breeding. Now these traits need further breeding improvement through advanced breeding.

Table 14.2 Lablab germplasm at several Lablab gene banks located throughout the world

S/n	Continent or region	Country or an institution hosting the gene-bank	Amount in number	Amount in the continent or region	References
1.	Asia	University of Agricultural Sciences (UAS), Bengaluru, India	650	3092	Ramesh and Byregowda (2016) and Missanga et al. (2021)
		National Bureau of Plant Genetic Resources (NBPGR), New Delhi, India	650		BI (2008)
		Plant gene banks in Bangladesh	551		ISLAM (2008)
		World Vegetable Research and Development Centre (AVRDC), Taiwan	447		AVRDC (2009) and Schreinemachers et al. (2014)
		Plant gene banks in China	410		BI (2008)
		Plant gene banks in the Philippines	209		Engle and Altoveros (2000)
		Plant gene banks in South Asia	93		BI (2008)
		Plant gene banks in South-east Asia (apart from India and Bangladesh)	82		BI (2008) and NIAS (2009)
2.	Africa	Nelson Mandela African Institutions of Science and Technology (NM-AIST), Tanzania	450	1216	Kirkhouse (2019) and Missanga et al. (2023a).
		World Vegetable Centre, Eastern and Southern Africa, Arusha-Tanzania	73		Ojiewo et al. (2010)
		Plant gene banks in Kenya	403		BI (2008)
		Ethiopian gene-banks including International Livestock Research Institute (ILRI)	223		BI (2008)
		Other sub-Saharan African countries including International Institute of Tropical Agriculture (IITA), Nigeria	67		BI (2008)
3.	America	Plant gene-banks in South America	134	186	BI (2008)
		United States, Department of Agriculture (USDA)	52		GRIN (2009)
4.	Oceania	Commonwealth Scientific and Industrial Research Organization (CSIRO), Australia	104	104	BI (2008)
5.	Europe	European research institutions	82	82	BI (2008) and VIR (2009)
Total amount				4680	

Source: Improved from Ramesh and Byregowda (2016), Missanga et al. (2021), and Missanga et al. (2023a)

14.3.3 The Crop with a Multiple Range of Benefits

14.3.3.1 Food and Nutritional Importance

Lablab is enhanced with various food components (Table 14.3). Protein, carbohydrates, lipid, nutrients, vitamins, and dietary fibers are part of these food ingredients that exhibit remarkable concentrations in Lablab (Ambrose et al. 2016; Singh and Abhilash 2019). Protein is an important food component in Lablab (Minde et al. 2021). Lablab leaves and seeds have 20–28% and 21–38% of protein, respectively (Naeem et al. 2020). Essential amino acids contained in Lablab are leucine, isoleucine, histidine, arginine, threonine, methionine, valine, and lysine (Ahmed et al. 2015; Purwanti et al. 2021). Lablab has also high contents of carbohydrate, fibers, minerals, and vitamins. The key minerals from Lablab include magnesium, phosphorus, calcium, iron, zinc, and copper while essential vitamins are thiamin and niacin (Shaahu et al. 2015). On top of that, Lablab is among the legumes with low fat content (Hossain et al. 2016). Therefore, Lablab could replace animal products as the least expensive source of protein (Qamar et al. 2020), helping to fight nutrient-deficiency disorders such as malnutrition (Minde et al. 2021).

14.3.3.2 Farming and Agronomic Potential

Production of Lablab in SSA involves smallholder farmers (Chawe et al. 2019; Foyer et al. 2016). These farmers produce Lablab primarily for conservation agriculture, enhanced soil fertility, animal and human consumption, and marketing (Chakoma et al. 2016; Bakari and Pauline 2020). While most farmers' cultivars are grown annually, Lablab is occasionally planted as a short-lived perennial. Lablab wild sp. thrive longer to become a perennial crop (FAO 2012). In production, Lablab is a predominantly self-pollinated crop with about 6–10% pollination by insects (Kukade and Tidke 2014). Most often, it is produced in the fields, gardens, or homesteads either alone or in combination with other crops. In mono-cropping, Lablab functions as a cover crop, pulse, and fodder. When farmers' main and minor crops, such as maize, sorghum, millet, sunflower, and groundnuts, are mixed, intercropped, or rotated with Lablab, they profit from soil nutrients, nutrients' recovery, decrease in weed competition, disease infestation, water dehydration, and wind influence (McDonald et al. 2001; Hoeschle-Zeledon 2019). Sudanese farmers intercrop Lablab with cotton (D'Alessandro and Molina 2021). Lablab differs in flowering and maturity time depending on cultivars. It begins to flower as soon as 40–45 days after sowing. Pods begin to form 60–70 days in advance and last 90–100 days. Harvesting mature seeds occurat least 150–210 days after sowing (Heuzé et al. 2016; Dhaliwal 2007). As a high-yield crop, potential in drought tolerance (Guretzki and Papenbrock 2014).

Lablab out-yields other relative crops such as common bean, cowpeas, and pigeon-peas (Nedumaran et al. 2015; Rai et al. 2018a). As much as 2.5–5.0 t ha^{-1} of green pods and 1.5–2.0 t ha^{-1} of dry seeds can be produced by the improved varieties of Lablab (Heuzé et al. 2016; Nord et al. 2020). Additionally, improved varieties of Lablab are capable of producing 6–13 t ha^{-1} of dry matter (Wangila et al. 2021). This is in contrast to the varieties of cowpeas and common beans, which can yield

Table 14.3 Food and nutritional values of Lablab seeds compared with other related legumes

S/n	Value in 100 g	Lablab	Common bean	Cowpeas	Soybean	Lentils	Chickpea	Bambara-nut
1	Protein (g)	24.2	21.7	22.5	28.21	9.02	16.4	18.8
2	Carbohydrate (g)	62.1	60	61.0	8.36	20.13	27.42	61
3	Lipid/ Fat (g)	1.0	1.5	1.4	8.97	0.38	2.59	6.2
4	Fiber (g)	8.6	4.4	5.4	6.0	7.9	7.6	4.8
5	Iron (mg)	9.0	8.2	N/A	5.14	3.33	2.89	12.2
6	Calcium (mg)	90	120	104	102	19	49	62
7	Phosphorus (mg)	330	323	416	245	180	170	276
8	β-Carotene (mg)	N/A	10.0	70.0	9.0	8.0	27.0	10.0
9	Ascorbic acid (mg)	[a]	1.0	2.0	1.7	1.5	1.3	[a]
10	Thiamin (mg)	0.5	0.4	N/A	0.1	0.155	0.116	0.169
11	Niacin (mg)	2.3	2.4	4.0	0.399	1.06	0.526	1.8
12	Riboflavin (mg)	0.1	0.2	0.9	0.285	0.073	0.063	0.1
13	Energy (Kcal)	350	341	336	446	N/A	378	N/A

Improved from Hossain et al. (2016), Wallace et al. (2016), Ganesan and Xu (2017), Kilonzi et al. (2017), Çakir et al. (2019), Purwanti et al. (2019), and Letting et al. (2021)

Note: [a]Trace amount; N/A: Not applicable; Units; Fiber: Total dietary fiber (TDF); Energy: Kilo-calories (Kcal); 1 cal: 4.184 Joules between 1.3 and 1.5 t ha^{-1} (Njonjo et al. 2019) and about 0.88 t ha^{-1} (CIAT; World Bank 2017) of dry seeds, respectively.

14.3.3.3 Socio-economic Benefits

Lablab offers a variety of socio-economic benefits to farmers (Raghu et al. 2018; Bakari and Pauline 2020). The presence of Lablab in cereal production system has enhanced dietary benefits of resource poor households (Pretty et al. 2012). It is gradually developing into a reliable source of income, especially in dry areas where rainfall is insufficient to support consistent production of other legumes (Athmaselvi et al. 2020). According to Forsythe (2019) and Nord et al. (2020), Lablab is becoming more popular even in places with marginal rainfall. For instance, Lablab has become an important cash crop with higher income among farmers in dry climatic regions of East Africa (Karanja 2016; Miller et al. 2018). Lablab has also developed into a high quality forage that enhances the digestion of inferior roughages since it surpasses other legumes in terms of nutritive ingredients, yield, and biomass. It provides fresh foliage when other pastures are no longer in the field (Ewansiha and Singh 2006). Aganga and Tshwenyane (2003) reported a great quality of the livestock products such as meat, and milk when feeding on Lablab forages. Despite all these benefits, Lablab is still considered as an underutilized crop with very few improved varieties in many countries including Tanzania.

14.3.3.4 Sustainable Production and Its Challenges

Overall challenges faced by farmers in Lablab production in SSA involve mainly biotic and abiotic factors (Kankwatsa and Muzira 2018; Miller et al. 2018) similarly to other Lablab growing areas in the world (Ramesh and Byregowda 2016). The main abiotic challenges that influence sustainable Lablab production in SSA are climatic and environmental factors (Ewansiha and Singh 2006). Within this umbrella, Lablab is affected by extreme temperatures, frost, heavy winds, salinity and toxicity of the soil, flooding, and water logging conditions, (Brown and Pengelly 2007; Grotelüschen 2014). The major biotic challenges are pests, and diseases, especially at flowering and podding stages. The main Lablab pests include arthropods, e.g., young millipedes; insect-pests, e.g., foliage insects (bruchids, aphids, and leaf miners); pod-sucking insects, e.g., coreid bugs and pod-eating larvae (pod borers); as well as parasitic weeds, e.g., striga, that can extract plant juice. The major diseases that affect Lablab include fungal infections, e.g., powdery mildew (*Erysiphe polygoni*), anthracnose (*Colletotrichum lindemuthianum*), and root rot (*Rhizoctonia* spp.); bacterial infections, e.g., bacterial leaf spot (*Xanthomonas* sp.), bacterial wilt (*Curtobacterium* sp.), and bacterial blight (*Pseudomonas* sp.); as well as viral diseases, e.g., Dolichos yellow mosaic virus (DoYMV) and Bean common mosaic virus (BCMV). Root-knot nematodes (RKN) can also affect Lablab sometimes with severe condition (Ewansiha et al. 2016). Bruchids is the main storage insect-pest for most of the legumes including Lablab (Letting et al. 2021). Other challenges that influence Lablab production in SSA are long production durations for the majority of Lablab cultivars. Many Lablab cultivars are indeterminate and highly photoperiod sensitive that requires short day length to switch from vegetative to reproductive phases (Ramtekey et al. 2019; Kamau 2022). As reported by Keerthi et al. (2014), normally it takes them longer time before meeting favorable condition for flowering. Due to overlapping behavior in their vegetative and reproductive phases, harvests in indeterminate Lablab cultivars are repetitive with very low yield (Sultana et al. 2001; Bakari and Pauline 2020). On contrary, the determinate cultivars have their main stems, and branches terminate their growth at the inflorescent stage. As a result, these cultivars require only constant number of days to transform from vegetative to reproductive stage, irrespective of day length. Thus, the determinate cultivars are suitable for economic production (Keerthi et al. 2018, b; Basanagouda et al. 2022). However, many of the naturally occurring Lablab cultivars are indeterminate cultivars (Ramesh and Byregowda 2016), and therefore, creating a need for their breeding and genetic improvement. Moreover, Lablab is given less preference among consumers when compared to other legumes due to the different anti-nutritional factors such as cyanogenic glucoside toxins, tannins, oxalates, phytates, and trypsin inhibitors (Kilonzi et al. 2017) and volatile compounds such as hydrocarbons, terpenes, esters, alcohols, and terpenoids. These compounds affect quality, nutritive value, and organoleptic properties of the crop such as flavor, taste, and odor as well as cooking interval (Kimani et al. 2019; Ngure et al. 2021). Poor agronomic practices among farmers, limited market channels (Miller et al. 2018), and policy factors (legumes given less emphasis compared to the staple food) (Ojiewo et al. 2019) have been associated with low production ofLablab in SSA. A use of local cultivars in Lablab production has also been associated with limited sources of quality seeds. Most of the Lablab farmers in SSA depend on their landraces for seeds. In order to address most of these challenges, especially lack of improved varieties in production, there should be more scientific studies on the genetic and genomic exploitation of the crop to develop drought tolerant - high yield varieties. To gratify on seed challenges , there should be a well-organized system of seed dissemination with active commercial supply chains where seed inventory tools in Lablab can be incorporated for quick registration, production, and circulation.

14.4 Drought Stress Tolerance in Lablab

Lablab has a number of characteristics that help it withstand drought stress (Rangaiah and D'Souza 2016). These characteristics include various morphological, biochemical, and molecular parameters (Missanga et al. 2021). There have been several studies on the morphological characteristics of Lablab, with little being conducted to connect these characteristics to drought stress during the seedling stage. Very few papers (D'Souza and Devaraj 2011; Devaraj et al. 2014a, b) have been documented to link drought tolerance and Lablab seedlings, however with a small number of accessions. On top of that, the relationship between drought and Lablab reproduction efficiency has not been thoroughly studied. According to Nadeem et al. (2019), water use efficiency (WUE) is an essential factor to consider during development of the drought-tolerant cultivars. Such types of cultivars should incorporate different traits tolerant to early, intermittent as well as terminal moisture stress (TMS) (Susmitha and Ramesh 2023). Farmers need these types of varieties because they perform well in drought-stressed environments, particularly at low soil moisture contents (MC) throughout their growing stages. Many crops usually begin to experience drought stress when the soil MC depletes to 35–50% in the top soil (30 cm) (Da Silva et al. 2019; Ding et al. 2021). This is different from Lablab where the lowest soil MC that can indicate drought stress ranges from 25 to 50% (Fening et al. 2009). Despite good knowledge available for MC that would elicit drought stress in crops, TMS-tolerant varieties have not been easily acquired in many crops such as Lablab due to the reason that drought tolerance is a complex process that is governed by polygenic traits (Serba and Yadav 2016; Missanga et al. 2021). So far, several methods have been applied to develop TMS-tolerant cultivars, although with some limitations. The first one involves yield evaluation of genotypes at moisture stress (MS) or moisture stress-free (MSF) conditions. This method is less preferred in selection of the best tolerant genotypes as it causes a strong interaction among genotypes, low genetic heritability, and low yield variations among the genotypes. The second method considers genetic manipulation of crops to tolerate TMS. Since there is still lack of sufficient knowledge on genetic mechanism for tolerance in many crops, this method is still also ineffective for TMS studies (Mitra 2001). From these shortcomings, an appropriate method to select TMS-tolerant genotypes considers yield evaluation of the genotypes at both TMS and MSF conditions. The yield obtained at two moisture conditions, i.e., MS condition relative to MSF, creates the relationship that generates ten empirical indices (Table 14.4) to use for computing the TMS-tolerant genotypes (Talebi et al. 2009; Gitore et al. 2021).

Table 14.4 Drought tolerance indices for quantifying the accessions' responses to TMS experiments

S/N	Drought tolerance indices	Code	Formula
1	Mean productivity	MP	$(Yp + Ys)/2$
2	Geometric mean productivity	GMP	$\sqrt{Yp \times Ys}$
3	Harmonic mean productivity	HMP	$2^* (Yp \times Ys) / (Yp + Ys)$
4	Stress tolerance index	STI	$(Yp \times Ys)/ \bar{Y}p^2)$
5	Abiotic tolerance index	ATI	$[(Yp - Ys)/(\bar{Y}p / \bar{Y}s) \times \sqrt{}(Yp \times Ys)]$
6	Stress non-stress production index	SNPI	$[^3\sqrt{}Yp + Ys/ (Yp - Ys)] \times [^3\sqrt{}Yp \times Ys]$
7	Yield index	YI	$(Yp/ \bar{Y}s)$
8	Modified STI 1	K_1STI	$Yp^2/ \bar{Y}s_2$
9	Modified STI 2	K_2STI	$Ys^2/ \bar{Y}s_2$
10	Drought resistance index	DI	$(Ys \times (Yp/Ys))/ \bar{Y}s$

Source: Missanga et al. (2023b), Susmitha and Ramesh (2023)
Yp and Ys: Seed yield $plant^{-1}$ of each accession under MSF and TMS conditions, respectively
$\bar{Y}p$ and $\bar{Y}s$: Mean seed yield $plant^{-1}$ of all accession under MSF and TMS conditions, respectively

These indices include mean productivity (MP), geometric mean productivity (GMP), stress tolerance index (STI), hormonic mean productivity (HMP), stress non-stress production index (SNPI), abiotic tolerance index (ATI), modified STI 1 and 2 (K_1STI and K_2STI), yield index (YI), and drought susceptibility index (DSI) (Bennani et al. 2017; Susmitha and Ramesh 2023).

Some findings on drought tolerance indices were reported in several crops (Talebi et al. 2009; Farshadfar and Elyasi 2012), however, with limited information in Lablab. The best TMS genotypes can be found by combining the drought indices with additional morphological and yield data acquired during a similar field evaluation procedure. Moreover, high temperatures in the field can produce reactive oxygen species (ROS), such as hydroxyl radicals ($OH^{\bullet}$), which can harm essential crop precursors (Foyer and Noctor 2012). According to Rangaiah and D'Souza (2016), Lablab has both enzymatic and non-enzymatic defense mechanisms against ROS biochemically. Polyphenol oxidase (PPOX), catalase (CAT), and other antioxidant enzymes are important biochemical compounds able to convert ROS into less dangerous compounds. Lablab stores non-enzymatic substances such as proline, flavonoid, and phenols to shield its cellular compartments from the effects of drought (Devaraj et al. 2014a, b). Genes that are resistant to drought stress and their transcription factors that are linked to the stress e.g. dehydrins, are well expressed when Lablab undergoes drought stress (Wang et al. 2018). By working together, UDOM and NM-AIST have been able to collect Lablab germplasm from both the farmers and different foreign gene-banks, screen them for drought tolerance at the seedling stage, and finally evaluate them for drought tolerance and improved yield in different dry farming systems based on drought tolerance indices (Missanga et al. 2021, 2023a, b). The study on biochemical analysis of different Lablab accessions has been conducted at UDOM (Fig. 14.1) to compare their seedling performance at stressed and non stressed conditions . However, the phenomena of gene expressions regulating drought stress have been studied in many crops using molecular markers. However, Lablab and some other legumes have seen slower advancements in marker technology than cereal crops, to the extent that they are now referred to as "orphan crops" (Vaijayanthi et al. 2018; Dhaliwal et al. 2020). In their steps of development, the first group of markers i.e.Amplified Fragment Length Polymorphism (AFLP), Restriction Fragment Length Polymorphism (RFLP), and Random Amplification of Polymorphic DNA (RAPD) were mainly employed to understand the diversity and genetic characterization. The second group of markers, namely, sequence-based markers including Single-Nucleotide Polymorphism (SNPs), Microsatellite or Simple Sequence Repeats (SSRs), and Expressed Sequence Tags (EST), were mostly utilized in mapping studies (Kimani et al. 2012; Kamotho et al. 2016; Keerthi et al. 2018a; Rai et al. 2018a, b; Vaijayanthi et al. 2018; Dholakia et al. 2019; Amkul et al. 2021; Kamau et al. 2021; Sserumaga et al. 2021). Despite the fact that molecular markers were applied multiple times in Lablab, little was done to utilize them in drought tolerance improvement in Lablab. Nonetheless, some Quantitative Trait Loci (QTLs) for drought tolerance found in other legumes (Table 14.5) may provide helpful information for the creation and transferability of markers in Lablab (Cheng et al. 2017; Lepcha et al. 2019; Jha et al. 2020).

(a) Screen-house covered by plastic materials

(b) Top soil and sandy soil (1:1) preparation

(c) The experiment involving two replicates of the experiment i.e. stressed and non-stressed experiments

(d) Pot experiment involving different Lablab genotypes screened for bio-chemical analysis three times

Fig. 14.1 Screening Lablab genotypes for drought tolerance by relating morphological, and biochemical assessment at seedling stage conducted at the University of Dodoma (UDOM), Tanzania (**a**) Screen-house covered by plastic materials (**b**) Top soil and sandy soil (1:1) preparation (**c**) The experiment involving two replicates i.e. stressed and non-stressed experiments (d) Pot experiment involving different Lablab genotypes screened for biochemical analysis three times Lablab genotypes

The study on "the expressed sequence tag-simple sequence repeat (EST-SSR) marker-based polymorphism contrasting for responses to drought stress" has recently been made possible by the collaboration between UDOM and NM-AIST. The findings from this study will be soon released to support scientific community in marker based drought tolerance research These research initiatives to develop drought-tolerant Lablab cultivars form an important step towards sustainable crop production and associated food security in SSA.

14.5 Future Prospects for Drought Tolerance and Improved Yield in Lablab

Many farmers and consumers in SSA are beginning to grasp numerous benefits from using Lablab. The crop has been the focus of a lot of scientific studies, however, many of which are conducted in Asia rather than in Africa, where the crop originated frome (Maass 2016; Missanga et al. 2021). There have been very few studies conducted in SSA that focus on advanced breeding and genomic development to release drought-tolerant high-yield Lablab varieties for market purposes. According to Ramesh and Byregowda (2016) and Missanga et al. (2021), Australia and Asia are home to several well-known commercial Lablab varieties , such as

Table 14.5 Molecular markers and QTLs for drought tolerance in some other legumes

S/n	Crop	Molecular markers/QTLs	Mapping population	Features in drought tolerance	Reference
1	Common beans	SSR; AFLP markers; 49 QTLs	82 Recombinant inbred lines (RILs)	Drought-responsive agronomic traits	Sedlar et al. (2020)
		SNP markers; 18 QTLs	97 RIL-from parent lines: Portillo × Red Hawk	Drought tolerance parameters	Onziga et al. (2019)
		SSRs and SNPs markers; 12 QTLs	F8; SEA 5 × AND 277	Drought (stress) conditions	Briñez et al. (2017)
		SSR; 69 QTLs SSR; 15 putative QTLs	DOR364 × BAT477	Drought tolerance parameters; Root-drought related traits	Asfaw and Blair (2012)
		AFLP, RAPD, SSR markers; 143 QTLs	100—RILs	Drought tolerance and related traits	Diaz et al. (2018)
		SNP markers; 14 QTLs	RILs; SEA5 × CAL96,	Phenology-yield-drought	Mukeshimana et al. (2014)
		53 SNP markers; 11 QTLs	128—F8 RILs	Drought stress conditions	Nabateregga et al. (2019)
		SNP markers; genes functioning	Pinto-Villa × Pinto Saltillo, F3:5 (289 genotypes)	Drought tolerance	Villordo-Pineda et al. (2015)
2	Cowpeas	184 genome-wide EST-derived SNP markers; drought tolerance QTLs	2 drought-tolerant lines (IT93K-503-1 and IT97K-499-35)	Green abilities and yield under water stress conditions	Batieno et al. (2016)
		412 DArT, 80 AFLP, 28 microsatellite markers; 2 QTLs	72-F1 derived double haploid (DH)	Stress tolerance (drought)	Fan et al. (2015)
		35 SNP markers; Tolerance index ranged between 69.19–142.01	305-F8 RILs	Tolerance to water-deficit conditions	Ravelombola et al. (2021)

3	Chickpeas	SNP markers and 21 major QTLs	232 RILs; ICC 4958 × ICC 1882	Drought tolerance parameters	Sivasakthi et al. (2018)
		828 SNPs; DREB and CAPS from QTL-hotspot region	264 RILs	Drought tolerance	Jaganathan et al. (2015)
		47, 53, and 46 SSRs markers in selection; QTL-hotspot region	Three elite cultivars	Drought tolerance and grain yield	Bharadwaj et al. (2021)
4	Soybeans	368 SSR including Satt277; QTLs	F2 mapping population (PK1180, SL 46 × UPSL 298, PK 1169)	Seedling survivability under drought conditions	Sreenivasa et al. (2020)
		8078 specific locus amplified fragments (SLAF) markers; 23 QTLs	RILs	Drought tolerance traits	Ren et al. (2020)
5	Mung bean	3690 SSR; 58 QTLs for plant parameters and 5 for drought tolerance	256 RIL population	Plant tolerance and associated parameters	Liu et al. (2017)
6	Asparagus bean	39 SNPs markers from GWAS	95 accessions	Tolerance to soil water stress	Xu et al. (2015)

Source: Improved from Missanga et al. (2021)

Hebbal Avare (HA). In SSA, Kenya has been able to commercialize several Lablab varieties (Eldo-KT: Black 1 and 2; Eldo-KT cream, and Eldo-KT Maridadi) earlier than other countries (Kirkhouse Trust 2015; KEPHIS 2017). The Nelson Mandela-Dolichos (NMD 19) and NMD 20 are the two newly Lablab released commercial varieties in Tanzania. Availability of a few Lablab commercial varieties in SSA compared to other Lablab production regions especially in Asia are primarily caused by research in Africa focusing more on morphological characterization, soil fertility and biomass assessment, cover crops and soil erosion, food testing, farmer selection, and certain production constraints like insect pests and diseases rather than breeding components of the crop and their associated advanced technologies. There is so far little use of genomic resources in Africa compared to other continents to enhance drought tolerance and yield improvement in Lablab. Research base in many countries in SSA lack modernized tools such as high-throughput and next-generation facilities to support gene editing and transcription analysis of the available germplasm toward securing the potential genotypes. The recent completion of the chromosomal scale genome by Njaci et al. (2023) has shown that African scientists are well-positioned to promote the commercial exploitation of untapped resources to address the region's concerns. Nonetheless, there is still a need for straightforward knowledge on extraction of these sources to address the problems of poor yield and drought stress in potential crops such as Lablab. Capacity building and research collaboration in advanced technology such as genomics and bioinformatics should also be integrated among the research systems in SSA for the efficient understanding, and applications .

14.6 Conclusion and Recommendations

Lablab is gaining more attention from researchers due to its various benefits. Compared to other closely related species, Lablab has demonstrated a stronger tolerance to drought stress along with high-yield features. Even with these advantages, there has not been much work done on genetic and genomic breeding of Lablab for drought tolerance and improved yield. The crop has therefore remained underutilized in many areas. Given Lablab's genetic and physiological potential, advanced breeding techniques are recommended to turn it into an economic crop. A few of these have already been put into practices to bring some modifications. Therefore, the crop can from an orphan legume to an industrial crop to serve the continent .

References

Abdallah OEM, Hamza NB, Dagash YMI (2015) Agronomic and molecular evaluation of six lablab bean (*Lablab Purpureus* L.) cultivars. Int J Sci Res Agric Sci 2(1):7–15

Aganga AA, Tshwenyane SO (2003) Lucerne, lablab and Leucaena leucocephala forages: production and utilization for livestock production. Pak J Nutr 2(2):46–53

Ambrose, M., Fondevilla, S., Mcgeough, E., Mcphee, K., & Thompson, R. (2016). Legume perspectives. Hyacinth bean: a gem among legumes state of the art in Lablab Purpureusresearch. https://www.legumesociety.org

Amkul K, Sookbang JM, Somta P (2021) Genetic diversity and structure of landrace of lablab (*Lablab purpureus* (L.) sweet) cultivars in Thailand revealed by SSR markers. Breed Sci 71(2):176–183. https://doi.org/10.1270/jsbbs.20074

Asfaw A, Blair MW (2012) Quantitative trait loci for rooting pattern traits of common beans grown under drought stress versus non- stress conditions. Mol Breed 30(2):681–695. https://doi.org/10.1007/s11032-011-9654-y

Athmaselvi KA, Sukumar A, Bhokarikar S (2020) Hyacinth beans. In: Manickavasagan A, Thirunathan P (eds) Pulses: processing and product development, pp 119–127. https://doi.org/10.1007/978-3-030-41376-7

AVRDC (2009) AVRDC Vegetable Genetic Resources Information System (AVGRIS), Shanhua. http://203.64.245.173/avgris/

Baa-Poku F (2018) Review on the role of underutilized crops in achieving food security in Ghana: implications for policy. Ghana J Agric Sci 52:113–120

Bakari AE, Pauline NM (2020) Trade-offs of Dolichos Lablab production in the context of the changing climate in semi-arid areas of Tanzania. Tanzania J Agric Sci 19(2):188–202

Basanagouda G, Ramesh S, Chandana BR, Siddu CB, Kirankumar R, Kalpana MP (2022) Inheritance of growth habit under photoperiod insensitive genetic background in dolichos bean [*Lablab purpureus* (L.) sweet]. Genet Resour Crop Evol 69(7):2535–2543

Batieno BJ, Danquah E, Tignegre J, Huynh B, Drabo I, Close TJ, Ofori K, Roberts P, Ouedraogo TJ (2016) Application of marker-assisted backcrossing to improve cowpea (*Vigna unguiculata* L. Walp) for drought tolerance. J Plant Breed Crop Sci 8(12):273–286. https://doi.org/10.5897/JPBCS2016.0607

Bationo-Kando P, Zongo JD, Siljak-Yakovlev S (2016) First genome size assessment, heterochromatin and rDNA chromosome mapping in the genus Sclerocarya (Anacardiaceae): insight into the new basic chromosome number. Bot Lett 163(1):11–17

Beltran-Peña A, Rosa L, D'Odorico P (2020) Global food self-sufficiency in the 21st century under sustainable intensification of agriculture. Environmental Research Letters 15(9):1–12.

Bennani S, Nsarellah N, Jlibene M, Tadesse W, Birouk A, Ouabbou H (2017) Efficiency of drought tolerance indices under different stress severities for bread wheat selection. Aust J Crop Sci 11(4):395–405

Bharadwaj C, Tripathi S, Soren KR, Thudi M, Singh RK, Sheoran S, Roorkiwal M, Patil BS, Chitikineni A, Palakurthi R, Venula A, Rathole A, Kumar Y, Chaturvedi SK, Mondal B, Shanmugavadivel PS, Srivastava SK, Dixit GP, Singh NP, Varshney RK (2021) Introgression of "QTL-hotspot" region enhances drought tolerance and grain yield in three elite chickpea cultivars. Plant Genome 14:e20076. https://doi.org/10.1002/tpg2.20076

BI (2008) Biodiversity Directory of germplasm collections. http://www.biodiversityinternational.org/information

Briñez B, Perseguini JMKC, Rosa JS, Bassi D, Gonçalves JGR, Almeida C, De Carvalho-Paulino JF, Blair MW, Chioratto AF, Carbonell SAM, Valdisser PAMR, Vianello RP, Benchimol-reis LL (2017) Mapping QTLs for drought tolerance in a SEA 5 x AND 277 common bean cross with SSRs and SNP markers. Genet Mol Biol 40(4):813–823

Brown S, Pengelly B (2007) Lablab Factsheet: Tropical Grasslands Society-Tropical Forages Database (SoFT). https://keys.lucidcentral.org

Byre-Gowda M, Girish G, Ramesh S, Mahadevu P, Keerthi CM (2015) Descriptors of Dolichos bean (Lablab purpureus L.). Journal of Food Legumes 28(3):203–214.

Çakir Ö, Uçarli C, Tarhan Ç, Pekmez M, Turgut-kara N (2019) Nutritional and health benefits of legumes and their distinctive genomic properties. Food Sci Technol 39(1):1–12

Chakoma I, Manyawu G, Gwiriri L, Moyo S, Sikhalazo D (2016) The Agronomy and use of Lablab Purpureus in Smallholder Farming Systems of Southern Africa. ILRI extenstion brief website: https://hdl.handle.net/10568/78513

Chang Y, Liu H, Liu M, Liao X, Sahu SK, Fu Y, Song B, Cheng S, Kariba R, Muthemba S, Hendre PS, Mayes S, Ho WK, Yssel AEJ, Kendable P, Wang S, Li L, Muchungi A, Jamnadass R, Liu X (2018) The draft genomes of five agriculturally important African orphan crops. GigaScience 8(3):1–16. https://doi.org/10.1093/gigascience/giy152

Chapman MA (2015) Transcriptome Sequencing and Marker Development for Four Underutilized Legumes. Applications in Plant Sciences 3(2):1–5.

Chawe KG, Venkataramana PB, Ndakidemi PA (2019) Assessment of farmers' indigenous knowledge and preferences: a tool for sustainable lablab bean (*Lablab purpureus*. L. Sweet) improvement and utilization in Northern Tanzania. J Adv Biol Biotechnol 21(1):1–14

Cheng A, Chai HH, Ho WK, Bamba ASA, Feldman A, Kendabie P, Halim RA, Tanzi A, Mayes S, Massawe F (2017) Molecular marker technology for genetic improvement of underutilised crops. In: Abdullah S, Chai-Ling H, Wagstaff C (eds) Crop improvement. https://doi.org/10.1007/978-3-319-65079-1_3

Chibarabada TP, Modi AT, Mabhaudhi T (2017) Expounding the value of grain legumes in the semi- and arid tropics. Sustain For 9:1–25

Chivenge P, Mabhaudhi T, Modi AT, Mafongoya P (2015) The potential role of neglected and under-utilized crop species as future crops under water scarce conditions in sub-Saharan Africa. Int J Environ Res Public Health 12:5685–5711

CTA (1991) SPORE Bulletin of the CTA No. 36. Technical Bulletin No. 36 of the Technical Centre for Agricultural and Rural Cooperation ACP-EU (CTA). https://hdl.handle.net/10568/45669

D'Alessandro C, Molina BP (2021) Fiwi (Lablab bean). In: Zocchi DM, Fontefrancesco MF (eds) The ark of taste in Tanzania: food, knowledge, and stories of gastronomic heritage. Università degli Studi di Scienze Gastronomiche, pp 58–59

D'Souza MR, Devaraj VR (2011) Specific and non-specific responses of Hyacinth bean (*Dolichos lablab*) to drought stress. Indian J Biotechnol 10(1):130–139

Da Silva AL, Coolong T, Diaz-Perez JC (2019) Principles of irrigation and scheduling for vegetable crops in Georgia. UGA Coop Ext Bull 1511:2–12

Devaraj VR, D'souza MR, Kokila S (2014a) Polyamine levels in leaves of hyacinth bean (*Lablab purpureus*) and their relation to drought-tolerance. World Appl Sci J 32(58):2398–2402. https://doi.org/10.5829/idosi.wasj.2014.32.12.1382

Devaraj VR, D'souza MR, Kokila S (2014b) Response of *Lablab purpureus* (Hyacinth bean) cultivars to drought stress. Asian J Plant Sci Res 4(5):48–55

Dhaliwal MS (2007) Legume vegetables: handbook of vegetable crops. Kalyani Publishers, pp 228–278

Dhaliwal SK, Talukdar A, Gautam A, Sharma P, Sharma V, Kaushik P (2020) Developments and prospects in imperative underexploited vegetable legumes breeding: a review. Int J Mol Sci 21:9615

Dholakia HP, Joshi MK, Delvadiya IR (2019) Molecular characterization of Indian bean (*Lablab purpureus* L.) genotypes. J Pharmacogn Phytochem 8(2):455–463

Diaz LM, Ricaurte J, Tovar E, Cajiao C, Teran H, Grajales M, Polania J, Rao I, Beebe S, Raatz B (2018) QTL analyses for tolerance to abiotic stresses in a common bean (*Phaseolus vulgaris* L.) population. PLoS One 13(8):e0202342

Ding Y, Gong X, Xing Z, Cai H, Zhou Z, Zhang D, Sun P, Shi H (2021) Attribution of meteorological, hydrological and agricultural drought propagation in different climatic regions of China. Agric Water Manag 255(2021):106996

Ewansiha SU, Singh BB (2006) Relative drought tolerance of important herbaceous legumes and cereals in the moist and semi-arid regions of West Africa. J Food Agric Environ 4(2):188–190

Ewansiha SU, Tarawali SA, Chiezey UF (2016) Observations on pests and diseases in *Lablab purpureus* in Nigerian northern Guinea savanna. J Int Legume Soc 13:23–25

Fan Y, Shabala S, Ma Y, Xu R, Zhou M (2015) Using QTL mapping to investigate the relationships between abiotic stress tolerance (drought and salinity) and agronomic and physiological traits. BMC Genomics 16:43. https://doi.org/10.1186/s12864-015-1243-8

FAO (2012) Grassland species index. Lablab Purpureus. A searchable catalogue of grass and forage legumes. FAO, Rome. www.fao.org

FAO (2017) The future of food and agriculture: trends and challenges. Food and Agriculture Organization (FAO) of the United Nations, Rome. www.fao.org/publications

FAO (2019) The state of food security and nutrition in the world 2019. Safeguarding against economic slowdowns and downturns. FAO, IFAD, UNICEF, WFP and WHO of The United Nations, Rome. https://doi.org/10.4060/ca9692en

Farshadfar E, Elyasi P (2012) Screening quantitative indicators of drought tolerance in bread wheat (*Triticum aestivum* L.) landraces. Eur J Exp Biol 2(3):577–584

Fening JO, Quansah C, Sarfo-Kantanka A (2009) Response of Three Forage Legumes To Soil. Journal of Science and Technology 29(3):24–30

Forsythe C (2019) Exploring the viability of re- introducing Lablab Purpureus (L.) sweet as a multifunctional legume in Northern Tanzania. (Swedish University of Agricultural Sciences. https://stud.epsilon.slu.se). Retrieved from https://stud.epsilon.slu.se

Foyer CH, Noctor G (2012) Managing the cellular redox hub in photosynthetic organisms. Plant Cell Environ 35(2):199–201

Foyer CH, Lam HM, Nguyen HT, Siddique KHM, Varshney RK, Colmer TD, Cowling W, Bramley H, Mori TA, Hodgson JM, Cooper JW, Miller AJ, Kunert K, Vorster J, Cullis C, Ozga JA, Wahlqvist ML, Liang Y, Shou H et al (2016) Neglecting legumes has compromised human health and sustainable food production. Nat Plants 2:16112

Ganesan K, Xu B (2017) Polyphenol-rich lentils and their health promoting effects. Int J Mol Sci 18(11):2390

Gitore SA, Danga B, Henga S, Gurmu F (2021) Evaluating drought tolerance indices for selection of drought tolerant Orange Fleshed Sweet Potato (OFSP) genotypes in Ethiopia. Int J Agric Sci Food Technol 7(2):249–254. https://doi.org/10.17352/2455-815X.000115

GRIN (2009) National plant germplasm system. Genetic resources. Information Network, Beltsville. http://www.ars-grin.gov/cgi-bin/npgs/acc/query.pl

Grotelüschen K (2014) Lablab purpureus (L.) sweet: a promising multipurpose legume for enhanced drought resistance and improved household nutritional status in smallholder farming systems of Eastern Kenya. Georg-August-University Göttingen, Germany

Gupta M, Rao KP, Rajwade VB (2017) Correlation study of floral traits, yield and nutritional parameters in dolichos bean (*Lablab purpureus* L.) genotypes under Allahabad agro-climatic zone. J Pharmacogn Phytochem 6(6):1585–1591

Guretzki S, Papenbrock J (2014) Characterization of Lablab purpureus regarding drought tolerance, trypsin inhibitor activity and cyanogenic potential for selection in breeding programmes. J Agron Crop Sci 200(1):24–35

Heuzé V, Tran G, Sauvant D, Renaudeau D, Bastianelli D, Lebas F (2016) Lablab (Lablab purpureus). https://www.feedipedia.org/node/297

Hoeschle-Zeledon I (2019) Africa research in sustainable intensification for the next generation: sustainable intensification of key farming systems in east and southern Africa. Technical Report, 01 Apr 2019–30 Sept 2019 Submitted to United States Agency for International Development (USAID). www.africa-rising.net

Hossain S, Ahmed R, Bhowmick S, Mamun AA, Hashimoto M (2016) Proximate composition and fatty acid analysis of *Lablab purpureus* (L.) legume seed: implicates to both protein and essential fatty acid supplementation. Springerplus 5(1899):1–10. https://doi.org/10.1186/s40064-016-3587-1

IPGRI (2002) Neglected and underutilized plant species: strategic action plan of the International Plant Genetic Resources Institute. https://scholar.google.com

Islam MT (2008) Morpho-agronomic diversity of hyacinth bean (*Lablab purpureus* (L.) sweet) accessions from Bangladesh. Bulletin de Ressources Phytogénétiques 73:73–78

Iwata A, Greenland CM, Jackson SA (2013) Cytogenetics of legumes in the Phaseoloid clade. Plant Genome 6(3):1–8

Jaganathan D, Thudi M, Kale S, Azam S, Roorkiwal M, Gaur PM, Kishor PBK, Nguyen H, Sutton T, Varshney RK (2015) Genotyping-by-sequencing based intra-specific genetic map refines a "QTL-hotspot" region for drought tolerance in chickpea. Mol Gen Genomics 290(2):559–571

Jha UC, Bohra A, Nayyar H (2020) Advances in "omics" approaches to tackle drought stress in grain legumes. Plant Breed 139:1–27. https://doi.org/10.1111/pbr.12761

Kamal N, Mun T, Reid D, Lin J, Akyol TY, Sandal N, Asp T, Hirakawa H, Stougaard J, Mayer KFX, Sato S, Andersen SU (2020) Insights into the evolution of symbiosis gene copy number and distribution from a chromosome-scale *Lotus japonicus* Gifu genome sequence. DNA Res 27(3):1–11

Kamau E (2022) Identification of new molecular markers for diversity analysis and breeding for early maturity and determinate lablab (*Lablab Purpureus* L.) varieties. University of Eldoret. URI: http://41.89.164.27:8080/xmlui/handle/123456789/1736

Kamau EM, Kinyua MG, Waturu CN, Kiplagat O, Wanjala BW, Kariba RK, David RK (2021) Diversity and population structure of local and exotic Lablab purpureus accessions in Kenya as revealed by microsatellite markers. Glob J Mol Biol 3(8):1–16

Kamotho GN, Kinyua MG, Muasya RM, Gichuki ST, Wanjala BW (2016) Assessment of genetic diversity of Kenyan Dolichos bean (*Lablab purpureus* L. sweet) using simple sequence repeat (SSR) markers. Int J Agric Environ Biores 1(01):26–43

Kang YJ, Kim SK, Kim MY, Lestari P, Kim KH, Ha B, Jun TH, Hwang WJ, Lee T, Lee J, Shim S, Yoon MY, Jang YE, Han KS, Taeprayoon P, Yoon N, Somta P, Tanya P, Kim KS, Lee SH (2014) Genome sequence of mungbean and insights into evolution within Vigna species. Nat Commun 5:5443

Kang YJ, Satyawan D, Shim S, Lee T, Lee J, Hwang WJ, Kim SK, Lestari P, Laosatit K, Kim KH, Ha TJ, Chitikineni A, Kim MY, Ko J, Gwag J, Moon J, Lee Y, Park B, Varshney RK, Lee S (2015) Draft genome sequence of adzuki bean, Vigna angularis. Sci Rep 5:8069. https://doi.org/10.1038/srep08069

Kankwatsa P, Muzira R (2018) Agronomic performance and sensory evaluation of lablab (*Lablab purpureus* L. sweet) accessions for human consumption in Uganda. Open Access Library J 05(03):1–23. https://doi.org/10.4236/oalib.1104481

Karanja D (2016) Pulses crops grown in Ethiopia, Kenya and United Republic of Tanzania for local and export markets. International Trade Centre, East Africa Grain Council, Geneva, Switzerland. https://intracen.org/file/kenyapulsesbooklet20201102webpdf

Kapoor D, Bhardwaj S, Landi M, Sharma A, Ramakrishnan M, Sharma A (2020) The Impact of Drought in Plant Metabolism: How to Exploit Tolerance Mechanisms to Increase Crop Production. Applied Sciences 10(5692):1–19.

Keerthi CM, Ramesh S, Byre Gowda M, Rao AM, Prasad BSR, Vaijayanthi PV (2014) Genetics of growth habit and photoperiodic response to flowering time in dolichos bean (*Lablab purpureus* (L.) sweet). J Genet 93(1):203–206. https://doi.org/10.1007/s12041-014-0336-5

Keerthi CM, Ramesh S, Byregowda M, Vaijayanthi PV (2018a) Simple sequence repeat (SSR) marker assay-based genetic diversity among Dolichos bean (*Lablab purpureus* L. sweet) advanced breeding lines differing for productivity per se traits. Int J Curr Microbiol App Sci 7(5):3736–3744. https://doi.org/10.20546/ijcmas.2018.705.433

Keerthi CM, Ramesh S, Byregowda M, Mohan Rao A, Reena GAM (2018b) Photo-thermal effects on time to flowering in dolichos bean (*Lablab purpureus* (L.) sweet) var. lignosus. Curr Sci 115(7):1320–1327

KEPHIS (2017) National crop variety list-Kenya. Kenya Plant Health Inspectorate Service, Nairobi. http://www.kephis.org/images/VarietyList/april20170525.pdf

Kilonzi SM, Makokha AO, Kenji GM (2017) Physical characteristics, proximate composition and anti-nutritional factors in grains of lablab bean (*Lablab purpureus*) genotypes from Kenya. J Appl Biosci 114:11289–11298

Kimani EN, Wachira FN, Kinyua M, G. (2012) Molecular diversity of Kenyan Lablab bean (*Lablab purpureus* (L.) sweet) accessions using amplified fragment length polymorphism markers. Am J Plant Sci 3:313–321

Kimani E, Matasyoh J, Kinyua M, Wachira FN (2019) Characterisation of volatile compounds and flavour attributes of *Lablab purpureus* bean accessions. Afr J Biotechnol 18(24):518–530

Kirkhouse Trust (2015) Dolichos improvement programme in Kenya. https://www.kirkhouse-trust.org

Kirkhouse Trust (2019) Supporting research and education in biological sciences: agricultural crop improvement for the relief of poverty, with a focus on legumes. https://www.kirkhousetrust.org/stolprojects

Kongjaimun A, Takahashi Y, Yoshioka Y, Tomooka N, Mongkol R, Somta P (2022) Molecular analysis of genetic diversity and structure of the Lablab (*Lablab purpureus* (L.) sweet) Gene Pool reveals two independent routes of domestication. Plan Theory 12(1):1–18

Kukade SA, Tidke JA (2014) Reproductive biology of Dolichos lablab L. (Fabaceae). Indian J Plant Sci 3:22–25

Lepcha P, Kumar PR, Sathyanarayana N (2019) Exploring genomics research in the context of some underutilized legumes – a review. In: Banerjee R, Kumar GV, Kumar SPJ (eds) OMICS-based approaches in plant biotechnology. Scrivener publishing, pp 1–18. https://doi.org/10.1002/9781119509967.ch1

Letting FK, Venkataramana PB, Ndakidemi PA (2021) Breeding potential of lablab [*Lablab purpureus* (L.) sweet]: a review on characterization and bruchid studies towards improved production and utilization in Africa. Genet Resour Crop Evol 2021:1–21

Liu C, Wu J, Wang L, Fan B, Cao Z, Su Q, Zhang Z, Wang Y, Tian J, Wang S (2017) Quantitative trait locus mapping under irrigated and drought treatments based on a novel genetic linkage map in mungbean (*Vigna radiata* L.). Theor Appl Genet 130:2375–2393. https://doi.org/10.1007/s00122-017-2965-6

Lonardi S, Munoz-Amatriain M, Liang Q, Shu S, Wanamaker SI, Lo S, Tanskanen J, Schulman AH, Zhu T, Luo M, Alhakami H, Ounit R, Hasan A, Verdier J, Roberts PA, Santos JRP, Ndeze A, Dolezel J, Vrana J, Hokin SA et al (2019) The genome of cowpea (*Vigna unguiculata* [L.] Walp.). Plant J 98:767–782. https://doi.org/10.1111/tpj.14349

Maass BL (2016) Origin, domestication and global dispersal of *Lablab purpureus* (L.) sweet (Fabaceae): current understanding. Legume Perspect 13:5–8

Maass BL, Chapman MA (2022) The lablab genome: recent advances and future perspectives. In: Underutilised crop genomes. Springer Nature, pp 229–253

Maass BL, Knox MR, Venkatesha SC, Angessa TT, Ramme S, Pengelly BC (2010) *Lablab purpureus*-A crop lost for Africa? Trop Plant Biol 3(3):123–135

Maass BL, Robotham O, Chapman MA (2017) Evidence for two domestication events of hyacinth bean (*Lablab purpureus* (L .) Sweet): a comparative analysis of population genetic data. Genet Resour Crop Evol 4(8):1–13

Massawe PI, Mtei KM, Munishi LK, Ndakidemi PA (2016) Effect of rhizobium and intercropping systems on soil nutrients and biological nitrogen fixation as influenced by legumes (*Phaseolus vulgaris* and *Dolichos lablab*). Int J Curr Microbiol App Sci 5(10):135–149

McDonald LM, Wright P, MacLeod DA (2001) Nitrogen fixation by lablab (*Lablab purpureus*) and lucerne (*Medicago sativa*) rotation crops in an irrigated cotton farming system. Aust J Exp Agric 41(2):219–225

Mihailović V, Mikić A, Ćeran M, Ćupina B, Dordevic V, Marjanovic-Jeromela A, Mikic S, Peric V, Savic A, Srebric M, Terzic S, Vazic M, Vasiljevic M, Vujic S (2016) Some aspects of biodiversity, applied genetics and agronomy in hyacinth bean (*Lablab purpureus*) research. J Int Legume Soc 13:9–15

Miller NR, Mariki W, Nord A, Snapp S (2018) Cultivar selection and management strategies for *Lablab purpureus* (L.) sweet in Africa. In: Filho WL (ed) Handbook of climate change resilience, pp 1–14. https://doi.org/10.1007/978-3-319-71025-9_102-1

Minde JJ, Venkataramana PB, Matemu AO (2021) Dolichos Lablab-an underutilized crop with future potentials for food and nutrition security: a review. Crit Rev Food Sci Nutr 61(13):2249–2261

Missanga JS, Venkataramana PB, Ndakidemi PA (2021) Recent developments in *Lablab purpureus* genomics: a focus on drought stress tolerance and use of genomic resources to develop stress-resilient varieties. Legume Sci 3(3):1–11

Missanga JS, Venkataramana PB, Ndakidemi PA (2023a) *Lablab purpureus*: analysis of landraces cultivation and distribution, farming systems, and some climatic trends in production areas in Tanzania. Open Agric 8(1):20220156

Missanga JS, Venkataramana PB, Ndakidemi PA (2023b) *Lablab purpureus:* evaluation and selection of drought-tolerant: high-yielding accessions in dry farming systems based on drought tolerance indices and multi- environmental yield trials. J Agric Sci (Tarim Bilimleri Dergisi) 29(2):690–709

Mitra J (2001) Genetics and genetic improvement of drought resistance in crop plants. Curr Sci 80:758–763

Mkonda M, He X (2017) Conservation agriculture in Tanzania. Sustain Agric Res 22:310–324. https://doi.org/10.1007/978-3-319-48006-0

Mugerwa S, Sserumaga JP, Kayondo SI, Kiggundu M, Kigozi A, Namazzi C et al (2024) Genetic variation of nutritional quality traits in grains of Lablab (Lablab purpureus L. sweet) accessions: implications for forage improvement program. J Crop Sci Biotechnol 27(3):339–348

Mukeshimana G, Butare L, Cregan PB, Blair MW, Kelly JD (2014) Quantitative trait loci associated with drought tolerance in common bean. Crop Sci 54(3):923–938

Murphy AM, Colucci PE (1999) A tropical forage solution to poor quality ruminant diets: A review of Lablab purpureus. Livestock Research for Rural Development. http://www.cipav.org.co/lrrd11/2/colu.htm.

Nabateregga M, Mukankusi C, Raatz B, Edema R, Nkalubo S, Alladassi BME (2019) Quantitative trait loci (QTL) mapping for intermittent drought tolerance in BRB 191 × SEQ 1027 Andean intra- gene cross recombinant inbred line population of common bean (*Phaseolus vulgaris* L.). Afr J Biotechnol 18(21):452–461

Nadeem M, Li J, Yahya M, Sher A, Ma C, Wang X, Qiu L (2019) Research progress and perspective on drought stress in legumes: a review. Int J Mol Sci 20(10):1–32. https://doi.org/10.3390/ijms20102541

Naeem M, Shabbir A, Ansari AA, Aftab T, Khan MMA, Uddin M (2020) Scientia Horticulturae Hyacinth bean (*Lablab purpureus* L.): an underutilised crop with future potential. Sci Hortic 272(2020):109551

Nedumaran S, Abinaya P, Jyosthnaa P, Shraavya B, Rao P, Bantilan C (2015) Grain legumes production, consumption and trade trends in developing countries. Working Paper Series No 60. ICRISAT Research Program, Markets, Institutions and Policies. http://oar.icrisat.org/8991/1/2015-101WPS60.pdf

Ngure D, Kinyua M, Kiplagat O (2021) Evaluation of cooking time and organoleptic traits of improved Dolichos (*Lablab purpureus* (L.) sweet) genotypes. Afr J Food Sci 15(5):218–225. https://doi.org/10.5897/AJFS2021.2098

NIAS (2009) Plant genetic resources databases. http://www.gene.affrc.go.jp/databases-plant_search_en.php

Njaci I, Waweru B, Kamal N, Muktar MS, Fisher D, Gundlach H, Muli C, Muthui L, Maranga M, Kiambi D, Maass BL, Emmrich PMP, Entfellner JD, Spannagly M, Chapman MA, Shorinola O, Jones CS (2023) Chromosome-level genome assembly and population genomic resource to accelerate orphan crop lablab breeding. Nat Commun 14(1):1915. https://doi.org/10.1038/s41467-023-37489-7

Njonjo MW, Muthomi JW, Mwang'Ombe AW, Carozzi M (2019) Production practices, postharvest handling, and quality of cowpea seed used by farmers in Makueni and Taita Taveta counties in Kenya. Int J Agron 2019(1607535):1–12. https://doi.org/10.1155/2019/1607535

Nord A, Miller NR, Mariki W, Drinkwater L, Snapp S (2020) Investigating the diverse potential of a multi-purpose legume, *Lablab purpureus* (L.) sweet, for smallholder production in East Africa. PLoS One 15(1):1–19

Ojiewo C, Tenkouano A, Oluoch M, Yang R (2010) The role of AVRDC: the world vegetable centre in vegetable value chains. Afr J Hortic Sci 3:1–23

Ojiewo C, Monyo E, Desmae H, Boukar O, Mukankusi-Mugisha C, Thudi M, Pandey MK, Saxena RK, Gaur PM, Chaturvedi SK, Fikre A, Rao NG, Sameer Kumar C, Okori P, Janila P, Rubyogo JC, Godfree C, Akpo E, Omoigui L, Varshney RK (2019) Genomics, genetics and breeding of tropical legumes for better livelihoods of smallholder farmers. Plant Breed 138:487–499. https://doi.org/10.1111/pbr.12554

Onziga ID, Nkalubo ST, Kelly JD (2019) Identification of QTL associated with drought tolerance in Andean common bean. Crop Sci 59:1–14

Padulosi S, Hoeschle-Zeledon I (2004) Underutilized plant species: what are they? Leisa-Leusden 20:5–6

Pandey MK, Pandey AK, Kumar R, Nwosu CV, Guo B, Wright GC, Bhat RS, Chen X, Bera SK, Yuan M, Jiang H, Faye I, Radhakrishnan T, Wang X, Liang X, Liao B, Zhang X, Varshey R, Zhuang W (2020) Translational genomics for achieving higher genetic gains in groundnut. Theor Appl Genet 133(5):1679–1702. https://doi.org/10.1007/s00122-020-03592-2

Pretty J, Toulmin C, Wiliams S (2012) Sustainable intensification in African agriculture. Int J Agric Sustain 9(1):5–24

Purwanti E, Prihanta W, Fauzi E (2019) Nutritional content characteristics of Dolichos lablab L. accessions in effort to investigate functional food source. In: Advances in social science, education and humanities research, vol 349, pp 166–170

Purwanti E, Hermanto FE, Souhaly JW, Prihanta W, Permana TI (2021) Exploring public health benefits of Dolichos lablab as a dietary supplement during the COVID-19 outbreak: a computational study. J Appl Pharm Sci 11(02):135–140. https://doi.org/10.7324/JAPS.2021.110217

Qamar S, Manrique YJ, Parekh H, Falconer JR (2020) Nuts, cereals, seeds and legumes proteins derived emulsifiers as a source of plant protein beverages: a review. Crit Rev Food Sci Nutr 2019:1–21

Raghu BR, Samuel DK, Mohan N, Aghora TS (2018) Dolichos bean: an underutilized and unexplored crop with immense potential. Int J Adv Multidiscip Res 05(12):4338–4341

Rai KK, Rai N, Rai SP (2018a) Investigating the impact of high temperature on growth and yield of *Lablab purpureus* L. inbred lines using integrated phenotypical, physiological, biochemical and molecular approaches. Indian J Plant Physiol 23(2):209–226

Rai KK, Rai N, Rai SP (2018b) Recent advancement in modern genomic tools for adaptation of *Lablab purpureus* L to biotic and abiotic stresses: present mechanisms and future adaptations. Acta Physiol Plant 40:1–29. https://doi.org/10.1007/s11738-018-2740-6

Ramesh S, Byregowda M (2016) Dolichos bean (*Lablab purpureus* L. sweet, var. lignosus) genetics and breeding-present status and future prospects. Mysore J Agric Sci 50:481–500

Ramtekey V, Bhuriya A, Ayer D, Parekh V, Modha K, Kale B (2019) Molecular tagging of photoperiod responsive flowering in Indian bean [*Lablab purpureus* (L.) sweet]. J Genet 79(1):264–269. https://doi.org/10.31742/IJGPB.79S.1.17

Rana R, Sayem ASM, Sabuz AA, Rahman M, Hos-Sain A (2021) Effect of Lablab bean (Lablab purpureus L.,) seed flour on the physicochemical and sensory properties of biscuits. International Journal of Food Science and Agriculture 5:52–57.

Rangaiah DV, D'Souza M (2016) Hyacinth bean (*Lablab purpureus*): an adept adaptor to adverse environments. J Int Legume Soc 13:20–22

Ravelombola W, Shi A, Huynh B (2021) Loci discovery, network-guided approach, and genomic prediction for drought tolerance index in a multi-parent advanced generation intercross (MAGIC) cowpea population. Hortic Res 8:1–13

Ren H, Han J, Wang X, Zhang B, Yu L, Gao H, Hong H, Sun R, Tian Y, Qi X, Liu Z, Wu X, Qiu L (2020) QTL mapping of drought tolerance traits in soybean with SLAF sequencing. Crop J 8(6):977–989

Robotham O, Chapman M (2017) Population genetic analysis of hyacinth bean (*Lablab purpureus* (L.) sweet, Leguminosae) indicates an East African origin and variation in drought tolerance. Genet Resour Crop Evol 64(1):139–148

Schmutz J, McClean PE, Mamidi S, Wu GA, Cannon SB, Grimwood J, Jenkins J, Shu S, Song Q, Chavarro C, Torres-Torres M, Geffroy V, Moghaddam SM, Gao D, Abernathy B, Barry K, Blair M, Brick MA, Chovatia M, Jackson SA (2014) A reference genome for common bean and genome-wide analysis of dual domestications. Nat Genet 46(7):707–716. https://pubmed.ncbi.nlm.nih.gov/24908249/-full-view-affiliation-10

Schreinemachers P, Ebert AW, Wu M (2014) Costing theex situconservation of plant genetic resourcesat AVRDC: the world vegetable center. Genet Resour Crop Evol 61:757–773. https://doi.org/10.1007/s10722-013-0070-5

Sedlar A, Zupin M, Maras M, Razinger J, Šuštar-vozliˇ J, Pipan B, Megliˇc V (2020) QTL mapping for drought-responsive agronomic traits associated with physiology, phenology, and yield in an Andean intra-Gene Pool common. Agronomy 10:1–18
Sennhenn A, Njarui DMG, Maass BL, Whitbread AM (2017) Exploring niches for short-season grain legumes in semi-arid eastern Kenya: coping with the impacts of climate variability. Front Plant Sci 8(699):1–17
Serba DD, Yadav RS (2016) Genomic tools in pearl millet breeding for drought tolerance: status and prospects. Front Plant Sci 7:1–10
Shaahu DT, Kaankuka FG, Okpanachi U (2015) Proximate, amino acid, anti-nutritional factor and mineral composition of different varieties of raw *Lablab purpureus* seeds. Int J Sci Technol Res 4(04):157–161
Shiferaw B, Tesfaye K, Kassie M, Abate T, Prasanna BM, Menkir A (2014) Managing vulnerability to drought and enhancing livelihood resilience in sub-Saharan Africa: Technological, insti tutional and policy options. Weather and Climate Extremes 3:67–79. https://doi.org/10.1016/j.wace.2014.04.004
Shivashankar G, Kulkarni RS (1989) Lablab purpureus. In: van der Maesen LJG, Somaatmadja S (eds) Plant resources of South-East Asia (PROSEA). No. 1. Pulses. Pudoc, Wageningen, pp 48–50
Singh A, Abhilash PC (2019) Varietal dataset of nutritionally important *Lablab purpureus* (L.) sweet from Eastern Uttar Pradesh, India. Data Brief 24:103935. https://doi.org/10.1016/j.dib.2019.103935
Singh VK, Saxena RK, Varshney RK (2017) Sequencing Pigeonpea Genome. In: Varshney R, Saxena R (eds) The Pigeonpea genome. Compendium of plant genomes. Springer, Cham. https://doi.org/10.1007/978-3-319-63797-6_9
Sivasakthi K, Thudi M, Tharanya M, Kale SM, Kholová J, Halime MH, Jaganathan D, Baddam R, Thirunalasandari T, Gaur PM, Varshney RK, Vadez V (2018) Plant vigour QTLs co-map with an earlier reported QTL hotspot for drought tolerance while water saving QTLs map in other regions of the chickpea genome. BMC Plant Biol 18:29. https://doi.org/10.1186/s12870-018-1245-1
Sreenivasa V, Lal SK, Babu PK, Swamy HKM, Yadav RR, Talukdar A, Rathod DR (2020) Inheritance and mapping of drought tolerance in soybean at seedling stage using bulked segregant analysis. In: Plant genetic resources: characterization and utilization, pp 1–8. https://doi.org/10.1017/S1479262120000052
Sserumaga JP, Kayondo SI, Kigozi A, Kiggundu M, Namazzi C, Walusimbi K, Bugeza J, Molly A, Mugerwa S (2021) Genome-wide diversity and structure variation among lablab [Lablab purpureus (L.) sweet] accessions and their implication in a forage breeding program. Genet Resour Crop Evol 68:2997–3010
Sultana N, Ozaki Y, Okubo H (2001) Early identification of determinate growth habit in lablab bean (*Lablab purpureus* (L.) sweet). J Fac Agric Kyushu Univ 46(1):31–38
Susmitha B, Ramesh S (2023) Identification of indices for empirical selection of dolichos bean [*Lablab purpureus* (L.) var. Lignosus] genotypes for tolerance to terminal moisture stress. Legume Res 46(6):695–699
Talebi R, Fayaz F, Naji AM (2009) Effective selection criteria for assessing drought stress tolerance in durum wheat (*Triticum durum* Desf.). Gen Appl Plant Physiol 35(1–2):64–74
Tian Y, Zeng Y, Zhang J, Yang CG, Yan L, Wang XJ, Shi CY, Xie J, Dai TY, Peng L, Huan YZ, Xu A, Huang Y, Zhang J, Ma X, Dong Y, Hao S, Sheng J (2015) High quality reference genome of drumstick tree (*Moringa oleifera* Lam.), a potential perennial crop. Sci China Life Sci 58(7):627–638. https://doi.org/10.1007/s11427-015-4872-x
UN (2019) World population prospects. https://population.un.org/wpp/
UNCC (2020) Climate change is an increasing threat to Africa. In United Nations climate change-news report on 27 October, 2020. Retrieved from https://unfccc.int/news/climate-change-is-an-increasing-threat-to-africa?
Vaijayanthi PV, Ramesh S, Gowda MB, Rao AM, Keerthi CM (2018) Genome-wide marker-trait association analysis in a core set of Dolichos bean germplasm. Plant Genet Resour 17(1):1–11

Vaijayanthi PV, Chandrakant Ramesh S (2019) Hyacinth bean (Lablab purpureus L. Sweet): Genetics, breeding and genomics. Advances in Plant Breeding Strategies: Legumes 7:287-318.

Valliyodan B, Lee SH, Nguyen HT (2017) Sequencing, assembly, and annotation of the soybean genome. In: Nguyen H, Bhattacharyya M (eds) The soybean genome. Compendium of plant genomes. https://doi.org/10.1007/978-3-319-64198-0_5

Venkatesha SC, Byregowda M, Mahadevu P, Mohan Rao A, Kim DJ, Ellis THN, Knox MR (2007) Genetic diversity within *Lablab purpureus* and the application of gene-specific markers from a range of legume species. Plant Genet Resour 5(3):154–171

Venkatesha SC, Ganapathy KN, Byre Gowda M, Ramanjini Gowda PH, Mahadevu P, Girish G, Ajay BC (2013) Variability and genetic structure among lablab bean collections of India and their relationship with exotic accessions. Vegetos 26(Special):121–130. https://doi.org/10.5958/j.2229-4473.26.2s.130

Vidigal V, Duarte B, Cavaco A, Caçador I, Figueiredo A, Mato A, Viegas W, Monteiro F (2018) Preliminary diversity assessment of an undervalued tropical bean (*Lablab purpureus* (L.) sweet) through fatty acid profiling. Plant Physiol Biochem 134:508–514. https://doi.org/10.1016/j.plaphy.2018.10.001

Villordo-Pineda E, González-Chavira MM, Giraldo-Carbajo P, Acosta-Gallegos JA, Caballero-Pérez J (2015) Identification of novel drought-tolerant-associated SNPs in common bean (*Phaseolus vulgaris*). Front Plant Sci 6:546. https://doi.org/10.3389/fpls.2015.00546

VIR (2009) Passport data of germplasm collections. N. I. Vavilov All-Russian Scientific Research Institute of Plant Industry, St. Petersberg

Wallace TC, Murray R, Zelman KM (2016) The nutritional value and health benefits of chickpeas and hummus. Nutrients 8(12):766

Wang B, Zhao M, Yao L, Joao MS, Babu V, Wu T, Nguyen HT (2018) Identification of drought-inducible regulatory factors in Lablab purpureus by a comparative genomic approach. Crop Pasture Sci 69:632–641

Wang X, Chen Y, Yang K, Duan F, Liu P, Wang Z, Wang J (2021) Effects of legume intercropping and nitrogen input on net greenhouse gas balances, intensity, carbon footprint and crop productivity in sweet maize cropland in South China. Journal of Cleaner Production 314(2021):127997. https://doi.org/10.1016/j.jclepro.2021.127997

Wangila AJ, Gachuiri CK, Muthomi JW, Ojiem JO (2021) Biomass yield and quality of fodder from selected varieties of Lablab (*Lablab purpureus*) in Nandi south sub-county of Kenya. Online J Anim Feed Res 11(1):28–35. https://doi.org/10.51227/ojafr.2021.6

World Bank (2017) Climate-smart agriculture in Tanzania. CSA Country Profiles for Africa Series. https://ccafs.cgiar.org/resources/publications/climate-smart-agriculture-tanzania

Xu P, Moshelion M, Wu X, Halperin O, Wang B, Luo J, Wallach R et al (2015) Natural variation and gene regulatory basis for the responses of asparagus beans to soil drought. Frontiers in Plant Sci 6:1–14.

Yang K, Tian Z, Chen C, Luo L, Zhao B, Wang Z, Yu L, Li Y, Sun Y, Weiyu L, Chen Y, Li Y, Zhang Y, Ai D, Zhao J, Shang C, Ma Y, Wu B, Wang M et al (2015) Genome sequencing of adzuki bean (*Vigna angularis*) provides insight into high starch and low fat accumulation and domestication. Proc Natl Acad Sci USA 112:13213–13218

Young ND, Zhou P (2019) The sequenced genomes of Medicago truncatula. In: The model legume Medicago Truncatula, vol 2019, pp 828–834

Young ND, Cannon SB, Sato S, Kim D, Cook DR, Town CD, Roe BA, Tabata S (2005) Sequencing the genespaces of *Medicago truncatula* and *Lotus japonicus*. Plant Physiol 137(4):1174–1181. https://doi.org/10.1104/pp.104.057034

Breeding Advancements in Neglected Legume Species for Drought Tolerance

15

Suman Dutta, Debashis Paul, Sampa Saha, and Sandeep Bagri

Abstract

In the realm of agricultural research, agricultural practices delve into the innovative strategies aimed at developing the drought resilience of overlooked legume varieties, shedding light on their historical neglect and the imperative for transformative interventions. Drawing upon a rich tapestry of global and regional insights, it navigates through the intricate landscape of neglected legumes, laying the groundwork for a comprehensive examination of breeding advancements in enhancing their adaptive capacities. This chapter delves into the breeding advancements aimed at enhancing the drought tolerance of neglected legume species, addressing their underutilisation and limited research focus. It begins with an exploration of neglected legumes, elucidating their origin, chromosome number, distribution, and nutritional quality, while also examining the global scenario and the influence of initiatives like the International Year of Pulses 2016. Specific attention is given to the Indian context, detailing species-wise production and associated challenges. The effect of drought and heat stress on neglected legume production is discussed, underlining the urgency for innovative breeding strategies. Conventional breeding approaches for drought tolerance are examined, paving way for the exploration of advanced methods, particularly OMICS interventions. Genomics, transcriptomics, and metabolomics approaches are evaluated for their efficacy in enhancing drought tolerance in neglected

S. Dutta (✉)
School of Agriculture and Rural Development, Ramakrishna Mission Vivekananda Educational and Research Institute, Howrah, West Bengal, India

D. Paul · S. Bagri
ICAR-Central Institute for Cotton Research, Regional Station, Sirsa, Haryana, India

S. Saha
ICAR-National Bureau of Plant Genetic Resources, New Delhi, India

© The Author(s), under exclusive license to Springer Nature Singapore Pte Ltd. 2025
M. A. Nadeem et al. (eds.), *Marker-Assisted Breeding in Legumes for Drought Tolerance*, https://doi.org/10.1007/978-981-96-4112-3_15

legumes, with an illustrated overview of these advanced breeding approaches showcasing their interconnectedness and synergy. In conclusion, the abstract emphasises the necessity for intra-governmental and public-private partnerships to foster improvements in neglected legumes, highlighting their transformative potential in addressing global issues pertaining to food security and climate change.

Keywords

Drought tolerance · Breeding advancements · OMICS interventions · Climate resilience · Food security

15.1 Introduction

Legumes belonging to Fabaceae or Leguminaceae family are the third largest among all cultivated crop species worldwide. Legumes are regarded as the second-most pertinent plant resource for human and animal dietetics, after cereals (Bhat and Karim 2009), especially in drought and famine conditions (Haq 2002). However, some of these species are neglected or underutilised. Legume crops that are not cultivated commercially and are not traded extensively are considered neglected. These are often minor crops of regional significance that are disregarded by industry and researchers because of their low economic significance in the global marketplace (Foyer et al. 2016). However, they have the potential to contribute to food security, health, revenue creation, environmental services, and sustainable agriculture (Jaenicke and Höschle-Zeledon 2006). They may be targeted for desired qualities with enhanced nutrition and tolerance to stress (Kumar and Bhalothia 2020). The neglected legumes are mostly unknown and confined to regions with particular cultural or historical importance. Despite improvements in modern medicine, substantial population, particularly in developing nations, still rely on these crops for their daily medical requirements in addition to being used as food and feed (Bagri and Janeja 2023). They frequently occupy ecological niches unoccupied by major crop plants, increasing genetic diversity and flexibility (Foyer et al. 2016). Due to the overexploitation of non-renewable natural resources, it is essential to introduce diversification in agriculture. Thus, the current progressive goals of sustainability emphasise agricultural diversification with neglected legumes (Samal et al. 2023).

The aim of this chapter is to provide a comprehensive overview of neglected legumes, focusing on their significance and potential in agriculture and sustainability. This chapter will begin with a brief description of some neglected legume species, detailing their unique characteristics and regional importance. It will then explore the global scenario of neglected legumes, highlighting their distribution and significance across different regions. This chapter will also address the problems associated with the production of these legume species, identifying the challenges that hinder their widespread cultivation and commercialisation. Furthermore, it will review conventional breeding approaches aimed at enhancing drought tolerance in neglected legumes and emphasise the necessity of genome sequencing to unlock the genetic potential of underutilised pulses. The importance of quantitative trait mapping in identifying and developing desirable traits in these legumes will also be discussed. Finally, the chapter will highlight the role of wild relatives as reservoirs of novel traits that can be utilised to improve the resilience and nutritional value of neglected legumes. By addressing these key points, the chapter aims to advocate for increased research and attention toward neglected legumes to enhance agricultural diversification and sustainability.

15.2 Importance of Neglected Legumes

Neglected legumes are a crucial source of micronutrients, proteins, and vitamins, making them an essential component of the human diet, especially in regions where food security is a concern (Samal et al. 2023). These legumes are rich in essential micronutrients such as iron, zinc, and magnesium, which are vital for various bodily functions, including immune system support and cognitive development (Samal et al. 2023). Additionally, the high protein content of legumes provides a significant dietary protein source, particularly important in areas where animal protein is scarce or expensive. Vitamins, including B vitamins and folate, found in legumes contribute to overall health and well-being, supporting energy production and cell metabolism. Moreover, legumes contain anti-nutrients like phytic acids, tannins, polyphenols, saponins, protease inhibitors, and amylase inhibitors, which play a role in increasing the disease resistance of plants (Pati et al. 2019). These compounds can deter pests and pathogens, contributing to the plant's defense mechanisms and reducing the need for chemical pesticides. While anti-nutrients can interfere with nutrient absorption in humans, traditional processing methods such as soaking, fermenting, and cooking can reduce these effects, making legumes a nutritious and safe food source. The symbiotic relationship between legumes and nitrogen-fixing bacteria is another significant benefit. Through this natural process, legumes improve soil fertility, reducing the need for synthetic fertilizers and promoting sustainable agricultural practices. This soil-enrichment feature is particularly valuable in low-input farming systems, where resources are limited. Neglected legumes are also utilised in the production of environmentally friendly products, animal feed, and natural manures (Hughes and Ebert 2011). Their biomass can be used to create organic fertilizers, enhancing soil health and productivity without harming the environment. Additionally, legumes serve as a valuable source of animal feed, providing livestock with essential nutrients and promoting sustainable livestock production. The versatility and ecological benefits of neglected legumes make them an important resource for promoting food security, environmental sustainability, and agricultural diversification (Singh et al. 2007; Hughes and Ebert 2011; De la Peña et al. 2011; Jain and Gupta 2013).

15.3 Brief Description of Some Neglected Legumes

The neglected legumes are renowned for their nutritional benefits mostly in their localised niches. The winged bean is a perennial legume plant with immature seeds, broad green pods, leaves, and tubers and often referred to as the "One Species Supermarket" (Tanzi et al. 2019). Azuki bean, also known as little red beans (Gohara et al. 2016), is the second most significant legume. Sword bean also known as jack or beach beans have sword blade-like pods (Luo et al. 2015). Scarlet runner bean is utilised for food consumption (green and dried pods) and as ornamental due to its red flower varieties (Łabuda 2010). Yam bean is known for its delicious and firm tubers (Kumalasari et al. 2014). Potato bean is a tuberous legume, consumed both as raw and cooked. Drought-tolerant legume is tepary bean used for domestic consumption. Agathi is green manuring multipurpose crop with good fibre quality. Velvet bean and moth bean are the potential reserve of high protein. Velvet bean also known as itching plant due to mucunain. Horse gram is consumed by humans (dal or rasam) and is also used as animal feed. It is generally grown in spaces between citrus plants. Yardlong bean is basically grown for its long pods that are delicious stringless, consumed as raw green beans. Similarly, African yam bean is grown for its seeds and tubers. Ground bean is hardy crop adaptable to harsh conditions (Savita 2023b). Table 15.1 presents a comprehensive overview of underutilised legume crops, detailing their species, scientific names, chromosome numbers, origins, distribution, nutritional quality, and associated health benefits.

15.4 Global Scenario of Neglected Legumes

Neglected legumes are mostly found in tropical parts in terms of global distribution. On the other hand, some species are extremely adaptive to different ecogeographical situations, allowing them to survive or be less altered by variations in environments (Popoola et al. 2019). Adzuki bean, winged bean, and sword bean are widely distributed in humid tropics and subtropics of South and Southeast Asia (Kaga et al. 2008; Moteetee 2016; Lepcha et al. 2017). Jack bean, potato bean, moth bean, scarlet runner bean, catjang bean, and velvet bean are cultivated in the tropics and subtropics of Asia, America, Africa, and the Pacific (Li et al. 2014; Tiamiyu et al. 2016; Kochhar 2016). Yam bean is widely cultivated in Hawaii, China, Singapore, India, the Philippines, Mexico, Nepal, Indonesia, and Burma. Tepary bean is distributed in Australia, Asia, Africa, and Madagascar. Agathi is widely spread from South America to Southeast Asia, mainly India. Africa, Australia, Burma, India, Nepal, Malaysia, Mauritius, and Sri Lanka are among the countries that cultivate horse gram for food (Asha et al. 2006). Yardlong bean is extensively cultivated in Africa and southeast Asia, basically in India for its pods. African locust beans, ground beans, and African yam beans are limited to tropical regions of the African continent (Akande et al. 2010; Savita 2023a). Marama beans are being grown in Africa, Australia, Israel, and America (Uzabakiriho 2016). Therefore, neglected legumes,

Table 15.1 A list of underutilised legume crops with species-wise origin, chromosome number, distribution and nutritional quality along with health benefits

Species	*Scientific name*	Chromosome no.	Origin	Nutritional quality	References
Winged bean	*Psophocarpus tetragonolobus* (L.)	18	India	34.3–40.7% protein, 31% carbohydrate, 19.1% oil	Vatanparast et al. (2016) and Martín-Cabrejas (2019)
Ground bean or bambara bean	*Vigna subterranea* (L.) Verdc.	22	West Africa	15–25% protein, 72% carbohydrate; 5.2–6.4% oil content	Foyer et al. (2016) and Samal et al. (2023)
Jack bean	*Canavalia ensiformis* (L.)	22	West Indies	36% protein, 41.8% carbohydrate, 2% fat	Tiamiyu et al. (2016)
Moth bean	*Phaseolus aconitifolius*	22	India, Pakistan, and Myanmar	27% protein, 71% carbohydrate, 2% fat	Kochhar (2016) and Diakostefani et al. (2024)
Scarlet runner bean	*Phaseolus coccineus*	22	Mexico and Central America	21% protein, 55% carbohydrate, dietary fibre (9 g)	Guerra-García et al. (2017)
Yam bean	*Pachyrhizus erosus* Urbn.	22	River Amazon of Tropical America	32% protein, 34% carbohydrate, 26% fat, dietary fibre (7.5 g)	Pati et al. (2019)
Potato bean	*Apios americana*	22	North America	25–30% protein	Belamkar et al. (2016)
Tepary bean	*Posphocarpus acutifolius*	22	Northern Mexico and the United States (Sonoran Desert)	25% protein, 38% carbohydrate, dietary fibre (11.5 g)	Bhardwaj and Hamama (2004) and Emmrich et al. (2020)

(continued)

Table 15.1 (continued)

Species	*Scientific name*	Chromosome no.	Origin	Nutritional quality	References
Adzuki bean	*Vigna angularis* (Wild.) Wright	22	Tropical Asia (China) and Africa	27% protein, 56% carbohydrate, dietary fibre (7.3 g)	Foyer et al. (2016) and Savita (2023a)
Sword bean	*Canavalia gladiata*	22, 24	Indo-Malayan (primary) and Tropical Africa (secondary)	31.8% protein, 5.20% oil content, 70.6% carbohydrate	Moteetee (2016) and Diakostefani et al. (2024)
Yardlong bean or asparagus bean or cowpea	*Vigna unguiculata* ssp. sesquipedalis	22	Southeast Asia	17% protein, 50% carbohydrate, dietary fibre (5 g)	Foyer et al. (2016) and Suma et al. (2021)
Agathi	*Sesbania grandiflora* Poir.	24	India or Indonesia	34% protein, 49% carbohydrate, 4% fat, dietary fibre (2 g), calcium (1.13 g), phosphorus (0.08 g), iron (0.004 g)	Karmakar et al. (2016)
Catjang bean	*Vigna unguiculata* var. *cylindrical* or *biflora*	22	India or Africa	25% protein, 63% carbohydrate, dietary fibre (10.7 g)	Zuluaga et al. (2021)
Red clover	*Trifolium pralense*	14	European origin	Protein (127.9–208.4 g/kg), digestible fibre (379.2 to 594.6 g/kg)	Asci (2012)
Horse gram	*Dolichus uniflorus* Lam. (*Dolichus biflorus* Auct. L.)	22, 24	India	20.06–24.22% protein, 57.2% carbohydrate, 0.33–0.75% oil content	Savita (2023a)
African yam bean	*Sphenotylis stenocarpa*	22	Africa	1% protein, 89% carbohydrate, 9% fat	Waweru et al. (2023)

(continued)

Table 15.1 (continued)

Species	*Scientific name*	Chromosome no.	Origin	Nutritional quality	References
African locust bean	*Parkia biglobosa*	24	West Africa	6.56% protein, 30–40% carbohydrate, 1.80% oil content	Uyoh et al. (2011)
Cluster bean	*Cyamopsis tetragonoloba*	14	West Africa, India	Crude ash (5.29 to 8.30%), oil (0.6 to 0.8%), fibre (14 to 16%	Akçura et al. (2020)
Velvet bean	*Mucuna deeringiana* (Bor.) Merr.	22	South Asia	39% protein, 59% carbohydrates, 6.3–7.4% lipid	Hao et al. (2022)
Marama bean	*Tylosema esculentum* (Burchell) Schreiber	44	South Africa	34.71% protein, 34 g oil content of 100 g seed	Li and Cullis (2023)
Common vetch	*Vicia sativa*	14	Middle East	Protein (283.63 g/kg), carbohydrates (435 g/kg), fat (15.8 g/kg)	Huang et al. (2017)
White lupin	*Lupinus albus*	50	Mediterranean region	water (71.1 g), protein (15.6 g), carbohydrates (9.88 g), fibre (2.8 g), lipids (2.92 g)	Pereira et al. (2022)
Grass pea	*Lathyrus sativus*	14	Central Asia, Abyssinia origin	Protein (22.6–28.1%), carbohydrates (51.8–58.5%), fat (1.0–1.6%), fibre (7.2–8.3%)	Urga et al. (1995), Foyer et al. (2016) and Samal et al. (2023)
Broad bean/ Faba bean	*Vicia faba*	12	Europe and Asia	20–41% protein, 1.2–1.9% oil content	Rana and Vilas (2017)

predominantly found in different ecogeographical regions, exhibit remarkable adaptability to various ecogeographical conditions.

15.5 Impact of International Year of Pulses 2016 on Neglected Legumes

The UN Food and Agriculture Organization declared 2016 to be the International Year of Pulses (IYP) to promote links across the food chain that would enhance the use of proteins derived from pulses, increase global production of pulses, improve crop rotation efficiency, and address trade issues (Foyer et al. 2016; Murrell 2016). With a focus on food security and nutrition and sustainable food production, the goal of IYP 2016 is to raise public knowledge of the nutritional advantages of pulse crops, which have greater protein content than other crops (Amanullah 2016). The primary initiatives of the plan were global and regional awareness campaigns via online forums, the Web, and social media; consultations with policymakers, and experts on pulses, producer and trade organisations; creation of documents and scientific studies (Sherasia et al. 2018). This plan has considerably boosted the level of cooperation and partnerships among significant players and raised the profile and knowledge of pulses across all segments of the population (Calles et al. 2019). The International Year of Pulses 2016 significantly heightened global awareness of neglected legumes, spotlighting their nutritional benefits, agricultural resilience, and potential to enhance food security. This initiative, led by the United Nations, aimed to educate the public and policymakers about the diverse and underutilised legume species that are rich in proteins, fibres, vitamins, and minerals. The campaign promoted the inclusion of these crops in diets worldwide, highlighting their role in sustainable agriculture due to their nitrogen-fixing properties, which improve soil health. Additionally, it encouraged research and development efforts to improve cultivation practices and yield. As a result, the year-long focus spurred greater interest in integrating these legumes into mainstream agriculture and diets, contributing to better health outcomes and environmental sustainability.

15.6 Species-Wise Area and Production in India

The cultivation of various legume species across India varies significantly in terms of area and production, reflecting the country's diverse agro-climatic zones and agricultural practices. Only Kerala, Tamil Nadu, Karnataka, Maharashtra, Goa, Orissa, West Bengal, Tripura, Mizoram, and Manipur are home to winged beans in India (Chandel et al. 1984). Many states, including Tamil Nadu, Andhra Pradesh, Kerala, Assam, Gujarat and Bengal, cultivate 'agathi'. It may spread like weeds and thrives well in hot, humid climates (Karmakar et al. 2016). The sword bean is used as fodder and is widely grown in Northern and Peninsular India, as well as in the Eastern and Western Ghats of South India. The primary distribution regions for adzuki beans are high-altitude regions in Nagaland, Arunachal Pradesh, and

Mizoram (Bhatt et al. 2021). Yam bean is widely cultivated in Odisha, Bihar, Jharkhand, Eastern Uttar Pradesh, West Bengal, Tripura, and Assam. Jack bean is found in southern areas of India (Vadivel and Janardhanan 2001). Moth bean is majorly cultivated in Rajasthan and then in Haryana, Punjab, Jammu and Kashmir, Gujarat, Maharashtra, Madhya Pradesh, and Uttar Pradesh (Kumar 2002).

15.7 Problems Associated with the Production of Neglected Legume Species

Even with the advent of advanced agricultural practices, the production of neglected legumes faces numerous challenges. The Food and Agriculture Organization (FAO) highlighted several issues in 1998 that continue to hinder the development of these valuable crops. One major problem is the restricted availability of germplasm or seed material. This limitation makes it difficult for farmers to access diverse and high-quality seeds necessary for cultivating robust legume varieties that can withstand environmental stresses and diseases (Dickstein 2019). Inadequate cultivation knowledge also poses a significant barrier. Many farmers lack the necessary information on optimal growing practices for neglected legumes, including planting techniques, pest management, and harvest timing. This knowledge gap leads to suboptimal yields and lower crop quality, discouraging farmers from investing in these crops. Additionally, there is little interest and limited policy support for the development of neglected legumes (Dickstein 2019). Agricultural policies and research funding often prioritise major crops like wheat, rice, and maize due to their high economic value and global demand. As a result, neglected legumes receive minimal attention from researchers and policymakers, stunting their development and potential. Distribution and marketing support for neglected legumes are also insufficient. Without established supply chains and market networks, farmers struggle to sell their produce at fair prices. This lack of market access reduces the profitability of growing neglected legumes, further disincentivising farmers from cultivating them. Furthermore, these crops receive little recognition at the global level. Neglected legumes are often overshadowed by more widely known and commercially successful crops, leading to a lack of awareness about their nutritional and ecological benefits (Savita 2023a). This lack of global recognition means, fewer international initiatives, and collaborations aimed at promoting their cultivation and use. In conclusion, the production of neglected legumes is hindered by restricted germplasm availability, inadequate cultivation knowledge, limited policy support, insufficient distribution and marketing infrastructure, and a lack of global recognition. Addressing these challenges through targeted research, policy interventions, and awareness campaigns is essential to unlock the potential of neglected legumes in contributing to food security and sustainable agriculture (Williams and Haq 2000; Savita 2023a).

15.8 Impact of Drought and Heat Stress

Drought and heat stress significantly impact the growth, yield, and nutritional quality of neglected legumes, which are crucial for food security and sustainable agriculture in arid and semi-arid regions (Khatun et al. 2021). These abiotic stresses lead to a reduction in leaf area, shoot and root growth, chlorophyll content, stomatal conductance, and water-use efficiency, ultimately resulting in lower yields (Khatun et al. 2021). The physiological responses to these stresses include the accumulation of reactive oxygen species (ROS), which cause cellular damage and impair metabolic functions (Fig. 15.1). Additionally, heat and drought stress during critical growth stages such as flowering and pod development can severely disrupt reproductive processes, further diminishing yield. Despite these challenges, legumes exhibit various adaptive mechanisms, such as altered gene expression, improved root architecture, and enhanced antioxidant defense systems, which can be harnessed through breeding and agronomic practices to develop more resilient crops (Popoola et al. 2022). These strategies are essential for mitigating the adverse effects of climate change on legume production and ensuring their contribution to global food security. Despite significant advances in understanding the effects of drought and heat stress on neglected legumes, several research gaps remain. Firstly, there is a need for more comprehensive studies on the physiological and molecular mechanisms that underlie stress tolerance in a wider range of neglected legume species. Most existing research focuses on well-known legumes like chickpeas and lentils, while many underutilised species remain understudied (Farooq et al. 2018; Karalija et al. 2022). Secondly, the interaction effects of simultaneous drought and heat stress on these legumes are not well understood (Awasthi et al. 2014). Most studies examine these stresses in isolation, which does not reflect the combined stresses plants often experience in the field. Additionally, there is limited information on the

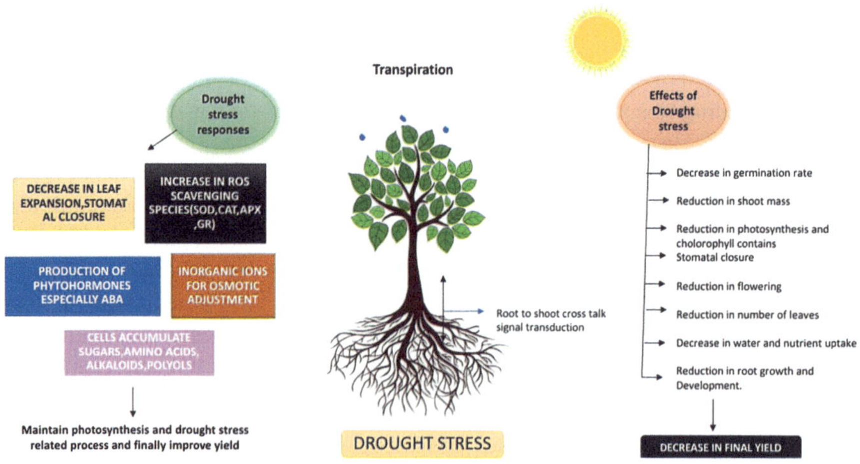

Fig. 15.1 Impact of drought and heat stress on production of neglected legumes

role of root architecture and below-ground traits in conferring stress tolerance, despite their critical importance in water and nutrient uptake under stress conditions . Finally, while some agronomic and biotechnological approaches have been proposed to mitigate stress effects, their practical implementation and effectiveness in diverse agroecological zones need further validation through field trials and farmer-participatory research. Addressing these gaps is crucial for developing robust strategies to enhance the resilience of neglected legumes to climate change.

Even though limited research was done on neglected legumes, significant pathways related to drought and heat response have been studied to develop adaptive strategies to overcome drought. Neglected legumes, despite being under-researched, hold immense potential for future agriculture due to their remarkable tolerance to harsh environments. Table 15.2 details the impact of drought and heat stress on the growth parameters of various neglected legumes, listing specific crops affected and referencing relevant studies.

Hailed as a "climate-smart crop," the winged bean (*Psophocarpus tetragonolobus*) thrives in drought conditions, making it a valuable addition to agricultural systems in arid and semi-arid regions. Lone et al. (2022) pinpointed heat-shock proteins (Hsps) and stress-tolerant metabolites in the root tubers of winged beans, which act as defense mechanisms against drought stress. These proteins and metabolites help stabilise cellular structures and mitigate the damaging effects of reactive oxygen species (ROS), thereby enhancing the plant's ability to withstand prolonged periods of water deficiency. Heat stress is the primary culprit behind yield loss in lima beans, particularly affecting its pollination process (Ernest and Wisser 2024). Night-time heat disrupts pollen viability and development, leading to reduced seed production. Bambara groundnut showcases all three drought-coping mechanisms such as avoidance, escape, and tolerance under adverse environments (Khan et al. 2022). Studies predict significant yield and water productivity increases for Bambara

Table 15.2 Impact of drought and heat stress on growth parameters in neglected legumes

S.n.	Crop	Impact	References
1	Winged bean	Increased heat-shock proteins (Hsps)	Lone et al. (2022)
2.	Lima bean	Reduced photosynthetic ability, reduced yield	Ernest and Wisser (2024)
3.	Bambara groundnut	Reduced stomatal conductance and shortened leaf length; activation of stress avoidance mechanisms	Chai et al. (2016) and Khan et al. (2022)
5.	Marama bean	Leaf reduction, stem elongation reduction, and reduced taproot development	Cullis et al. (2018)
6.	Horse gram	Valine degradation and gluconeogenesis, pathways such as glucosinolate biosynthesis, flavonoid biosynthesis, and sucrose degradation	Bhardwaj et al. (2013)
7.	Lablab beans	Altered stomatal behaviour	Kokila et al. (2014)
8.	Adzuki bean	Reduced growth and development in early stages, limited root and shoot growth	Chun et al. (2021)
9.	*Canavalia rosea*	41 *CrHsp20* genes upregulated	Zhang et al. (2022)

groundnut under anticipated climate shifts in South Africa (Soumare et al. 2022; Mayes et al. 2019). The marama bean boasts a unique tuberous root system that stores water and nutrients, enabling it to survive in scorching temperatures of up to 50 °C (Cullis et al. 2018). This adaptation allows the marama bean to thrive in extremely arid conditions, where other crops might fail. The water-storing capability of its tubers provides a crucial buffer during periods of drought, ensuring the plant's survival and sustained growth. It can even undergo complete dieback during cooler periods (Cullis et al. 2018). Additionally, multiple adaptations like reduced leaf area, deeper taproots, and stomatal closure minimise water loss during dry spells. Horse gram utilises various pathways like valine degradation and glucosinolate biosynthesis to combat drought stress (Bhardwaj et al. 2013). Transcription factors also play a crucial role in its drought tolerance. Studies by Kokila et al. (2014), Grotelüschen et al. (2014), and Guretzki and Papenbrock (2014) in Lablab beans highlight the diverse morphological responses of lablab bean to drought stress at different growth stages. While valued for its protein content, drought stress significantly impacts its growth and yield in Adzuki bean, especially during early stages (Wang et al. 2018; Chun et al. 2021). Aquaporins (AQPs) are believed to play a role in its stress tolerance. *Canavalia rosea* thrives in high-temperature environments. Zhang et al. (2022) identified heat shock protein 20 (Hsp20) genes as crucial for cellular protection during heat stress. These genes were significantly upregulated, suggesting their role in adaptation. Neglected legumes, despite being relatively unexplored, showcase a remarkable arsenal of mechanisms to combat drought and heat stress. Further research into these underutilised powerhouses holds immense promise for developing climate-resilient crops for future generations.

15.9 Conventional Breeding Approaches for Drought Tolerance

In terms of architectural traits, maturity duration, crop production, and nutritional composition, the genetic enhancement and breeding endeavours aimed at improving underutilised and neglected leguminous crops have fallen short of anticipated success (Gedil et al. 2016). Despite efforts involving traditional hybridisation and other breeding techniques targeting specific traits, the desired outcomes have not been fully realised. Limited successful crosses have been recorded for underutilised legumes, with notable challenges documented in *Cajanus cajan* and certain species (Hickey et al. 2017). Sinha et al. (2020) identified 17 pigeonpea (*Cajanus cajan* L.) accessions with beneficial haplotypes in three drought-responsive genes, which are useful for breeding drought-tolerant varieties. Evaluation by genotype-by-environment interaction (GEI) helped in pigeonpea crop improvement programmes to identify and recommend high-performing and stable genotypes for breeding or production purposes against abiotic stresses (Yohane et al. 2021). Development of core using Geographic and Qualitative Morphological Descriptors could help significantly in selection of parental lines in further development of drought-resistant lines (Reddy et al. 2005). In the case of African yam bean, successful breeding lines have yet been reported (Bekkering and Tian 2019). However, initial-level studies such as morphometric analysis and characterisation (Aremu et al. 2019) and yield stability (Aremu et al. 2020) have a potential in crop improvement programme regarding drought tolerance. Interaction due to genotype × water stress levels have been studied in Dolichos beans for studying drought (Dhanuja 2024). Study by Guretzki and Papenbrock (2014) compared *L. purpureus* genotypes to identify those best suited to evolving climate conditions. Four genotypes were assessed for drought tolerance using conventional growth indicators, infrared thermal imaging, and stomatal responses. Variability in drought response was observed among genotypes at different growth stages, with inconsistent rankings across methods. Trypsin inhibitor activity and cyanogenic potential (HCNp) also exhibited genotype variations, independent of each other. To guide future breeding programmes, it is recommended to employ multiple methods for drought assessment for selection of genotypes. Studies by Missanga et al. (2023) used three indices that were shown to be useful tools for selecting drought-tolerant lines: the stress tolerance index, the geometric mean productivity index, and the mean productivity index with stronger correlation coefficients. A collection of techniques compiled for assessing drought tolerance in plants was used by Guretzki and Papenbrock (2013), where methods encompass traditional growth metrics such as measures of plant height, dry weight biomass, leaf size, and plant water content (PWC). Additionally, stomatal behaviour is evaluated through leaf conductance and infrared thermal imaging, completing the toolkit for identifying drought tolerance. The outcomes from these methods are used to establish a ranking of the tested genotypes based on their drought tolerance levels in Lablab beans. Selection criteria for identifying drought-tolerant cultivars of winged beans (*Psophocarpus tetragonolobus*) in arid zones prioritise traits such as reduced ratios of leaf weight, inadequate nodulation, narrow stems, and dense pulvuni. Additionally, characteristics like stem coloration, size and shape of terminal trifoliate leaves, phyllotaxy, petiole/internode ratios, and lengths are considered in the selection process. These criteria aid in the identification of winged bean varieties that can thrive in dry conditions (Karikari 1977). In the context of a drought study, these findings could suggest that certain morpho-agronomic and phenological traits of lima beans are particularly relevant indicators of drought tolerance. For instance, if the number of emitted and aborted flowers shows a strong genetic correlation with seed and pod production under drought conditions, it implies that these traits may serve as early indicators of drought stress in lima beans. Similarly, positive correlations between pod dimensions and seed characteristics may suggest that certain traits related to pod size could be indicative of drought tolerance, potentially influencing seed development and yield under water-limited conditions. Thus, understanding these relationships can aid in the selection and breeding of drought-tolerant lima bean varieties by focusing on traits that demonstrate strong correlations with drought resilience and yield stability (Machado et al. 2022). Landraces identified as promising cultivar can be used as a parental line for drought breeding. For example, when subjected to temperature and drought stress, the 'Pintat' landrace demonstrated significant competitiveness in comparison to the commercial cultivar (Martínez-Nieto et al. 2020). Strategic breeding approaches as suggested by

Massawe et al. (2005) for bambara groundnut include assessing genetic diversity, developing stable pure lines under varying environments, constructing multi-lines for specific conditions, and conducting multi-locational trials to evaluate drought tolerance. Utilising pure lines in crossing programmes aids in introducing desirable traits, while genetic mapping identifies drought-related quantitative trait loci (QTLs). Assessment of a segregating population for the improvement of drought tolerance in bambara groundnut was also recorded by Chai et al. (2014). Many underutilised legumes encounter reproductive barriers such as embryo abortion, impeding genetic progress. Tissue culture and micro-propagation methods can be utilised to generate viable haploid plants. *Vigna subterranea, V. vexillata*, *Vigna aconitifolia,* and *Canavalia* sp. have a huge potential to overcome these barriers giving more attention toward in vitro study and understanding characters contributing to drought tolerance (Koné et al. 2007; D'Cunha and Sridhar 2011; Bhadkaria et al. 2021).

15.10 Requirement of Genome Sequencing in Underutilised Pulses

The ability to democratise research has been one of the features of DNA sequencing technological advancements over the past 20 years (Das et al. 2022). It was too expensive to do molecular genetic studies on any species other than fruit flies and *Arabidopsis* prior to the development of next-generation sequencing. There is optimism that crop species that lag behind others in terms of genetic resources will benefit from the knowledge gained from other species. Comparative genomics has a special promise to expedite research in understudied species in legumes, where there is considerable genome synteny throughout the whole family (Ren et al. 2019). This work has been simpler with the creation of databases to collect genetic data from legumes that are grown more extensively (Berendzen et al. 2021). Quick progress in the sequencing technologies of genomes over the past 10 years have enhanced the genomic resources, which now include simple sequence repeats (SSRs), whole genome assemblies, genome-wide distributed high-throughput molecular markers, transcriptomes, and single nucleotide polymorphisms (SNPs) of various underutilised legumes (Adhikari et al. 2017; Dutta et al. 2020). Various underutilised legumes have had their trait gene(s) analysed, tagged, and mapped using hybridisation-based molecular markers such as restriction fragment length polymorphisms (RFLP) and PCR-based molecular markers, including RAPD and SSR markers (Bohra et al. 2014). However, conventional PCR-based molecular markers for genotyping have been supplanted by next-generation sequencing technology (NGS)-based high-throughput (HTP) markers, particularly SNPs. Through whole-genome sequencing and transcriptome sequencing initiatives in a variety of crops, including underutilised legumes, second- and third-generation sequencing technologies have made it possible to mine enormous quantities of SSRs and SNPs (Edwards and Batley 2010). Similarly, the development of NGS-based HTP genotyping platforms like Illumina's HiSeq 4000 platform, GoldenGate assay, and

Infinium SNP array has made it possible to find a large number of SNPs across several genomes, which has facilitated a variety of studies, including the diversity of genebank collections (Sokolkova et al. 2020).

De novo genome sequencing of underutilised legumes is the process of sequencing the whole genome of a species of legume that has not been widely researched or is not grown for commercial purposes (Jha et al. 2022). DNA is separated from the plant material at the beginning of this procedure, and then it is fragmented, sequenced, assembled, and annotated. Underutilised legumes frequently contain priceless genetic resources that can support sustainable agriculture, crop innovation, and biodiversity conservation (Upadhyaya et al. 2011). Through genome sequencing, scientists may find genetic variants, pinpoint the genes causing desired characteristics like resilience to disease or drought, and investigate their evolutionary connections with other bean species (Dutta et al. 2024a). Breeding programmes targeted at improving the agronomic features and nutritional value of underutilised legumes can be informed by the insights this technique offers about the genetic diversity and evolutionary history of these crops (Samal et al. 2023). Additionally, it makes it easier to create molecular markers for breeding and to support conservation initiatives that protect genetic variety. De novo genome sequencing of neglected legumes also advances our knowledge of plant biology and evolution and creates new avenues for collaboration and multidisciplinary study in the domains of genomics, genetics, plant breeding, and agricultural science. Table 15.3 presents the compiled genome sequence assemblies of various underutilised legume crops, offering a comprehensive overview of the genomic resources available for these lesser-known species. This information serves as a valuable reference for researchers and breeders aiming to harness the genetic potential of underutilised legumes for drought-sensitive regions.

15.11 Challenges and Future Prospects

Breeding advancements aimed at enhancing drought tolerance in neglected legumes face numerous challenges. Despite their vital importance for global food security, neglected legumes suffer from limited research and investment, resulting in insufficient knowledge and resources for breeding programmes. Moreover, the global scenario highlights the disparity in attention and support compared to major crops, necessitating collaboration with international partners to overcome research gaps and accelerate progress. While initiatives like the International Year of Pulses 2016 have raised awareness, neglected legumes continue to encounter production challenges such as limited access to quality seeds and susceptibility to pests and diseases. Furthermore, the increasing prevalence of drought and heat stress poses significant threats to agricultural productivity, emphasising the urgency of developing resilient varieties. Conventional breeding approaches are hindered by genetic constraints and lengthy breeding cycles, underscoring the need for advanced genomic tools like marker-assisted selection and genome sequencing. However, genome sequencing initiatives are hampered by high costs and complexity,

Table 15.3 A list of underutilised legume crops with genomic sequence assemblies. (Modified form Jha et al. 2022)

Crop name	Genotype	Size of genome	No. of protein coding genes	Sequencing platform used	References
Trifolium pratense	Tatra	314.6	47,398	Illumina HiSeq 2000	Ištvánek et al. (2014)
Vigna radiata	VC 1973A	579 Mb	22,427	Illumina HiSeq 2000, GS FLX +	Kang et al. (2014)
Vigna angularis var. angularis	IT213134	612 Mb	26,857	Illumina HiSeq 2000	Kang et al. (2015)
Trifolium pratense	Milvus B	309 Mb	40,868	Illumina HiSeq 2000	De Vega et al. (2015)
Vigna angularis	Jingnong 6	450 Mb	34,183	HiSeq 2000	Yang et al. (2015)
T. subterraneum L.	Daliak	471.8 Mb	42,706	Illumina MiSeq and HiSeq 2000	Hirakawa et al. (2016)
Lupinus angustifolius	Tanjil	609 Mb	33,076	Illumina	Hane et al. (2017)
T. subterraneum L.	TSUd_r1.1	512 Mb	31,272	Illumina HiSeq 2000	Kaur et al. (2017)
Cyamopsis tetragonoloba (L.)	Vaviloskij 130	1.2 Gb		Illumina and Oxford Nanopore	Grigoreva et al. (2019)
Vigna umbellata	VRB3	414 Mb	31,276	Illumina and PacBio platform	Kaul et al. (2019)
V. subterranea		535.05 Mb	31,707	HiSeq 2000 platform (Illumina)	Chang et al. (2019)
L. purpureus		395.47 Mb	20,946	HiSeq 2000 platform (Illumina)	Chang et al. 2019
Cyamopsis tetragonoloba (L.)	RGC-936	550.31 Mbp	34,680	Illumina, 10× Chromium and Oxford Nanopore	Gaikwad et al. (2020)
Lupinus albus	AMIGA	451 Mb	38,258	PacBio Sequel platform	Hufnagel et al. (2020)
Medicago polymorpha	Huaiyang Jinhuacai	441.83 Mb	36,087	Illumina, PacBio and Hi-C technologies	Cui et al. (2021)
Phaseolus acutifolis A.	wild accession	676 Mb	27,095	Illumina HiSeq platforms	Moghaddam et al. (2021)
Lupinus angustifolius	Tanjil	615.8 Mb	33,907	PacBio Sequel II platform	Wang et al. (2021a)

(continued)

Table 15.3 (continued)

Crop name	Genotype	Size of genome	No. of protein coding genes	Sequencing platform used	References
Vigna mungo	Pant U-31	475 Mb	18,655	Illumina and Nanopore sequencing	Jegadeesan et al. (2021)
M. ruthenica		903.56 Mb	50,268	Illumina, PacBio, and Hi-C	Yin et al. (2021)
Macrotyloma uniflorum	PHG-9	279.1 Mb	24,521	Illumina HiSeq	Mahesh et al. (2021)
Macrotyloma uniflorum	HPK-4	259.2 Mb		Illumina HiSeq 2000	Shirasawa et al. (2021a)
M. ruthenica		904.13 Mb	50,162	PacBio, Illumina, 10 × Genomics, and Hi-C	Wang et al. (2021b)
Phaseolus lunatus L.	G27455	512 Mbp	28,326	Illumina HiSeq	Garcia et al. (2021)
Phaseolus acutifolis A.	Frijol Bayo	684 Mb	27,538	Illumina HiSeq platforms	Moghaddam et al. (2021)
Vicia sativa	KSR5	1.5 Gb	31,146	HiSeq2000	Shirasawa et al. (2021b)
Vigna radiata	VC 1973A	475 Mb	30,958	PacBio RS II platform	Ha et al. (2021)
Vigna mungo	Chai Nat 80	499 Mb	29,411	Illumina HiSeq × Ten	Pootakham et al. (2021)

especially for underutilised pulses with limited research investment. Addressing these challenges requires concerted efforts to advocate for sustained investment, collaboration, and innovation in breeding programmes aimed at enhancing drought tolerance in neglected legumes.

Despite the challenges, the future prospects for enhancing drought tolerance in neglected legumes are promising. Neglected legume species encompass a rich reservoir of genetic diversity that holds immense potential for breeding resilient crops capable of withstanding harsh environmental conditions. Advancements in genomic technologies, including high-throughput sequencing and genome editing, offer unprecedented opportunities to decipher the genetic basis of drought tolerance and accelerate breeding efforts. By identifying key genes and alleles associated with drought resistance, breeders can develop targeted molecular markers for efficient selection of desirable traits, thereby expediting the development of drought-tolerant varieties. Moreover, genomic tools enable the integration of diverse genetic resources from wild relatives and landraces, further enriching the breeding pool and enhancing the adaptive potential of cultivated legumes (Upadhyaya et al. 2011; Dutta et al. 2024b). Collaborative initiatives and partnerships are essential for harnessing the full potential of neglected legumes. International consortia and networks facilitate the exchange of germplasm, knowledge, and expertise, enabling breeders to access diverse genetic resources and leverage complementary strengths across regions. Furthermore, interdisciplinary approaches that integrate genomics with phenomics, physiology, and agronomy provide holistic insights into the complex mechanisms underlying drought tolerance, guiding the development of multifaceted breeding strategies tailored to diverse agroecological environments. Public awareness and support are crucial drivers for advancing neglected legume research and breeding. Initiatives such as the International Year of Pulses have raised awareness of the nutritional, environmental, and economic benefits of legume crops, fostering greater appreciation and demand for neglected species. Increased investment from governments, research institutions, and private sectors is essential to strengthen research infrastructure, capacity building, and technology transfer in neglected legume breeding programmes. By prioritising neglected legumes in agricultural policies and development agendas, stakeholders can unlock their full potential to enhance food security, promote sustainable agriculture, and mitigate the impacts of climate change on vulnerable farming communities (Maity et al. 2023). Through concerted efforts and collective action, neglected legumes can emerge as resilient and indispensable crops for a changing world.

15.12 Conclusion

Neglected legumes hold immense potential for future food security in a drought-prone world. These underutilised cousins of common beans and peas often possess natural drought tolerance, allowing them to thrive in harsh conditions. By incorporating modern breeding techniques like marker-assisted selection and CRISPR, it is possible to unlock these inherent stress-resistant traits. This would allow us to develop new legume varieties specifically designed for drier regions. Furthermore, neglected legumes often boast high nutritional value and symbiotic nitrogen fixation, making them ideal candidates for sustainable agriculture. Investing in research and breeding programmes focused on these forgotten crops could be a game-changer in the fight against drought and malnutrition.

Acknowledgements The authors gratefully acknowledge the support and resources provided by the Ramakrishna Mission Vivekananda Educational & Research Institute, ICAR—Central Institute for Cotton Research, Regional Station, Sirsa, and ICAR—National Bureau of Plant Genetic Resources.

Authors' Contribution SD and DP: Conceptualisation and data management; SD, DP, SS, and SB: Writing of the first draft of the manuscript; DP, SD and SB: Revision of the first draft; SD and DP: Editing and proofreading of the manuscript.

References

Adhikari S, Saha S, Biswas A, Rana TS, Bandyopadhyay TK, Ghosh P (2017) Application of molecular markers in plant genome analysis: a review. Nucleus 60:283–297

Akande FB, Adejumo OA, Adamade CA, Bodunde J (2010) Processing of locust bean fruits: challenges and prospects. Afr J Agric Res 5(17):2268–2271

Akçura M, Müftüoğlu NM, Kaplan M, Türkmen C (2020) Nutrient potential and mineral contents of some vegetable cluster bean genotypes. Cereal Chem 97(6):1193–1203

Amanullah (2016) International year of pulses 2016. EC Agric ECO 01:05–07

Aremu CO, Ojuederie OB, Ayo-Vaughan F, Dahunsi O, Adekiya AO, Olayanju A, Adebiyi OT, Sunday I, Inegbedion H, Asaleye AJ, Abolusoro S (2019) Morphometric analysis and characterization of the nutritional quality in African yam bean accessions. Plant Physiol Rep 24:446–459

Aremu C, Ige SA, Ibirinde D, Raji I, Abolusoro S, Ajiboye B, Obaniyi S, Adekiya A, Asaleye A (2020) Assessing yield stability in African yam bean (*Sphenostylis stenocarpa*) performance using year effect. Open Agric 5(1):202–212

Asci OO (2012) Biodiversity in red clover (*Trifolium pratense* L.) collected from Turkey. II: Nutritional values. Afr J Biotechnol 11(18):4248–4257

Asha KI, Latha M, Abraham Z, Jayan PK, Nair MC, Mishra SK (2006) Genetic resources. In: Kumar D (ed) Horsegram in India. Scientific Publisher, Jodhpur, pp 11–28

Awasthi R, Kaushal N, Vadez V, Turner NC, Berger J, Siddique KH, Nayyar H (2014) Individual and combined effects of transient drought and heat stress on carbon assimilation and seed filling in chickpea. Funct Plant Biol 41(11):1148–1167

Bagri S, Janeja HS (2023) Seed production of underutilized vegetables. In: Production technology of underutilized vegetable crops. Springer, Cham, pp 325–337

Bekkering CS, Tian L (2019) Thinking outside of the cereal box: breeding underutilized (pseudo) cereals for improved human nutrition. Front Genet 10:495433

Belamkar V, Farmer AD, Weeks NT, Kalberer SR, Blackmon WJ, Cannon SB (2016) Genomics-assisted characterization of a breeding collection of *Apios americana*, an edible tuberous legume. Sci Rep 6(1):34908

Berendzen J, Brown AV, Cameron CT, Campbell JD, Cleary AM, Dash S, Hokin S, Huang W, Kalberer SR, Nelson RT, Redsun S (2021) The legume information system and associated online genomic resources. Legum Sci 3(3):e74

Bhadkaria A, Srivastava N, Bhagyawant SS (2021) A prospective of underutilized legume moth bean (*Vigna aconitifolia* (Jacq.) Marechàl): phytochemical profiling, bioactive compounds and in vitro pharmacological studies. Food Biosci 42:101088

Bhardwaj HL, Hamama AA (2004) Protein and mineral composition of tepary bean seed. HortScience 39(6):1363–1365

Bhardwaj J, Chauhan R, Swarnkar MK, Chahota RK, Singh AK, Shankar R, Yadav SK (2013) Comprehensive transcriptomic study on horse gram (*Macrotyloma uniflorum*): De novo assembly, functional characterization and comparative analysis in relation to drought stress. BMC Genomics 14:1–18

Bhat R, Karim AA (2009) Exploring the nutritional potential of wild and underutilized legumes. Compr Rev Food Sci Food Saf 8(4):305–331. https://doi.org/10.1111/j.1541-4337.2009.00084.x

Bhatt KC, Malav PK, Gore PG, Tripathi K, Rathi RS, Tiwari UL, Ahlawat SP (2021) A note on distribution and potential of Japanese wild adzuki bean [*Vigna angularis* var. *nipponensis* (Ohwi) Ohwi and H. Ohashi] in India. Genet Resour Crop Evol 68:2157–2166

Bohra A, Jha UC, Kishor PK, Pandey S, Singh NP (2014) Genomics and molecular breeding in lesser explored pulse crops: current trends and future opportunities. Biotechnol Adv 32(8):1410–1428

Calles T, Castello DR, Baratelli M, Xipsiti M, Navarro DK (2019) International year of pulses. Food and Agriculture Organization of the United Nations

Chai HH, Massawe F, Mayes S (2014) Assessment of a segregating population for the improvement of drought tolerance in Bambara groundnut. In: XXIX International horticultural congress on horticulture: sustaining lives, livelihoods and landscapes (IHC2014): 1127, pp 339–346
Chai HH, Massawe F, Mayes S (2016) Effects of mild drought stress on the morpho-physiological characteristics of a bambara groundnut segregating population. Euphytica 208:225–236
Chandel KPS, Pant KC, Arora RK (1984) Winged bean in India. NBPGR Sci Monogr 8:31
Chang Y, Liu H, Liu M, Liao X, Sahu SK, Fu Y, Song B, Cheng S, Kariba R, Muthemba S, Hendre PS (2019) The draft genomes of five agriculturally important African orphan crops. GigaScience 8(3):giy152
Chun HC, Sanghun LEE, Choi YD, Gong DH, Jung KY (2021) Effects of drought stress on root morphology and spatial distribution of soybean and adzuki bean. J Integr Agric 20(10):2639–2651
Cui J, Lu Z, Wang T, Chen G, Mostafa S, Ren H, Liu S, Fu C, Wang L, Zhu Y, Lu J (2021) The genome of *Medicago polymorpha* provides insights into its edibility and nutritional value as a vegetable and forage legume. Hortic Res 8:47
Cullis C, Chimwamurombe P, Barker N, Kunert K, Vorster J (2018) Orphan legumes growing in dry environments: Marama bean as a case study. Front Plant Sci 9:404670
Das A, Singh S, Islam Z, Munshi AD, Behera TK, Dutta S, Weng Y, Dey SS (2022) Current progress in genetic and genomics-aided breeding for stress resistance in cucumber (*Cucumis sativus* L.). Sci Hortic 300:111059
D'Cunha M, Sridhar K (2011) Micropropagation of the wild legume *Canavalia rosea* (Sw.) DC. from coastal sand dunes. Biol Lett 48(1):85–96
De La Peña RC, Ebert AW, Gniffke PA, Hanson P, Symonds RC (2011) Genetic adjustment to changing climates: vegetables. In: Yadav SS, Redden RJ, Hatfield JL, Lotze-Campen H, Hall AE, editors. Crop adaptation to climate change. WileyChichester, pp. 396–410
De Vega JJ, Ayling S, Hegarty M, Kudrna D, Goicoechea JL, Ergon Å, Rognli OA, Jones C, Swain M, Geurts R, Lang C (2015) Red clover (*Trifolium pratense* L.) draft genome provides a platform for trait improvement. Sci Rep 5(1):17394
Dhanuja GS (2024) Studies on drought tolerance in dolichos bean (*Lablab purpureus* L. Sweet). Int J Adv Biochem Res 8(2):635–639. https://doi.org/10.33545/26174693.2024.v8.i2h.644
Diakostefani A, Velissaris R, Cvijanovic E, Bulgin R, Pantelides A, Leitch IJ, Chapman MA (2024) Genome resources for underutilised legume crops: genome sizes, genome skimming and marker development. Genet Resour Crop Evol 71(1):427–438
Dickstein R (2019) The future of legume genetic data resources: challenges, opportunities, and priorities. Legum Sci 1(1):e16
Dutta S, Muthusamy V, Chhabra R, Zunjare RU, Hossain F (2020) Two-step method for isolation of high-quality RNA from stored seeds of maize rich in starch. 3 Biotech 10:1–8
Dutta S, Paul D, Das A, Kumar S, Roy A, Barua A, Chinnusamy V (2024a) Directed mutagenesis in plants through genome editing using guide RNA library. Nucleus 67:61–73
Dutta S, Zunjare RU, Sil A, Mishra DC, Arora A, Gain N, Chand G, Chhabra R, Muthusamy V, Hossain F (2024b) Prediction of matrilineal specific patatin-like protein governing in-vivo maternal haploid induction in maize using support vector machine and di-peptide composition. Amino Acids 56(1):20
Edwards D, Batley J (2010) Plant genome sequencing: applications for crop improvement. Plant Biotechnol J 8(1):2–9
Emmrich PM, Sarkar A, Njaci I, Kaithakottil GG, Ellis N, Moore C, Edwards A, Heavens D, Waite D, Cheema J, Trick M (2020) A draft genome of grass pea (Lathyrus sativus), a resilient diploid legume. BioRxiv:2020–2004. https://doi.org/10.1101/2020.04.24.058164
Ernest EG, Wisser RJ (2024) Multiple physiological mechanisms involved in reproduction could be targeted for breeding heat tolerance in lima bean. Crop Sci 64(1):179–191
Farooq M, Hussain M, Usman M, Farooq S, Alghamdi SS, Siddique KH (2018) Impact of abiotic stresses on grain composition and quality in food legumes. J Agric Food Chem 66(34):8887–8897
Foyer CH, Lam HM, Nguyen HT, Siddique KH, Varshney RK, Colmer TD, Cowling W, Bramley H, Mori TA, Hodgson JM, Cooper JW (2016) Neglecting legumes has compromised human

health and sustainable food production. Nat Plants 2(8):1–10. https://doi.org/10.1038/nplants.2016.112
Gaikwad K, Ramakrishna G, Srivastava H, Saxena S, Kaila T, Tyagi A, Sharma P, Sharma S, Sharma R, Mahla HR, Amitha Mithra SV (2020) Chromosome scale reference genome of Cluster bean (*Cyamopsis tetragonoloba* (L.) Taub.). bioRxiv:2020–2005
Garcia T, Duitama J, Zullo SS, Gil J, Ariani A, Dohle S, Palkovic A, Skeen P, Bermudez-Santana CI, Debouck DG, Martínez-Castillo J (2021) Comprehensive genomic resources related to domestication and crop improvement traits in Lima bean. Nat Commun 12(1):702
Gedil M, Ferguson M, Girma G, Gisel A, Stavolone L, Rabbi I (2016) Perspectives on the application of next-generation sequencing to the improvement of Africa's staple food crops. In: Next generation sequencing–advances, applications and challenges. InTechOpen, pp 287–321
Gohara AK, Souza AHPD, Gomes STM, Souza NED, Visentainer JV, Matsushita M (2016) Nutritional and bioactive compounds of adzuki beans cultivars using chemometric approach. Ciência e Agrotecnologia 40(1):104–113
Grigoreva E, Ulianich P, Ben C, Gentzbittel L, Potokina E (2019) First insights into the guar (*Cyamopsis tetragonoloba* (L.) Taub.) genome of the 'Vavilovskij 130'accession, using second and third-generation sequencing technologies. Russ J Genet 55:1406–1416
Grotelüschen K, Sennhenn A, Whitbread AM, Maass BL, Njarui DM (2014) *Lablab Purpureus* (L) Sweet: a promising multipurpose legume for enhanced drought resistance in smallholder farming-systems of eastern Kenya
Guerra-García A, Suárez-Atilano M, Mastretta-Yanes A, Delgado-Salinas A, Piñero D (2017) Domestication genomics of the open-pollinated scarlet runner bean (*Phaseolus coccineus* L.). Front Plant Sci 8:299557
Guretzki S, Papenbrock J (2013) Comparative analysis of methods analyzing effects of drought on the herbaceous plant *Lablab purpureas*. J Appl Bot Food Qual 86(1):47–54
Guretzki S, Papenbrock J (2014) Characterization of *Lablab purpureus* regarding drought tolerance, trypsin inhibitor activity and cyanogenic potential for selection in breeding programmes. J Agron Crop Sci 200(1):24–35
Ha J, Satyawan D, Jeong H, Lee E, Cho KH, Kim MY, Lee SH (2021) A near-complete genome sequence of mungbean (*Vigna radiata* L.) provides key insights into the modern breeding program. Plant Genome 14(3):e20121
Hane JK, Ming Y, Kamphuis LG, Nelson MN, Garg G, Atkins CA, Bayer PE, Bravo A, Bringans S, Cannon S, Edwards D (2017) A comprehensive draft genome sequence for lupin (*Lupinus angustifolius*), an emerging health food: insights into plant–microbe interactions and legume evolution. Plant Biotechnol J 15(3):318–330
Hao S, Ge Q, Shao Y, Tang B, Fan G, Qiu C, Lee SMY (2022) Chromosomal-level genome of velvet bean (*Mucuna pruriens*) provides resources for L-DOPA synthetic research and development. DNA Res 29(5):dsac031
Haq N (2002) ICUC activities in South Asia. In: Fruits for the future in Asia. Proceedings of a regional consultation meeting on utilization of tropical fruit trees in Asia, Bangkok
Hickey JM, Chiurugwi T, Mackay I, Powell W (2017) Genomic prediction unifies animal and plant breeding programs to form platforms for biological discovery. Nat Genet 49(9):1297–1303
Hirakawa H, Kaur P, Shirasawa K, Nichols P, Nagano S, Appels R, Erskine W, Isobe SN (2016) Draft genome sequence of subterranean clover, a reference for genus Trifolium. Sci Rep 6:30358
Huang YF, Gao XL, Nan ZB, Zhang ZX (2017) Potential value of the common vetch (*Vicia sativa* L.) as an animal feedstuff: a review. J Anim Physiol Anim Nutr 101(5):807–823
Hufnagel B, Marques A, Soriano A, Marquès L, Divol F, Doumas P, Sallet E, Mancinotti D, Carrere S, Marande W, Arribat S (2020) High-quality genome sequence of white lupin provides insight into soil exploration and seed quality. Nat Commun 11(1):492
Hughes JDA, Ebert AW (2011) Research and development of underutilized plant species: the role of vegetables in assuring food and nutritional security. In: II International symposium on underutilized plant species: crops for the future-beyond food security
Ištvánek J, Jaroš M, Křenek A, Řepková J (2014) Genome assembly and annotation for red clover (*Trifolium pratense*; Fabaceae). Am J Bot 101(2):327–337

Jaenicke H, Höschle-Zeledon I (2006) Strategic framework for underutilized plant species research and development: with special reference to Asia and the Pacific, and to Sub-Saharan Africa. Bioversity International

Jain SM, Gupta SD (2013) Biotechnology of neglected and underutilized crops. Springer

Jegadeesan S, Raizada A, Dhanasekar P, Suprasanna P (2021) Draft genome sequence of the pulse crop blackgram [*Vigna mungo* (L.) Hepper] reveals potential R-genes. Sci Rep 11(1):11247

Jha UC, Nayyar H, Parida SK, Bakır M, von Wettberg EJ, Siddique KH (2022) Progress of genomics-driven approaches for sustaining underutilized legume crops in the post-genomic era. Front Genet 13:831656

Kaga A, Isemura T, Tomooka N, Vaughan DA (2008) The genetics of domestication of the azuki bean (*Vigna angularis*). Genetics 178(2):1013–1036

Kang YJ, Kim SK, Kim MY, Lestari P, Kim KH, Ha BK, Jun TH, Hwang WJ, Lee T, Lee J, Shim S (2014) Genome sequence of mungbean and insights into evolution within Vigna species. Nat Commun 5(1):5443

Kang YJ, Satyawan D, Shim S, Lee T, Lee J, Hwang WJ, Kim SK, Lestari P, Laosatit K, Kim KH, Ha TJ (2015) Draft genome sequence of adzuki bean, *Vigna angularis*. Sci Rep 5(1):8069

Karalija E, Vergata C, Basso MF, Negussu M, Zaccai M, Grossi-de-Sa MF, Martinelli F (2022) Chickpeas' tolerance of drought and heat: current knowledge and next steps. Agronomy 12(10):2248

Karikari SK (1977) Characters in the selection of winged beans (*Psophocarpus tetragonolobus* (L) DC) tolerant to drought. In: V Africa Symposium on Horticultural Crops 84, pp 19–22

Karmakar P, Singh V, Yadava RB, Singh B, Singh R, Kushwaha M (2016) Agathi [*Sesbania grandiflora* L.(Agast)]: Current status of production, protection and genetic improvement. In: National symposium on vegetable legumes for soil and human health, 153–161

Kaul T, Eswaran M, Thangaraj A, Meyyazhagan A, Nehra M, Raman NM, Gayacharan BJ, Badapanda C, Balamurali B (2019) Rice Bean (*Vigna umbellata*) draft genome sequence: unravelling the late flowering and unpalatability related genomic resources for efficient domestication of this underutilized crop. bioRxiv:816595

Kaur P, Bayer PE, Milec Z, Vrána J, Yuan Y, Appels R, Edwards D, Batley J, Nichols P, Erskine W, Doležel J (2017) An advanced reference genome of *Trifolium subterraneum* L. reveals genes related to agronomic performance. Plant Biotechnol J 15(8):1034–1046

Khan MMH, Rafi MY, Ramlee SI, Jusoh M, Mamun A (2022) Bambara groundnut (*Vigna Subterranea* L. Verdc): a model underutilized legume, its resiliency to drought and excellency of food & nutrient values-a review. Malays J Genet (MJG) 2:12

Khatun M, Sarkar S, Era FM, Islam AM, Anwar MP, Fahad S, Islam AA (2021) Drought stress in grain legumes: effects, tolerance mechanisms and management. Agronomy 11(12):2374

Kochhar SL (2016) Mat or Moth bean [*Phaseolus aconitifolius*]. In: Kochhar (ed) Economic botany: a comprehensive study, 5th edn. Cambridge University Press, p 430

Kokila S, Myrene RD, Devaraj VR (2014) Response of *Lablab purpureus* (Hyacinth bean) cultivars to drought stress. Asian J Plant Sci Res 4(5):48–55

Koné M, Patat-Ochatt EM, Conreux C, Sangwan RS, Ochatt SJ (2007) In vitro morphogenesis from cotyledon and epicotyl explants and flow cytometry distinction between landraces of Bambara groundnut [*Vigna subterranea* (L.) Verdc], an under-utilised grain legume. Plant Cell Tissue Organ Cult 88:61–75

Kumalasari ID, Nishi K, Harmayani E, Raharjo S, Sugahara T (2014) Immunomodulatory activity of Bengkoang (*Pachyrhizus erosus*) fiber extract in vitro and in vivo. Cytotechnology 66(1):75–85

Kumar D (2002) Production technology for moth bean in India

Kumar B, Bhalothia P (2020) Orphan crops for future food security. J Biosci 45:1–8

Łabuda H (2010) Runner bean (*Phaseolus coccineus* L.)-biology and use. Acta Sci Polonorum-Hortorum Cultus 9(3):117–132

Lepcha P, Egan AN, Doyle JJ, Sathyanarayana N (2017) A review on current status and future prospects of winged bean (*Psophocarpus tetragonolobus*) in tropical agriculture. Plant Foods Hum Nutr 72:225–235. https://doi.org/10.1007/s11130-017-0627-0

Li J, Cullis C (2023) Comparative analysis of 84 chloroplast genomes of *Tylosema esculentum* reveals two distinct cytotypes. Front Plant Sci 13:1025408

Li J, Jiang J, Stel HV, Homkes A, Corajod J, Brown K, Chen Z (2014) Phylogenetics and biogeography of Apios (Fabaceae) inferred from sequences of nuclear and plastid genes. Int J Plant Sci 175(7):764–780

Lone RA, Sarvendra K, Singh V, Bano N, Bag SK, Mohanty CS, Barik SK (2022) Adaptation of winged bean (*Psophocarpus tetragonolobus* (L.) DC.) to drought stress is mediated by root-tuber heat-shock proteins and specific metabolites. Curr Plant Biol 32:100266

Luo SF, Hu LH, Chen YY, Li PX (2015) The effects of modified atmosphere packaging on the quality of sword bean storage and antioxidant enzyme activities (in Chinese). Food Sci 17:1–8

Machado EO, Ferraz GV, Almeida RDC, Lopes ACDA, Gomes RLF, Silva VBD (2022) Evaluation of Lima Bean Accessions at High Temperatures. Rev Caatinga 35:791–798

Mahesh HB, Prasannakumar MK, Manasa KG, Perumal S, Khedikar Y, Kagale S, Soolanayakanahally RY, Lohithaswa HC, Rao AM, Hittalmani S (2021) Genome, Transcriptome, and Germplasm Sequencing Uncovers Functional Variation in the Warm-Season Grain Legume Horsegram (*Macrotyloma uniflorum* (Lam.) Verdc.). Front Plant Sci 12:75811

Maity A, Paul D, Lamichaney A, Sarkar A, Babbar N, Mandal N, Dutta S, Maity PP, Chakrabarty Chakrabarty SK (2023) Climate change impacts on seed production and quality: current knowledge, implications, and mitigation strategies. Seed Sci Technol 51(1):7–38

Martín-Cabrejas MÁ (ed) (2019) Legumes: nutritional quality, processing and potential health benefits (vol. 8). Royal Society of Chemistry

Martínez-Nieto MI, Estrelles E, Prieto-Mossi J, Roselló J, Soriano P (2020) Resilience capacity assessment of the traditional Lima Bean (*Phaseolus lunatus* L.) landraces facing climate change. Agronomy 10(6):758

Massawe FJ, Mwale SS, Azam-Ali SN, Roberts JA (2005) Breeding in bambara groundnut (*Vigna subterranea* (L.) Verdc.): strategic considerations. Afr J Biotechnol 4(6):463–471

Mayes S, Ho WK, Chai HH, Gao X, Kundy AC, Mateva KI, Azam-Ali SN (2019) Bambara groundnut: an exemplar underutilised legume for resilience under climate change. Planta 250:803–820

Missanga J, Ndakidemi P, Venkataramana P (2023) Lablab purpureus: evaluation and selection of drought-tolerant – high-yielding accessions in dry farming systems based on drought tolerance indices and multi-environmental yield trials. J Agric Sci 29(2):690–709. https://doi.org/10.15832/ankutbd.1169256

Moghaddam SM, Oladzad A, Koh C, Ramsay L, Hart JP, Mamidi S, Hoopes G, Sreedasyam A, Wiersma A, Zhao D, Grimwood J (2021) The tepary bean genome provides insight into evolution and domestication under heat stress. Nat Commun 12(1):2638

Moteetee AN (2016) Canavalia (Phaseoleae, Fabaceae) species in South Africa: naturalized and indigenous. S Afr J Bot 103:6–16

Murrell D (2016) Global research and funding survey on pulse productivity and sustainability. Global Pulse Confederation, Dubai, p 71

Pati K, Zhang F, Batley J (2019) First report of genome size and ploidy of the underutilized leguminous tuber crop Yam Bean (*Pachyrhizus erosus* and P. tuberosus) by flow cytometry. Plant Genet Res 17(5):456–459

Pereira A, Ramos F, Sanches Silva A (2022) Lupin (*Lupinus albus* L.) seeds: balancing the good and the bad and addressing future challenges. Molecules 27(23):8557

Pootakham W, Nawae W, Naktang C, Sonthirod C, Yoocha T, Kongkachana W, Sangsrakru D, Jomchai N, U-thoomporn S, Somta P, Laosatit K (2021) A chromosome-scale assembly of the black gram (*Vigna mungo*) genome. Mol Ecol Resour 21(1):238–250

Popoola J, Ojuederie O, Omonhinmin C, Adegbite A (2019) Neglected and underutilized legume crops: improvement and future prospects. In: Recent advances in grain crops research. IntechOpen

Popoola JO, Aworunse OS, Ojuederie OB, Adewale BD, Ajani OC, Oyatomi OA, Eruemulor DI, Adegboyega TT, Obembe OO (2022) The exploitation of orphan legumes for food, income, and nutrition security in Sub-Saharan Africa. Front Plant Sci 13:782140

Rana MK, Vilas CA (2017) Broad Bean. In: Vegetable crop science. CRC Press, pp 683–692

Reddy LJ, Upadhyaya HD, Gowda CLL, Singh S (2005) Development of core collection in pigeonpea [*Cajanus cajan* (L.) Millspaugh] using geographic and qualitative morphological descriptors. Genet Resour Crop Evol 52:1049–1056

Ren L, Huang W, Cannon SB (2019) Reconstruction of ancestral genome reveals chromosome evolution history for selected legume species. New Phytol 223(4):2090–2103

Samal I, Bhoi TK, Raj MN, Majhi PK, Murmu S, Pradhan AK, Kumar D, Paschapur AU, Joshi DC, Guru PN (2023) Underutilized legumes: nutrient status and advanced breeding approaches for qualitative and quantitative enhancement. Front Nutr 10:1110750. https://doi.org/10.3389/fnut.2023.1110750

Savita VV (2023a) Underutilized vegetables introduction and identification. In: Production technology of underutilized vegetable crops. Springer, pp 1–10

Savita (2023b) Production technology of underutilized vegetables of leguminosae family. In: Production technology of underutilized vegetable crops. Springer, pp 25–99

Sherasia PL, Garg MR, Bhanderi BM (2018) Pulses and their by-products as animal feed. UN

Shirasawa K, Chahota R, Hirakawa H, Nagano S, Nagasaki H, Sharma T, Isobe S (2021a) A chromosome-scale draft genome sequence of horsegram (*Macrotyloma uniflorum*). Gigabyte

Shirasawa K, Kosugi S, Sasaki K, Ghelfi A, Okazaki K, Toyoda A, Hirakawa H, Isobe S (2021b) Genome features of common vetch (*Vicia sativa*) in natural habitats. Plant Direct 5(10):e352

Singh RJ, Chung GH, Nelson RL (2007) Landmark research in legumes. Genome 50(6):525–537

Sinha P, Singh VK, Saxena RK, Khan AW, Abbai R, Chitikineni A, Varshney RK (2020) Superior haplotypes for haplotype-based breeding for drought tolerance in pigeonpea (*Cajanus cajan* L.). Plant Biotechnol J 18(12):2482–2490

Sokolkova A, Burlyaeva M, Valiannikova T, Vishnyakova M, Schafleitner R, Lee CR, Ting CT, Nair RM, Nuzhdin S, Samsonova M, von Wettberg E (2020) Genome-wide association study in accessions of the mini-core collection of mungbean (*Vigna radiata*) from the World Vegetable Gene Bank (Taiwan). BMC Plant Biol 20:1–9

Soumare A, Diedhiou AG, Kane A (2022) Bambara groundnut: a neglected and underutilized climate-resilient crop with great potential to alleviate food insecurity in sub-Saharan Africa. J Crop Improv 36(5):747–767

Suma A, Latha M, John JK, Aswathi PV, Pandey CD, Ajinkya A (2021) Yard-long bean. In: The beans and the peas. Woodhead Publ, pp 153–172

Tanzi AS, Eagleton GE, Ho WK, Wong QN, Mayes S, Massawe F (2019) Winged bean (*Psophocarpus tetragonolobus* (L.) DC.) for food and nutritional security: synthesis of past research and future direction. Planta 250(3):911–931

Tiamiyu LO, Okomoda VT, Akpa PO (2016) Nutritional profile of toasted *Canavalia ensiformis* seed and its potential as partially replacement for soybean in the diet of *Clarias gariepinus*. Braz J Aquat Sci Technol 20(2):12–17

Upadhyaya HD, Dwivedi SL, Ambrose M, Ellis N, Berger J, Smýkal P, Debouck D, Duc G, Dumet D, Flavell A, Sharma SK (2011) Legume genetic resources: management, diversity assessment, and utilization in crop improvement. Euphytica 180:27–47

Urga K, Fite A, Kebede B (1995) Nutritional and antinutritional factors of grass pea (*Lathyrus sativus*) germplasms. Bull Chem Soc Ethiop 9(1)

Uyoh EA, Urua IS, Ntui VO, Okpako EC (2011) Flow cytometric analysis of nuclear DNA content, mitotic chromosome number and protein separation by SDS-PAGE in three accessions of African locust bean (*Parkia biglobosa* Benth.). J Crop Sci Biotechnol 14:227–232

Uzabakiriho JD (2016) Diversity and distribution of *Tylosema esculentum* (Marama bean) endophytic bacteria communities in Omitara, Harnas and Otjinene, eastern Namibia. [Doctoral dissertation]. University of Namibia

Vadivel V, Janardhanan K (2001) Diversity in nutritional composition of wild jack bean (*Canavalia ensiformis* L. DC) seeds collected from south India. Food Chem 74(4):507–511

Vatanparast M, Shetty P, Chopra R, Doyle JJ, Sathyanarayana N, Egan AN (2016) Transcriptome sequencing and marker development in winged bean (*Psophocarpus tetragonolobus*; Leguminosae). Sci Rep 6(1):1–14

Wang L, Zhu J, Li X, Wang S, Wu J (2018) Salt and drought stress and ABA responses related to bZIP genes from *V. radiata* and *V. angularis*. Gene 651:152–160

Wang P, Zhou G, Jian J, Yang H, Renshaw D, Aubert MK, Clements J, He T, Sweetingham M, Li C (2021a) Whole-genome assembly and resequencing reveal genomic imprint and key genes of rapid domestication in narrow-leafed lupin. Plant J 105(5):1192–1210

Wang T, Ren L, Li C, Zhang D, Zhang X, Zhou G, Gao D, Chen R, Chen Y, Wang Z, Shi F (2021b) The genome of a wild Medicago species provides insights into the tolerant mechanisms of legume forage to environmental stress. BMC Biol 19(1):96

Waweru B, Njaci I, Murungi E, Paliwal R, Muli C, Maranga M, Shorinola O (2023) Chromosome-scale assembly of the African yam bean genome. bioRxiv:10

Williams JT, Haq N (2000) Global research on underutilized crops–an assessment of current activities and proposals for enhanced cooperation. International Centre for Underutilized Crops, Southampton. International Standard Book Number, 92-9043-545-3

Yang K, Tian Z, Chen C, Luo L, Zhao B, Wang Z, Yu L, Li Y, Sun Y, Li W, Chen Y (2015) Genome sequencing of adzuki bean (*Vigna angularis*) provides insight into high starch and low fat accumulation and domestication. Proc Natl Acad Sci 112(43):13213–13218

Yin M, Zhang S, Du X, Mateo RG, Guo W, Li A, Wang Z, Wu S, Chen J, Liu J, Ren G (2021) Genomic analysis of *Medicago ruthenica* provides insights into its tolerance to abiotic stress and demographic history. Mol Ecol Resour 21(5):1641–1657

Yohane EN, Shimelis H, Laing M, Mathew I, Shayanowako A (2021) Genotype-by-environment interaction and stability analyses of grain yield in pigeonpea [*Cajanus cajan* (L.) Millspaugh]. Acta Agric Scand B Soil Plant Sci 71(3):145–155. https://doi.org/10.1080/09064710.2020.1859608

Zhang M, Jian S, Wang Z (2022) Comprehensive analysis of the Hsp20 gene family in *Canavalia rosea* indicates its roles in the response to multiple abiotic stresses and adaptation to tropical coral islands. Int J Mol Sci 23(12):6405

Future Perspectives in Legume Breeding for Sustainable Agriculture

16

Muhammad Tanveer Altaf, Amjad Ali, Waqas Liaqat, Parnaz Mortazavi, Sarmad Ali Qureshi, Muhammed Tatar, Amna Jamil, Muhammad Sheraz, Muhammad Aneeq Ur Rahman, Sabeen Rehman Soomro, Salma Naimatullha Soomro, Muhammad Azhar Nadeem, and Faheem Shehzad Baloch

Abstract

This chapter investigates the multifaceted challenges and opportunities in legume breeding under sustainable agriculture. It explores the impacts of elevated CO_2 levels and temperature on legume production and nodulation, emphasizing the necessity for enhanced stress tolerance. A comprehensive review of traditional breeding methods, alongside recent advancements and OMICS approaches, is presented. This chapter highlights successful cultivar development using both

The original version of the chapter has been revised. A correction to this chapter can be found at https://doi.org/10.1007/978-981-96-4112-3_17

M. T. Altaf (✉)
Department of Field Crops, Faculty of Agriculture, Recep Tayyip Erdoğan University, Rize/Pazar, Türkiye
e-mail: muhammadtanveer.altaf@erdogan.edu.tr

A. Ali · M. Tatar
Department of Plant Protection, Faculty of Agricultural Sciences and Technologies, Sivas University of Science and Technology, Sivas, Turkey

W. Liaqat
Department of Field Crops, Faculty of Agriculture, Institute of Natural and Applied Sciences, Çukurova University, Adana, Turkey

P. Mortazavi · S. A. Qureshi · S. R. Soomro · S. N. Soomro
Department of Plant Production and Technologies, Faculty of Agricultural Sciences and Technologies, Sivas University of Science and Technology, Sivas, Turkey

A. Jamil
Department of Horticultural Sciences, Faculty of Agriculture and Environment, The Islamia University of Bahawalpur, Bahawalpur, Pakistan

© The Author(s), under exclusive license to Springer Nature Singapore Pte Ltd. 2025, Corrected Publication 2026

M. A. Nadeem et al. (eds.), *Marker-Assisted Breeding in Legumes for Drought Tolerance*, https://doi.org/10.1007/978-981-96-4112-3_16

conventional and modern techniques and identifies critical adaptive traits essential for resilient pulse crop production. Addressing genetic limitations of pulses, it discusses emerging technologies and tools, such as high-throughput phenotyping and genome editing, which are pivotal for overcoming these challenges. This synthesis aims to provide a forward-looking perspective on leveraging innovative breeding strategies to ensure the sustainability and productivity of legume crops in the face of global climate challenges.

Keywords

Legumes · Genome editing · OMICS approaches · Participatory approaches

16.1 Introduction

The effects of climate change on agricultural production have become more evident, with legume crops being especially susceptible to these changes. Fluctuations in temperature, patterns of precipitation, and the occurrence of extreme weather changes have a direct impact on the development, productivity, and standard of legume crops (Nadeem et al. 2023). Higher temperatures accelerate phenological development, shortening growth and lowering crop yields. Changes in precipitation patterns can cause drought stress, which hinders photosynthesis and nitrogen fixation, which legumes need to grow. Floods and storms can also damage crops, threatening their stability and production. To ensure food security and agricultural sustainability, legume breeding must adapt to climate change (Altaf et al. 2023). Legumes crops use rhizobia bacteria to change the atmospheric nitrogen into usable form, making them crucial to global agriculture. This strategy improves soil fertility and reduces synthetic nitrogen fertilizer use, supporting sustainable farming. Legumes are rich in protein, amino acids, vitamins, and minerals. They are essential to human nutrition, especially in meat-restricted environments. Legumes also improve crop diversification and pest and disease resistance, ensuring food security and sustainable agriculture (Abd-Alla et al. 2023). Modern legume breeding uses old and modern approaches to boost crop yield.

Traditional breeding involves selecting and interbreeding plants with desirable traits in order to develop novel cultivars. Nevertheless, these techniques can require a significant amount of time and manual effort. Legume breeding techniques, such

M. Sheraz
Fodder and Forrage Program, Crop Sciences Institute, NARC, Islamabad, Pakistan

M. A. U. Rahman
Biotechnology Research Institute, Graduate School of Chinese Academy of Agricultural Sciences, Beijing, China

M. A. Nadeem
Department of Biotechnology, Faculty of Science, Mersin University, Mersin, Türkiye

F. S. Baloch
Department of Biotechnology, Faculty of Science, Mersin University, Mersin, Türkiye

Department of Plant Resources and Environment, Jeju National University, Jeju, South Korea as marker-assisted selection (MAS) and genomic selection (GS), have recently made significant progress, providing more accurate and faster methods (Ali et al. 2022). These strategies facilitate the identification and selection of particular genes linked to favorable characteristics, hence expediting the breeding process. The breeding of legumes has experienced a considerable transition because of the application of OMICS technologies like genomes, metabolomics, transcriptomics, and proteomics (Patra et al. 2024). These innovative approaches of breeding provide a comprehensive understanding of the genetic, metabolic, and physiological underpinnings of significant characteristics. There are notable examples of cultivar production that demonstrate the efficacy of mixing traditional and modern breeding approaches to achieve significant advances in genetics. These improvements include the development of cultivars that are resistant to disease and abiotic stresses (Jha et al. 2023; Majhi et al. 2024).

The future of legume breeding is being influenced by emerging technology and techniques, which provide new opportunities for speeding up breeding efforts and improving genetic gains. The breeding landscape is being revolutionized by high-throughput phenotyping, CRISPR/Cas9 genome editing technologies, and powerful bioinformatics tools (Chinnusamy et al. 2023; Naik et al. 2024; Younas et al. 2023). By allowing for the precise manipulation of genetic material, the rapid screening of phenotypic characteristics, and the efficient management of enormous datasets, these tools make it possible to optimize the breeding process. It is vital to utilize particular breeding procedures in order to generate legume varieties that are adaptive to climate conditions in order to address the challenges that are brought about by climate change. It is of the utmost importance to selectively develop legume crops through breeding in order to acquire characteristics such as resilience to biotic stressors, heat tolerance, and drought tolerance. In light of the fact that the environmental circumstances are always shifting, this is essential in order to preserve the consistency and productivity of these crops. The significance of these approaches is shown by case studies that illustrate successful breeding efforts to mitigate the effects of climate change on legume crops. These approaches include the development of chickpea variations that are able to survive high temperatures and cowpea variants that are resistant to drought (Ngalamu et al. 2023; Singh et al. 2023).

Participation in breeding and involvement of stakeholders are essential for the effective advancement and acceptance of novel bean cultivars. Engaging farmers, breeders, and other stakeholders in the breeding process guarantees that the resulting varieties align with the requirements and preferences of end-users, thereby amplifying the significance and effectiveness of breeding programs. Modern breeding programs also aim to improve legume crops' nutritional quality and agronomic features (Altaf et al. 2024a). Enhancing the protein content, increasing the bioavailability of micronutrients, and maintaining consistent crop yields are crucial in order to address the nutritional requirements of a rapidly expanding population and guarantee food security. This chapter intends to examine the difficulties and possibilities in legume breeding for sustainable agriculture, focusing on the effects of climate change, improving stress resistance, and utilizing sophisticated breeding methods and emerging technology for future progress.

16.2 Addressing Climate Change Challenges

16.2.1 Impact of Climate Change (Higher CO_2 and Temperature) on Legume Production

The potential use of grain legumes as a climate-smart crop may appear promising; nevertheless, the negative impact of climate change on the productivity of legumes poses inquiries and necessitates strategies to address these issues. The presence of climatic anomalies and the increase in greenhouse gases (GHGs) and temperature pose a greater challenge and complexity in reaching the desired crop production and guaranteeing nutritional security. Studies have demonstrated that the increasing concentration of CO_2 and temperature has both direct and indirect effects on the physiological and biochemical features of grain legumes. Comprehensive investigations have clearly demonstrated a notable and favorable response of elevated carbon dioxide levels (eCO_2) in several types of grain legumes.

In a study conducted by Srinivasarao et al. (2016), it was shown that the shoot and total biomass of pigeon pea and black gram increased when exposed to elevated levels of carbon dioxide (550 ppm CO_2), compared to the ambient levels (350 ppm CO_2). Pearl millet and sorghum, which are C4 crops, did not exhibit improved performance under eCO_2, as reported by Srinivasarao et al. (2016). Other grain legumes, such as pigeon pea, chickpea, field pea, lentil, and green gram, have also shown a similar beneficial response to eCO_2 in terms of growth and yield qualities (Jin et al. 2013; Lam et al. 2012; Nasser et al. 2008; Pandey et al. 2016; Saha et al. 2011).

Undoubtedly, grain legumes are unequivocally the most environmentally intelligent crops. However, in order to do this, legumes necessitate an optimal provision of essential nutrients such as nitrogen (N), phosphorus (P), and potassium (K), as well as adequate soil moisture. A meta-analysis of multiple FACE experiments revealed the intricate correlation between eCO_2, crop performance, and the utilization of applied inputs (Kimball 2016). The results demonstrated that C3 legumes, such as soybean and clover, exhibited a significant rise of up to 25% in shoot biomass when exposed to both N and water. In contrast, C3 cereals, including rice, wheat, and barley, showed a response of 19%, while the effect was essentially minimal in C4 crops like sorghum and maize. Curiously, when N is not present, the C3 cereals had just a 10% rise in shoot biomass. In contrast, legumes, which benefit from biological nitrogen fixation (BNF), demonstrated a positive growth rate without any negative impact on yield (Kimball 2016).

An extended FACE experiment conducted in Australia, with eCO_2 content (550 ppm), demonstrated a significant increase in grain yield and biomass. Field pea showed a 26% increase in grain output and a 28% increase in biomass, while lentil exhibited a 60% increase in grain yield and a 62% increase in biomass (Bourgault et al. 2016, 2017). Improved coordination between the source and sink, reduced harvest index, continuous growth pattern, and larger reproductive organs enable legumes to adapt effectively to changing climate conditions and emerge as a promising crop for the future (Ainsworth et al. 2006). A study conducted by Mishra and Agrawal (2015) found that green gram experienced a 20–34% increase in biomass when exposed to eCO_2 levels (700 ppm). Similar higher yield responses were observed in various grain legumes, including soybean, black gram, lentil, desi chickpea, pigeon pea, and kabuli chickpea (Bhatia et al. 2021; Bourgault et al. 2017; Drag et al. 2020; Lenka et al. 2020; Saha et al. 2011; Sathish et al. 2014; Vanaja et al. 2010-).

Lamichaney et al. (2021) conducted an experiment to evaluate the impact of eCO_2 (600 parts per million) on chickpea plants of the JG 14 variety. The application of eCO_2 resulted in a significant increase in plant height, leaf and stem biomass, and seed yield, compared to the ambient CO_2 conditions. The increase in yield was attributed to a higher number of pods and seeds per plant. Nevertheless, eCO_2 resulted in a decrease in total phenol, seed protein, and thiobarbituric acid reactive compounds, while simultaneously boosting starch content and water absorption. The presence of eCO_2 did not affect seed germination and vigor. The results indicate that eCO_2 levels of up to 600 ppm can improve the growth and productivity of chickpeas in a sub-tropical climate, which has potential implications for enhancing its nutritional content.

The precise effects of future atmospheric conditions on legume output and quality are currently unknown. Although legumes may demonstrate enhanced yield advantages and decreased protein loss at eCO_2 levels compared to non-N_2-fixing plants (Singer et al. 2020). In contrast to eCO_2, elevated temperature (eT) negatively affects the productivity of grain legumes. The severity of harm caused by eT is determined on the developmental stage and intensity of temperature. The detrimental effect of eT is seen in various legumes, such as lentil. For every degree Celsius increase above 32 °C, the quantity of lentil seeds decreases by 0.10–0.20% (Delahunty et al. 2018). The heat stress in legumes is caused by factors such as stomata closure, flower and pod abortion, and disruption of normal physiological functions like respiration or electron transport (Saha et al. 2016). A study by Bindumadhava et al. (2018) found that green gram cultivars that are resilient to high temperatures exhibited a 25% increase in total biomass and a 31% increase in seed yield under eCO_2 conditions.

In their study, Ahmed et al. (2023) demonstrated that legumes utilizing the C3 fixation pathway exhibited an elevated rate of photosynthesis, a decrease in photorespiration, increased biomass output, and improved water usage efficiency in the presence of eCO_2. Nevertheless, as the temperature increases, plants exhibit an accelerated rate of growth, a reduced lifespan, a shorter period for grain filling, and a decreased yield. Similarly, the beneficial effect of eCO_2 on nodulation was hindered by an increase in temperature. Legumes have the ability to tolerate eCO_2, even up to 1000 ppm, by allocating carbohydrates in the form of sucrose and storing them as starch. In addition, besides the process of starch mobilization, legumes also undergo protein synthesis to aid in their adaptation to the dynamic climate. To mitigate the adverse effects of climate extremes, it is necessary to implement various adaptation strategies. These include the creation of new legume crop varieties that are resilient to climate change, designing ideal crop characteristics using process-based models, adjusting sowing dates, utilizing short-duration cultivars, employing precision agriculture tools for precise irrigation and fertilization, practicing intercropping, transitioning to legume cultivars that are better suited to the changing climate, and diversifying crops (Ahmed et al. 2023).

Jumrani et al. (2024) found that eT and water stress led to a decrease in CO_2 exchange, resulting in an adverse impact on nitrogen fixation in soybean. The activity of nitrogenase in the nodules decreased dramatically as the temperature increased. The high temperatures caused a decrease in the levels of leghemoglobin, heme-chrome, and ureids in the nodules. This decrease ultimately led to a drop in nitrate reductase activity, chlorophyll content, and total free amino acid. The decrease in nitrogen and carbon fixation ultimately led to a decrease in both the quantity and quality of seeds produced. Water stress, when applied at varying temperatures, exacerbated the impact of temperature. Additionally, the combination of water stress and high temperature had a more deleterious effect. The indeterminate growth habit and high flexibility in phenology of legumes make them well suited for future conditions (Liu et al. 2003).

16.2.2 Impact of Climate Change (Higher CO_2 and Temperature) on Nodulation

Considerable focus has been dedicated to understanding the effects of climate change on the physiological processes of legumes both above and below ground. However, there has been comparatively less interest in studying the growth and performance of nodules (Dutta et al. 2022). The process of nodule formation and nitrogen fixation consists of several complex processes. However, it is challenging to accurately evaluate the specific effects of eCO_2 on each individual step. Overall, eCO_2 has been found to have a beneficial effect on the growth and physiological functions of nodules in many bean species. The influence of eCO_2 on nodulation is strongly correlated with the specific crop species, bacterial symbiont, plant nutritional state, physicochemical soil conditions, and environmental factors.

The duration of exposure to eCO_2 has a significant and varied effect on the formation and development of nodules. In general, brief exposure enhances the functions of nodules, such as nitrogenase activity, N_2 fixation, and the overall nitrogen status of the plant. However, extended exposure leads to changes in nodule development, specifically in terms of nodule biomass and the number of nodules per plant (Dutta et al. 2022). The inquiry led by Saha et al. (2016) examined the effects of eCO_2 (580 ppm) on chickpea. The results demonstrated a significant rise in both nodule number and biomass, with a respective growth of 39% and 76% compared to the control group. Furthermore, they found that the concentration of leghemoglobin was 66% higher under eCO_2 (580 ppm) compared to normal atmospheric conditions.

Pandey et al. (2016) found that the weight of fresh nodules in green gram rose by 38% when exposed to eCO_2 (550 ppm) compared to normal settings. The quantity and biomass of nodules in various leguminous plant species, including lentil, bean, soybean, alfa-alfa, and field pea, showed a significant increase when exposed to eCO_2 (Nasser et al. 2008; Haase et al. 2007; Guo et al. 2013; Prévost et al. 2010).

Additionally, Cabrerizo et al. (2001) observed positive outcomes in field pea, while Pandey et al. (2016) reported a 1.6-fold increase in Nase activity in green gram under elevated carbon dioxide levels (550 ppm). The relationship between legumes and rhizobium, known as legume-rhizobium interaction, is significantly affected by eT, which poses a substantial risk to nodule metabolism and biochemical properties (Yuksel and Vardar 2023). The eT disrupts the metabolism of the nodule by controlling the allocation of carbon, increasing the likelihood of oxidative stress, hindering normal nitrogen metabolism, and ultimately exposing the bacteroids to oxygen. This combination of factors impairs the smooth functioning of Nase (Aranjuelo et al. 2014; King and Purcell 2005; Pla et al. 2003). In addition, reduces the photosynthetic rate by forcibly sealing guard cells, reducing the overall amount of soluble proteins in leaves, inhibiting carboxylase activity, and causing water stress (Aranjuelo et al. 2014). Nevertheless, Aranjuelo et al. (2014) found no reliable association between photosynthetic downregulation and Nase activity, indicating the need for additional research before reaching a definitive conclusion.

Elevated temperature stress is expected to affect the regulation of genes associated with nodulation and rhizobium infection. Rhizobia are very responsive to temperature, with their activity being most efficient at temperatures ranging from 20 to 25 °C. Even a slight change in soil temperature, less than 2 °C, can severely damage the symbiotic connection by hindering root growth and the rate of infection (Aranjuelo et al. 2014). In addition, when exposed to conditions, the root system experiences hindered growth of root hairs due to the buildup of harmful nitrogenous compounds and inadequate supply of carbohydrates. This leads to the deterioration of nod gene activity and eventually weakens the delicate symbiotic relationship between the crop and microorganism (Aranjuelo et al. 2014). Although eCO_2 has a favorable effect on nodulation, its impact is likely to be influenced by environmental factors such as temperature stress, soil nitrogen status, and moisture availability (Dutta et al. 2022). Experimental investigations have shown that legumes are capable of tolerating up to 1000 ppm of eCO_2 levels. They achieve this by maintaining nodule growth and Nase activity by a careful regulation of carbohydrate levels. Specifically, they balance these levels by allocating sucrose for immediate use and storing excess carbohydrates in the form of starch. In addition to the mobilization of starch, the enhanced synthesis of proteins likely facilitated the adaptation of legumes to the changing climate. In summary, it can be stated that, in addition to natural adaptations such as carbohydrate allocation, the development of climate resilient agro-technologies for appropriate agro-ecosystems would aid legumes in adapting and meeting the grain requirement for humanity.

16.2.3 Enhancing Stress Tolerance in Legumes

Enhancing the root systems of crops is widely acknowledged as a crucial approach to assure the long-term viability and yield of crops, especially under challenging environmental situations. The genetic diversity present in legumes can be effectively utilized to tackle the difficulties posed by climatic unpredictability (Foyer et al. 2019).

The study conducted by Prince et al. (2018) achieved the necessary genetic variety to comprehend the mechanisms governing root system architecture (RSA) by examining 397 soybean accessions. Genetic loci for RSA were detected using a genome-wide association mapping technique. Among the genes that were identified, Glycine Lateral Root 1 (GLR1) was determined to regulate the quantity of lateral roots. The different forms of this gene at a specific location enhanced the rate of long-range neuronal connections by controlling the shape and size of cells in the cortex. In addition, soybean accessions possessing the particular allelic variations of this locus exhibited enhanced root branching capacity under both water-limited and ideal field conditions (Prince et al. 2018). Elevated diurnal temperatures exert a detrimental influence on the reproductive efficacy of soybean and other leguminous plants (Foyer et al. 2019). Therefore, it is crucial to identify and characterize genotypes that have a high tolerance for high temperatures. The study conducted by Djanaguiraman et al. (2018) revealed that elevated daytime temperatures during the transition from gametogenesis to full bloom in soybean plants growing in the field result in reduced pod-set across various genotypes. This decrease in pod-set is attributed to a decline in pollen viability. Moreover, these authors discovered soybean genotypes that can withstand high temperatures through a rapid and efficient screening process.

Changing climatic patterns are expected to cause an increase in flood episodes in various places. Flooded soils commonly experience an absence of oxygen, leading to a significant stress for plants known as root oxygen deprivation. Munir et al. (2018) examined the reactions of two cool-season grain legumes, chickpea and faba bean, to root-zone applications of carbon dioxide and ethylene. Root respiration was suppressed, and growth was diminished by the presence of carbon dioxide. Ethylene enhanced the gas-filled porosity of roots. This plant hormone is widely recognized for its ability to stimulate the development of aerenchyma in many species (Yamauchi et al. 2018). The study demonstrated that faba bean exhibited much higher levels of root gas-filled porosity compared to chickpea when exposed to ethylene. While high levels of carbon dioxide had a negative effect on root growth in both species, acute lack of oxygen was even more harmful. Faba bean exhibited greater tolerance than chickpea to development in an oxygen-deficient root zone, underscoring the need of selecting the appropriate grain legume species for cultivation in flood-prone fields.

Soil waterlogging presents a significant danger to the productivity of important legume crops such as chickpea, pea, and mung bean. These crops are especially vulnerable to this type of stress (Foyer et al. 2019). Zaman et al. (2018) assessed the ability of three distinct pea genotypes to germinate under varying durations of waterlogging. They also used RNA-Seq and gene expression analyses to discover genes that were expressed differently and to understand the mechanisms connected to stress tolerance. To summarize, the authors propose two distinct tolerance mechanisms associated with pea germination in response to waterlogging stress. One technique is the quiescence strategy, which involves preserving the seed reserve energy during waterlogging. The second technique is escape technique, which involves fast germination, which relies on energy derived from both protein and lipid metabolism to withstand waterlogging stress. Additional research is necessary to explore the genes linked to these pathways in order to successfully employ them in future crop breeding programs for enhancement.

The study conducted by Ricter et al. (2018) established a correlation between salt stress tolerance and metabolic responses, as well as the buildup of Na^+, K^+, or Cl^- in *Vicia faba*. The yield of lentil seeds and their nutritional quality were reduced by exposure to drought and/or heat stress (Sehgal et al. 2018). Nevertheless, these authors discovered a lentil genotype that is resistant to drought and also shows tolerance to heat stress. Gaining insight into the impact of heat stress on reproductive organs is crucial for enhancing grain legume yields in upcoming climates. Jiang et al. (2018) demonstrated that heat stress exposure in pea plants results in a more significant suppression of pollen formation and viability compared to ovules. A heat-tolerant pea genotype was discovered to display a more consistent lipid content in the stamens.

16.3 Current Status of Legume Breeding

16.3.1 Review of Traditional Breeding Methods for Legumes (Conventional Breeding Approach)

Traditional breeding methods for legumes typically emphasize phenotypic selection, hybridization, and backcrossing to develop new cultivars with traits such as enhanced yield, resistance to diseases, and drought endurance. This conventional approach involves selecting parent plants that exhibit favorable characteristics and cross-pollinating them to produce offspring that inherit these desirable traits. The process is repeated across multiple generations to stabilize these traits in the new cultivar. While this method can be effective, it is often time-consuming and labor-intensive, and its success depends significantly on the breeder's skill and experience (Majidian 2022).

16.4 Discussion on Recent Advancements in Legume Breeding Techniques

Recent advancements in legume breeding techniques have significantly accelerated the development of improved cultivars. Marker-assisted selection (MAS) has become a powerful tool, allowing breeders to identify and select plants with specific genetic markers linked to desirable traits (Baloch et al. 2023a). This technique reduces the reliance on phenotypic selection and increases the efficiency and precision of breeding programs. Genomic selection (GS) further enhances breeding efficiency by using genome-wide markers to predict the performance of breeding lines, enabling the selection of superior candidates at an early stage. GS integrates high-throughput genotyping and advanced statistical models to improve the accuracy of selection decisions (Crossa et al. 2017).

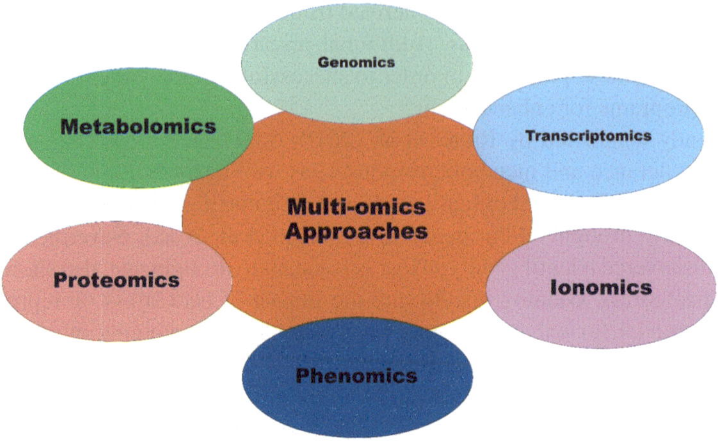

Fig. 16.1 Multi-omics approaches for the improvement of legumes

16.4.1 Advanced Breeding Approaches—OMICS Approaches

OMICS approaches, including genomics, metabolomics, transcriptomics, and proteomics, have revolutionized legume breeding (Fig. 16.1). Genomics provides insights into the genetic architecture of complex traits and facilitates the identification of candidate genes for breeding (Nadeem et al. 2022). Metabolomics analyzes the biochemical pathways and metabolite profiles associated with specific traits, offering a deeper understanding of plant physiology and stress responses. Transcriptomics examines gene expression patterns under various conditions, revealing the regulatory networks that control important traits. Proteomics studies the protein composition of plants, linking genotype to phenotype. Integrating these OMICS approaches can enhance the precision and scope of legume breeding programs.

16.4.2 Successful Examples of Cultivar Development Using Conventional and Modern Breeding Approaches

Numerous successful examples demonstrate the effectiveness of both conventional and modern breeding approaches in legume cultivar development. For instance, the development of disease-resistant soybean varieties through traditional hybridization and selection has significantly improved crop resilience and productivity (Tripathi et al. 2022). More than 200 chickpeas have been developed by conventional breeding (Chaturvedi et al. 2014). Modern techniques such as MAS have been employed to develop chickpea cultivars with enhanced drought tolerance and nutritional quality (Kundu et al. 2023). Genomic selection has been successfully used in pea breeding programs to accelerate the development of high-yielding, disease-resistant varieties. These examples highlight the synergy between traditional and advanced breeding methods in achieving significant improvements in legume crops. In summary, the current status of legume breeding reflects a dynamic field that combines time-tested conventional methods with cutting-edge biotechnological advances. The integration of traditional and modern approaches holds great promise for meeting the challenges of sustainable agriculture and food security in the face of global climate change.

16.4.3 Critical Adaptive Traits for Resilient Pulse Crop Production

The resilience of crop plants seems to be achieved in nature through the rearrangement and reassortment of genes across various locations, resulting in the development of new adaptive characteristics that ultimately lead to improved suitability for a few individuals in the altered environmental conditions (Kumar et al. 2019). McKersie (2015) has explored many approaches to lessen the impact of climate change on crop output. One significant strategy for addressing future climate change is the development of genotypes that are well suited to the changing conditions. The process entails integrating innovative traits into agricultural crops to improve food yield and stability. In order to develop pulse cultivars that are resilient to climate conditions, it is crucial to make genetic enhancements for adaptive characteristics (Cutforth et al. 2007; Choudhary et al. 2016). Climate change poses a significant risk to the future production of both warm-season and cool-season pulses. Therefore, it is crucial to extract germplasm-derived traits for adaptability and incorporate them into emerging cultivars. This approach is extremely desirable (Kumar et al. 2019). The cultivation of adaptive traits is necessary to enhance the resistance of crops to the existing conditions of climate change, in order to ensure the continuity of productivity. Adaptive traits demonstrate their ability to adjust to changing environmental conditions and assist crop plants in their survival and reproduction amidst biotic and abiotic stress situations (Cullis and Kunert 2017). These adaptive qualities can manifest as agromorphological as well as biochemical attributes (Bargaz et al. 2015; Huang et al. 2017; Shunmugam et al. 2018). Breeders have endeavored to enhance numerous characteristics in pulses to suit the specific target environment. Hence, it is necessary to include particular adaptation traits into the enhanced genotypes for every distinct growth condition.

The reproductive phase significantly impacts the number of seeds produced in agricultural plants. Drought stress during the pod-filling stage in chickpea plants leads to pod loss and decreases the number of seeds per plant. On the other hand, if the drought occurs during the initial podding phase, chickpea yield is reduced by 85% (Pang et al. 2017). Therefore, the capacity to fill pods can be specifically focused on as an agro-morphological characteristic in order to design cultivars that are tolerant to drought in settings with limited moisture. Quan et al. (2016) have identified specific leaf features, including leaf weight, leaf area, and leaf growth rate that are associated with tolerance to drought stress in Mediterranean climatic conditions. Physiological features, which exhibit a high degree of consistency regardless of the environment, offer increased breeding value (Reynolds and Langridge 2016).

Several physiological characteristics, such as leaf variables, seed yield, pod abscisic acid content, and root attributes, are known to provide drought resistance in chickpea (Chen et al. 2017; Pang et al. 2017). Prince et al. (2017) conducted a groundbreaking investigation to unravel the processes behind drought tolerance in legumes. They found that root xylem plasticity plays a crucial role in enhancing water-use efficiency in soybean plants when they are exposed to water stress.

A well-developed root system is a beneficial physiological characteristic that aids in preserving seed yield in pulses during drought conditions by improving the extraction of soil water (Turner et al. 2001). Therefore, it is advantageous to utilize root and leaf characteristics when selecting for drought resistance in pulse crops. Research conducted by Shunmugam et al. (2018) has confirmed that crop plants produce proteins and metabolites in many tissues as a reaction to environmental challenges. Accurately identifying, confirming, and describing various proteins/metabolites found in particular organs and tissues during periods of stress can serve as valuable biochemical indicators for developing crops that are adaptable to climate change. Ramalingam et al. (2015) investigated the differential expression of proteins under different stress conditions at various growth phases in many pulse crops, such as chickpea, pea, green gram, and common bean.

Several morphophysiological features that provide resistance to these pressures have been recognized in pulse crops, enabling them to adjust to the increasing global temperatures observed in recent years (Fig. 16.2) (Choudhary et al. 2016; Gaur et al. 2015). Photosynthetic activity is a crucial physiological feature that can be affected by low and high temperatures, leading to photodamage in photosystem-II (Murata et al. 2007). The discovery of heat-tolerant genotypes in lentil can be facilitated by examining pollen and leaf features (Sita et al. 2017). Chickpea breeders have utilized early flowering characteristics in their breeding operations, resulting in the creation of novel chickpea varieties that are well suited for warmer, short-season conditions. This has led to a significant transformation in chickpea cultivation in southern India (Berger et al. 2006). The creation of rapid-maturing cultivars has boosted the potential for growing lentil crops in rice-fallow fields by reducing yield losses due to forced maturity (Kumar et al. 2018). Hence, it is evident that the morphophysiological and biochemical traits that contribute to adaptability differ depending on the specific abiotic stress and plant species. In order to breed intelligent pulses for a certain scenario, it is necessary to utilize specific adaptive traits (Kumar et al. 2019).

16.5 Genetic Limitations of Pulses

Multiple genes regulate numerous significant quantitative traits, including grain yield, nutritional components, and certain types of tolerance to environmental and biological challenges. Genetic restrictions on adaptive development refer to genetic factors that hinder or decrease the probability of natural selection directly raising the intermediate phenotype to its optimal level (Richard and Houle 2009). A quantitative restriction refers to a situation where there is a capacity to respond to

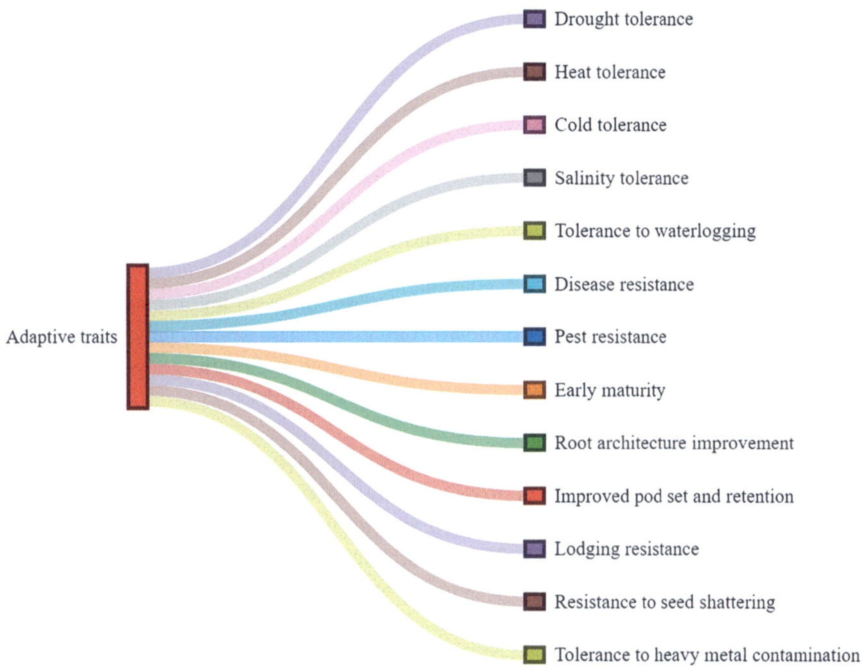

Fig. 16.2 A list of adaptive traits that need to be exploited in important pulses for sustainable agriculture

selection, but this capacity is restricted or constrained (Walsh and Blows 2009). There may be limitations in the genetic diversity of natural populations due to the presence of multiple genetic factors. Genetic differences can be observed in almost all traits when there is strong natural and sexual selection. However, there is a contradiction between natural and sexual selection, which leads to a reduction in genetic variation (Pilson 2000). Chevin (2012) reported that when directional selection acts consistently throughout time, it is likely to result in negative genetic correlations between pairs of traits that are adaptively significant. Genetic correlations across traits can limit the ability to respond to natural selection, but the degree to which these correlations are influenced by adaptation to patterns of directional selection is not well understood (Connallon and Hall 2018). The requirements for maintaining stable polymorphism are limited, particularly when selection is minimal (Kawecki and Ebert 2004).

Polymorphism can result in a diminished level of genetic and evolutionary autonomy among different traits, genetic trade-offs among different aspects of fitness, and less adaptability due to pleiotropic genetic variation (Arnqvist et al. 2010). Therefore, if the necessary circumstances for sustained coexistence of several forms of a gene are met by antagonistic pleiotropy, it is anticipated that there will be a significant amount of dominant variance in one or both aspects of fitness. However, such observations are infrequently reported in experimental studies (Abobatta et al.

2021). The strong genetic similarity between wild relatives and leguminous relatives' germplasm groups is utilized to enhance gene maps and identify specific genes responsible for improving agricultural traits and increasing leguminous productivity (Varshney et al. 2017).

Genetic limitation in a population refers to a situation where there is a low level of genetic variation, limited availability of sources of resistance to both biotic and abiotic pressures in cultivated gene pools, and the potential for introducing new genes and alleles from wild species and landraces into cultivated species. The challenges associated with hybridizing farmed species with wild species are regarded as a significant obstacle in the development of genetic maps in legume crops such as peanut, pigeon pea, and chickpea (Ojiewo et al. 2019). The limited genetic diversity of pulses remains a major obstacle to achieving higher genetic gains, as well as hindering accurate and efficient selection processes (Baloch et al. 2023b; Kumar et al. 2020).

Wild species and landraces, although important, have not been fully utilized to enhance crops due to obstacles in crossbreeding between different species, limited information on specific traits, and limited availability of advanced breeding tools, among others (Großkinsky et al. 2015). Enhancing these characteristics through traditional breeding methods is a challenging and time-consuming endeavor. Novel approaches, rooted in classical genetics, such as linkage disequilibrium (LD)-based correlation mapping and advanced back-cross QTL (AB-QTL) analysis, present the potential to surmount some obstacles (Ojiewo et al. 2019). The significance of genetic diversity for crop breeders lies in its potential to facilitate the creation of novel and enhanced varieties possessing desirable qualities, including those chosen by farmers and breeders, such as yield and its various components (Govindaraj et al. 2015). To enhance the quality, productivity, and profitability of legume crops in the future, it is crucial to harness the vast genetic variety they possess (Abobatta et al. 2021). The utilization of molecular and genetic engineering techniques is currently being applied to the genetic and genomic resources of legume crops, which are considered crucial strategic tools. Plant breeding is a powerful technique for enhancing the traits of legume crops, promoting a balance between the environment and agriculture, and improving the variety for sustainable agricultural practices. This leads to crops that are more appealing in terms of health and economics.

16.6 Emerging Technologies and Tools for Addressing the Limitations

The future of legume breeding is set to be revolutionized by emerging technology, which will provide novel possibilities to expedite breeding advancements and improve genetic gains. High-throughput phenotyping enables breeders to analyze massive numbers of legumes with higher efficiency and accuracy by rapidly and precisely measuring plant traits. This technique facilitates the detection of favorable characteristics, such as resistance to drought and diseases, with enhanced velocity and accuracy (Ahn et al. 2023). The advent of genome editing tools, specifically CRISPR/Cas9, has provided unparalleled opportunities for accurate genetic alterations (Ali et al. 2023, 2024). These technologies provide the precise modification of particular genes linked to desirable characteristics, thereby expediting the advancement of enhanced various crop cultivars. Advanced bioinformatics technologies are revolutionizing legume breeding by enabling the analysis and interpretation of large volumes of genetic data. These approaches facilitate the identification of genetic markers associated with significant features and enhance the efficiency of marker-assisted selection (MAS) and genomic selection (GS). These emerging technologies have the potential to make the breeding process more efficient, decrease the time needed to generate new varieties, and achieve major genetic improvements in legume crops (Altaf et al. 2024b). This will ultimately contribute to the development of more sustainable and productive agricultural systems.

16.6.1 High-Throughput Phenotyping in Legume Breeding

High-throughput phenotyping, often known as HPP, is revolutionizing the breeding of legumes by making it possible to evaluate plant characteristics on a broad range in a timely and reliable way. By utilizing cutting-edge imaging techniques, sensors, and robots, HTP is able to evaluate essential features such as the pace of growth, the amount of biomass, the efficiency with which water is used, and other stress reactions (Mishra et al. 2024; Sabag et al. 2024). Using this technology, breeders are able to analyze thousands of plants under a wide range of environmental circumstances, thereby selecting superior genotypes in a precise and expedient manner. Through the provision of exhaustive phenotypic data, HTP speeds up the selection process, which in turn makes it possible to rapidly advance the plants that perform the best through breeding programs (Ball et al. 2020; Bhat and Yu 2021). The development of legume varieties that are capable of meeting the problems posed by climate change and ensuring food security requires this higher efficiency with great importance.

16.6.2 Genome Editing in Legume Breeding

The use of genome editing, and more specifically CRISPR/Cas9, is a powerful technique in the breeding of legumes because it enables precise alterations to be made in particular genes (Table 16.1). Through the application of this technology, it is possible to produce legumes that have improved characteristics, such as increased resilience to disease and drought, as well as higher nutritional quality (Bhowmik et al. 2021). In contrast to conventional breeding practices, which can take a significant number of generations to obtain the required characteristics, genome editing can bring about these modifications in a far shorter amount of time. The production of superior bean varieties can be sped up by the use of genome editing, which involves directly targeting and modifying genes linked with major characteristics. This rapid progress is essential for enhancing the production and resilience of legumes, as well as adapting them to changing environmental conditions (Baloglu et al. 2022).

Table 16.1 Gene-editing technology in different legume crops

Legume	Target gene	Technique	Result	References
Medicago truncatula	*FMO1-like*, *ERDJ2*, *MEL1*, and *FBL1-like*	CRISPR/Cas 9	Statistically significant effects on nodule	Curtin et al. (2017)
	MtPDS		Albino plants	Meng et al. (2017)
	MtPDS		70% mutation efficiency	Wolabu et al. (2020)
	MtSUP		MtSUP was found to be orthologous of *AtSUP*	Rodas et al. (2021)
	NPD genes		Smaller nodules, earlier onset of nodule, senescence, ineffective nodules production	Trujillo et al. (2019)
	MtPDS		Homozygous and biallelic mutants	Zhang et al. (2020)
	MtHen1		Efficient mutation	Curtin et al. (2018)
	CYP93E2 and *CYP72A61*		51 *CYP93E2* mutant plant lines	Confalonieri et al. (2021)
Glycine max	*GmAGO7a* and *GmAGO7b*	CRISPR/Cas 9	Inherited mutation until T_2 lines	Zhang et al. (2020)
	KAS1		Deletion and insertion mutations	Virdi et al. (2020)
	GmLox1, *2*, and *3*		Loss of lipoxygenase activity	Wang et al. (2020)
	GmFAD2-1A, *GmFAD2-1B*		Increased oleic acid content	Do et al. (2019)
	GmFT2a and *GmFT5a*		ft2a, ft5a, and ft2aft5a mutants	Cai et al. (2020)
	Glycine max Dicer-like2		*GmDicer-like2* mutant plants	Curtin et al. (2018)
	Glyma14g04180		Insertion and deletion mutations	Di et al. (2019)
	GmFAD2-1A, *GmFAD2-2A*		Increased oleic acid content	Wu et al. (2020)
	GmSPL9		Altered plant architecture	Bao et al. (2019)
	FAD2-2		Increased oleic acid content	Al Amin et al. (2019)
	Glyma03g36470		Insertion and deletion mutations	Di et al. (2019)
	Pooled platform-102		Multiplex mutations	Bai et al. (2020)
	Glyma06g136900		Insertion and deletion mutations	Di et al. (2019)
	E1		Early flowering	Han et al. (2019)
	GmDrb2a and *GmDrb2b*		Biallelic double mutant	Curtin et al. (2018)
	Conglycinin		Mutations in three of nine genes	Li et al. (2019)
	GmFT2a and *GmFT5a*		Deletion mutations	Cai et al. (2020)
	GmPDS11 and *GmPDS18*		Albino and dwarf buds	Du et al. (2016)

Mechanism of Gene Editing CRISPR/Cas9 technology has gained popularity in recent years for its ability to modify genomes in a wide range of organisms, including plants. Furthermore, it has expanded the range of agricultural studies by enabling the development of new plant types through the removal of detrimental traits or the introduction of notable characteristics. CRISPR is an advancing approach that enables various genetic manipulations, such as producing knockouts, making accurate alterations, performing multiplex genome engineering, or activating and suppressing genes (Arora and Narula 2017). The CRISPR/Cas9 system consists of two primary components: the Cas9 protein and the guide RNA (gRNA). Cas9 is an enzyme that breaks DNA and is dependent on RNA. It forms a complex with gRNA. The gRNA is composed of 20 nucleotides that are a perfect match to the target DNA sequence and functions as an indicator to attract Cas9. CRISPR/Cas9 utilizes RNA-DNA interactions rather than DNA-protein interactions to identify the target DNA sequence, distinguishing it from previous genome-editing tools. Zinc Finger Nucleases (ZFNs) and Transcription Activator-Like Effector Nucleases (TALENs) require two distinct DNA-binding domains for each target site. These domains facilitate the DNA-protein interactions necessary to precisely target specific sequences. This procedure is somewhat cumbersome. CRISPR/Cas9, which relies on DNA-RNA interactions, only requires the creation of an 18–20 base pair. Cas9 and gRNA must bind to a certain protospacer in order to be effective. The adjacent motif (PAM) is a pattern located at the 3′ end of target sequences. The sequence 5′-NGG-3′ serves as the protospacer adjacent motif (PAM) for Cas9 derived from *Streptococcus pyogenes*, which is the most frequently used PAM sequence that is commonly chosen and employed for genome editing. Cas9 stimulates DNA repair by introducing double-stranded breaks (DSBs) in the DNA target. The repair mechanism involves non-homologous end joining (NHEJ) for creating gene knockouts and homology-directed repair (HDR) for producing gene modification and gene insertion (Huang and Puchta 2019). A frameshift mutation arises when non-homologous end joining (NHEJ) randomly introduces an insertion or deletion of a DNA strand inside a coding region, leading to a complete inactivation of the gene. Using a homologous repair template is not required for non-homologous end joining (NHEJ). Hence, the non-homologous end joining (NHEJ) repair mechanism is widely recognized as the predominant and highly efficient approach for DNA damage repair in all plant species, including legumes. The HDR approach, however, has the ability to precisely include preset sequences derived from a donor DNA template. The limited editing efficiency of HDR has constrained its use in plants (Atkins and Voytas 2020). The CRISPR technique enables several prospective uses, such as the examination of gene expression, augmentation of gene function, and reduction of gene function. CRISPR has been applied in legume farming to utilize its genome-editing capabilities for producing superior and environmentally friendly agricultural goods, specifically legumes.

16.7 Bioinformatics in Legume Breeding

In contemporary legume breeding, bioinformatics plays a crucial role because it supplies the computational tools that are required to handle and interpret the enormous volumes of data that are produced by high-throughput phenotyping and genome editing methodologies. Through the use of sophisticated bioinformatics software and algorithms, it is possible to identify genetic markers that are associated with desired characteristics, which in turn enables selection processes to be more effective (Khalid et al. 2021; Tan et al. 2022). Bioinformatics provides breeders with the ability to better comprehend the complex relationships that determine plant performance. This is accomplished by integrating genetic, phenotypic, and environmental data. The most potential prospects for further development are identified through this in-depth study, which speeds up the process of breeding (Kehinde et al. 2023). Therefore, the application of bioinformatics helps to improve genetic gains in legume crops, which in turn helps to encourage the development of varieties that are both resilient and high yielding.

16.8 Participatory Breeding and Stakeholder Engagement

Overview of participatory breeding approaches involving farmers, breeders, and other stakeholders in the breeding process. It is challenging to define participation uniformly because its application and associated theories vary widely (Chanie 2015). World Bank (2007) defined the crucial role of stakeholder engagement for the successful implementation and effectiveness of new varieties in diverse agro-ecosystem, participation is described as the process through which stakeholder influence and exert control over the establishment of priorities, developing policies, allocating resources and accessing public goods and policies. It denotes "granting people the authority to gather their talents, serve as active agents rather than passive observers, administer resources, make choices and regulate the actions that impact their lives." Participation refers to the committed involvement of residents in predetermined programs or projects. This can include activities like contributing to a survey, acting as primary informant, or taking part in researcher-led trials. Pearse and Stiefel (1979) stated that participation refers to the intentional actions taken by previously marginalized groups and movements to gain control over resources and regulatory institutions within certain social contexts. Therefore, involvement in this deskwork is defined as the voluntary support, teamwork, involvement of faming families, and development schemes or projects. According to Narayan (1993), participatory development is defined as the inclusion of users and communities in every stage of the advancement procedure. Levels of engagement: The level of participation might vary depending on the research topic, the researchers' facilitation abilities, the farmers' expertise in agricultural trials, and the degree of trust between scientists and farmers. Participation is generally accessed through a scale that features categories such as associative, informative, farmer directed, and researcher controlled. Participatory plant breeding is a process in which farmers actively engage in a plant breeding program, consistently having opportunities to make decisions at various stages. The participation of farmers in the participatory breeding plan (PPB) encompasses various activities such as establishing breeding objectives, choosing sources of germplasm, conducting trials, selecting lines for crossbreeding, analyzing results, organizing activities, proposing methodological modifications, and multiplying and marketing seeds from stable lines. According to Narayan (1993), participatory development is defined as the inclusion of participants and community members in every stage of the developmental framework. Participatory plant breeding is a process in which farmers actively engage in a plant breeding program, consistently having opportunities to make decisions at various stages. Participatory Plant Breeding (PPB) typically requires a greater and more intricate level of farmer participation, due to their role in decision-making during formative and more essential phases of the variety breeding process. Consequently, PPB has a greater impact on empowering farmers compared to Participatory Variety Selection (PVS) (Witcombe et al. 2005).

16.8.1 Participatory Varietal Selection

Participatory variety selection (PVS) is a method where a few finished or nearly finished varieties are field-tested with partners' participation (Mkumbira et al. 2003). This selection process is essential in breeding programs and is part of participatory plant breeding, helping identify farmer-acceptable varieties to replace old or obsolete ones (Ogliari et al. 2013). PVS serves as a preliminary trial to assess the land and time needed for a full participatory plant breeding program (Osiru et al. 2010). Farmers evaluate traits important to them, aiding in varietal diversity, faster replacement, and scaling up. PVS can also assess quality traits such as milling percentage, cooking and keeping quality, taste, and market price, which are challenging to evaluate in conventional trials. PVS is becoming popular for determining breeding program materials and testing them in farming environments before release (Qazi et al. 2014). Identifying farmers' needs, testing relevant materials, and experimenting on their fields are phases of PVS. Farmers' involvement is crucial due to their knowledge and skills in selecting varieties that fit local environments and meet consumer satisfaction, which the current breeding system often overlooks (Rahman et al. 2015). Climate variability and smallholder farmers' indigenous knowledge highlight the need for their participation in variety developments. PVS integrates farmers' knowledge with breeders' quantitative methods, enhancing cultivar diversity and reducing research costs. Participatory plant breeding through decentralized selection in target environments improves varietal diversity and addresses stakeholders' needs. Benefits of participatory research include enhancing farmers' knowledge, increasing varietal diversity, and empowering local communities. Involving farmers in selection and testing novel cultivars is critical for small-scale and resource-poor farmers to achieve progress tailored to their conditions and needs.

16.8.2 Participatory Plant Breeding

Participatory plant breeding (PPB) involves farmers and trained breeders collaborating at various stages of the breeding process, often using farmers' fields instead of research stations to select for traits suited to specific needs. This method emerged from criticism of development initiatives from the 1950s aimed at introducing modern agriculture to resource-limited areas. Plant breeding has ancient roots, dating back to the domestication of crops around 10,000 years ago (Tarekegne et al. 2019) and became a scientific field with Mendel's discoveries in the early 1900s. Described as a "science-based technology," plant breeding seeks to improve cultivars by selecting among genetically diverse populations, with PPB seen as an extension of participatory varietal selection (Thapa et al. 2009).

PPB methodologies vary but consistently include meaningful farmer involvement. Farmers can engage at different stages, such as setting breeding goals, making crosses, selecting progeny, evaluating varieties, and disseminating seeds (Probst and Jones 2016). There are distinctions between "formal participatory plant breeding," led by public research scientists, and "farmer-led participatory plant breeding," where scientists support farmers' plans (Tracy 2004). Projects can be centralized at research stations or decentralized in farmers' fields, aiming to improve varieties for marginalized areas, preserve biodiversity, empower disadvantaged populations (especially women), and reduce breeding costs and time (Fig. 16.3).

16.8.3 Farmers' Selection Criteria

Each farmer may have distinct selection considerations that necessitate the presence of multiple cultivars within a community to fulfill various needs, based on their own conditions (Thapa et al. 2009). As the number of selection criteria increases, the farmer will need to plant a greater number of landraces in order to fulfill all of their requirements. Farmers typically consider various factors when selecting crops, including their involvement in technology development, the cropping system, and

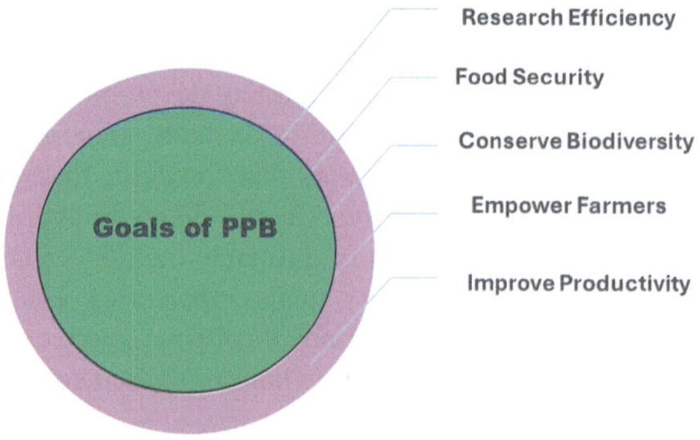

Fig. 16.3 Possible major goals of participatory plant breeding (PPB) projects

the crop's usefulness for their family and the market. However, these criteria may vary depending on the gender, age, and socioeconomic status of the farmers, including their access to credit. Farmers commonly utilize crop features such as the large size, shape, and color of the harvested component, the absence of pest damage, and the duration of maturity (Voss 1996). Additional cultivars are selected and cultivated as contingency measures in the event that one fails as a result of environmental pressures. Farmers can capitalize on various markets by choosing cultivars that are highly sought after.

16.9 Achievements, Farmer's Evaluations, and Experiences of Participatory Approaches in Different Countries

Insights from PPB and PVS reveal that farmers who take part in these practices function as researchers alongside plant breeders. As noted by Smith and Weltzien (2000), some farmers are skilled in seed selection and saving, making them valuable to participatory breeding teams. Despite doubts about farmers' selection skills, many projects have shown farmers to possess extensive expertise. Research by Sthapit et al. (1996) indicated that inclusive rice breeding in Nepal elevated farmer's efforts and time investment, with joint selections by farmers and breeders producing successful lines (Gyawali et al. 2007). The study by Ceccarelli et al. (2001) revealed that Syrian farmers were more proficient than breeders at choosing high-quality barley genotypes in their fields. Ashby (2009) highlighted the impact of PPB and PVS on various crops in multiple countries, citing different authors. In the Burkina Faso, farmers planted cultivars with varying growth periods based on field availability and bird control. In Rwanda, bean experts' selections outperformed local mixtures in yield 64–89% of the case, while breeder selections exceeded local mixtures 41–51% of the time across the country (Sperling et al. 1993). In Ethiopia, efforts have been made to develop and popularize common bean varieties through PPB and PVS (Asfaw et al. 2004), and bread wheat (Asaye Demelash et al. 2014).

16.10 PVS in Common Bean in Sidama Zone of Southern Ethiopia

Farmers typically prioritize qualitative characteristics of common beans, including seed color, drought tolerance, disease resistance, marketability, seed size, shattering tolerance, flavor, and cooking time. These criteria imply that farmers consider factors beyond yield when selecting common bean varieties. According to these criteria, every farmer who took part within the common bean mother experiment favored the Ibado variety as their top choice. This preference was due to its red speckled seed color, large seed size, high market demand, and 2 tons per hectare early maturity. Based on its seed color (light red), marketability, and flavor, the local variety was ranked second (Table 16.2). The researchers chose Awash-1 and Omo-95 based on their grain output, as shown in Table 16.2. Additionally, they emphasized that

Table 16.2 Preference rankings of common bean varieties by farmers in the Umbullo Watershed Mother Trial

Varieties	F1	F2	F3	F4	F5	F6	F7	F8	F9	F10	F11	F12	F13	F14	F15	Av.R	Av. R
Awash M.	2	3	3	4	3	2	1	5	3	3	4	3	2	2	3	2.9	3
Ibado	1	1	1	2	1	1	3	1	2	1	1	1	2	1	1	1.3	1
Awash-1	4	4	3	3	4	2	4	4	4	4	3	3	2	2	3	3.3	4
Local	3	2	2	1	2	3	3	3	1	2	2	2	1	3	2	2.1	2
Omo-95	5	2	4	6	5	3	5	2	3	5	5	3	4	5	2	3.9	5
Roba-1	6	6	4	5	5	4	2	2	5	3	5	4	3	4	4	4	6

Awash M. Awash Melka, *Av. R* Average rank, *F* Framer

NB: Traits: disease, drought, pest resistance, seed color, seed size, taste and cooking time, marketability, shattering tolerance, (Source: Fekadu 2013)

farmers possess a strong awareness of the selection criteria and are adept at choosing and prioritizing different types. Certain criteria are consistent with breeder's standards, while others exceed the breeder's expectations.

16.11 PVS in Faba Bean in Dabat District in Ethiopia

Mulualem et al. (2012) conducted a field study to assess the performance of PVS at four growth stages: vegetative, flowering, maturity, and harvesting (Table 16.3). The evaluation was based on criteria preferred by farmers, including number of branches (NOB), stem strength (STS), seed size (SS), plant establishment (PES), overall performance (OAP), and grain yield. The results are presented in Table 16.3. They additionally mentioned that farmers and researchers employed distinct metrics and methodologies to assess the tested genotypes. Therefore, researchers must consider the characteristics preferred by farmers, namely, seed production, seed diameter, and comprehensive filed efficiency, when developing new varieties. The ongoing selection process has further validated the farmers' ability to choose significant characteristics that contribute to grain output. Based on these features, they have successfully identified superior varieties that are well suited to their specific region. In the study conducted by Mulualem et al. (2012), it was shown that PVS (participatory varietal selection) proved to be a successful and dependable method for determining suitable faba bean cultivars in collaboration with farmers who have limited resources.

16.12 Preliminary Visual Survey (PVS) Conducted in the Pawe District in North-Western Ethiopia Focused on the Soybean Crop

The farmers executed a joint review of soybean genotypes at the stage of initial growth and ripening. They determined that suitable selection criteria were pods per plant, seed dimension, plant height, shattering susceptibility, conformity, and consumer need. The most individuals involved favored the Awassa-95 variety from the initial set, the Gishama variety from the med-tier set, and the Wegayen variety from the late flowering soybean types. According to typical output, researchers selected the Wegayen variety for the late set, the Gishama variety for the medium set, and the Awassa-95 variety for the early set, all of which showed high yields (Table 16.4). The farmers' preferences align with the breeders' selection in this particular scenario (Getahun et al. 2016). Olaoye et al. (2009) elucidated that the limited involvement of farmers in the process of selecting varieties results in a meager proportion of the varieties created by breeders being ultimately adopted. Ceccarelli and Grando (2007) emphasized that farmer's participatory varietal selection is a method that addresses the constraints of traditional breeding. It allows farmers to select the varieties that best meet their requirements and adapt to their specific environment.

Table 16.3 Aggregate scores for traits across three farmer sites, overall mean values of selection criteria, and genotype rankings

Variety	PE	SS	STS	NOB	OP	Total	Mean	Rank
WAYU	12	12	15	12	12	63	12.6	3
HOLETTA-2	9	13	8	10	8	48	9.6	6
WOLKI	13	12	15	13	13	66	13.2	2
EH99051–3	11	8	8	10	10	47	9.4	7
MOTI	13	15	8	10	8	54	10.8	5
CS20DK	8	13	7	7	8	43	8.6	8
GEBELCHO	8	10	10	8	7	43	8.6	8
DEGAGA	8	7	8	10	8	41	8.2	9
SELALE	11	7	15	12	12	57	11.4	4
DOSHA	14	15	12	15	12	68	13.6	1

OP Overall Performance, *SS* Seed Size, *PE* Plant Establishment, *NB* Number of Branches, *STS* Stem Strength, Criteria: 1 = very poor, 2 = poor, 3 = average, 4 = good, 5 = very good
(Source: Mulualem et al. 2012).

Table 16.4 Selection parameters and variety preference among farmers

Variety	PH	NPP	SS	Sh	U	MD	Total	mean	Rank
AFDAT	18	14	16	13	17	16	94	11.75	2
Belase-95	9	14	13	16	18	15	85	10.625	3
Crowford	6	4	14	10	15	14	63	7.875	3
Ethio-Yugoslavia	16	15	16	14	16	15	92	11.5	2
Willams	6	9	18	12	17	17	79	9.875	2
Gishama	15	13	17	19	17	17	98	12.25	1
Wegayen	17	14	17	12	18	17	95	11.875	1
Awassa 95	15	16	16	16	18	17	98	12.25	1
Gizo	10	13	14	12	16	15	80	10	3

PH plant height, *Sh* Shattering, *SS* Seed Size, *U* uniformity, *NPP* No. of pods per plant; market demand, Criteria: 1 = very poor, 2 = poor, 3 = average, 4 = good, 5 = very good
(Source: Mulualem et al. 2012; Getahun et al. 2016)

16.13 Future Directions and Challenges

The future of legume breeding for sustainable agriculture lies in the integration of multiple traits to create resilient and high-yielding varieties. Breeding programs must focus on combining traits such as disease resistance, drought tolerance, and nutritional quality to meet the demands of a growing population and changing climate. This approach will ensure that legumes remain a vital component of sustainable agricultural systems. Additionally, harnessing big data and digital agriculture technologies will revolutionize legume breeding. Advanced genomic tools, coupled with artificial intelligence and machine learning, can accelerate the identification of beneficial traits and their incorporation into new varieties. Digital platforms will facilitate data sharing and collaboration among researchers, enhancing the efficiency of breeding programs.

Socioeconomic considerations and farmer adoption are critical to the success of these innovations. Breeding programs must engage with farmers to understand their needs and preferences, ensuring that new legume varieties are not only scientifically advanced but also practical and acceptable to end-users. This requires participatory breeding approaches and effective extension services to educate farmers about the benefits and proper management of new varieties. Moreover, addressing regulatory frameworks and biosafety is essential for the successful deployment of genetically improved legumes. Clear and science-based regulations must be established to assess and manage the risks associated with genetically modified organisms (GMOs) and other biotechnological innovations. This will facilitate the safe introduction of new legume varieties while maintaining public trust and environmental safety.

16.14 Conclusion

In conclusion, the future of legume breeding for sustainable agriculture involves a multi-faceted approach that integrates advanced genetic techniques, digital tools, farmer engagement, and robust regulatory frameworks. Overcoming these challenges will require collaboration among scientists, policymakers, and farmers to develop and adopt legume varieties that contribute to food security, environmental sustainability, and economic viability. By addressing these future directions and challenges, legume breeding can play a pivotal role in achieving sustainable agricultural systems worldwide.

Authors Contribution M.T.A., A.A., W.L., P.M., S.A.Q., M.T., A.J., M.S., M.A.R., S.R.S., and S.N.S. contributed to the writing of the original draft, as well as the preparation of tables and figs. M.T.A., M.A.N., and F.S.B. assisted with the review and editing of the manuscript and revised the book chapter. All authors have read and approved the final version of the book chapter and are willing to submit it.

References

Abd-Alla MH, Al-Amri SM, El-Enany AWE (2023) Enhancing rhizobium–legume symbiosis and reducing nitrogen fertilizer use are potential options for mitigating climate change. Agriculture 13(11):2092

Abobatta WF, El-Hashash EF, Hegab RH (2021) Challenges and opportunities for the global cultivation and adaption of legumes. J Appl Biotechnol Bioeng 8(5):160–172

Ahmed M, Sameen A, Parveen H, Ullah MI, Fahad S, Hayat R (2023) Climate change impacts on legume crop production and adaptation strategies. In: Global agricultural production: resilience to climate change. Springer, Cham, pp 149–181

Ahn E, Botkin J, Curtin SJ, Zsögön A (2023) Ideotype breeding and genome engineering for legume crop improvement. Curr Opin Biotechnol 82:102961

Ainsworth EA, Rogers A, Vodkin LO, Walter A, Schurr U (2006) The effects of elevated CO_2 concentration on soybean gene expression. An analysis of growing and mature leaves. Plant Physiol 142:135–147

Al Amin N, Ahmad N, Wu N, Pu X, Ma T, Du Y et al (2019) CRISPR-Cas9 mediated targeted disruption of FAD2–2 microsomal omega-6 desaturase in soybean (Glycine max. L). BMC Biotechnol 19:1–10

Ali A, Altaf MT, Nadeem MA, Karaköy T, Shah AN, Azeem H et al (2022) Recent advancement in OMICS approaches to enhance abiotic stress tolerance in legumes. Front Plant Sci 13:952759

Ali A, Ölmez F, Altaf MT, Liaqat W, Umar UUD, Iqbal J (2023) Advanced and sustainable approaches in sugarcane crop improvements with reference to environmental stresses. In: Aasim M, Baloch FS, Nadeem MA, Habyarimana E, Ahmed S, Chung G (eds) Biotechnology and omics approaches for bioenergy crops. Springer, Singapore. https://doi.org/10.1007/978-981-99-4954-0_8

Ali A et al (2024) Biotechnological advancements toward sugarcane crop improvement. In: Kumar N (ed) Industrial crop plants. Interdisciplinary biotechnological advances. Springer, Singapore. https://doi.org/10.1007/978-981-97-1003-4_13

Altaf MT, Liaqat W, Iqbal J, Baig MMA, Ali A, Nadeem MA, Baloch FS (2023) Exploring omics approaches to enhance stress tolerance in soybean for sustainable bioenergy production. In: Biotechnology and omics approaches for bioenergy crops. Springer, Singapore, pp 123–154

Altaf MT, Liaqat W, Ali A, Jamil A, Bedir M, Nadeem MA et al (2024a) Conventional and biotechnological approaches for the improvement of industrial crops. In: Industrial crop plants. Springer, Singapore, pp 1–48

Altaf MT, Liaqat W, Bedir M, Ali A, Nadeem MA, Baloch FS (2024b) Wheat biofortification: a promising approach to improve public health. In: Advances in wheat breeding: towards climate resilience and nutrient security. Springer, Singapore, pp 623–651

Aranjuelo I, Cabrerizo PM, Aparicio-Tejo PM, Arrese-Igor C (2014) Unravelling the mechanisms that improve photosynthetic performance of N_2- fixing pea plants exposed to elevated [CO_2]. Environ Exp Bot 1:167–174

Arnqvist G, Dowling DK, Eady P, e al. (2010) Genetic architecture of metabolic rate: environment specific epistasis between mitochondrial and nuclear genes in an insect. Evolution 64:3354–3363

Arora L, Narula A (2017) Gene editing and crop improvement using CRISPR-Cas9 system. Front Plant Sci 8:1932. https://doi.org/10.3389/fpls.2017.01932

Asaye Demelash L, Desalegn T, Alemayehu G (2014) Participatory varietal selection of bread wheat (Triticum aestivum L.) genotypes at Marwold Kebele, Womberma Woreda, west Gojam, Ethiopia. Afr J Agric 9(3):327–333

Asfaw A, Dauro D, Kimani PM (2004) Decentiralized participatory bean breeding in southern and Mathematics 3, 1., ISSN (Online): 2394–2894

Ashby JA (2009) The impact of participatory plant breeding. In: Plant breeding and farmer participation, 649–671

Atkins PA, Voytas DF (2020) Overcoming bottlenecks in plant gene editing. Curr Opin Plant Biol 54:79–84. https://doi.org/10.1016/j.pbi.2020.01.002

Bai M, Yuan J, Kuang H, Gong P, Li S, Zhang Z et al (2020) Generation of a multiplex mutagenesis population via pooled CRISPR -Cas9 in Soya Bean. Plant Biotechnol. J. 18(3):721–731. Epub 2019 Sep 9. PMID: 31452351; PMCID: PMC7004907. https://doi.org/10.1111/pbi.13239

Ball KR, Power SA, Brien C, Woodin S, Jewell N, Berger B, Pendall E (2020) High-throughput, image-based phenotyping reveals nutrient-dependent growth facilitation in a grass-legume mixture. PLoS One 15(10):e0239673

Baloch FS, Altaf MT, Liaqat W, Bedir M, Nadeem MA, Cömertpay G et al (2023a) Recent advancements in the breeding of sorghum crop: current status and future strategies for marker-assisted breeding. Front Genet 14:1150616

Baloch FS, Altaf MT, Bedir M, Nadeem MA, Tatar M, Karaköy T, Aasim M (2023b) iPBS-retrotransposons variations: DNA fingerprinting and the evaluation of genetic diversity and population structure in international cowpea germplasm. Genet Resour Crop Evol 70(6):1867–1877

Baloglu MC, Celik Altunoglu Y, Baloglu P, Yildiz AB, Türkölmez N, Özden Çiftçi Y (2022) Gene-editing technologies and applications in legumes: Progress, evolution, and future prospects. Front Genet 13:859437

Bao A, Chen H, Chen L, Chen S, Hao Q, Guo W et al (2019) CRISPR/Cas9- mediated targeted mutagenesis of GmSPL9 genes alters plant architecture in soybean. BMC Plant Biol 19:131. https://doi.org/10.1186/s12870-019-1746-6

Bargaz A, Zaman-Allah M, Farissi M, Lazali M, Drevon JJ, Maougal RT, Georg C (2015) Physiological and molecular aspects of tolerance to environmental constraints in grain and forage legumes. Int J Mol Sci 16:18976–19008

Berger JD, Ali M, Basu PS, Chaudhary BD, Chaturvedi SK, Deshmukh PS, Dharmaraj PS, Dwivedi SK, Gangadhar GC, Gaur PM et al (2006) Genotype by environment studies demonstrate the critical role of phenology in adaptation of chickpea (Cicer arietinum L.) to high and low yielding environments of India. Field Crop Res 98:230–244

Bhat JA, Yu D (2021) High-throughput NGS-based genotyping and phenotyping: role in genomics-assisted breeding for soybean improvement. Legum Sci 3(3):e81

Bhatia A, Mina U, Kumar V, Tomer R, Kumar A, Chakrabarti B, Singh RN, Singh B (2021) Effect of elevated ozone and carbon dioxide interaction on growth, yield, nutrient content and wilt disease severity in chickpea grown in northern India. Heliyon 7:06049

Bhowmik P, Konkin D, Polowick P, Hodgins CL, Subedi M, Xiang D et al (2021) CRISPR/Cas9 gene editing in legume crops: opportunities and challenges. Legum Sci 3(3):e96

Bindumadhava H, Sharma L, Nair RM, Nayyar H, Riley JJ, Easdown W (2018) High-temperature-tolerant mungbean (Vignaradiata L.) lines produce better yields when exposed to higher CO_2 levels. J Crop Imp 4:418–430

Bourgault M, Brand J, Tausz M, Fitzgerald GJ (2016) Yield, growth and grain nitrogen response to elevated CO_2 of five field pea (Pisumsativum L.) cultivars in a low rainfall environment. Field Crop Res 196:1–9

Bourgault M, Brand J, Tausz-Posch S, Armstrong RD, O'Leary GL, Fitzgerald GJ, Tausz M (2017) Yield, growth and grain nitrogen response to elevated CO_2 in six lentil (Lens culinaris) cultivars grown under free air CO_2 enrichment (FACE) in a semi-arid environment. Eur J Agron 87:50–58

Cabrerizo PM, González EM, Aparicio-Tejo PM, Arrese-Igor C (2001) Continuous CO_2 enrichment leads to increased nodule biomass, carbon availability to nodules and activity of carbon-metabolising enzymes but does not enhance specific nitrogen fixation in pea. Physiol Plant 113:33–40

Cai Y, Wang L, Chen L, Wu T, Liu L, Sun S et al (2020) Mutagenesis of GmFT2a and GmFT5a mediated by CRISPR/Cas9 contributes for expanding the regional adaptability of Soybean. Plant Biotechnol. J. 18(1):298–309. Epub 2019 Jul 5. PMID: 31240772; PMCID: PMC6920152. https://doi.org/10.1111/pbi.13199

Ceccarelli S, Grando S (2007) Decentralized-participatory plant breeding: an example of demand driven research. Euphytica 155:349–360

Ceccarelli S, Grando S, Bailey E, Amri A, El-Felah M, Nassif F et al (2001) Farmer participation in barley breeding in Syria, Morocco and Tunisia. Euphytica 122:521–536

Chanie Y (2015) Review of participatory agricultural research approach and its importance in Ethiopia journal of biology. Agric Healthcare 5(17):192–200

Chaturvedi SK, Singh NP, Gaur PM, Varshney RK, Mishra N (2014) New challenges in breeding chickpea under changing climate. In: Pulses: challenges & opportunities. Indian Society of Pulses Research and Development, pp 102–113

Chen Y, Ghanem ME, Siddique KHM (2017) Characterising root trait variability in chickpea (Cicer arietinum L.) germplasm. J Exp Bot 68:1987–1999

Chevin LM (2012) Genetic constraints on adaptation to a changing environment. Evolution 67(3):708–721

Chinnusamy V, Schepler-Luu V, Mangrauthia SK, Ramesh SV (2023) Genome editing in plants: a tool for precision breeding and functional genomics. J Plant Biochem Biotechnol 32(4):657–660

Choudhary AK, Kumar J, Gupta S, Sultana R, Singh IS (2016, August) Breeding for adaptive traits in pulses. In Proceedings of the National Conference on bringing self-sufficiency in pulses for eastern India, Kanpur, pp 5–6

Confalonieri M, Carelli M, Gianoglio S, Moglia A, Biazzi E, Tava A (2021) CRISPR/Cas9-mediated targeted mutagenesis of CYP93E2 modulates the triterpene Saponin biosynthesis in Medicago Truncatula. Front Plant Sci 12:690231. https://doi.org/10.3389/fpls.2021.690231
Connallon T, Hall MD (2018) Genetic constraints on adaptation: a theoretical primer for the genomics era. Ann NY Acad Sci 1422:1–23
Crossa J, Pérez-Rodríguez P, Cuevas J, Montesinos-López O, Jarquín D, De Los Campos G et al (2017) Genomic selection in plant breeding: methods, models, and perspectives. Trends Plant Sci 22(11):961–975
Cullis C, Kunert KJ (2017) Unlocking the potential of orphan legumes. J Exp Bot 68:1895–1903
Curtin SJ, Tiffin P, Guhlin J, Trujillo DI, Burghardt LT, Atkins P et al (2017) Validating genome-wide association candidates controlling quantitative variation in nodulation. Plant Physiol. 173(2):921–931. Epub 2017 Jan 5. PMID: 28057894; PMCID: PMC5291020. https://doi.org/10.1104/pp.16.01923
Curtin SJ, Xiong Y, Michno J-M, Campbell BW, Stec AO, Čermák T et al (2018) CRISPR/Cas9 and TALENs generate heritable mutations for genes involved in small RNA processing of glycine maxand Medicago Truncatula. Plant Biotechnol J 16:1125–1137. https://doi.org/10.1111/pbi.12857
Cutforth HW, McGinn SM, McPhee KE, Miller PR (2007) Adaptation of pulse crops to the changing climate of the northern Great Plains. Agron J 99:1684–1699
Delahunty A, Nuttall J, Nicolas M, Brand J (2018) Response of lentil to high temperature under variable water supply and carbon dioxide enrichment. Crop Pasture Sci 69:1103–1112
Di YH, Sun XJ, Hu Z, Jiang QY, Song GH, Zhang B et al (2019) Enhancing the CRISPR/Cas9 system based on multiple GmU6 promoters in soybean. Biochem Biophys Res Commun 519(4):819–823
Djanaguiraman M, Schapaugh W, Fritschi F, Nguyen H, Prasad PVV (2018) Reproductive success of soybean (Glycine max L. Merril) cultivars and exotic lines under high daytime temperature. Plant Cell Environ 42:321–336
Do PT, Nguyen CX, Bui HT, Tran LTN, Stacey G, Gillman JD et al (2019) Demonstration of highly efficient dual gRNA CRISPR/Cas9 editing of the Homeologous GmFAD2-1A and GmFAD2-1B genes to yield a high oleic, low linoleic and α-linolenic acid phenotype in soybean. BMC Plant Biol 19:311. https://doi.org/10.1186/s12870-019-1906-8
Drag DW, Slattery R, Siebers M, DeLucia EH, Ort DR, Bernacchi CJ (2020) Soybean photosynthetic and biomass responses to carbon dioxide concentrations ranging from pre-industrial to the distant future. J Exp Bot 71:3690–3700
Du H, Zeng X, Zhao M, Cui X, Wang Q, Yang H et al (2016) Efficient targeted mutagenesis in soybean by TALENs and CRISPR/Cas9. J Biotechnol 217:90–97. https://doi.org/10.1016/j.jbiotec.2015.11.005
Dutta A, Trivedi A, Nath CP, Gupta DS, Hazra KK (2022) A comprehensive review on grain legumes as climate-smart crops: challenges and prospects. Environ Chall 7:100479
Fekadu G (2013) Assessment of Farmers' Criteria for Common Bean Variety Selection, the case of Umbullo Watershed in Sidama Zone of the Southern Region of Ethiopia. Ethiopian e-journal for research and innovation sight. 5(2):4–13
Foyer CH, Nguyen H, Lam HM (2019) Legumes—the art and science of environmentally sustainable agriculture. Plant Cell Environ 42(1):1–5
Gaur P, Saminen S, Krishnamurthy L, Kumar S, Ghane M, Beebe S, Rao I, Chaturvedi S, Basu P, Nayyar H et al (2015) High temperature tolerance in grain legumes. Leg Perspect 7:23–24
Getahun A, Atnaf M, Abady S, Degu T, Dilnesaw Z (2016) Participatory variety selection of soybean (Glycine max (L.) Merrill) varieties under rain fed condition of Pawe District, North-Western Ethiopia. Int J Appl Sci Math 3(1) ISSN (Online): 2394-2894
Govindaraj M, Vetriventhan M, Srinivasan M (2015) Importance of genetic diversity assessment in crop plants and its recent advances: an overview of its analytical perspectives. Genet Res Int 2015:431487

Großkinsky DK, Svensgaard J, Christensen S et al (2015) Plant phenomics and the need for physiological phenotyping across scales to narrow the genotype–to–phenotype knowledge gap. J Exp Bot 66(18):5429–5440
Guo H, Sun Y, Li Y, Liu X, Ren Q, Zhu-Salzman K, Ge F (2013) Elevated CO_2 modifies N acquisition of Medicagotruncatula by enhancing N fixation and reducing nitrate uptake from soil. PLoS One 8:81373
Gyawali S, Sunwar S, Subedi M et al (2007) Collaborative breeding with farmers can be effective. Field Crop Res 101:88–95
Haase S, Neumann G, Kania A, Kuzyakov Y, Römheld V, Kandeler E (2007) Elevation of atmospheric CO_2 and N-nutritional status modify nodulation, nodule-carbon supply, and root exudation of Phaseolus vulgaris L. Soil Biol Biochem 39:2208–2221
Han J, Guo B, Guo Y, Zhang B, Wang X, Qiu L-J (2019) Creation of early flowering germplasm of soybean by CRISPR/Cas9 technology. Front Plant Sci 10:1446. https://doi.org/10.3389/fpls.2019.01446
Huang T-K, Puchta H (2019) CRISPR/Cas-mediated gene targeting in plants: finally a turn for the better for homologous recombination. Plant Cell Rep 38:443–453. https://doi.org/10.1007/s00299-019-02379-0
Huang S, Gali KK, Tar'an B, Warkentin TD, Bueckert RA (2017) Pea phenology: crop potential in a warming environment. Crop Sci 57:1540–1551
Jha UC, Nayyar H, Chattopadhyay A, Beena R, Lone AA, Naik YD et al (2023) Major viral diseases in grain legumes: designing disease resistant legumes from plant breeding and OMICS integration. Front Plant Sci 14:1183505
Jiang Y, Lahlali R, Karunakaran C, Warkentin TD, Davis AR, Bueckert RA (2018) Pollen, ovules, and pollination in pea: success, failure, and resilience in heat. Plant Cell Environ 42:354–372
Jin J, Tang C, Armstrong R, Butterly C, Sale P (2013) Elevated CO_2 temporally enhances phosphorus immobilization in the rhizosphere of wheat and chickpea. Plant Soil 368:315–328
Jumrani K, Bhatia VS, Kataria S, Rastogi A (2024) The interactive effect of high temperature and water deficit stress on nitrogen fixation, photosynthesis, chlorophyll fluorescence, seed yield and quality in soybean (Glycine max). Plant Physiol Rep 29(1):125–140
Kawecki TJ, Ebert D (2004) Conceptual issues in local adaptation. Ecol Lett 7(12):1225–1241
Kehinde O, Emmanuel O, Wong LS, Siti Sarah M (2023) Genetic algorithm for forecasting Bioinformatic outcomes of mutation-induced cowpeas for sustainable development. J Data Sci 2023(12):1–13
Khalid MN, Abdullah A, Ijaz Z, Naheed N, Hamad A, Sheir MA et al (2021) Application and potential use of advanced bioinformatics techniques in agriculture and animal sciences. Ind J Pure App Biosci 9(3):237–246
Kimball BA (2016) Crop responses to elevated CO_2 and interactions with H_2O, N, and temperature. Curr Opin Plant Biol 31:36–43
King CA, Purcell LC (2005) Inhibition of N_2 fixation in soybean is associated with elevated ureides and amino acids. Plant Physiol 137:1389–1396
Kumar J, Gupta S, Biradar RS, Gupta P, Dubey S, Singh NP (2018) Association of functional markers with flowering time in lentil. J Appl Genet 59:9–21
Kumar J, Choudhary AK, Gupta DS, Kumar S (2019) Towards exploitation of adaptive traits for climate-resilient smart pulses. Int J Mol Sci 20(12):2971
Kumar S, Gupta P, Choukri H, Siddique KH (2020) Efficient breeding of pulse crops. Accelerated plant breeding, volume 3: food legumes, 1–30
Kundu A, Sanwal SK, Kumar A, Singh V, Nain A (2023) Genetic improvement of chickpea through MAS for abiotic stresses. J Soil Salinity Water Qual 15(2):146–157
Lam SK, Hao X, Lin E, Han X, Norton R, Mosier AR, Chen D (2012) Effect of elevated carbon dioxide on growth and nitrogen fixation of two soybean cultivars in northern China. Biol Fertil Soils 48:603–606
Lamichaney A, Tewari K, Basu PS, Katiyar PK, Singh NP (2021) Effect of elevated carbon-dioxide on plant growth, physiology, yield and seed quality of chickpea (Cicer arietinum L.) in indo-Gangetic plains. Physiol Mol Biol Plants 27:251–263

Lenka NK, Lenka S, Thakur JK, Yashona DS, Shukla AK, Elanchezhian R, Singh KK, Biswas AK, Patra AK (2020) Carbon dioxide and temperature elevation effects on crop evapotranspiration and water use efficiency in soybean as affected by different nitrogen levels. Agric Water Manag 230:105936

Li C, Nguyen V, Liu J, Fu W, Chen C, Yu K et al (2019) Mutagenesis of seed storage protein genes in soybean using CRISPR/Cas9. BMC Res Notes 12:176. https://doi.org/10.1186/s13104-019-4207-2

Liu PH, Gan Y, Warkentin T, McDonald C (2003) Morphological plasticity of chickpea in a semi-arid environment. Crop Sci 43:426–429

Majhi PK, Bhoi TK, Sharma P, Joshi A, Samal I, Patra B (2024) Genetic diversity of food legumes and OMICS prospective to ensure nutritional security. In: Sustainable utilization and conservation of plant genetic diversity, pp 107–125

Majidian P (2022) Legume breeding: from conventional method to modern technique. In: Legumes research-volume 1. IntechOpen

McKersie B (2015) Planning for food security in a changing climate. J Exp Bot 66:3435–3450

Meng Y, Hou Y, Wang H, Ji R, Liu B, Wen J et al (2017) Targeted mutagenesis by CRISPR/Cas9 system in the model legume Medicago Truncatula. Plant Cell Rep 36:371–374. https://doi.org/10.1007/s00299-016-2069-9

Mishra AK, Agrawal SB (2015) Cultivar specific response of CO_2 fertilization on two tropical mung bean (Vignaradiata L.) cultivars: ROS generation, antioxidant status, physiology, growth, yield and seed quality. J Agron Crop Sci 252:797–811

Mishra A, Mani DBKV, Dash M (2024) Recent advances in plant breeding for pulse improvement. Smart Breed 36:329–369

Mkumbira J, Chiwona-Karltun L, Lagercrantz U, Mahungu NM, Saka J, a. (2003) Classification of cassava into 'bitter'and 'cool'in Malawi: from farmers' perception to characterization by molecular markers. Euphytica 132:7–22

Mulualem T, Dessalegn T, Dessalegn Y (2012) Participatory varietal selection of faba bean (Vicia faba L.) for yield and yield components in Dabat district, Ethiopia. Wudpecker. J Agric Res 1(7):270–274

Munir R, Konnerup D, Khan HA, Siddique KHM, Colmer TD (2018) Sensitivity of chickpea and faba bean to root-zone hypoxia, elevated ethylene, and carbon dioxide. Plant Cell Environ 42:85–97

Murata N, Takahashi S, Nishiyama Y, Allakhverdiev SI (2007) Photoinhibition of photosystem II under environmental stress. Biochim Biophys Acta 1767:414–421

Nadeem MA, Altaf MT, Zia MAB, Karaköy T, Baloch FS, Aasim M (2022) Breeding advancements in fenugreek for environmental stresses. In: Environmental challenges and medicinal plants: sustainable production solutions under adverse conditions. Springer, Cham, pp 449–460

Nadeem MA, Baloch FS, Fiaz S, Aasim M, Habyarimana E, Sönmez O, Zencirci N (2023) Legumes biofortification. Springer, Cham, pp 1–549. https://doi.org/10.1007/978-3-031-33957-8

Naik YD, Zhao C, Channale S, Nayak SN, Bhutia KL, Gautam A et al (2024) Bioinformatics for plant genetics and breeding research. In: Frontier Technologies for Crop Improvement. Springer, Singapore, pp 35–64

Narayan D (1993) Focus on participation: evidence from 121 rural water supply projects. UNDPWorld Bank Water Supply and Sanitation Program, Washington, DC, World Bank

Nasser RR, Fuller MP, Jellings AJ (2008) Effect of elevated CO_2 and nitrogen levels on lentil growth and nodulation. Agron Sustain Dev 28:175–180

Ngalamu T, Galla JO, Ofori K, Meseka SK (2023) Genetic improvement for development of a climate resilient food legume crops: relevance of cowpea breeding approach in improvement of food legume crops for future. In: Climate change and legumes. CRC Press, pp 97–120

Ogliari JB, Kist V, Canci A (2013) The participatory genetic enhancement of a local maize variety in Brazil. In: Community biodiversity management. Routledge

Ojiewo C, Monyo E, Desmae H et al (2019) Genomics, genetics and breeding of tropical legumes for better livelihoods of smallholder farmers. Plant Breed 138:487–499

Olaoye G, Ajala SO, Adedeji SA (2009) Participatory selection of a maize (Zea mays L.) variety for the control of stem borers in a southeastern Nigeria location

Osiru DSO, Balyejusa-Kizito E, Bisikwa J, Baguma Y, Turyagyenda L (2010) Participatory selection and development of drought tolerant cassava varieties for farmers in marginal areas. In second ruforum biennial regional conference on" building capacity for food security in Africa", Entebbe, Uganda, 1009–1012

Pandey R, Meena S, Krishnapriya V, Khetarpal S, Pal M (2016) Interactive effects of phosphorus nutrition and atmospheric carbon dioxide levels on growth, nitrogen fixation and yield of green gram. Ind J Fertil 12:56–64

Pang J, Turner NC, Khan T, Du YL, Xiong JL, Colmer TD, Devilla R, Stefanova K, Siddique KHM (2017) Response of chickpea (Cicer arietinum L.) to terminal drought: leaf stomatal conductance, pod abscisic acid concentration, and seed set. J Exp Bot 68:1973–1985

Patra S, Chakraborty A, Saikia P (2024) Role of omics tools in understanding the stress tolerance in legumes. In: Current omics advancement in plant abiotic stress biology. Academic, pp 215–226

Pearse and Stiefel (1979. Quoted in Kanji, Nazneen, and Greenwood, Laura. 2001) Participatory approaches to Research and Development in IISD: learning from experience. IIED, London, p 8

Pilson D (2000) Herbivory and natural selection on flowering phenology in wild sunflower. Helianthus Annuus Oecologia 122(1):72–82

Pla CL, Ocaña A, Carrero JJ, Tejera N (2003) Interaction plant-soil microorganisms: nitrogen-fixing symbioses. In: Plant ecophysiology: a science of synthesis thomsonparaninfo, pp 331–360

Prévost D, Bertrand A, Juge C, Chalifour FP (2010) Elevated CO_2 induces differences in nodulation of soybean depending on bradyrhizobial strain and method of inoculation. Plant Soil 331:115–127

Prince SJ, Murphy M, Mutava RN, Durnell LA, Valliyodan B, Shannon JG, Nguyen HT (2017) Root xylem plasticity to improve water use and yield in water-stressed soybean. J Exp Bot 68:2027–2036

Prince SJ, Vallivodam B, Ye H, Yang M, Tai S, Hu W et al (2018) Understanding genetic control of root system architecture in soybean: insights into the genetic basis of lateral root number. Plant Cell Environ 42:212–229

Probst J, Jones K (2016) Looking ahead: rural-urban differences in anticipated

Qazi HA, Rao PS, Kashikar A, Suprasanna P, Bhargava S (2014) Alterations in stem sugar content and metabolism in sorghum genotypes subjected to drought stress. Funct Plant Biol 41:954–962

Quan W, Liu X, Wang H, Chan Z (2016) Comparative physiological and transcriptional analyses of two contrasting drought tolerant alfalfa varieties. Front Plant Sci 6:1256

Rahman MA, Thant AA, Win M, Tun MS, Moet P (2015) Participatory varietal selection (PVS): a "bottom-up" breeding approach helps rice farmers in the Ayeyarwady Delta, Myanmar. SABRAO J Breed Genet 47:299–314

Ramalingam A, Kudapa H, Pazhamala LT, Weckwerth W, Varshney RK (2015) Proteomics and metabolomics: two emerging areas for legume improvement. Front Plant Sci 6:1116

Reynolds M, Langridge P (2016) Physiological breeding. Curr Opin Plant Biol 31:162–171

Richard G, Houle D (2009) Demographic and genetic constraints on evolution. Am Nat 174(6):18–29

Ricter JA, Behr JH, Erban A, Kopta J, Zörb C (2018) Ion -dependent metabolic responses of Vicia faba L. to salt stress. Plant Cell Environ 42:295–309

Rodas AL, Roque E, Hamza R, Gómez-Mena C, Minguet EG, Wen J et al (2021) MtSUPERMAN plays a key role in compound inflorescence and flower development in Medicago Truncatula. Plant J 105:816–830. https://doi.org/10.1111/tpj.15075

Sabag I, Bi Y, Sahoo MM, Herrmann I, Morota G, Peleg Z (2024) Leveraging genomics and temporal high-throughput phenotyping to enhance association mapping and yield prediction in sesame. Plant Genome 17(3):e20481

Saha S, Chakraborty D, Pal M, Nagarajan S (2011) Impact of elevated CO_2 on utilization of soil moisture and associated soil biophysical parame ters in pigeon pea (Cajanuscajan L.). Agric Ecosys Environ 142:213–221.jin

Saha S, Chakraborty D, Sehgal VK, Nain L, Pal M (2016) Long-term atmospheric CO_2 enrichment impact on soil biophysical properties and root nodule biophysics in chickpea (Cicer arietinum L.). Eur J Agron 75:1–11

Sathish P, Vijay Kumar G, Jyothi Lakshmi N, Vanaja M, Yadav SK, Vagheera P (2014) Impact of CO_2 enhancement on photosynthesis and protein profile-response studies with a CO_2 responsive black gram genotype. Int J Appl Biol Pharm Technol 5:441–450

Sehgal A, Sita K, Bhandari K, Kumar S, Kumar J, Prasad PVV et al (2018) Influence of drought and heat stress, applied independently or in combination during seed development, on qualitative and quantitative aspects of seeds of lentil (Lens culinaris Medikus) genotypes, differing in drought sensitivity. Plant Cell Environ 42:198–211

Shunmugam A, Kannan U, Jiang Y, Daba K, Gorim L (2018) Physiology based approaches for breeding of next-generation food legumes. Plan Theory 7:72

Singer SD, Chatterton S, Soolanayakanahally RY, Subedi U, Chen G, Acharya SN (2020) Potential effects of a high CO_2 future on leguminous species. Plant-Environ Interact 1(2):67–94

Singh C, Kumar R, Sehgal H, Bhati S, Singhal T, Gayacharan et al (2023) Unclasping potentials of genomics and gene editing in chickpea to fight climate change and global hunger threat. Front Genet 14:1085024

Sita K, Sehgal A, Kumar J, Kumar S, Singh S, Siddique KH, Nayyar H (2017) Identification of high-temperature tolerant lentil (Lens culinaris Medik) genotypes through leaf and pollen traits. Front Plant Sci 8:744

Smith M, Weltzien E (2000) Scaling-up in participatory plant breeding. In: Almekinders C, De Boef

Sperling L, Loevinsohn ME, Ntabomvura B (1993) Rethinking the farmer's role in plant breeding: local bean experts and on-station selection in Rwanda. Exp Agric 29:509–519

Srinivasarao C, Kundu S, Shanker AK, Naik RP, Vanaja M, Venkanna K, Rao VUM (2016) Continuous cropping under elevated CO_2: differential effects on C4 and C3 crops, soil properties and carbon dynamics in semi-arid alfisols. Agric Ecosyst Environ 15:73–86

Sthapit B, Joshi KD, Witcombe JR (1996) Farmer participatory crop improvement. III. Participatory plant breeding, a case study for Nepal. Exp Agric 32:479–496

Tan YC, Kumar AU, Wong YP, Ling APK (2022) Bioinformatics approaches and applications in plant biotechnology. J Genet Eng Biotechnol 20(1):106

Tarekegne W, Mekbib F, Dessalegn Y (2019) Performance and participatory variety evaluation of finger millet [Eleusine coracana (L.) Gaertn] varieties in west Gojam zone. Northwest Ethiopia East Afr J Sci 13:27–38

Thapa DB, Sharma RC, Mudwari A, Ortiz-Ferrara G, Sharma S et al (2009) Identifying superior wheat cultivars in participatory research on resource poor farms. Field Crop Res 112:124–130

Tracy WF (2004) What is plant breeding? In seeds and breeds for 21st century agriculture. In: Proceedings of a summit conference, Washington, DC, 23–30

Tripathi N, Tripathi MK, Tiwari S, Payasi DK (2022) Molecular breeding to overcome biotic stresses in soybean: update. Plan Theory 11(15):1967

Trujillo DI, Silverstein KAT, Young ND (2019) Nodule-specific PLAT domain proteins are expanded in the Medicago lineage and required for nodulation. New Phytol 222:1538–1550. Epub 2019 Feb 27. https://doi.org/10.1111/nph.15697

Turner NC, Wright GC, Siddique KHM (2001) Adaptation of grain legumes (pulses) to water-limited environments. Adv Agron 71:193–231

Vanaja M, Ram Reddy PR, Lakshmi NJ, Abdul Razak SK, Vagheera P, Archana G, Yadav SK, Maheswari M, Venkateswarlu B (2010) Response of seed yield and its components of red gram (Cajanuscajan L. Mill sp.) to elevated CO_2. Plant Soil Environ 56:458–462

Varshney RK, Hoisington DA, Upadhyaya HD et al (2017) Molecular genetics and breeding of grain legume crops for the semi–arid tropics. Chapter 10. In: Varshney R, Tuberosa R (eds) Genomic assisted crop improvement: vol 2: genomics applications in crops, pp 207–241

Virdi KS, Spencer M, Stec AO, Xiong Y, Merry R, Muehlbauer GJ et al (2020) Similar seed composition phenotypes are observed from CRISPR- generated in-frame and knockout alleles of a soybean KASI Ortholog. Front Plant Sci 11:1006. https://doi.org/10.3389/fpls.2020.01005

Voss J (1996) Participatory breeding and IDRC's biodiversity programme. In Participatory plant breeding: proceedings of a workshop on participatory plant breeding, 26–29
Walsh B, Blows MW (2009) Abundant genetic variation+strong selection = multivariate genetic constraints: a geometric view of adaptation. Annu Rev Ecol Evol Syst 40:41–59
Wang L, Sun S, Wu T, Liu L, Sun X, Cai Y et al (2020) Natural variation and CRISPR/Cas9-mediated mutation in GmPRR37 affect photoperiodic flowering and contribute to regional adaptation of soybean. Plant Biotechnol J 18:1869–1881. https://doi.org/10.1111/pbi.13346
Witcombe JR, Joshi KD, Gyawali S, Musa AM, Johansen C, Virkand DS, Sthapit BR (2005) Participatory plant breeding is better described as highly client-oriented plant breeding. I. Four indicators of client-orientation in plant breeding. Exp Agric 41(03):299–319
Wolabu TW, Park JJ, Chen M, Cong L, Ge Y, Jiang Q et al (2020) Improving the genome editing efficiency of CRISPR/Cas9 in Arabidopsis and Medicago Truncatula. Planta 252:15. https://doi.org/10.1007/s00425-020-03415-0
World Bank (2007) World development report 2008: agriculture for development. The World Bank, Washington, DC, p 365
Wu N, Lu Q, Wang P, Zhang Q, Zhang J, Qu J et al (2020) Construction and analysis of GmFAD2-1A and GmFAD2-2A soybean fatty acid desaturase mutants based on CRISPR/Cas9 technology. Int J Mol Sci 21(3):1104. https://doi.org/10.3390/ijms21031104
Yamauchi T, Colmer TD, Pedersen O, Nakazono M (2018) Regulation of root traits for internal aeration and tolerance to soil waterlogging-flooding stress. Plant Physiol 176:1118–1130
Younas A, Yousaf Z, Rashid M, Riaz N, Aftab A, Shahid MJ et al (2023) Contribution of biotechnological approaches to micronutrient improvements in legumes. In: Legumes biofortification. Springer, Cham, pp 131–176
Yuksel B, Vardar F (2023) Impact of high temperature stress and its alleviation in Fabaceae. In: Climate change and legumes. CRC Press, pp 79–95
Zaman MSU, Malik AI, Erskine W, Kaur W (2018) Changes in gene expression during germination reveal pea genotypes with either quiescence or escape mechanisms of waterlogging tolerance. Plant Cell Environ 42:245–258
Zhang H, Cao Y, Zhang H, Xu Y, Zhou C, Liu W et al (2020) Efficient generation of CRISPR/Cas9-mediated homozygous/Biallelic Medicago Truncatula mutants using a hairy root system. Front Plant Sci 11:294. https://doi.org/10.3389/fpls.2020.00294

Correction to: Future Perspectives in Legume Breeding for Sustainable Agriculture

Muhammad Tanveer Altaf, Amjad Ali, Waqas Liaqat,
Parnaz Mortazavi, Sarmad Ali Qureshi, Muhammed Tatar,
Amna Jamil, Muhammad Sheraz,
Muhammad Aneeq Ur Rahman, Sabeen Rehman Soomro,
Salma Naimatullha Soomro, Muhammad Azhar Nadeem,
and Faheem Shehzad Baloch

Correction to:
Chapter 16 in: M. A. Nadeem et al. (eds.),
Marker-Assisted Breeding in Legumes for Drought Tolerance,
https://doi.org/10.1007/978-981-96-4112-3_16

The first name of the author, Muhammed Tatar, was unfortunately published with an incorrect spelling. The initially published version has now been corrected.

The updated version of this chapter can be found at
https://doi.org/10.1007/978-981-96-4112-3_16

© The Author(s), under exclusive license to Springer Nature Singapore Pte Ltd. 2026
M. A. Nadeem et al. (eds.), *Marker-Assisted Breeding in Legumes for Drought Tolerance*, https://doi.org/10.1007/978-981-96-4112-3_17

The manufacturer's authorised representative in the EU is Springer
Nature Customer Service Centre GmbH, Europaplatz 3, 69115 Heidelberg,
Germany. If you have any concerns regarding our products, please
contact ProductSafety@springernature.com

Printed and bound by CPI Group (UK) Ltd, Croydon, CR0 4YY
01/10/2026
02225929-0011